Advances in Mechanics and Mathematics

Volume 23

T0214473

For more titles in this series, go to
http://www.springer.com/series/5613

Advances in Mechanics and Mathematics

Volume 27

Guang-Ren Duan

Analysis and Design
of Descriptor Linear Systems

 Springer

Guang-Ren Duan
Harbin Institute of Technology
Center for Control Theory and Guidance Technology
Harbin, 150001
P. R. China
g.r.duan@hit.edu.cn

ISSN 1571-8689   e-ISSN 1876-9896
ISBN 978-1-4614-2684-4   ISBN 978-1-4419-6397-0 (eBook)
DOI 10.1007/978-1-4419-6397-0
Springer New York Dordrecht Heidelberg London

Mathematics Subject classification (2010): 58E25, 93B52, 93C05, 93C35

Springer is part of Springer Science+Business Media (www.springer.com)

To Shichao and Jiefu

Preface

Descriptor linear systems theory is an important part in the general field of control systems theory, and has attracted much attention in the last two decades. In spite of the fact that descriptor linear systems theory has been very rich in content, there have been only a few comprehensive books on this topic, e.g., Campbell (1980), Campbell (1982), and Dai (1989b). There do exist some other books and some PhD thesises related to descriptor systems, but they are all focused on very special topics.

This book aims to provide a relatively systematic introduction to the basic results in descriptor linear systems theory. The whole book has 11 chapters, and focuses on the analysis and design problems on continuous-time descriptor linear systems. Materials about analysis and design of discrete-time descriptor linear systems are not included. Besides most of the fundamental context, it also contains some of the author's research work, which are reflected in the topics of response analysis, regularization, dynamical order assignment, eigenstructure assignment, and parametric approaches for observer design, etc.

Many researchers in the world have made great contribution to descriptor linear systems theory. Owing to length limitation and the structural arrangement of the book, many of their published results are not included or even not cited. I would extend my apologies to these researchers.

Most of the materials of the book have been lectured by the author himself in the spring terms of 2002~2005 in a postgraduate course at Harbin Institute of Technology. My colleagues, Prof. Zhi-Bin Yan and Dr Cang-Hua Jiang, assisted me in lecturing this course in the spring terms of 2006~2008, respectively, and have helped a lot in proofreading the manuscripts. Prof. Zhi-Bin Yan, Prof. Xian Zhang and Dr Ai-Guo Wu have all coauthored with me a few papers, which have been included in this book. Here, I would like to express my heartfelt appreciation of their contribution.

All my graduate and PhD students and those who took the graduate course "Descriptor Linear Systems" at Harbin Institute of Technology in the spring terms of 2002~2008 have offered tremendous help in finding the errors and typos in the manuscripts. Their help has greatly improved the quality of the manuscripts, and is indeed very much appreciated. Dr Hai-Hua Yu, Dr Ai-Guo Wu, Dr Bing Liang, Dr Yan-Ming Fu, Dr Ying Zhang, Dr Liu Zhang, Dr Yong-Zheng Shan, and Dr Hong-Liang Liu, who were really my PhD students years ago, have helped me with the indices, the references, and the parts of the revision of the book. My present Ph.D students, Mr. Da-Ke Gu, Mr. Shi-Jie Zhang, Ms. Ling-Ling Lv, Mr. Yan-Jiang Li, Ms. Shi Li, and Mr. Guang-Bin Cai, all helped me with the examples of the book. Particularly, Dr Hai-Hua Yu, besides all the above, has helped me with the whole formatting of the book. I would extend my great thanks to all of them. My thanks would also go to my colleague, Prof. Hui-Jun Gao, who once was in 2003 a student in my class of the course, has proofread several chapters of the book as well.

I would also like to thank my wife, Ms Shi-Chao Zhang, for her continuous support in every aspect. Sincere thanks also go to my secretary, Ms Ming-Yan Liu, for helping me in typing a few chapters of the manuscripts. Part of the book was written when I was with the Queen's University of Belfast, UK, from September 1998 to October 2002. I would like to thank Professor G. W. Irwin and Dr S. Thompson for their help, suggestions, and support. The reviewers of the book have given some real valuable and helpful comments and suggestions, which are indeed very much appreciated.

The author would like to gratefully acknowledge the financial support kindly provided by the many sponsors, including NSFC, the National Natural Science Foundation of China (National Science Fund for Distinguished Young Scholar's Grant No.60474015), the Ministry of Education (Program of The New Century Excellent Talents in University and the Chang Jiang Scholars Program), and also EPSRC, the UK Engineering and Physical Science Research Council (GR/K83861/01).

At the last, let me thank in advance all the readers for choosing to read this book. I would be indeed very grateful if readers could possibly provide, via email: g.r.duan@hit.edu.cn, feedback about any problems found. Your help will certainly make any future editions of the book much better.

Harbin Institute of Technology,   *Guang-Ren Duan*
12 December 2009

Contents

List of Notation

Notations Related to Subspaces

$\mathbb{R}$ Set of all real numbers

$\mathbb{R}^+$ Set of all positive real numbers

$\mathbb{R}^-$ Set of all negative real numbers

$\mathbb{C}$ Set of all complex numbers

$\mathbb{C}^+$ The open right half complex plane

$\mathbb{C}^-$ The open left half complex plane

$\bar{\mathbb{C}}^+$ The closed right half complex plane, $\bar{\mathbb{C}}^+ = \{s \mid s \in \mathbb{C}, \mathrm{Re}(s) \geq 0\}$

$\bar{\mathbb{C}}^-$ The closed left half complex plane, $\bar{\mathbb{C}}^- = \{s \mid s \in \mathbb{C}, \mathrm{Re}(s) \leq 0\}$

$\mathbb{R}^n$ Set of all real vectors of dimension n

$\mathbb{C}^n$ Set of all complex vectors of dimension n

$\mathbb{R}^{m \times n}$ Set of all real matrices of dimension $m \times n$

$\mathbb{C}^{m \times n}$ Set of all complex matrices of dimension $m \times n$

$\mathbb{R}_r^{m \times n}$ Set of all $m \times n$ real matrices with rank r

$\mathbb{C}_r^{m \times n}$ Set of all $m \times n$ complex matrices with rank r

$\mathbb{R}^{m \times n}[s]$ Set of all polynomial matrices of dimension $m \times n$ with real coefficients

$\mathbb{C}^{m \times n}[s]$ Set of all polynomial matrices of dimension $m \times n$ with complex coefficients

$\mathbb{R}^{m \times n}(s)$ Set of all rational matrices of dimension $m \times n$ with real coefficients

$\mathbb{C}^{m \times n}(s)$ Set of all rational matrices of dimension $m \times n$ with complex coefficients

$\mathbb{C}_p^k$ Set of all k times piecewise continuously differentiable functions

$\emptyset$ Null set

$\dim(V)$ Dimension of a vector space V

$V^\perp$ Orthogonal complement of subspace V

ker(T)	Kernel of transformation or matrix T
Image(T)	Image of transformation or matrix T
span(T)	Subspace spanned by the columns of matrix T

Notations Related to Vectors and Matrices

0_n	Zero vector in $\mathbb{R}^n$
$0_{m \times n}$	Zero matrix in $\mathbb{R}^{m \times n}$
I_n	Identity matrix of order n
$[a_{ij}]_{m \times n}$	Matrix of dimension $m \times n$ with the i-th row and j-th column element being a_{ij}
A^{-1}	Inverse matrix of matrix A
A^{T}	Transpose of matrix A
$\bar{A}$	Complex conjugate of matrix A
A^*	Transposed complex conjugate of matrix A
Re(A)	Real part of matrix A
Im(A)	Imaginary part of matrix A
det(A)	Determinant of matrix A
adj(A)	Adjoint of matrix A
tr(A)	Trace of matrix A
rank(A)	Rank of matrix A
$\rho(A)$	Spectral radius of matrix A
$A > 0$	A is symmetric positive definite
$A \geq 0$	A is symmetric semi-positive definite
$A > B$	$A - B > 0$
$A \geq B$	$A - B \geq 0$
$A^{\frac{1}{2}}$	The symmetric matrix Z satisfying $Z^2 = A > 0$
$\lambda_i(A)$	i-th eigenvalue of matrix A
$\sigma(A)$	Set of all eigenvalues of matrix A
$\sigma(E, A)$	Set of all finite eigenvalues of the regular matrix pencil (E, A)
$\lambda_{\min}(A)$	Minimum eigenvalue of matrix A
$\lambda_{\max}(A)$	Maximum eigenvalue of matrix A
$\sigma_i(A)$	i-th singular value of matrix A
$\sigma_{\max}(A)$	Maximum singular value of matrix A
$\sigma_{\min}(A)$	Minimum singular value of matrix A
$\|A\|_{\mathrm{F}}$	Frobenius norm of matrix A
$\|A\|_1$	Row-sum norm of matrix A

$\|A\|_2$	Spectral norm of matrix A
$\|A\|_\infty$	Column-sum norm of matrix A
$\mu_i(A)$	Measure induced from $\|A\|_i$, $i = 1, 2, \infty$ and F
$\dot{x}(t)$	First order derivative of vector x with respective to t
$\ddot{x}(t)$	Second order derivative of vector x with respective to t
$x^{(i)}(t)$	i-th order derivative of vector x with respective to t

Notations of Relations and Manipulations

$\Rightarrow$	Imply
$\Leftrightarrow$	If and only if
$\in$	Belong to
$\notin$	Not belong to
$\subset$	Subset
$\cap$	Intersection
$\cup$	Union
$\forall$	Arbitrarily chosen
$\sim$	Equivalent to
$\oplus$	Internal or external direct sum of two linear spaces of vectors
$\equiv$	Identical, equal everywhere
$\mathscr{L}[\cdot]$	Laplace transformation of a function
$\mathbb{B}\backslash\mathbb{A}$	Complement of set $\mathbb{A}$ in $\mathbb{B}$

Other Notations

δ_{ij}	Kronecker function
$\delta(t - \tau)$	Dirac (or Delta) function
$Q_c[A, B]$	Controllability matrix of the matrix pair (A, B)
$Q_o[A, C]$	Observability matrix of the matrix pair (A, C)
$\deg(\cdot)$	Degree of a polynomial
$\max\{c_1, c_2, \ldots, c_n\}$	Maximum value among real scalars $c_1, c_2, \ldots, c_n$
$\min\{c_1, c_2, \ldots, c_n\}$	Minimum value among real scalars $c_1, c_2, \ldots, c_n$
$\mathrm{span}\{x_1, x_2, \ldots, x_n\}$	Vector space spanned by the vectors $x_1, x_2, \ldots, x_n$
$\mathrm{diag}(d_1, d_2, \ldots, d_n)$	Diagonal matrix with diagonal elements $d_1, d_2, \ldots, d_n$

Chapter 1
Introduction

This chapter gives a brief introduction to descriptor systems. Section 1.1 discusses the state space representation of descriptor systems, with an emphasis on the time-invariant descriptor linear systems, which will be mainly discussed in this book. Section 1.2 presents examples of descriptor systems, specifically, the electrical circuit systems, the interconnected large-scale systems, the constrained mechanical systems, and the robotic systems. The purpose of this section is to assure the readers about the existence of descriptor systems in various fields. In Sect. 1.3, various feedback laws in descriptor linear systems are first introduced and then, a general introduction to the basic problems involving descriptor linear systems analysis and design is presented. An overview of the book is given in Sect. 1.4 and some notes and comments follow in Sect. 1.5.

1.1 Models for Descriptor Systems

It is well known from modern control theory that two main mathematical representations for dynamical systems are the transfer matrix representation and the state space representation. The former describes only the input–output property of the system, while the latter gives further insight into the structural property of the system. In this section, we describe the models for descriptor systems. Since the content of this book is mainly based on the state space approach, here only the state space representation is presented.

1.1.1 State Space Representation

State space approach was developed at the end of the 1950s and the beginning of the 1960s, which has the advantage that it not only provides us with an efficient method for control system analysis and synthesis, but also offers us more understanding about the various properties of the systems.

G.-R. Duan, *Analysis and Design of Descriptor Linear Systems*, Advances
in Mechanics and Mathematics 23, DOI 10.1007/978-1-4419-6397-0_1,
© Springer Science+Business Media, LLC 2010

State space models of systems are obtained mainly using the so-called state space variable method. To obtain a state space model of a practical system, we need to choose some physical variables such as speed, weight, temperature, or acceleration, which are sufficient to characterize the system. Then, by the physical relationships among the variables or by some model identification techniques, a set of equations can be established. Naturally, this set of equations are usually differential and/or algebraic equations, which form the mathematical model of the system. By properly defining a state vector $x(t)$ and an input vector $u(t)$, which are formed by the physical variables of the system, and an output vector $y(t)$, whose elements are properly chosen measurable variables of the system, this set of equations can be arranged into two equations: one is the so-called state equation, which is of the following general form:

$$f(\dot{x}(t), x(t), u(t), t) = 0, \tag{1.1}$$

and the other is the output equation, or the observation equation, which is in the form of

$$g(x(t), u(t), y(t), t) = 0, \tag{1.2}$$

where f and g are vector functions of appropriate dimensions with respect to $\dot{x}(t)$, $x(t)$, $u(t)$, $y(t)$ and t. Equations (1.1) and (1.2) give the state space representation for a general nonlinear dynamical system.

A special form of (1.1)–(1.2) is the following:

$$\begin{cases} E(t)\dot{x}(t) = H(x(t), u(t), t) \\ y(t) = J(x(t), \dot{x}(t), u(t), t) \end{cases}, \tag{1.3}$$

where $t \geq t_0$ is the time variable, H and J are appropriate dimensional vector functions, $x(t) \in \mathbb{R}^n$ is the state vector, $u(t) \in \mathbb{R}^r$ is the control input vector, $y(t) \in \mathbb{R}^m$ is the measured output vector. The matrix $E(t)$ may be singular for some $t \geq t_0$. Equation (1.3) is the general form for the so-called (nonlinear) descriptor systems.

When H is a linear function of $x(t)$ and $u(t)$, and J a linear function of $x(t)$, $\dot{x}(t)$ and $u(t)$, the general nonlinear descriptor system (1.3) reduces to the following form:

$$\begin{cases} E(t)\dot{x}(t) = A(t)x(t) + B(t)u(t) \\ y(t) = C_1(t)x(t) + C_2(t)\dot{x}(t) + D(t)u(t) \end{cases}, \tag{1.4}$$

where $E(t)$, $A(t) \in \mathbb{R}^{\tilde{n} \times n}$, $B(t) \in \mathbb{R}^{\tilde{n} \times r}$ and $C_1(t)$, $C_2(t) \in \mathbb{R}^{m \times n}$ are matrix functions of time t, and they are called the coefficient matrices of system (1.4). Equation (1.4) describes a general (time-varying) descriptor linear system.

The descriptor linear system (1.4) has a single observation equation, which observes a linear combination of both the state and the state derivative. In practical applications, the state variables and the state derivative variables are often separately measured. In this case, a special form of (1.4) is obtained as follows:

$$\begin{cases} E(t)\dot{x}(t) = A(t)x(t) + B(t)u(t) \\ y_p(t) = C_p(t)x(t) + D_p(t)u(t) \\ y_d(t) = C_d(t)\dot{x}(t) + D_d(t)u(t) \end{cases} \tag{1.5}$$

where $x(t) \in \mathbb{R}^n$ is the state vector, $u(t) \in \mathbb{R}^r$ is the control input vector, $y_p(t) \in \mathbb{R}^{m_p}$ and $y_d(t) \in \mathbb{R}^{m_d}$ are the measured proportional output vector and derivative output vector, respectively; $E(t)$, $A(t) \in \mathbb{R}^{\tilde{n} \times n}$, $B(t) \in \mathbb{R}^{\tilde{n} \times r}$, $C_p(t) \in \mathbb{R}^{m_p \times n}$ and $C_d(t) \in \mathbb{R}^{m_d \times n}$ are matrix functions of time t. Equation (1.5) describes a (time-varying) descriptor linear system, which is often encountered in practice.

1.1.2 Time-Invariant Descriptor Linear Systems

If all the coefficient matrices of the system (1.4) are constant ones, system (1.4) becomes the following time-invariant descriptor linear system

$$\begin{cases} E\dot{x}(t) = Ax(t) + Bu(t) \\ y(t) = C_1 x(t) + C_2 \dot{x}(t) + Du(t) \end{cases},$$

where $x(t) \in \mathbb{R}^n$ is the state vector, $u(t) \in \mathbb{R}^r$ is the control input vector, $y(t) \in \mathbb{R}^m$ is the measured output vector. The constant matrices E, $A \in \mathbb{R}^{\tilde{n} \times n}$, $B \in \mathbb{R}^{\tilde{n} \times r}$ and $C_1, C_2 \in \mathbb{R}^{m \times n}$ and $D \in \mathbb{R}^{m \times r}$ are called the coefficient matrices of the system.

Similarly, if all the coefficient matrices of the system (1.5) are constant ones, system (1.5) becomes the following time-invariant descriptor linear system

$$\begin{cases} E\dot{x}(t) = Ax(t) + Bu(t) \\ y_p(t) = C_p x(t) + D_p u(t) \\ y_d(t) = C_d \dot{x}(t) + D_d u(t) \end{cases} \tag{1.6}$$

where $x(t) \in \mathbb{R}^n$ is the state vector, $u(t) \in \mathbb{R}^r$ is the control input vector, $y_p(t) \in \mathbb{R}^{m_p}$ is the measured (proportional) output vector, and $y_d(t) \in \mathbb{R}^{m_d}$ is the measured derivative output vector. The constant matrices E, $A \in \mathbb{R}^{\tilde{n} \times n}$, $B \in \mathbb{R}^{\tilde{n} \times r}$, $C_p \in \mathbb{R}^{m_p \times n}$, $C_d \in \mathbb{R}^{m_d \times n}$, and $D_p \in \mathbb{R}^{m_d \times r}$ and $D_d \in \mathbb{R}^{m_d \times r}$ are called the coefficient matrices of the system.

The above descriptor linear system (1.6) is the main system to be studied in this book. For future convenience, we here clarify some basic concepts related to this system.

In accordance with the sizes of the coefficient matrices E and A, the system (1.6) is said to have dimension $\tilde{n} \times n$. When $\tilde{n} = n$, the system is said to be square and to have order n.

The system (1.6) is called a normal system if it is square and the matrix E is nonsingular. Specially, when the matrix E is an identity matrix, the system is called a standard normal system, which has been intensively studied in linear systems theory (Chen 1970; Blackman 1977; Kailath 1980; Furuta et al. 1988; Brogan 1991; Fairman 1998; Duan 2004a). Obviously, any normal system in the form of (1.6) can be written in the following standard normal form:

$$\begin{cases} \dot{x}(t) = E^{-1}Ax(t) + E^{-1}Bu(t) \\ y_{\mathrm{p}}(t) = C_{\mathrm{p}}x(t) + D_{\mathrm{p}}u(t) \\ y_{\mathrm{d}}(t) = C_{\mathrm{d}}\dot{x}(t) + D_{\mathrm{d}}u(t) \end{cases} . \qquad (1.7)$$

Therefore, theory for standard normal linear systems analysis and synthesis can be readily applied to a linear system in the form of (1.6) by first converting it into the standard form (1.7). However, we point out that this is not a good approach since the matrix inversion may subject to numerical problems and the computational error produced in this very first step (the conversion) may make the final analysis or synthesis results inaccurate. It is strongly recommended that theory for descriptor linear systems analysis and design be used to deal with normal systems in the form of (1.6) due to numerical considerations.

For the descriptor linear system (1.6), there is also a concept called the dynamical order, which is defined to be $n_0 = \mathrm{rank}E$. Clearly, for normal linear systems, the system dynamical order coincides with the usual system order n, while for a general description linear system, the system dynamical order is generally less than the system order n.

The corresponding (E, A) is called the matrix pair associated with the system (1.6). The corresponding polynomial matrix

$$P_{(E, A)}(s) = sE - A$$

is called the matrix pencil of system (1.6). It is uniquely determined by the matrix pair (E, A), and is called square when $\tilde{n} = n$.

When the descriptor linear system (1.6) is square,

$$\det(sE - A) = 0 \qquad (1.8)$$

is called the characteristic equation of the system, and any finite $s \in \mathbb{C}$ satisfying (1.8) is called a finite pole or a finite eigenvalue of the system, or the matrix pair (E, A). The set of all finite eigenvalues of the matrix pair (E, A) is denoted by $\sigma(E, A)$, that is

$$\sigma(E, A) = \{\lambda \mid \lambda \in \mathbb{C}, \lambda \text{ finite}, \det(\lambda E - A) = 0\} .$$

Obviously, when $E = I$, the identical matrix, $\sigma(E, A) = \sigma(I, A)$ reduces to the set of eigenvalues of matrix A, and is usually denoted by $\sigma(A)$.

In practical system analysis and control system design, many system models may be established in the general descriptor system form of (1.6), while they cannot be described by the standard normal system form of (1.7). In the next section, we present some examples of descriptor linear systems to assure the readers about the existence of descriptor linear systems.

1.2 Examples of Descriptor Linear Systems

Descriptor systems appear in many fields, such as power systems, electrical networks, aerospace engineering, chemical processes, social economic systems, network analysis, biological systems, time-series analysis, and so on. In this section, some examples of descriptor systems are presented, from which readers can indeed see the existence of descriptor linear systems in our real world.

1.2.1 Electrical Circuit Systems

Many electrical circuit systems can be described by descriptor linear systems. Here, let us consider two such simple circuit networks.

Example 1.1 (Dai 1989b). Consider a simple circuit network as shown in Fig. 1.1, where R, L, and C_0 stand for the resistor, inductor, and capacity, respectively, and their voltages are denoted by $V_R(t)$, $V_L(t)$, $V_C(t)$, respectively. $V_S(t)$ is the voltage source which is taken as the control input. Following basic circuit theory and the Kirchoff's law (Smith 1966), we have the following equations, which describe the system:

$$L\dot{I}(t) = V_L(t), \tag{1.9}$$

$$\dot{V}_C(t) = \frac{1}{C_0}I(t), \tag{1.10}$$

$$RI(t) = V_R(t), \tag{1.11}$$

$$V_L(t) + V_C(t) + V_R(t) = V_S(t). \tag{1.12}$$

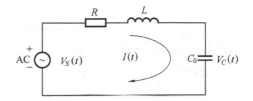

Fig. 1.1 A single-loop circuit network

By choosing the state vector x, the input vector u and the output vector y as follows:

$$x = \begin{bmatrix} I(t) \\ V_L(t) \\ V_C(t) \\ V_R(t) \end{bmatrix}, \ u = V_S(t), \ y = V_C(t),$$

the above (1.9)–(1.12) can be written in the following descriptor linear system form:

$$\begin{cases} E\dot{x} = Ax + Bu \\ y = Cx \end{cases} \tag{1.13}$$

with

$$E = \begin{bmatrix} L & 0 & 0 & 0 \\ 0 & 0 & 1 & 0 \\ 0 & 0 & 0 & 0 \\ 0 & 0 & 0 & 0 \end{bmatrix}, \ B = \begin{bmatrix} 0 \\ 0 \\ 0 \\ -1 \end{bmatrix}$$

and

$$A = \begin{bmatrix} 0 & 1 & 0 & 0 \\ \frac{1}{C_0} & 0 & 0 & 0 \\ -R & 0 & 0 & 1 \\ 0 & 1 & 1 & 1 \end{bmatrix}, \ C = \begin{bmatrix} 0 & 0 & 1 & 0 \end{bmatrix}.$$

Example 1.2 (Dai 1989b). Consider the circuit system that is shown in Fig. 1.2, where u_{c_1}, u_{c_2} are the voltages of C_1, C_2, and I_1, I_2 are the amperages of the currents flowing over them. Choose the state vector as

$$x = \begin{bmatrix} u_{c_1} & u_{c_2} & I_2 & I_1 \end{bmatrix}^{\mathrm{T}}.$$

Further, choose the control input u, and the output variable y as

$$u = u_e \ \text{and} \ y = u_{c_2}.$$

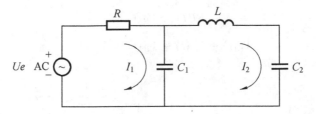

Fig. 1.2 A two-loop circuit network

Then, again according to the Kirchoff's second law we can establish the following state space representation for the system:

$$\begin{cases} \begin{bmatrix} C_1 & 0 & 0 & 0 \\ 0 & C_2 & 0 & 0 \\ 0 & 0 & -L & 0 \\ 0 & 0 & 0 & 0 \end{bmatrix} \dot{x} = \begin{bmatrix} 0 & 0 & 0 & 1 \\ 0 & 0 & 1 & 0 \\ -1 & 1 & 0 & 0 \\ 1 & 0 & R & R \end{bmatrix} x + \begin{bmatrix} 0 \\ 0 \\ 0 \\ -1 \end{bmatrix} u \\ y = \begin{bmatrix} 0 & 1 & 0 & 0 \end{bmatrix} x \end{cases}$$

1.2.2 Large-Scale Systems with Interconnections

Example 1.3 (Dai 1989b). Consider a class of interconnected large-scale systems with the following subsystems

$$\begin{cases} \dot{x}_i(t) = A_i x_i(t) + B_i a_i(t) \\ b_i(t) = C_i x_i(t) + D_i a_i(t), \quad i = 1, 2, \dots, N \end{cases}, \qquad (1.14)$$

where $x_i(t)$, $a_i(t)$, and $b_i(t)$ are the substate, the control input, and the output, respectively, of the i-th subsystem. By denoting

$$x(t) = \begin{bmatrix} x_1(t) \\ x_2(t) \\ \vdots \\ x_N(t) \end{bmatrix}, \quad a(t) = \begin{bmatrix} a_1(t) \\ a_2(t) \\ \vdots \\ a_N(t) \end{bmatrix}, \quad b(t) = \begin{bmatrix} b_1(t) \\ b_2(t) \\ \vdots \\ b_N(t) \end{bmatrix},$$

$$A = \mathrm{diag}(A_1, A_2, \dots, A_N), \quad B = \mathrm{diag}(B_1, B_2, \dots, B_N),$$
$$C = \mathrm{diag}(C_1, C_2, \dots, C_N), \quad D = \mathrm{diag}(D_1, D_2, \dots, D_N),$$

the equations in (1.14) can be rewritten as

$$\begin{cases} \dot{x}(t) = Ax(t) + Ba(t) \\ b(t) = Cx(t) + Da(t) \end{cases}. \qquad (1.15)$$

Assume that the subsystem interconnection is linear and is given by

$$\begin{cases} a(t) = L_{11}b(t) + L_{12}u(t) + R_{11}a(t) + R_{12}y(t) \\ y(t) = L_{21}b(t) + L_{22}u(t) + R_{21}a(t) + R_{22}y(t) \end{cases}, \qquad (1.16)$$

where $u(t)$ is the overall input of the large-scale system; $y(t)$ is the overall measured output; $L_{ij}, R_{ij}, i, j = 1, 2$, are constant matrices of appropriate dimensions. Then by letting

$$x_e(t) = [x^T(t) \; a^T(t) \; b^T(t) \; y^T(t)]^T, \; y_e(t) = y(t),$$

(1.15) and (1.16) can be converted into the form of

$$\begin{cases} E_e \dot{x}_e(t) = A_e x_e(t) + B_e u(t) \\ y_e(t) = C_e x_e(t) \end{cases}$$

with

$$E_e = \begin{bmatrix} I & 0 & 0 & 0 \\ 0 & 0 & 0 & 0 \\ 0 & 0 & 0 & 0 \\ 0 & 0 & 0 & 0 \end{bmatrix}, \; A_e = \begin{bmatrix} A & B & 0 & 0 \\ C & D & -I & 0 \\ 0 & R_{11} - I & L_{11} & R_{12} \\ 0 & R_{21} & L_{21} & R_{22} - I \end{bmatrix},$$

$$B_e = \begin{bmatrix} 0 \\ 0 \\ L_{12} \\ L_{22} \end{bmatrix}, \; C_e = [0 \; 0 \; 0 \; I],$$

which is in the form of the descriptor linear system (1.6).

1.2.3 Constrained Mechanical Systems

Constrained linear mechanical systems can be described as follows:

$$M\ddot{z} + D\dot{z} + Kz = Lf + J\mu, \tag{1.17}$$

$$G\dot{z} + Hz = 0, \tag{1.18}$$

where $z \in \mathbb{R}^n$ is the displacement vector, $f \in \mathbb{R}^n$ is the vector of known input forces, $\mu \in \mathbb{R}^q$ is the vector of Lagrangian multipliers, M is the inertial matrix, which is usually symmetric and positive definite, D is the damping and gyroscopic matrix, K is the stiffness and circulator matrix, L is the force distribution matrix, J is the Jacobian of the constraint equation, G and H are the coefficient matrices of the constraint equation. All matrices in (1.17)–(1.18) are known and constant ones of appropriate dimensions.

Equation (1.17) is the dynamical equation, while (1.18) is the constraint equation.

Assume that a linear combination of displacements and velocities is measurable, then, the output equation is of the form

$$y = C_p z + C_v \dot{z}, \tag{1.19}$$

where $C_p, C_v \in \mathbb{R}^{m \times n}$. By further choosing the state vector and the input vector as

$$x = \begin{bmatrix} z \\ \dot{z} \\ \mu \end{bmatrix} \text{ and } u = f,$$

respectively, then the above (1.17)–(1.19) can be written in the following descriptor linear system form:

$$\begin{cases} E\dot{x} = Ax + Bu \\ y = Cx \end{cases} \tag{1.20}$$

with

$$E = \begin{bmatrix} I & 0 & 0 \\ 0 & M & 0 \\ 0 & 0 & 0 \end{bmatrix}, \quad B = \begin{bmatrix} 0 \\ L \\ 0 \end{bmatrix}$$

and

$$A = \begin{bmatrix} 0 & I & 0 \\ -K & -D & J \\ H & G & 0 \end{bmatrix}, \quad C = \begin{bmatrix} C_p & C_v & 0 \end{bmatrix}.$$

Example 1.4 (Schmidt 1994). Consider a mechanical system shown in Fig. 1.3. This system consists of two one-mass oscillators connected by a dashpot element. Let

$$m_1 = m_2 = 1\,\text{kg}, \ d = 1\,\text{Ns/m}, \ k_1 = 2\,\text{N/m}, \ k_2 = 1\,\text{N/m}.$$

Then the three equations (1.17)–(1.19) for this system can be obtained as follows:

$$\begin{bmatrix} 1 & 0 \\ 0 & 1 \end{bmatrix}\begin{bmatrix} \ddot{z}_1 \\ \ddot{z}_2 \end{bmatrix} + \begin{bmatrix} 1 & 1 \\ 1 & 1 \end{bmatrix}\begin{bmatrix} \dot{z}_1 \\ \dot{z}_2 \end{bmatrix} + \begin{bmatrix} 2 & 0 \\ 0 & 1 \end{bmatrix}\begin{bmatrix} z_1 \\ z_2 \end{bmatrix} = \begin{bmatrix} -1 \\ 1 \end{bmatrix} f + \begin{bmatrix} 1 \\ 1 \end{bmatrix} \mu,$$

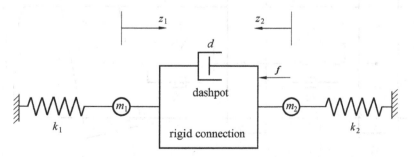

Fig. 1.3 Two connected one-mass oscillators

$$[1\ 1]\begin{bmatrix} z_1 \\ z_2 \end{bmatrix} = 0,$$

$$y = [1\ 0]\begin{bmatrix} z_1 \\ z_2 \end{bmatrix}.$$

Therefore, we have

$$M = \begin{bmatrix} 1 & 0 \\ 0 & 1 \end{bmatrix}, \ D = \begin{bmatrix} 1 & 1 \\ 1 & 1 \end{bmatrix}, \ K = \begin{bmatrix} 2 & 0 \\ 0 & 1 \end{bmatrix},$$

$$L = \begin{bmatrix} -1 \\ 1 \end{bmatrix}, \ J = \begin{bmatrix} 1 \\ 1 \end{bmatrix},$$

$$G = [0\ 0], \ H = [1\ 1],$$

and

$$C_p = [1\ 0], \ C_v = [0\ 0].$$

Based on these matrices, the descriptor linear system form (1.20) for the system can be readily written out.

Remark 1.1. This example is actually a modified version of Schmidt (1994). In Schmidt (1994), the mechanical system is free of external forces, and eventually the L matrix, and hence the B matrix, in the system models are zero ones. While here we have added the external force f, and have derived a descriptor linear system in the form of (1.20) with a nonzero matrix B.

Example 1.5. Rolling ring drive (Schmidt 1994, see also Hou 1995) Fig. 1.4 shows a simplified model of a rolling ring drive, which converts rotary motion of a plain

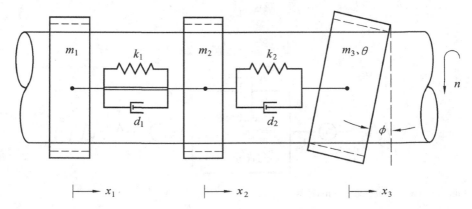

Fig. 1.4 A rolling ring drive

round shaft into linear movement. It operates like a nut on a threaded bar, however the pitch of both left-hand and right-hand is variable depending on the angle of the rolling rings. The connection between the rolling ring drive (m_3, θ) and the pay load (m_2) as well as that between (m_1) and (m_2) has been modeled as spring-dashpot elements with coefficients k_2 and d_2 as well as k_1 and d_1, respectively. In addition, the masses m_1 and m_2 are connected by a rigid massless bar. The translational velocity $\dot{x}_3$ depends linearly on the angular deflection ϕ of the rolling rings, as long as small angles ϕ are regarded. This rolling ring drive represents an example of a dynamical system with both holonomic and nonholonomic constraints.

This system is described by the following equations:

$$\theta\ddot{\phi} = u,$$

$$m_1\ddot{x}_1 + d_1(\dot{x}_1 - \dot{x}_2) + k_1(x_1 - x_2) = \mu_2,$$

$$m_2\ddot{x}_2 + d_1(\dot{x}_2 - \dot{x}_1) + d_2(\dot{x}_2 - \dot{x}_3) + k_1(x_2 - x_1) + k_2(x_2 - x_3) = -\mu_2,$$

$$m_3\ddot{x}_3 + d_2(\dot{x}_3 - \dot{x}_2) + k_2(x_3 - x_2) = \mu_1,$$

$$x_1 = x_2, \tag{1.21}$$

$$\dot{x}_3 = c\phi, \tag{1.22}$$

$$y = \begin{bmatrix} x_2 \\ x_3 \end{bmatrix}.$$

Equations (1.21) and (1.22) represent the constraints of the system; μ_2 and μ_1 are the Lagrangian multipliers correspondingly.

For simplicity, let

$$m_1 = m_2 = m_3 = 1\,\text{kg}, \ \theta = 1, \ c = 1\,\text{m/srad},$$

$$d_1 = d_2 = 1\,\text{Ns/m}, \ k_1 = k_2 = 1\,\text{N/m},$$

and denote

$$z = \begin{bmatrix} \phi & x_1 & x_2 & x_3 \end{bmatrix}^{\text{T}}, \ \mu = \begin{bmatrix} \mu_1 & \mu_2 \end{bmatrix}^{\text{T}}.$$

Then the system can be arranged into the form of (1.17)–(1.19) with $M = I_4$ and

$$D = K = \begin{bmatrix} 0 & 0 & 0 & 0 \\ 0 & 1 & -1 & 0 \\ 0 & -1 & 2 & -1 \\ 0 & 0 & -1 & 1 \end{bmatrix}, \ L = \begin{bmatrix} 1 \\ 0 \\ 0 \\ 0 \end{bmatrix}, \ J = \begin{bmatrix} 0 & 0 \\ 0 & 1 \\ 0 & -1 \\ 1 & 0 \end{bmatrix},$$

$$G = \begin{bmatrix} 0 & 0 & 0 & 0 \\ 0 & 0 & 0 & 1 \end{bmatrix}, \ H = \begin{bmatrix} 0 & -1 & 1 & 0 \\ -1 & 0 & 0 & 0 \end{bmatrix},$$

$$[C_p \quad C_v] = \begin{bmatrix} 0 & 0 & 1 & 0 & 0 & 0 & 0 & 0 \\ 0 & 0 & 0 & 1 & 0 & 0 & 0 & 0 \end{bmatrix},$$

and can thus be further arranged into the descriptor system form (1.20).

1.2.4 Robotic System–A Three-Link Planar Manipulator

In this subsection, let us consider the modeling of a three-link planar manipulator, which is a simplified model of a cleaning robot. The motion of a cleaning robot is restricted by the cleaning surface. This makes cleaning robot a good example of descriptor systems.

Task Description

Figure 1.5 shows a large mobile manipulator cleaning the facade of a high building (Wanner et al. 1986). The development of this kind of manipulator is part of several research projects in which some German research institutes and companies are involved (Wanner and Kong 1990; Hiller 1994). The mobile manipulator belongs to an important type of service robots widely utilized in disaster and emergency relief, construction, public services, and environmental protection (Schraft et al. 1993). The cleaning robot named SKYWASH already works in applications (Schraft and Wanner 1993).

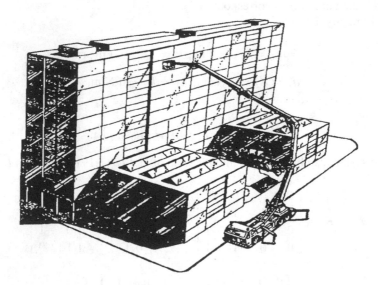

Fig. 1.5 A cleaning mobile manipulator

Fig. 1.6 A tree-link planar manipulator

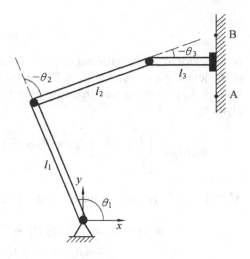

Figure 1.6 shows a three-link planar manipulator whose task is to clean the region between the points A and B. The robot fulfills the task by moving the end-effector of the manipulator from point A to point B repeatedly with the specified contact force within the given time. This is a simplified model and task description of the large mobile manipulator cleaning the facade of a large building illustrated in Fig. 1.5.

Assume that the cleaning flat surface is a rigid body and the end of the third arm is a smooth and rigid plate. Thus, there are two constraints on the motion of the robot:

- The restriction on the motion in the x direction, usually given as $x \leq 1$.
- The orthogonality of the third arm to the cleaning surface, which can be described by $\theta_1 + \theta_2 + \theta_3 = 0$.

Obviously, during the cleaning process these two constraints should be kept active.

The Nonlinear Model

The motion of constrained robots can be easily modeled as a descriptor system. One needs only to add the constraint equations and the corresponding generalized constrained force to the motion equation of a free robot. The systematic modeling procedure for the motion of a free robot is well established, see, for example, Craig (1986).

The dynamics of the three-link planar manipulator in Fig. 1.6 is described by the following equations in the joint coordinates:

$$M_\theta(\theta)\ddot{\theta} + C_\theta(\theta, \dot{\theta}) + G_\theta(\theta) = u_\theta + F_\theta^T \mu, \qquad (1.23)$$

$$\psi_\theta(\theta) = 0, \qquad (1.24)$$

where
$$\theta = \begin{bmatrix} \theta_1 & \theta_2 & \theta_3 \end{bmatrix}^{\mathrm{T}}$$

is the vector of joint displacements, $u_\theta \in \mathbb{R}^3$ is the vector of control torques applied at the joints. $F_\theta = \partial \varphi_\theta / \partial \theta$, $\mu \in \mathbb{R}^2$ represents the vector of the Lagrangian multipliers, $F_\theta^{\mathrm{T}} \mu$ is the generalized constraint force. The constraint function $\psi_\theta(\theta)$ is given by

$$\psi_\theta(\theta) = \begin{bmatrix} l_1 \cos \theta_1 + l_2 \cos(\theta_1 + \theta_2) + l_3 \cos(\theta_1 + \theta_2 + \theta_3) - l \\ \theta_1 + \theta_2 + \theta_3 \end{bmatrix}.$$

$M_\theta(\theta) \in \mathbb{R}^{3 \times 3}$ is the mass matrix, and is given by $M_\theta(\theta) = [m_{ij}(\theta)]_{3 \times 3}$ with

$$
\begin{aligned}
m_{11}(\theta) &= m_1 l_1^2 + m_2 \left(l_1^2 + l_2^2 + 2 l_1 l_2 \cos \theta_1 \right) \\
&\quad + m_3 (l_1^2 + l_2^2 + l_3^2 + 2 l_1 l_2 \cos \theta_2) \\
&\quad + m_3 \left(2 l_2 l_3 \cos \theta_3 + 2 l_2 l_3 \cos(\theta_2 + \theta_3) \right), \\
m_{12}(\theta) &= m_2 \left(l_2^2 + l_1 l_2 \cos \theta_2 \right) \\
&\quad + m_3 (l_2^2 + l_3^2 + l_1 l_2 \cos \theta_2 + 2 l_2 l_3 \cos \theta_3) \\
&\quad + m_3 \left(l_1 l_3 \cos(\theta_2 + \theta_3) \right), \\
m_{22}(\theta) &= m_2 l_2^2 + m_3 \left(l_2^2 + l_3^2 + 2 l_2 l_3 \cos \theta_3 \right), \\
m_{23}(\theta) &= m_3 (l_3^2 + l_2 l_3 \cos \theta_3), \\
m_{33}(\theta) &= m_3 l_3^2.
\end{aligned}
$$

$C_\theta(\theta, \dot{\theta}) \in \mathbb{R}^3$ is the centrifugal and Coriolis vector, and is given by

$$C_\theta(\theta, \dot{\theta}) = C_{\mathrm{I}}(\theta) \Theta_{\mathrm{N}} + C_{\mathrm{II}}(\theta) \Theta_{\mathrm{S}},$$

where

$$\Theta_{\mathrm{N}} = \begin{bmatrix} \dot{\theta}_1 \dot{\theta}_2 & \dot{\theta}_1 \dot{\theta}_3 & \dot{\theta}_2 \dot{\theta}_3 \end{bmatrix}^{\mathrm{T}},$$
$$\Theta_{\mathrm{S}} = [\dot{\theta}_1^2 \ \dot{\theta}_2^2 \ \dot{\theta}_3^2]^{\mathrm{T}},$$

and
$$C_{\mathrm{I}}(\theta) = \begin{bmatrix} c_{\mathrm{I},ij}(\theta) \end{bmatrix}_{3 \times 3}, \quad C_{\mathrm{II}}(\theta) = \begin{bmatrix} c_{\mathrm{II},ij}(\theta) \end{bmatrix}_{3 \times 3}$$

with

$$
\begin{aligned}
c_{\mathrm{I},11}(\theta) &= -2 m_2 l_1 l_2 \sin \theta_2 - 2 m_3 l_1 (l_2 \sin \theta_2 + l_3 \sin(\theta_2 + \theta_3)), \\
c_{\mathrm{I},12}(\theta) &= -2 m_3 l_3 (l_2 \sin \theta_3 + l_1 \sin(\theta_2 + \theta_3)), \\
c_{\mathrm{I},13}(\theta) &= c_{\mathrm{I},12}(\theta),
\end{aligned}
$$

$$c_{I,21}(\theta) = c_{I,32}(\theta) = c_{I,33}(\theta) = 0,$$
$$c_{I,22}(\theta) = c_{I,23}(\theta) = -c_{I,31}(\theta) = -2m_3 l_2 l_3 \sin\theta_3,$$
$$c_{II,11}(\theta) = c_{II,22}(\theta) = c_{II,33}(\theta) = 0,$$
$$c_{II,21}(\theta) = -c_{II,12}(\theta) = (m_2 + m_3)l_1 l_2 \sin\theta_2 + m_3 l_1 l_3 \sin(\theta_2 + \theta_3),$$
$$c_{II,31}(\theta) = -c_{II,13}(\theta) = m_3 l_3 (l_2 \sin\theta_3 + l_1 \sin(\theta_2 + \theta_3)),$$
$$c_{II,32}(\theta) = -c_{II,23}(\theta) = m_3 l_2 l_3 \sin\theta_3.$$

$G_\theta(\theta) \in \mathbb{R}^3$ is the vector of gravity, which is given by

$$G_\theta^T(\theta) = \begin{bmatrix} g_1(\theta) & g_2(\theta) & g_3(\theta) \end{bmatrix}$$

with

$$g_1(\theta) = g m_1 l_1 \cos\theta_1 + g m_2 (l_1 \cos\theta_1 + l_2 \cos(\theta_1 + \theta_2))$$
$$+ g m_3 (l_1 \cos\theta_1 + l_2 \cos(\theta_1 + \theta_2) + l_3 \cos(\theta_1 + \theta_2 + \theta_3)),$$
$$g_2(\theta) = g m_2 l_2 \cos(\theta_1 + \theta_2)$$
$$+ g m_3 (l_2 \cos(\theta_1 + \theta_2) + l_3 \cos(\theta_1 + \theta_2 + \theta_3)),$$
$$g_3(\theta) = g m_3 l_3 \cos(\theta_1 + \theta_2 + \theta_3).$$

The Linearized Model

Before linearizing the system, let us first express the dynamics of the manipulator in Cartesian coordinates because the constraint of the environment as well as the statement of the task is often easily described in these coordinates. Let

$$z = \begin{bmatrix} x & y & \phi \end{bmatrix}^T$$

be the Cartesian vector representing the position and the orientation of the end-effector. In Cartesian coordinates, the description (1.23)–(1.24) becomes

$$M_z(\theta)\ddot{z} + C_z(\theta, \dot{\theta}) + G_z(\theta) = u_z + F_z^T \mu, \tag{1.25}$$

$$\psi_z(\theta) = 0, \tag{1.26}$$

with

$$M_z(\theta) = J^{-T}(\theta)M_\theta(\theta)J^{-1}(\theta),$$
$$G_z(\theta) = J^{-T}(\theta)G_\theta(\theta),$$
$$C_z(\theta, \dot{\theta}) = J^{-T}(\theta)\left[C_\theta(\theta, \dot{\theta}) - M_\theta(\theta)J^{-1}(\theta)\dot{J}(\theta)\dot{\theta} \right]$$

and

$$u_z = J^{-T}(\theta)u_\theta,$$

where the Jacobian $J(\theta)$ satisfies $\dot{z} = J(\theta)\dot{\theta}$ and is given by

$$J(\theta) = \begin{bmatrix} -l_1s_1 - l_2s_{12} - l_3s_{123} & -l_2s_{12} - l_3s_{123} & -l_3s_{123} \\ l_1c_1 + l_2c_{12} + l_3c_{123} & l_2c_{12} + l_3c_{123} & l_3c_{123} \\ 1 & 1 & 1 \end{bmatrix}$$

with

$$s_1 = \sin\theta_1, \quad s_{12} = \sin(\theta_1 + \theta_2), \quad s_{123} = \sin(\theta_1 + \theta_2 + \theta_3),$$

$$c_1 - \cos\theta_1, \quad c_2 = \cos(\theta_1 + \theta_2), \quad c_{123} = (\theta_1 + \theta_2 + \theta_3).$$

Noticing the relations

$$x = l_1c_1 + l_2c_{12} + l_3c_{123} \text{ and } \phi = \theta_1 + \theta_2 + \theta_3,$$

we can obtain that

$$\psi_z(z) = F_0 z - L_0$$

with

$$F_0 = \begin{bmatrix} 1 & 0 & 0 \\ 0 & 0 & 1 \end{bmatrix}, \quad L_0 = \begin{bmatrix} l \\ 0 \end{bmatrix}.$$

Therefore,

$$F_z = \frac{\partial \psi_z(z)}{\partial z} = F_0.$$

This means that in Cartesian coordinates the considered robot model has linear constraints. This outcome is favorable in linearizing the model.

Let the system parameters be

$$m_1 = 20 \, \text{kg}, m_2 = 10 \, \text{kg}, m_3 = 5 \, \text{kg},$$

$$l_1 = 2 \, \text{m}, l_2 = 1.5 \, \text{m}, l_3 = 1 \, \text{m},$$

and

$$l = 2 \, \text{m}, \Delta l = 1 \, \text{m}, g = 0.98 \, \text{m/s}^2.$$

Choose the working point of linearization as

$$z_\omega = \begin{bmatrix} l & l + \dfrac{\Delta l}{2} & 0 \end{bmatrix}^{\text{T}}, \quad \dot{z}_\omega = \begin{bmatrix} 0 & \dot{y}_\omega & 0 \end{bmatrix}^{\text{T}}$$

with

$$\dot{y}_\omega = 0.3 \, \text{m/s}, \text{ and } \ddot{z}_\omega = \begin{bmatrix} 0 & 0 & 0 \end{bmatrix}^{\text{T}}.$$

Further, denote

$$\begin{cases} \delta z = z - z_\omega \\ \delta \dot{z} = \dot{z} - \dot{z}_\omega \\ \delta \ddot{z} = \ddot{z} - \ddot{z}_\omega \end{cases}.$$

Then, the linearized model of (1.25)–(1.26) can be obtained as follows:

$$M_0 \delta \ddot{z} + D_0 \delta \dot{z} + K_0 \delta z = S_0 \delta u + F_0^T \delta \mu, \tag{1.27}$$

$$F_0 \delta z = 0 \tag{1.28}$$

with

$$M_0 = M_z|_{z=z_\omega} = \begin{bmatrix} 18.75320 & -7.94493 & 7.94494 \\ -7.94493 & 31.81820 & -26.81820 \\ 7.94494 & -26.81820 & 26.81820 \end{bmatrix},$$

$$D_0 = \left.\frac{\partial C_z}{\partial \dot{z}}\right|_{\substack{z=z_\omega \\ \dot{z}=\dot{z}_\omega}} = \begin{bmatrix} -1.52143 & -1.55168 & 1.55168 \\ 3.22064 & 3.28467 & -3.28467 \\ -3.22064 & -3.28467 & 3.28467 \end{bmatrix},$$

$$K_0 = \left.\frac{\partial C_z}{\partial z}\right|_{\substack{z=z_\omega \\ \dot{z}=\dot{z}_\omega}} + \left.\frac{\partial G_z}{\partial z}\right|_{z=z_\omega} = \begin{bmatrix} 67.48940 & 69.23930 & -69.23920 \\ 69.81240 & 1.68624 & -1.68617 \\ -69.81230 & -1.68617 & -68.27070 \end{bmatrix},$$

$$S_0 = J^{-T}|_{\theta=\theta_\omega} = \begin{bmatrix} -0.216598 & -0.338060 & 0.554659 \\ 0.458506 & -0.845153 & 0.386648 \\ -0.458506 & 0.845153 & 0.613353 \end{bmatrix},$$

$$F_0 = F_z = \begin{bmatrix} 1 & 0 & 0 \\ 0 & 0 & 1 \end{bmatrix},$$

and

$$\delta u = u - u_\omega \text{ and } \delta \mu = \mu - \mu_\omega.$$

Here, θ_ω is determined by z_ω through inverse kinematics (Craig 1986). μ_ω is a two-dimensional vector. The first element of μ_ω is chosen to be equal to the desired contact force in the x direction, i.e., $\mu_{\omega 1} = -20N$, and the second element is chosen to be zero. Then, u_ω is suitably determined such that (1.27) remains balanced.

Define the state vector

$$x^T = \begin{bmatrix} \delta z^T & \delta \dot{z}^T & \delta \mu^T \end{bmatrix}$$

with

$$\delta z = \begin{bmatrix} \delta x & \delta y & \delta \phi \end{bmatrix}^T \text{ and } \delta \mu = \begin{bmatrix} \delta \mu_1 & \delta \mu_2 \end{bmatrix}^T.$$

Choose $\delta y, \delta \mu_1$, and $\delta \mu_2$ as the tracking outputs, then, the system (1.27)–(1.28) can be written in the following descriptor linear system form

$$\begin{cases} E\dot{x} = Ax + Bu \\ y = Cx + Du \end{cases} \tag{1.29}$$

with

$$E = \begin{bmatrix} I & 0 & 0 \\ 0 & M_0 & 0 \\ 0 & 0 & 0 \end{bmatrix}, \quad A = \begin{bmatrix} 0 & I & 0 \\ -K_0 & -D_0 & F_0^{\mathrm{T}} \\ F_0 & 0 & 0 \end{bmatrix}, \quad B = \begin{bmatrix} 0 \\ S_0 \\ 0 \end{bmatrix}, \tag{1.30}$$

and

$$C = \begin{bmatrix} 0 & 1 & 0 & 0 & 0 & 0 & 0 & 0 \\ 0 & 0 & 0 & 0 & 0 & 0 & 1 & 0 \\ 0 & 0 & 0 & 0 & 0 & 0 & 0 & 1 \end{bmatrix}, \quad D = 0. \tag{1.31}$$

1.3 Problems for Descriptor Linear Systems Analysis and Design

The purpose of this section is to outline the basic problems for descriptor linear systems analysis and design. To do this, we first need to introduce the concept of feedback in descriptor linear systems.

1.3.1 Feedback in Descriptor Linear Systems

To control a system means to alter the system in certain ways through a controller such that the system will "behave" as desired. Therefore, controller design is essential for controlling a system. In order to realize online adjustment, most controllers take the form of feedback. Since the context of this book is restricted to linear systems, here we only give a brief overview of linear feedback, which can be classified into two categories: the static feedback and the dynamical feedback. In this subsection, we introduce this concept for system (1.6).

Static Feedback

Static feedback involves no dynamical behavior and is thus represented by only algebraic equations. Static feedback for descriptor systems includes proportional feedback and derivative feedback. Let us first start with the former.

Proportional Feedback The most simple static feedback for the system (1.6) is the following state feedback

$$u = Kx + u_e, \tag{1.32}$$

where $K \in \mathbb{R}^{r \times n}$ is a parameter matrix to be designed, and is often called the feedback gain matrix, u_e is an external input.

When this state feedback controller is applied to system (1.6), the system is changed into

$$\begin{cases} E\dot{x} = A_c x + B u_e \\ y_p = C_p^c x + D_p u_e \\ y_d = C_d \dot{x} + H x + D_d u_e \end{cases} \tag{1.33}$$

with

$$\begin{cases} A_c = A + BK \\ C_p^c = C_p + D_p K \\ H = D_d K \end{cases} \tag{1.34}$$

The above system (1.33)–(1.34) is called the closed-loop system of system (1.6) under the state feedback controller (1.32). The gain matrix K needs to be so designed that the closed-loop system (1.33)–(1.34) meets certain desired requirements.

When the state vector $x(t)$ is not available, the following output feedback controller can be designed for system (1.6):

$$u = K y_p + u_e, \quad K \in \mathbb{R}^{r \times m_p}.$$

When $D_p = 0$ and $D_d = 0$, the closed-loop system is in the form of (1.33) with

$$\begin{cases} A_c = A + BK C_p \\ C_p^c = C_p \\ H = 0 \end{cases}.$$

Derivative Feedback It is well known from classical control theory that derivative feedback is essential in improving the system performance. Yet it has been pointed out by Dai (1989b) and Mukundan and Dayawansa (1983) that the use of derivative feedback in a normal state space system does not contribute any improvement over that which could be obtained using proportional feedback. However, for descriptor systems in the form of (1.6), derivative feedback can alter many features of the system, such as the regularity and the dynamical order of the system, the infinite and the finite frequencies, and the susceptibility to noise. Due to this reason, derivative feedback has been widely used in descriptor linear systems.

A pure general derivative feedback for descriptor linear system (1.6) is in the following form:

$$u = -K y_d + u_e, \quad K \in \mathbb{R}^{r \times m_d}. \tag{1.35}$$

Since it uses only a linear combination of the state derivative variables, the above controller is also often referred to as partial-state derivative feedback. When $D_p = 0$ and $D_d = 0$, the resulted closed-loop system is clearly given by

$$E_c \dot{x} = Ax + Bu_e, \tag{1.36}$$

where

$$E_c = E + BKC_d.$$

In the special case of $C_d = I$, the above partial-state derivative feedback (1.35) becomes the following full-state derivative feedback

$$u = -K\dot{x} + u_e, \quad K \in \mathbb{R}^{r \times n}.$$

The corresponding closed-loop system also takes the form of (1.36), but with

$$E_c = E + BK.$$

P-D Feedback A general static feedback for the system (1.6) is the following output plus partial-state derivative feedback:

$$\begin{aligned} u &= K_p y_p - K_d y_d + u_e \\ &= K_p C_p x - K_d C_d \dot{x} + u_e, \end{aligned} \tag{1.37}$$

where $u_e \in \mathbb{R}^r$ is an external input, $K_p \in \mathbb{R}^{r \times m_p}$ and $K_d \in \mathbb{R}^{r \times m_d}$ are two parameter matrices to be designed, and are often called the feedback gain matrices of the controller.

When the controller (1.37) is applied to the system (1.6) with $D_p = 0$ and $D_d = 0$, the closed-loop system is obtained as

$$\begin{cases} E_c \dot{x} = A_c x + B u_e \\ y_p = C_p x \\ y_d = C_d \dot{x} \end{cases},$$

where

$$\begin{cases} E_c = E + BK_d C_d \\ A_c = A + BK_p C_p \end{cases}.$$

The general output plus partial-state derivative feedback (1.37) clearly contains the pure proportional feedback and the pure derivative feedback discussed above as special cases. Furthermore, it also has the following several special forms.

(a) When $C_p = C_d$, it reduces to the following output proportional plus derivative feedback

$$\begin{aligned} u &= K_p y - K_d \dot{y} + u_e \\ &= K_p C_p x - K_d C_p \dot{x} + u_e. \end{aligned}$$

(b) When $C_p = I$ or $C_d = I$, it reduces to the following state plus partial-state derivative feedback

$$u = K_p x - K_d y_d + u_e,$$

or the output plus full-state derivative feedback

$$u = K_p y_p - K_d \dot{x} + u_e.$$

(c) When $C_p = C_d = I$, it reduces to the following state plus full-state derivative feedback

$$u = K_p x - K_d \dot{x} + u_e.$$

Dynamical Feedback

A general dynamical compensator for the descriptor linear system (1.6) takes the following form:

$$\begin{cases} \tilde{E}\dot{z} = Fz + H_p y_p - H_d y_d \\ u = Nz + M_p y_p - M_d y_d + u_e \end{cases}, \quad (1.38)$$

where $z \in \mathbb{R}^p$ is the state vector of the dynamical controller, $u_e \in \mathbb{R}^r$ is an external input. $\tilde{E}$, F, N, H_p, H_d, M_p, M_d are coefficient matrices of appropriate dimensions, and are to be designed.

When the matrix $\tilde{E}$ is square and nonsingular, dynamical compensator (1.38) is called a normal dynamical compensator. Otherwise, it is called a descriptor dynamical compensator.

For convenience, let us assume that the output y_p and y_d are free from the input effect, that is,

$$D_p = 0, \quad D_d = 0.$$

In this case, we have

$$y_p = C_p x, \quad y_d = C_d \dot{x}.$$

Substituting these two equations into (1.38) yields

$$\begin{cases} \tilde{E}\dot{z} = Fz + H_p C_p x - H_d C_d \dot{x} \\ u = Nz + M_p C_p x - M_d C_d \dot{x} + u_e \end{cases}. \quad (1.39)$$

Further, substituting the second equation in (1.39) into (1.6) gives

$$\begin{cases} (E + BM_d C_d)\dot{x} = (A + BM_p C_p)x + BNz + Bu_e \\ y_p = C_p x \\ y_d = C_d \dot{x} \end{cases}. \quad (1.40)$$

Combining (1.39) and (1.40) produces the closed-loop system

$$\begin{cases} E_c \dot{x}_c = A_c x_c + B_c u_e \\ y_p = \begin{bmatrix} C_p & 0 \end{bmatrix} x_c \\ y_d = \begin{bmatrix} C_d & 0 \end{bmatrix} \dot{x}_c \end{cases},$$

where

$$x_c^T = \begin{bmatrix} x^T & z^T \end{bmatrix},$$

and

$$\begin{cases} E_c = \begin{bmatrix} E + B M_d C_d & 0 \\ H_d C_d & \tilde{E} \end{bmatrix} \\ A_c = \begin{bmatrix} A + B M_p C_p & B N \\ H_p C_p & F \end{bmatrix} \\ B_c = \begin{bmatrix} B \\ 0_{p \times r} \end{bmatrix} \end{cases}.$$

The aim of the design is to select the parameter matrices $\tilde{E}$, F, N, H_p, H_d, M_p, M_d such that the closed-loop system (1.40) possesses certain desired properties.

1.3.2 Problems for Descriptor Linear Systems Analysis

Concerning analysis of descriptor linear systems, there are many problems. The following gives a brief description of the most basic ones.

The Equivalence Relation Section 1.1.2 introduces the state space representations for descriptor systems. It is easily understood that the state space representation of a descriptor system is not unique since the choice of the state vector has some flexibility. For normal linear systems, this phenomenon is well described by the concept of algebraic equivalence between systems, which involves mainly a similarity transformation between the coefficient matrices of two systems. For descriptor linear systems, again, the same idea is used to handle the lack of uniqueness of the state space representation, and the concept of restricted system equivalence (r.s.e.) is introduced. However, due to the general form of descriptor linear systems, the equivalence relation between two descriptor linear systems is no longer determined by a similarity transformation, but a transformation which involves two nonsingular matrices. Basic problems related to the restricted system equivalence relation between two descriptor linear systems include the following:

- Exploring the common features of two equivalent systems. This includes the finite and infinite poles of the two systems, the transfer functions, and the different types of controllability and observability of the system, etc.

- Finding for a certain type of descriptor linear systems some equivalent canonical forms, which offer easy understandings about this type of systems and provide convenience for analysis and control of this type of systems.

Normalizability As we have found out in Sect. 1.1.2, normal linear systems are special descriptor linear systems. Normal linear systems are well understood from linear systems theory. A normal linear system has a dynamical order which coincides with the system order n, and therefore possesses n poles. It does not involve infinite frequencies, and its system response is impulse-free. Due to these reasons, it is desirable, in descriptor linear systems design, to make the designed feedback control system be normal. This leads to the problem of normalizability: Given a descriptor linear system, find the conditions of existence of a type of controllers for the system such that the closed-loop system is normal.

Regularity For a normal linear system, there always exists a unique solution to the system. However, for a general descriptor linear system, the case can be much more complicated. As we will soon find out, a descriptor linear system may not have a solution, or may have more than one solution. Investigation of the condition of existence and uniqueness of solution to a descriptor system results in the concept of regularity. Since it guarantees the existence and uniqueness of solutions to a descriptor linear system, regularity is a very important property for descriptor linear systems. On the one hand, problems of gaining better understanding about regularity will be investigated, on the other hand, problems of finding out the special features that a regular descriptor linear system has will also be explored.

Regularizability Since regularity guarantees the existence and uniqueness of solution to a descriptor linear system, and offers many nice features as well, it is natural to include regularity as one of the basic requirements for descriptor system design. Therefore, regularizability of a descriptor linear system becomes a natural basic topic in descriptor linear systems theory. The problem is to present conditions for the existence of a certain type of controllers for a descriptor linear system, such that the closed-loop system is regular. The types of controllers involved are basically those static linear feedback ones discussed in Sect. 1.3.1.

Stability Stability of a dynamical system describes the response behavior of the system at infinity with respect to initial condition disturbances, and is well regarded as one of the most important properties of dynamical systems. The celebrated Lyapunov stability theory is suitable to analyze the stability of a given descriptor system. For normal linear systems, the stability is totally determined by the system poles. Such a fact is certainly strongly desirable to be extended to descriptor linear systems.

Controllability and Observability It is well known from normal linear systems theory that controllability and observability are a pair of dual concepts, and have close relation with normal linear systems design. Specifically, the poles of a linear system can be arbitrarily assigned by state feedback if and only if the system is controllable; and a full-order state observer for a linear system exists and the observer poles can be arbitrarily assigned if and only if the system is observable. These basic concepts and facts are desired to be generalized to the descriptor linear system case. Basic questions to be answered include the following:

- How to define controllability and observability for descriptor linear systems? As we will find out, for descriptor linear systems, there are several types of controllability and observability.
- How to determine the controllability and observability of a given descriptor linear system? This involves presenting various criteria for controllability and observability, and gives better understanding about the two concepts.
- What do controllability and observability offer in descriptor linear systems analysis and design? As we will find out, like the normal linear system case, all controllable and observable descriptor linear systems admit certain special structures, which are convenient for control systems design. Specifically, controllability and observability offer the conditions for the problems of pole (eigenstructure) assignment and observer design.

Stabilizability and Detectability Again, it is well known from normal linear system theory that stabilizability and detectability are a pair of dual concepts. They are, respectively, weaker than controllability and observability. As a matter of fact, stabilization is the minimum requirement for a normal linear system to be stabilized by state feedback, while detectability is the minimum condition for a normal linear system to have a full-order state observer. These concepts and facts can also be generalized to the case of descriptor linear systems. Problems involving these two concepts are the following:

- Defining stabilizability and detectability for descriptor linear systems.
- Establishing criteria for identifying stabilizability and detectability of a given descriptor linear system.
- Exploring the relations between these two concepts and descriptor linear systems analysis and design.

1.3.3   Problems for Descriptor Linear Systems Design

The meaning of control systems design could be very broad. Within the linear systems context, control systems design often means to design a controller of a certain type such that the closed-loop system, and sometimes the controller itself, meet certain design criteria. The different design criteria often define different design problems. In a practical control system design, often several criteria are required to be met simultaneously, this gives rise to the topic of multiobjective design. However, as an introduction work this book only considers some single-objective design problem. A brief description of the basic single-objective design problems in descriptor linear systems theory is given below.

Regularization As we have mentioned in the above subsection, regularity is a very important property for descriptor linear systems. Therefore, it should be one of the design requirements in a descriptor control systems design. This has two meanings:

- When the open-loop system is regular, we need to design a controller such that the closed-loop system maintains regularity together with some other requirements.
- When the open-loop system is not regular, the first thing we should look into is whether the system can be made regular by certain types of feedback controllers. This raises the problem of regularizability. When the system is verified to be regularizable by the desired type of feedback controllers, then a controller needs to be designed to assure the closed-loop regularity.

Dynamical Order Assignment As we have seen in Sect. 1.1.2, the dynamical order of a general descriptor linear system in the form of (1.6) is defined to be $n_0 = \text{rank}\,E$. It is clear that any proportional static feedback cannot alter the dynamical order of the system since the matrix E remains the same in both the open-loop and the closed-loop systems. However, when a controller involving derivative feedback is applied to the system, then the matrix E is changed, and the dynamical order of the system can then be changed consequently. With this understanding, it is clear to see that dynamical order assignment is only related with derivative feedback. The problems involved include the following:

- Given a typical descriptor linear system, determine the range of the dynamical order assignable by a certain type of controllers involving derivative feedback.
- For any given desired dynamical order p within the allowable range, find the whole set of controllers, S_p, of the prescribed type such that the closed-loop system has dynamical order p.
- In certain applications, feedback with smaller gain is more preferable. Therefore, when the set of controllers, S_p, which assign a desired allowable dynamical order p to the closed-loop system is determined, we can further find a subset $\tilde{S}_p$ of S_p such that every controller in $\tilde{S}_p$ assigns the desired dynamical order p to the system, and meanwhile, possesses gains with the minimum Frobenius norm.

Normalization A general descriptor linear system in the form of (1.6) is called normal if it is square and $\text{rank}\,E = n$ (the system order). Normalization is concerned with designing a controller for a descriptor linear system such that the closed-loop system is normal. Like the case of regularization, problems include the following:

- Normalizability analysis: to determine whether a given descriptor linear system can be made normal by a certain type of controllers.
- Normalizing controller design: to find all the controllers of the prescribed type for a normalizable descriptor linear system, such that the corresponding closed-loop systems are normal.

It is easily observed that the problem of normalization is a special case of the problem of dynamical order assignment. Set the desired dynamical order p to be assigned equal to the system order n, then the dynamical order assignment problem is actually a normalization problem. Therefore, results for normalization can be readily deduced from those for dynamical order assignment.

Impulse Elimination Different from that of a normal linear system, the response of a descriptor linear system may contain impulse terms. These impulse terms may cause saturation of control and may even destroy the system, and hence are usually not expected to exist in most practical applications. Therefore, eliminating the impulsive behavior of a descriptor linear system via certain feedback control is an important fundamental problem in descriptor linear systems theory. As we will find out, this problem is closely related with normalization of descriptor linear systems.

Eigenstructure Assignment The eigenstructure of a descriptor linear system includes both the finite and infinite poles as well as their corresponding eigenvectors, and we will see in the later chapters that the eigenstructure of a system is closely related with many properties of the system, such as the stability and the dynamical response. Because of this, when we design a controller for a descriptor linear system, it is often desirable to let the closed-loop system possess desired eigenstructure, that is, desired finite and infinite closed-loop eigenvalues and their corresponding eigenvectors. A general eigenstructure assignment problem can be described as follows:

Given a descriptor linear system, a type of controllers and a desired closed-loop Jordan structure, determine all the controllers of this type such that the closed-loop systems possess the desired Jordan structure, and meanwhile, give the whole set of permissible closed-loop eigenvectors.

It follows from the above description that solving the above eigenstructure assignment problem would consequently give a set of solutions instead of one pair of solutions (one gain matrix and one closed-loop eigenvector matrix). The set of solutions are often presented via parametric expressions, where the parameters represent the design degrees of freedom. A very important fact lying in the parametric approach for eigenstructure assignment is that the established design degrees of freedom, represented by the set of parameters, can be further utilized to achieve many additional design requirements, such as, closed-loop regularity, desired dynamical order and various robustness requirements.

Stabilization Stability is one of the most important properties for dynamical systems, and is consequently one of the most important requirements for control system design. In the descriptor linear systems context, stabilization is concerned with designing a controller of a prescribed type such that the closed-loop system is stable. Problems involving this topic include the following basic ones:

- Stabilizability analysis: to determine whether the descriptor system can be made stable, or simply stabilized, by a certain type of controllers.
- Stabilizing controller design: to determine a controller of the prescribed type when the system is stabilizable by this type of controllers, such that the closed-loop system is stable.

Optimal Control For control systems design, there may be many types of design criteria. Stabilization and regularization take stability and regularity, respectively, as the design criteria; eigenstructure assignment aims to achieve desired closed-loop eigenstructure. While optimal control is related to a different type of design criteria, which are represented by certain objective functions, or indices, the design purpose is to determine a certain type of controllers for a given system such that the objective function is minimized. Similar to the standard normal case, an important type of objective functions involved in the optimal control of descriptor linear systems often take the following quadratic form:

$$J = x^{\mathrm{T}}(0)Sx(0) + \int_0^{\mathrm{T}} (x^{\mathrm{T}}Qx + u^{\mathrm{T}}Ru)\mathrm{d}t,$$

where Q, R, and S are some properly chosen matrices, and T is a time instant. This leads to the linear quadratic optimal regulation of descriptor linear systems. When the terminal time T is infinity, the problem is called infinite time regulation, in which case the first term in the objective vanishes. Another type of objective functions take the following simple form:

$$J = t,$$

which represents the time period taken to drive the state of the system from one position to another. The problem is to seek, from a type of constrained controllers, a controller, which minimizes this objective. This leads to the problem of time-optimal control, or the well-known bang–bang control of descriptor linear systems.

Observer Design It is well known that full-state feedback contains the most "information" about the system and therefore has a bigger capacity for control than output feedback. This has been reflected, as known from the standard normal linear systems theory, by the pole assignment problem, in which state feedback can arbitrarily assign the whole set of closed-loop eigenvalues (under the controllability condition) while output feedback generally cannot. Therefore, state feedback should be applied whenever all the state variables are available. However, for systems with large dimensions, often not all the state variables are measured, or can be measured. In this case, to realize state feedback, we need to construct the state vector based on the system model and the output variables. This gives rise to the problems of state observer design.

A state observer is a constructed dynamical system whose output asymptotically approaches the state of the given system. Different from the standard normal system case, a state observer for a descriptor linear system may take the form of either a standard normal system form or a general descriptor system form, and the purpose is to properly determine the coefficient matrices involved in the constructed system such that the output of the constructed system approaches asymptotically the state vector of the given descriptor linear system.

Observer design problems do arise in applications which need only a linear combination of the state variables, Kx, rather than the whole state vector. In this case, it is more natural and often more economical to observe Kx directly. This leads to the well-known Luenberger-type of function observers. This type of observers is certainly also an important issue in the descriptor systems context.

To end this section, we finally make the following remarks.

Remark 1.2. Besides the problems described above, there are still many other problems related to descriptor linear systems design, such as, H_2/H_∞ control (Ishihara and Terra 1999; Yue and Lam 2004; Feng et al. 2005), decentralized stabilization (Lohmann and Labibi 2000), disturbance rejection (Chen and Cheng 2004), robust fault detection and isolation (Duan et al. 2002a; Yeu et al. 2005), fault tolerant control (Liang and Duan 2004; Gao and Ding 2007a,c), filtering (Ho et al. 2005), model reduction (Olsson and Ruhe 2006; Stykel 2006a,b), decentralized control (Chen et al. 2006; Zhai 2005), and various other kinds of design problems in descriptor linear systems. All these problems are not considered in this book.

Remark 1.3. Some researchers have used descriptor linear systems theory as a tool to treat some more complicated normal systems design problems, for example, Shaked and Fridman (2002) considered H_∞ control using descriptor systems approach, Cao and Lin (2004) studied robust stabilizability of the polytopic systems based on descriptor systems approach, Gao and Ding (2007b) introduced a descriptor system approach to investigate sensor fault reconstruction and sensor compensation for a class of nonlinear state-space systems with Lipschitz constraints, and Tanaka et al. (2007) presented a descriptor system approach to fuzzy control system design using fuzzy Lyapunov functions.

Remark 1.4. The above problems do arise in both continuous-time and discrete-time descriptor linear systems. However, this book only concentrates on the solutions to these basic problems for continuous-time descriptor linear systems.

1.4 Overview of the Book

Besides this chapter, this book contains the following four parts:

- Introduction
- Part I. Descriptor Linear Systems Analysis
- Part II. Descriptor Linear Systems Design
- Appendices

The general structure of Part I is as follows:

$$\text{Part I} \begin{cases} \text{Equivalence and solutions of descriptor linear systems (Chap. 2)} \\ \text{Regular descriptor linear systems (Chap. 3)} \\ \text{Controllability and observability (Chap. 4)} \end{cases}$$

The equivalence relation between descriptor linear systems introduced in Chap. 2 reveals the structural property of descriptor linear systems and, as we will soon see, allows us to deal with analysis and control of a general descriptor linear system by studying some of its equivalent canonical forms. Based on the so-called Kronecker canonical form, the solution to a general descriptor linear system is further investigated, and the condition for the existence and uniqueness of solution is presented, which actually gives rise to the term of regularity.

Chapter 3 further looks at the various properties of regular systems. These include the special equivalent canonical forms, the eigenstructure, the classical and distributional state responses, the stability as well as the transfer functions.

As we all know, controllability and observability are a pair of very important concepts for linear systems. In Chap. 4, various types of controllability and observability are introduced. The relation between controllability and observability, revealed by the so-called dual principle, is also presented. The problems of system decomposition according to controllability and observability minimal realization are also discussed.

The general structure of Part II is as follows:

$$
\text{Part II} \begin{cases}
\text{Regularization of descriptor linear systems (Chap. 5)} \\
\text{Dynamical order assignment and normalization (Chap. 6)} \\
\text{Impulse elimination (Chap. 7)} \\
\text{Pole assignment and stabilization(Chap. 8)} \\
\text{Eigenstructure assignment (Chap. 9)} \\
\text{Optimal control (Chap. 10)} \\
\text{Observer design (Chap. 11)}
\end{cases}
$$

Since regularity is one of the most important and basic properties for descriptor linear systems, regularization is first investigated in Part II. Conditions for regularizability under various feedback control laws are proposed, and the important fact that, a regularizable descriptor linear system can be regularized by "almost all" feedback controllers, is given and proven. In Chap. 6, another basic problem, namely, dynamical order assignment, is studied. As a special case of dynamical order assignment, normalization is also investigated. Besides these, all the typical design problems, such as stabilization, eigenstructure assignment, optimal control, and observer design, are treated in this part.

Some technical materials closely related to certain chapters are given in the appendix part, which is structured as follows:

$$
\text{Appendices} \begin{cases}
A. \text{ Some mathematical results} \\
B. \text{ Rank-constrained matrix matching} \\
\quad \text{and least square problems} \\
C. \text{ Generalized Sylvester matrix equations}
\end{cases}
$$

1.5 Notes and References

In this chapter, we introduced models for descriptor linear systems, presented some practical examples, and also outlined the structure of the book. It is worth pointing out that more practical examples of descriptor systems may be found in Campbell (1982), Campbell and Rose (1982), Haggman and Bryant (1984), Kiruthi et al. (1980), Lewis (1986), Luenberger (1977), Martens et al. (1984), Newcomb (1981), Newcomb (1982), Petzold (1982), Singh and Liu (1973), Stengel et al. (1979), Wang and Dai (1986), and Duan (2004d).

The studies of descriptor systems began at the end of the 1970s, although they were first mentioned in 1973 (Singh and Liu 1973). In many articles, descriptor systems are also called singular systems, generalized state space systems, semi-state systems, differential-algebraic systems, degenerate systems, constrained systems, etc.

Since 1970s, descriptor systems have been intensively studied. To date, descriptor systems theory has been very rich in content. However, as a textbook on descriptor linear systems theory, this book contains only the basic results in the analysis and design of descriptor linear systems.

Besides many published papers, the celebrated comprehensive work of Dai (1989b) has been a main reference. The PhD thesis of Hou (1995), and the books by Campbell (1982), Mehrmann (1991), and Kunkel and Mehrmann (2006) have also been referenced in the writing of this book. In addition to these, there are also some other books on descriptor linear systems. As a matter of fact, Campbell (1980) and Campbell (1982) are two separate volumes of one book treating singular systems of differential equations, which covers linear time-varying systems as well as nonlinear singular systems. Mehrmann (1991) mainly focuses on linear quadratic control problem theory and numerical solution, and Kunkel and Mehrmann (2006) is more from a mathematical point of view and mainly looks at the solutions to linear and nonlinear differential-algebraic equations. Speaking of books on this topic, there are at least three in Chinese: Liu and Wen (1997), Zhang (1997), and Zhang and Yang (2003). The first one studies the flexible structure control of descriptor linear systems, while the second and the third investigate decentralized control of large-scale descriptor systems and robust control of uncertain descriptor linear systems, respectively.

In the descriptor linear systems literature, there are also quite a few PhD theses. For instances, Wang (1984) examines disturbance resistance and output regulation in linear singular systems, Bender (1985) presents a geometric control theory of descriptor linear systems, and Hou (1995) studies the problems of observer design and fault diagnosis in descriptor linear systems. The more recent one, Stykel (2002a), establishes a Lyapunov equation theory for the stability of descriptor linear systems.

This book is a combination of some of the reported works and the author's own research results in the field of descriptor linear systems. The author's work are reflected in several chapters of the book:

- Chapter 2 introduces a few canonical equivalent forms for descriptor linear systems, and the canonical equivalent form for derivative feedback is due to the author (Duan and Zhang 2002a,b), which has performed important functions in the later chapters.
- Chapter 3 examines certain properties of regular descriptor linear systems, where the distributional and classical solutions of regular descriptor linear systems in Sects. 3.4 and 3.5 have adopted the ideas in Yan and Duan (2005), while

the eigenvectors and generalized eigenvectors in Sect. 3.6 as well as the eigenstructure decomposition in Sect. 3.7 are based on the work in Duan (1998a). Particularly, regarding distributional solutions of regular descriptor linear systems, a mistake in some existing results, e.g., Cobb (1983a) and Dai (1989b), has been corrected.

- Chapter 4 discusses the various types of controllability of descriptor linear systems. The indirect criteria based on canonical equivalent form for derivative feedback are given by the author, which turn to be very simple, convenient, and also very numerically reliable.
- Chapter 5 is based on the author's work in Duan and Zhang (2002c, 2003) on regularization of descriptor linear systems. Regularization of descriptor linear systems has been studied by many researchers, e.g. Bunse-Gerstner et al. (1992). However, most of the reported results are based on certain transformations, while our result gives direct simple criteria for regularizability based on the original system coefficient matrices.
- Chapter 6 is based on the author's work in Duan and Zhang (2002a,b), on dynamical order assignment in descriptor linear systems. The special feature of this chapter is giving the direct and complete parametric representation of all the solutions to the problems, including the problem of dynamical order assignment via derivative feedback with minimum Frobenius norm.
- Chapter 7 deals with the problem of impulse elimination. Some results in Sects. 7.1, 7.2, and 7.5 are based on the author's work (Duan and Wu 2005b,c,d). We have introduced the concept of I-controllablizability, and have given simple and convenient conditions for I-controllablizability and for a system to be impulse-free. Furthermore, simple parametric solutions to the problems of impulse elimination by state feedback and state P-D feedback are derived. These results are obtained based on the canonical equivalent form for derivative feedback given in Chap. 2.
- Chapter 9 is based on the author's work on eigenstructure assignment in descriptor linear systems. The author has been working on eigenstructure assignment in descriptor linear systems since 1989 and has made a few contributions to this topic (see Duan 1992b, 1995, 1998b, 2003a,b, 2004d; Duan and Patton 1997, 1998a; Duan and Liu 2002; Duan et al. 1999b, 2000c, 2001a, 2003; Duan and Wang 2003). However, as an introduction book on descriptor linear systems, only the work in Duan (1998b) is introduced in this chapter.
- Chapter 11 treats the problem of observer design. This chapter partially contains the author's work on parametric design approaches for the Luenberger-type observers (see Duan et al. 2002a; Duan and Wu 2005a).

This book is, to some extent, parallel in structure with the one written by the author in Chinese on normal linear systems theory (Duan 1996a, 2004a). Clarity has been one of the goals in the writing of the book.

Part I
Descriptor Linear Systems Analysis

Chapter 2
Equivalence and Solutions of Descriptor Linear Systems

Modeling is probably the most basic topic in any system theory. In Sect. 1.1, we introduced the state space representation of descriptor systems which depends very much on the state vector selected. Since there exist different choices of the state vector variables, the state space representation of a given descriptor system is certainly not unique. In normal linear systems theory, this phenomenon has been well described by the concept of algebraic equivalence (see Duan 1996a, 2004a). The purpose of this chapter is to generalize this concept into the case of descriptor linear systems, and based on which to examine some of the basic equivalent structures and the solution of a general descriptor linear system.

In Sect. 2.1, the concept of restricted system equivalence for descriptor linear systems is introduced, and some basic common properties of restricted equivalent systems are discussed. Some important canonical equivalent forms for descriptor linear systems are introduced in Sect. 2.2. Based on the Kronecker equivalent form for general descriptor linear systems introduced in Sect. 2.2, solutions to descriptor linear systems are discussed in Sect. 2.3. Some notes and comments follow in Sect. 2.4.

2.1 Restricted System Equivalence

In this section, we define the restricted system equivalence for descriptor linear systems, which is a generalization of the algebraic equivalence relation in standard normal linear systems theory. For standard normal linear systems, the equivalence relation involves mainly a similarity transformation between the coefficient matrices of two systems. We will soon find out that, for descriptor linear systems, the equivalence relation between two descriptor linear systems is determined by a transformation, which involves two nonsingular matrices.

To a certain extent, this equivalence relation to be introduced in this section is resulted in by different choices of the state vector when establishing the state representation of a practical system. Therefore, two equivalent systems must possess certain common properties. Some of the very basic common properties of equivalent systems are also revealed in this section.

G.-R. Duan, *Analysis and Design of Descriptor Linear Systems*, Advances in Mechanics and Mathematics 23, DOI 10.1007/978-1-4419-6397-0_2,

2.1.1 The Definition

For simplicity, let us first make the following conventions:

1. The time variable t in the descriptor linear system (1.6) will be omitted in our discussion unless it is necessary to be included.
2. The initial time for the system considered is taken as $t_0 = 0$ unless otherwise specified.
3. The derivative output equation in the system (1.6) is omitted unless when dealing with some problems involving derivative feedback.
4. The coefficient matrix D_p in the descriptor linear system (1.6) is also taken as a zero matrix unless otherwise specified.

Based on the above conventions, a descriptor linear system would often appear in the following form:

$$\begin{cases} E\dot{x} = Ax + Bu \\ y = Cx \end{cases}, \tag{2.1}$$

where $E, A \in \mathbb{R}^{\tilde{n} \times n}$, $B \in \mathbb{R}^{\tilde{n} \times r}$, and $C \in \mathbb{R}^{m \times n}$ are the system coefficient matrices and $x \in \mathbb{R}^n$, $u \in \mathbb{R}^r$, and $y \in \mathbb{R}^m$ are, respectively, the state vector, the input vector and the output vector of the system. For convenience, this system is also often denoted by the matrix quadruple (E, A, B, C).

In order to define the equivalence relation between two descriptor linear systems in the form of (2.1), let us recall the following definition about equivalence relations.

Definition 2.1. Let $\mathcal{F}$ be a set, in which a certain relationship π is established between any two elements Σ_1, $\Sigma_2 \in \mathcal{F}$. Denote the π relationship between Σ_1 and Σ_2 by "$\Sigma_1 \overset{\pi}{\sim} \Sigma_2$". Then π is called an equivalence relation if it satisfies the following:

1. Reflexivity: $\forall \Sigma_1 \in \mathcal{F}$, $\Sigma_1 \overset{\pi}{\sim} \Sigma_1$;
2. Transitivity: $\forall \Sigma_1$, Σ_2, $\Sigma_3 \in \mathcal{F}$, if $\Sigma_1 \overset{\pi}{\sim} \Sigma_2$ and $\Sigma_2 \overset{\pi}{\sim} \Sigma_3$, then $\Sigma_1 \overset{\pi}{\sim} \Sigma_3$;
3. Invertibility: $\forall \Sigma_1$, $\Sigma_2 \in \mathcal{F}$, if $\Sigma_1 \overset{\pi}{\sim} \Sigma_2$, then $\Sigma_2 \overset{\pi}{\sim} \Sigma_1$.

Let

$$\begin{cases} \tilde{E}\dot{\tilde{x}} = \tilde{A}\tilde{x} + \tilde{B}u \\ y = \tilde{C}\tilde{x} \end{cases} \tag{2.2}$$

with $\tilde{E}, \tilde{A} \in \mathbb{R}^{\tilde{n} \times n}$, $\tilde{B} \in \mathbb{R}^{\tilde{n} \times r}$ and $\tilde{C} \in \mathbb{R}^{m \times n}$, be another system which has a relationship with system (2.1) defined as follows:

There exist two nonsingular matrices $Q \in \mathbb{R}^{\tilde{n} \times \tilde{n}}$ and $P \in \mathbb{R}^{n \times n}$ such that

1. The state vectors of the two systems are related by

$$x = P\tilde{x}. \tag{2.3}$$

2. The coefficient matrices of the two systems satisfy the following relations

$$QEP = \tilde{E}, \quad QAP = \tilde{A}, \quad QB = \tilde{B}, \quad CP = \tilde{C}. \tag{2.4}$$

Then it can be easily shown that this relation between systems (2.1) and (2.2) possesses reflexivity, transitivity and invertibility, and is thus an equivalence relation following from Definition 2.1. For this particular relation, we give it a more specific term.

Definition 2.2. Given two systems (2.1) and (2.2), if there exist two nonsingular matrices $Q \in \mathbb{R}^{\tilde{n} \times \tilde{n}}$ and $P \in \mathbb{R}^{n \times n}$ such that (2.3) and (2.4) hold, then the two systems (2.1) and (2.2) are called restricted system equivalent (r.s.e.) under transformation (P, Q), and this equivalence relation is denoted by

$$\text{System(2.1)} \overset{(P,Q)}{\Longleftrightarrow} \text{System(2.2)},$$

or

$$(E, A, B, C) \overset{(P,Q)}{\Longleftrightarrow} (\tilde{E}, \tilde{A}, \tilde{B}, \tilde{C}).$$

For convenience, we often call the matrix Q the left transformation matrix, and the matrix P the right transformation matrix. To demonstrate the above definition, let us consider an example.

Example 2.1. Consider the circuit network shown in Fig. 1.1 again. As is shown in Example 1.1, when the state vector x, the input vector u, and the output vector y are chosen as

$$x = \begin{bmatrix} I(t) \\ V_L(t) \\ V_C(t) \\ V_R(t) \end{bmatrix}, \quad u = V_S(t), \quad y = V_C(t),$$

the system is in the form of (2.1) with

$$E = \begin{bmatrix} L & 0 & 0 & 0 \\ 0 & 0 & 1 & 0 \\ 0 & 0 & 0 & 0 \\ 0 & 0 & 0 & 0 \end{bmatrix}, \quad A = \begin{bmatrix} 0 & 1 & 0 & 0 \\ \frac{1}{C_0} & 0 & 0 & 0 \\ -R & 0 & 0 & 1 \\ 0 & 1 & 1 & 1 \end{bmatrix},$$

and

$$B^T = \begin{bmatrix} 0 & 0 & 0 & -1 \end{bmatrix}, \quad C = \begin{bmatrix} 0 & 0 & 1 & 0 \end{bmatrix}.$$

On the other hand, if we choose the state vector as

$$\tilde{x} = \begin{bmatrix} I(t) \\ V_R(t) + V_L(t) \\ V_L(t) + V_C(t) \\ V_R(t) + V_C(t) \end{bmatrix},$$

it can be easily verified that the system is in the form of (2.2) with

$$\tilde{E} = \begin{bmatrix} 2L & 0 & 0 & 0 \\ 0 & -1 & 1 & 1 \\ L & 0 & 0 & 0 \\ L & 0 & 0 & 0 \end{bmatrix}, \quad \tilde{A} = \begin{bmatrix} 0 & 1 & 1 & -1 \\ \frac{2}{C_0} & 0 & 0 & 0 \\ -R & 1 & 0 & 0 \\ 0 & 0 & 0 & -1 \end{bmatrix},$$

$$\tilde{B}^T = \begin{bmatrix} 0 & 0 & 0 & 1 \end{bmatrix}, \quad \tilde{C} = \frac{1}{2} \begin{bmatrix} 0 & -1 & 1 & 1 \end{bmatrix}.$$

Both systems (E, A, B, C) and $(\tilde{E}, \tilde{A}, \tilde{B}, \tilde{C})$ are mathematical models that describe the circuit network in Fig. 1.1. As a matter of fact, these two systems can be easily verified to be r.s.e. under transformation (P, Q) with the transformation matrices

$$Q = \begin{bmatrix} 2 & 0 & 0 & 0 \\ 0 & 2 & 0 & 0 \\ 1 & 0 & 1 & 0 \\ 1 & 0 & 0 & -1 \end{bmatrix}, \quad P = \frac{1}{2} \begin{bmatrix} 2 & 0 & 0 & 0 \\ 0 & 1 & 1 & -1 \\ 0 & -1 & 1 & 1 \\ 0 & 1 & -1 & 1 \end{bmatrix}.$$

2.1.2 Common Properties

It is easily understood from the above Example 2.1 that restricted equivalent systems are actually the different state space representations of the same practical plant. Therefore, they must have many properties in common. Such properties will be introduced gradually in the following chapters. At this moment, due to limited knowledge about descriptor linear systems, we only give some very basic and simple facts. To be more strict, let us first introduce the following definition.

Definition 2.3. Consider the square descriptor linear system (2.1). Any finite $\alpha \in \mathbb{C}$ satisfying

$$\text{rank} \begin{bmatrix} \alpha E - A & B \end{bmatrix} < n \tag{2.5}$$

is called an input transmission zero of the system, and any finite $\beta \in \mathbb{C}$ satisfying

$$\text{rank} \begin{bmatrix} \beta E - A \\ C \end{bmatrix} < n$$

is called an output transmission zero of the system.

The following theorem reveals some of the basic common properties of restricted equivalent systems.

Theorem 2.1. *Two restricted equivalent systems (2.1) and (2.2) have*

1. *The same dynamical order.*
2. *The same characteristic equation when they are square, and hence the same finite poles.*
3. *The same input and output transmission zeros when they are square.*

Proof. It follows from (2.4) that

$$QEP = \tilde{E},$$

which clearly implies, in view of the nonsingularity of the matrices P and Q,

$$\text{rank}\, E = \text{rank}\, \tilde{E}.$$

Thus, the first conclusion holds.

It again follows from (2.4) that

$$s\tilde{E} - \tilde{A} = Q(sE - A)P.$$

When the matrices E, A, $\tilde{E}$ and $\tilde{A}$ are square, taking determinant over both sides of the above equation, gives

$$\det(s\tilde{E} - \tilde{A}) = \det Q \det P \det(sE - A). \tag{2.6}$$

Since $\det Q \det P \neq 0$, (2.6) clearly indicates that the two systems (2.1) and (2.2) possess the same characteristic equation, and hence the same finite poles.

The last conclusion of the theorem is an immediate corollary of the second conclusion and the following basic proposition.   □

Proposition 2.1. *Any input (output) transmission zeros of the square descriptor linear system (2.1) are the finite poles of the system.*

Proof. Let $\alpha \in \mathbb{C}$ be an input transmission zero of system (2.1), then (2.5) holds. Suppose α is not a finite pole of system (2.1), then, by definition,

$$\det(\alpha E - A) \neq 0,$$

that is ,

$$\text{rank}(\alpha E - A) = n.$$

This contradicts with (2.5). Therefore, α must be a finite pole of system (2.1).

Similarly, we can show that any output transmission zero of system (2.1) is a finite pole of the system.   □

2.2 Canonical Equivalent Forms

In this section, we will introduce some canonical r.s.e. forms for descriptor linear systems. These canonical forms all reveal, in a certain sense, the structures of descriptor linear systems, and may have convenient applications in systems analysis and/or design.

Consider again the following descriptor linear system

$$\begin{cases} E\dot{x} = Ax + Bu \\ y - Cx \end{cases} , \tag{2.7}$$

where $x \in \mathbb{R}^n$, $u \in \mathbb{R}^r$, and $y \in \mathbb{R}^m$ are, respectively, the state vector, the input vector, and the system output vector; E, $A \in \mathbb{R}^{\tilde{n} \times n}$, $B \in \mathbb{R}^{\tilde{n} \times r}$, and $C \in \mathbb{R}^{m \times n}$ are constant coefficient matrices. As mentioned before, this system (2.7) can be represented by the matrix quadruple (E, A, B, C). When the output equation disappears, the system can be represented by the matrix triple (E, A, B).

2.2.1 Dynamics Decomposition Form

Now let us introduce the so-called dynamics decomposition form, which is the most basic restricted equivalent form for descriptor linear systems.

Suppose $\text{rank} E = n_0$, then it follows from matrix theory that there exist nonsingular matrices Q and P such that

$$\tilde{E} = QEP = \begin{bmatrix} I_{n_0} & 0 \\ 0 & 0 \end{bmatrix}. \tag{2.8}$$

This immediately suggests the following result.

Theorem 2.2. *Given a descriptor linear system* (E, A, B, C) *with* E, $A \in \mathbb{R}^{\tilde{n} \times n}$, $B \in \mathbb{R}^{\tilde{n} \times r}$, $C \in \mathbb{R}^{m \times n}$, *and* $\text{rank} E = n_0$, *there always exist two nonsingular matrices* Q, P *such that*

$$(E, A, B, C) \overset{(P,Q)}{\Longleftrightarrow} (\tilde{E}, \tilde{A}, \tilde{B}, \tilde{C})$$

with the matrix $\tilde{E}$ *possessing the special form of (2.8).*

Decompose the state vector of the system $(\tilde{E}, \tilde{A}, \tilde{B}, \tilde{C})$ as follows:

$$\tilde{x} = P^{-1}x = \begin{bmatrix} x_1 \\ x_2 \end{bmatrix}, \ x_1 \in \mathbb{R}^{n_0}, \ x_2 \in \mathbb{R}^{n-n_0},$$

and let

$$\tilde{A} = QAP = \begin{bmatrix} A_{11} & A_{12} \\ A_{21} & A_{22} \end{bmatrix},$$

$$\tilde{B} = QB = \begin{bmatrix} B_1 \\ B_2 \end{bmatrix},$$

$$\tilde{C} = CP = \begin{bmatrix} C_1 & C_2 \end{bmatrix},$$

where the partitions are compatible, then the system $(\tilde{E}, \tilde{A}, \tilde{B}, \tilde{C})$ can be written as follows:

$$\begin{cases} \dot{x}_1 = A_{11}x_1 + A_{12}x_2 + B_1 u \\ 0 = A_{21}x_1 + A_{22}x_2 + B_2 u \\ y = C_1 x_1 + C_2 x_2 \end{cases} \tag{2.9}$$

which is a restricted equivalent form of system (2.7). Since this canonical form clearly separates the dynamical part (represented by the dynamical state equation) and the static part (represented by the algebraic equation), it is called the dynamics decomposition form.

The dynamics decomposition form (2.9) reveals the physical structure of descriptor linear systems. The differential equation is composed of the dynamical subsystems, and the algebraic one represents the connection between the dynamical subsystems. Thus, descriptor systems may be viewed as a composite system formed by several interconnected subsystems (see Example 1.1). Furthermore, substates x_1 and x_2 reflect a layer property in some descriptor systems: one layer has a dynamic property (described by the differential equation); the other has interconnection, constraint, and administration properties (described by the algebraic equation).

It should be pointed out that the dynamics decomposition form is not unique. More specifically, the transformation (P, Q) that determines the form is not unique, and this consequently results in the lack of uniqueness of the coefficient matrices of the canonical form (2.9).

We also point out that singular value decomposition (see Sect. A.7 in Appendices) may be used to obtain the dynamics decomposition form. Let the singular value decomposition of the matrix E be

$$QEP = \begin{bmatrix} \Sigma & 0 \\ 0 & 0 \end{bmatrix}, \tag{2.10}$$

where Σ is a diagonal positive definite matrix of dimension $n_0 \times n_0$. Then it is easy to see that under the transformation (P, Q), system (2.7) is restricted equivalent to the following dynamics decomposition form:

$$\begin{cases} \Sigma \dot{x}_1 = A_{11}x_1 + A_{12}x_2 + B_1 u \\ 0 = A_{21}x_1 + A_{22}x_2 + B_2 u \\ y = C_1 x_1 + C_2 x_2 \end{cases}$$

which can also be changed into the standard form of (2.9) by pre-multiplying both sides of the first equation by Σ^{-1}. The advantage of employing the singular value decomposition (2.10) is that its derivation is generally numerical reliable, furthermore, instead of getting two general nonsingular transformation matrices, we can get two orthogonal ones. Since the inverses of orthogonal matrices can be exactly obtained by taking their transposes, the coefficient matrices of the finally obtained dynamics decomposition form are very accurate.

Example 2.2. Consider the system (1.29) in Sect. 1.2.4 with the parameters given in (1.30) and (1.31). It can be easily verified that under the transformation (P, Q),

$$
Q = \begin{bmatrix}
0 & 0 & 0 & -0.03459 & 0.09226 & -0.08449 & 0 & 0 \\
0 & 0 & 0 & -0.24331 & -0.05618 & -0.03825 & 0 & 0 \\
0 & 0 & 0 & -0.02414 & 0.43397 & 0.48377 & 0 & 0 \\
1 & 0 & 0 & 0 & 0 & 0 & 0 & 0 \\
0 & 1 & 0 & 0 & 0 & 0 & 0 & 0 \\
0 & 0 & 1 & 0 & 0 & 0 & 0 & 0 \\
0 & 0 & 0 & 0 & 0 & 0 & 1 & 0 \\
0 & 0 & 0 & 0 & 0 & 0 & 0 & 1
\end{bmatrix}, \quad P = Q^{\mathrm{T}},
$$

the systems (1.29) is r.s.e. to the following dynamics decomposition form:

$$
\begin{cases}
\dot{x}_1 = A_{11}x_1 + A_{12}x_2 + B_1 u \\
0 = A_{21}x_1 \\
y = C_1 x_1 + C_2 x_2
\end{cases},
$$

where

$$
A_{11} = \begin{bmatrix}
-0.09060 & 0.21120 & 0.04660 & -10.00544 & 2.09672 & -8.00748 \\
-0.00962 & 0.02243 & 0.00495 & 23.01308 & 17.00564 & -14.32962 \\
0.01802 & -0.04200 & -0.00927 & 5.10624 & 1.75563 & 32.08770 \\
-0.03459 & -0.24331 & -0.02414 & 0 & 0 & 0 \\
0.09226 & -0.05618 & 0.43397 & 0 & 0 & 0 \\
-0.08449 & 0.03825 & 0.48377 & 0 & 0 & 0
\end{bmatrix},
$$

$$
A_{12} = \begin{bmatrix}
-0.03459 & -0.08449 \\
-0.24331 & 0.03825 \\
-0.02414 & 0.48377 \\
0 & 0 \\
0 & 0 \\
0 & 0
\end{bmatrix}, \quad A_{21} = \begin{bmatrix}
0 & 0 & 0 & 1 & 0 & 0 \\
0 & 0 & 0 & 0 & 0 & 1
\end{bmatrix},
$$

$$B_1 = \begin{bmatrix} 0.08853 & -0.13769 & -0.03533 \\ 0.00940 & 0.16206 & -0.13321 \\ -0.01761 & 0.05025 & 0.45112 \\ 0 & 0 & 0 \\ 0 & 0 & 0 \\ 0 & 0 & 0 \end{bmatrix},$$

$$C_1 = \begin{bmatrix} 0 & 0 & 0 & 0 & 1 & 0 \\ 0 & 0 & 0 & 0 & 0 & 0 \\ 0 & 0 & 0 & 0 & 0 & 0 \end{bmatrix}, \quad C_2 = \begin{bmatrix} 0 & 0 \\ 1 & 0 \\ 0 & 1 \end{bmatrix}.$$

2.2.2 The Kronecker Form

A very basic restricted equivalent form for descriptor linear systems is the Kronecker form, which gives full insight of the structure of a general descriptor linear system.

Theorem 2.3. *Given a descriptor linear system* (E, A, B, C) *with* E, $A \in \mathbb{R}^{\tilde{n} \times n}$, $B \in \mathbb{R}^{\tilde{n} \times r}$, $C \in \mathbb{R}^{m \times n}$, *there always exist two nonsingular matrices* Q, P *such that*

$$(E, A, B, C) \overset{(P,Q)}{\Longleftrightarrow} (\tilde{E}, \tilde{A}, \tilde{B}, \tilde{C})$$

with the matrices $\tilde{E}$, $\tilde{A}$ *possessing the following special forms:*

$$\tilde{E} = QEP = \mathrm{diag}(0_{\tilde{n}_0 \times n_0}, L_1, L_2, \ldots, L_p, L_1^\infty, L_2^\infty, \ldots, L_q^\infty, I, N), \quad (2.11)$$

$$\tilde{A} = QAP = \mathrm{diag}(0_{\tilde{n}_0 \times n_0}, J_1, J_2, \ldots, J_p, J_1^\infty, J_2^\infty, \ldots, J_q^\infty, A_1, I), \quad (2.12)$$

where $A_1 \in \mathbb{R}^{h \times h}$,

$$L_i = \begin{bmatrix} 1 & 0 & & \\ & 1 & 0 & \\ & & \ddots & \ddots \\ & & & 1 & 0 \end{bmatrix}, \quad J_i = \begin{bmatrix} 0 & 1 & & \\ & 0 & 1 & \\ & & \ddots & \ddots \\ & & & 0 & 1 \end{bmatrix} \in \mathbb{R}^{\tilde{n}_i \times (\tilde{n}_i + 1)}, \quad (2.13)$$

$$L_j^\infty = \begin{bmatrix} 1 & & \\ 0 & 1 & \\ & \ddots & \ddots \\ & & 1 \\ & & 0 \end{bmatrix}, \quad J_j^\infty = \begin{bmatrix} 0 & & \\ 1 & 0 & \\ & \ddots & \ddots \\ & & 0 \\ & & 1 \end{bmatrix} \in \mathbb{R}^{(n_j + 1) \times n_j}, \quad (2.14)$$

$$i = 1, 2, \ldots, p, \quad j = 1, 2, \ldots, q$$

and

$$N = \text{diag}(N_1, N_2, \ldots, N_l) \in \mathbb{R}^{g \times g}, \tag{2.15}$$

with

$$N_i = \begin{bmatrix} 0 & 1 & & & \\ & 0 & 1 & & \\ & & \ddots & \ddots & \\ & & & 0 & 1 \\ & & & & 0 \end{bmatrix} \in \mathbb{R}^{k_i \times k_i}, \ i = 1, 2, \ldots, l, \tag{2.16}$$

and the dimensions of the above matrices satisfy the relations

$$\begin{cases} \tilde{n}_o + \sum_i \tilde{n}_i + \sum_j (n_j + 1) + \sum_i k_i + h = \tilde{n} \\ n_o + \sum_j n_j + \sum_i (\tilde{n}_i + 1) + \sum_i k_i + h = n \ . \\ \sum_i k_i = g \end{cases} \tag{2.17}$$

Proof. The proof immediately follows from the lemma below.   □

Lemma 2.1 (Gantmacher 1974). *For any two matrices E, $A \in \mathbb{R}^{\tilde{n} \times n}$, there always exist two nonsingular matrices Q, P such that*

$$QEP = \tilde{E}, \quad QAP = \tilde{A}$$

with the matrices $\tilde{E}$ and $\tilde{A}$ given by (2.11)–(2.17).

The above Kronecker form is useful in some theoretical analysis. As will be seen in Sect. 2.3, this form plays an essential role in deriving the general solution of a descriptor linear system. Nevertheless, this form is not quite suitable for numeral analysis and design since the derivation process may not be numerically reliable. In other words, the transformation matrices P and Q may sometimes not be obtained to a required degree of precision. Moreover, since these two matrices are only nonsingular, their inverses, which are often involved in the analysis and design problems, may also possess poor precision. Due to this reason, we here will not give an algorithm for deriving the Kronecker form. Readers who are interested in an algorithm may refer to the proof of the above Lemma 2.1 in Gantmacher (1974).

2.2.3 Canonical Equivalent Forms for Derivative Feedback

In this subsection, we introduce another three canonical restricted equivalent forms for the general descriptor linear system (2.7). They are called canonical restricted equivalent forms for derivative feedback because, as we will find later in the second part of the book, they have close relations with derivative feedback designs.

The First Form

First let us start with the most simple one.

Noting $\operatorname{rank} B = r$, by performing the SVD on the matrix B, we have

$$Q_1 B U_1 = \begin{bmatrix} \Sigma_B \\ 0 \end{bmatrix}, \tag{2.18}$$

where $Q_1 \in \mathbb{R}^{\tilde{n} \times \tilde{n}}$ and $U_1 \in \mathbb{R}^{r \times r}$ are orthogonal matrices, and $\Sigma_B \in \mathbb{R}^{r \times r}$ is a diagonal positive definite matrix. Based on this fact, the following theorem can be immediately obtained.

Theorem 2.4. *Given a system (E, A, B) with $E, A \in \mathbb{R}^{\tilde{n} \times n}$, $B \in \mathbb{R}^{\tilde{n} \times r}$ and $\operatorname{rank} B = r > 0$, there exist orthogonal matrices $Q_1 \in \mathbb{R}^{\tilde{n} \times \tilde{n}}$, $P_1 \in \mathbb{R}^{n \times n}$ and $U_1 \in \mathbb{R}^{r \times r}$ such that*

$$(E, A, B) \overset{(P_1, Q_1)}{\Longleftrightarrow} (\tilde{E}, \tilde{A}, \tilde{B})$$

with the matrix $\tilde{B}$ possessing the following form:

$$\tilde{B} = Q_1 B = \begin{bmatrix} \Sigma_B \\ 0 \end{bmatrix} U_1^{\mathrm{T}}, \tag{2.19}$$

where $\Sigma_B \in \mathbb{R}^{r \times r}$ is a diagonal positive definite matrix.

The descriptor system $(\tilde{E}, \tilde{A}, \tilde{B})$, with the matrix $\tilde{B}$ possessing the form of (2.19), is called the first canonical restricted equivalent form for derivative feedback of the system (E, A, B). Although this form is very simple, it is sufficient to be utilized to solve many problems.

The Second Form

The following theorem gives a canonical form with both the matrices $\tilde{B}$ and $\tilde{E}$ possessing certain special forms.

Theorem 2.5. *Consider a system (E, A, B) with E, $A \in \mathbb{R}^{\tilde{n} \times n}$, $B \in \mathbb{R}^{\tilde{n} \times r}$, and $\operatorname{rank} B = r > 0$. Let*

$$l = \operatorname{rank}[E \ B] - r. \tag{2.20}$$

Then, there exist orthogonal matrices $Q_2 \in \mathbb{R}^{\tilde{n} \times \tilde{n}}$, $P_2 \in \mathbb{R}^{n \times n}$ and $U_2 \in \mathbb{R}^{r \times r}$ such that

$$(E, A, B) \overset{(P_2, Q_2)}{\Longleftrightarrow} (\tilde{E}, \tilde{A}, \tilde{B})$$

with the matrices $\tilde{E}$ and $\tilde{B}$ possessing the following forms:

$$\tilde{E} = Q_2 E P_2 = \begin{bmatrix} \Sigma_1 & 0 \\ E_{21} & E_{22} \\ 0 & 0 \end{bmatrix}, \tag{2.21}$$

and

$$\tilde{B} = Q_2 B = \begin{bmatrix} 0 \\ \Sigma_B \\ 0 \end{bmatrix} U_2^{\mathrm{T}}, \tag{2.22}$$

where $\Sigma_1 \in \mathbb{R}^{l \times l}$, $\Sigma_B \in \mathbb{R}^{r \times r}$ are diagonal positive definite matrices, while E_{21} and E_{22} are some matrices of dimensions $r \times l$ and $r \times (n - l)$, respectively.

Proof. It follows from Theorem 2.4 that there exist orthogonal matrices $Q_1 \in \mathbb{R}^{\tilde{n} \times \tilde{n}}$ and $U_1 \in \mathbb{R}^{r \times r}$ such that (2.19) or (2.18) holds.

Letting

$$Q_{1a} = \begin{bmatrix} 0 & I_{\tilde{n}-r} \\ I_r & 0 \end{bmatrix} Q_1, \tag{2.23}$$

then we have

$$Q_{1a} B U_1 = \begin{bmatrix} 0 \\ \Sigma_B \end{bmatrix}. \tag{2.24}$$

Further, let $l = \mathrm{rank} E_1$, and

$$Q_{1a} E = \begin{bmatrix} E_1 \\ E_2 \end{bmatrix}, \quad E_1 \in \mathbb{R}^{(\tilde{n}-r) \times n}, \ E_2 \in \mathbb{R}^{r \times n}. \tag{2.25}$$

Then, by carrying out the singular value decomposition of the matrix E_1, we obtain

$$Q_{1b} E_1 P_2 = \begin{bmatrix} \Sigma_1 & 0 \\ 0 & 0 \end{bmatrix}, \tag{2.26}$$

where $Q_{1b} \in \mathbb{R}^{(\tilde{n}-r) \times (\tilde{n}-r)}$ is an orthogonal matrix, while P_2 and Σ_1 are as described in the theorem. Partition the matrix E_2 as

$$E_2 P_2 = [E_{21} \ E_{22}], \ E_{21} \in \mathbb{R}^{r \times l}, \tag{2.27}$$

and let

$$Q_{1c} = \begin{bmatrix} I_l & 0 & 0 \\ 0 & 0 & I_r \\ 0 & I_{\tilde{n}-r-l} & 0 \end{bmatrix}. \tag{2.28}$$

It then follows from (2.24) to (2.27) that

$$Q_{1c} \begin{bmatrix} Q_{1b} & 0 \\ 0 & I_r \end{bmatrix} Q_{1a} E P_2 = \begin{bmatrix} \Sigma_1 & 0_{l \times (n-l)} \\ E_{21} & E_{22} \\ 0 & 0 \end{bmatrix} = \tilde{E}, \tag{2.29}$$

$$Q_{1c} \begin{bmatrix} Q_{1b} & 0 \\ 0 & I_r \end{bmatrix} Q_{1a} B U_1 = \begin{bmatrix} 0_{l \times r} \\ \Sigma_B \\ 0 \end{bmatrix}. \tag{2.30}$$

Putting

$$Q_2 = Q_{1c} \begin{bmatrix} Q_{1b} & 0 \\ 0 & I_r \end{bmatrix} Q_{1a}, \quad U_2 = U_1, \tag{2.31}$$

then, (2.29) and (2.30) obviously give the relations (2.21) and (2.22), respectively.

□

Based on the above proof, an algorithm for obtaining, for a given descriptor linear system, a second canonical form for derivative feedback can be given as follows.

Algorithm 2.1. The second canonical form for derivative feedback. [Given $E \in \mathbb{R}^{\tilde{n} \times n}$, $B \in \mathbb{R}^{\tilde{n} \times r}$ with $\text{rank} B = r > 0$, finding the orthogonal matrices $Q_2 \in \mathbb{R}^{\tilde{n} \times \tilde{n}}$, $P_2 \in \mathbb{R}^{n \times n}$ and $U_2 \in \mathbb{R}^{r \times r}$]

Step 1 Carry out the singular value decomposition (2.18) and obtain the orthogonal matrices $Q_1 \in \mathbb{R}^{\tilde{n} \times \tilde{n}}$, $U_1 \in \mathbb{R}^{r \times r}$ and the diagonal positive matrix Σ_B.

Step 2 If $r = \tilde{n}$, output the matrices

$$Q_2 = Q_1, \quad U_2 = U_1, \quad P_2 = I_n,$$

and terminate the algorithm. Otherwise, carry on with the following steps.

Step 3 Compute the orthogonal matrix $Q_{1a} \in \mathbb{R}^{\tilde{n} \times \tilde{n}}$ and the matrices $E_1 \in \mathbb{R}^{(\tilde{n}-r) \times n}$ and $E_2 \in \mathbb{R}^{r \times n}$ according to (2.23) and (2.25).

Step 4 Carry out the singular value decomposition (2.26) of the matrix E_1, and obtain the orthogonal matrices $Q_{1b} \in \mathbb{R}^{(\tilde{n}-r) \times (\tilde{n}-r)}$ and $P_2 \in \mathbb{R}^{n \times n}$ and the diagonal positive definite matrix Σ_1.

Step 5 Find $l = \text{rank} \Sigma_1 = \text{rank} E_1$ and compute the orthogonal matrix Q_{1c} according to (2.28).

Step 6 Compute the orthogonal matrices Q_2 and U_2 according to (2.31).

The above second canonical equivalent form for derivative feedback is very convenient in dealing with certain control systems design problems involving derivative feedback. It is numerically favorable because all the matrices P_2, Q_2 and U_2 are orthogonal.

The Third Form

For certain design problems, such as dynamical order assignment (see Chap. 6), the matrix Q_2 in the above second canonical equivalent form does not need to be orthogonal. The following theorem shows that when the orthogonality of the matrix Q_2 is removed, the forms of the matrices $\tilde{E}$ and $\tilde{B}$ can be further simplified.

Theorem 2.6. *Consider the descriptor linear system (E, A, B) with $E, A \in \mathbb{R}^{\tilde{n} \times n}$, $B \in \mathbb{R}^{\tilde{n} \times r}$, and $\text{rank} E = n_0$, $\text{rank} B = r > 0$. Let l be an integer defined in (2.20). Then there exist a nonsingular matrix $Q_3 \in \mathbb{R}^{\tilde{n} \times \tilde{n}}$ and two orthogonal matrices $P_3 \in \mathbb{R}^{n \times n}$ and $U_3 \in \mathbb{R}^{r \times r}$ such that*

$$(E, A, B) \overset{(P_3, Q_3)}{\Longleftrightarrow} (\tilde{E}, \tilde{A}, \tilde{B})$$

with the matrices $\tilde{B}$ and $\tilde{E}$ possessing the following special forms:

$$\tilde{B} = Q_3 B = \begin{bmatrix} 0_{l \times r} \\ I_r \\ 0 \end{bmatrix} U_3^{\mathrm{T}},$$

$$\tilde{E} = Q_3 E P_3 = \begin{bmatrix} I_l & 0 \\ 0 & M \\ 0 & 0 \end{bmatrix},$$

where

$$M = \mathrm{diag}(d_1, d_2, \ldots, d_{n_0 - l}, 0_{(r+l-n_0) \times (n-n_0)}) \in \mathbb{R}^{r \times (n-l)} \tag{2.32}$$

with

$$d_i > 0, \quad i = 1, 2, \ldots, n_0 - l.$$

Proof. It follows from Theorem 2.5 that there exist orthogonal matrices $Q_2 \in \mathbb{R}^{\tilde{n} \times \tilde{n}}$, $P_2 \in \mathbb{R}^{n \times n}$ and $U_2 \in \mathbb{R}^{r \times r}$ such that (2.21) and (2.22) hold.

Let

$$Q_{2a} = \begin{bmatrix} \Sigma_1^{-1} & 0 & 0 \\ -\Sigma_B^{-1} E_{21} \Sigma_1^{-1} & \Sigma_B^{-1} & 0 \\ 0 & 0 & I_{\tilde{n}-r-l} \end{bmatrix}. \tag{2.33}$$

Then, pre-multiplying both sides of (2.21) and (2.22) by the matrix Q_{2a} gives

$$Q_{2a} Q_2 E P_2 = Q_{2a} \begin{bmatrix} \Sigma_1 & 0 \\ E_{21} & E_{22} \\ 0 & 0 \end{bmatrix}$$

$$= \begin{bmatrix} I_l & 0 \\ 0 & \Sigma_B^{-1} E_{22} \\ 0 & 0 \end{bmatrix}, \tag{2.34}$$

and

$$Q_{2a} Q_2 B = Q_{2a} \begin{bmatrix} 0_{l \times r} \\ \Sigma_B \\ 0 \end{bmatrix} U_2^{\mathrm{T}}$$

$$= \begin{bmatrix} 0 \\ I_r \\ 0 \end{bmatrix} U_2^{\mathrm{T}}. \tag{2.35}$$

Further, by performing the singular value decomposition on the matrix $\Sigma_B^{-1} E_{22}$, we can find orthogonal matrices $Q_{2b} \in \mathbb{R}^{r \times r}$ and $P_{2b} \in \mathbb{R}^{(n-l) \times (n-l)}$ and a diagonal positive definite matrix Σ_2 of dimension $(n_0 - l) \times (n_0 - l)$, satisfying

$$Q_{2b} \left(\Sigma_B^{-1} E_{22} \right) P_{2b} = M, \tag{2.36}$$

where the matrix M is given by (2.32).

Let

$$Q_3 = \begin{bmatrix} I_l & 0 & 0 \\ 0 & Q_{2b} & 0 \\ 0 & 0 & I \end{bmatrix} Q_{2a} Q_2, \tag{2.37}$$

$$P_3 = P_2 \begin{bmatrix} I_l & 0 \\ 0 & P_{2b} \end{bmatrix}, \tag{2.38}$$

and

$$U_3 = U_2 Q_{2b}^{\mathrm{T}}. \tag{2.39}$$

It then follows from (2.34) to (2.39) that

$$\begin{aligned} Q_3 E P_3 &= \begin{bmatrix} I_l & 0 & 0 \\ 0 & Q_{2b} & 0 \\ 0 & 0 & I \end{bmatrix} \begin{bmatrix} I_l & 0 \\ 0 & \Sigma_B^{-1} E_{22} \\ 0 & 0 \end{bmatrix} \begin{bmatrix} I_l & 0 \\ 0 & P_{2b} \end{bmatrix} \\ &= \begin{bmatrix} I_l & 0 \\ 0 & M \\ 0 & 0 \end{bmatrix}, \end{aligned}$$

and

$$\begin{aligned} Q_3 B U_3 &= \begin{bmatrix} I_l & 0 & 0 \\ 0 & Q_{2b} & 0 \\ 0 & 0 & I \end{bmatrix} \left(\begin{bmatrix} 0 \\ I_r \\ 0 \end{bmatrix} U_2^{\mathrm{T}} \right) U_2 Q_{2b}^{\mathrm{T}} \\ &= \begin{bmatrix} 0 \\ I_r \\ 0 \end{bmatrix}. \end{aligned}$$

Therefore, the conclusion holds.   □

Based on the above proof, we can give an algorithm for deriving, for a given descriptor linear system, a third canonical equivalent form for derivative feedback.

Algorithm 2.2. The third canonical form for derivative feedback. [Given $E \in \mathbb{R}^{\tilde{n} \times n}$, $B \in \mathbb{R}^{\tilde{n} \times r}$ with rank $B = r > 0$, finding the matrices $Q_3 \in \mathbb{R}^{\tilde{n} \times \tilde{n}}$, $P_3 \in \mathbb{R}^{n \times n}$ and $U_3 \in \mathbb{R}^{r \times r}$]

Step 1 Obtain using Algorithm 2.1 the orthogonal matrices $Q_2 \in \mathbb{R}^{\tilde{n} \times \tilde{n}}$, $P_2 \in \mathbb{R}^{n \times n}$ and $U_2 \in \mathbb{R}^{r \times r}$, the diagonal positive definite matrices $\Sigma_1 \in \mathbb{R}^{l \times l}$ and $\Sigma_B \in \mathbb{R}^{r \times r}$, as well as the matrices E_{21} and E_{22} satisfying (2.21).

Step 2 Obtain based on singular value decomposition two orthogonal matrices $Q_{2b} \in \mathbb{R}^{r \times r}$ and $P_{2b} \in \mathbb{R}^{(n-l) \times (n-l)}$ and the matrix M in the form of (2.32) satisfying (2.36).

Step 3 Compute the matrix Q_{2a} according to (2.33), and then the matrices $Q_3 \in \mathbb{R}^{\tilde{n} \times \tilde{n}}$, $P_3 \in \mathbb{R}^{n \times n}$ and $U_3 \in \mathbb{R}^{r \times r}$ according to (2.37)–(2.39).

We will see in Chap. 6 that the above third canonical form for derivative feedback performs an essential role in solving the dynamical order assignment problem with minimum gains.

2.3 Solutions of Descriptor Linear Systems

When dealing with solutions of descriptor linear systems, we suffice only to consider the following model:

$$E \dot{x}(t) = A x(t) + B u(t), \tag{2.40}$$

where $E, A \in \mathbb{R}^{\tilde{n} \times n}$, $B \in \mathbb{R}^{\tilde{n} \times r}$ and $u(t) \in \mathbb{R}^r$. For solutions to a descriptor linear system we have the following definition which is a slightly altered from the form that in Kunkel and Mehrmann (2006).

Definition 2.4. Consider the descriptor linear system (2.40).

1 A function $x(t)$ is called a solution of (2.40), if it satisfies (2.40) pointwise.
2 The function $x(t)$ is called a solution of the initial value problem (2.40) with the initial condition.

$$x(0) = x_0, \tag{2.41}$$

if it furthermore satisfies (2.41).

For system (2.40), in this section we are mainly concerned with the existence, uniqueness, and structure of its solutions.

2.3.1 System Decomposition Based on the Kronecker Form

It follows from Theorem 2.3 for Kronecker equivalent form reduction that there exist two nonsingular matrices P and Q, such that the system (2.40) is restricted equivalent to the following Kronecker form:

$$\tilde{E} \dot{\tilde{x}}(t) = \tilde{A} \tilde{x}(t) + \tilde{u}(t), \tag{2.42}$$

where the state vector $\tilde{x}$ and the input vector $\tilde{u}$ are, respectively, defined by

$$\tilde{x}(t) = P^{-1}x(t) \tag{2.43}$$

and

$$\tilde{u}(t) = QBu(t),$$

and the matrices $\tilde{E}$ and $\tilde{A}$ possess the special forms described by (2.11)–(2.17).

It follows from Theorem 2.1 that the systems (2.40) and (2.42) have the same solution property. More specifically, due to the relation (2.43), once a solution to (2.42) is obtained, a corresponding one to (2.40) can be immediately obtained.

Partition the state vector $\tilde{x}$ and the input vector $\tilde{u}$ of system (2.42) according to the structures of the matrices $\tilde{E}$ and $\tilde{A}$ as follows

$$\tilde{x}(t) = \left[x_{n_o}^{\mathrm{T}} \ x_{\tilde{n}_1}^{\mathrm{T}} \cdots x_{\tilde{n}_p}^{\mathrm{T}} \ x_{n_1}^{\mathrm{T}} \cdots x_{n_q}^{\mathrm{T}} \ x_h^{\mathrm{T}} \ x_{k_1}^{\mathrm{T}} \cdots x_{k_l}^{\mathrm{T}} \right]^{\mathrm{T}},$$

$$\tilde{u}(t) = \left[u_{n_o}^{\mathrm{T}} \ u_{\tilde{n}_1}^{\mathrm{T}} \cdots u_{\tilde{n}_p}^{\mathrm{T}} \ u_{n_1}^{\mathrm{T}} \cdots u_{n_q}^{\mathrm{T}} \ u_h^{\mathrm{T}} \ u_{k_1}^{\mathrm{T}} \cdots u_{k_l}^{\mathrm{T}} \right]^{\mathrm{T}}.$$

Then, by expanding, system (2.42) can be equivalently decomposed into the following five sets of equations:

$$0_{\tilde{n}_0 \times n_0} \dot{x}_{n_0}(t) = u_{\tilde{n}_0}(t), \tag{2.44}$$

$$L_i \dot{x}_{\tilde{n}_i}(t) = J_i x_{\tilde{n}_i}(t) + u_{\tilde{n}_i}(t), \quad i = 1, 2, \ldots, p, \tag{2.45}$$

$$L_j^{\infty} \dot{x}_{n_j}(t) = J_j^{\infty} x_{n_j}(t) + u_{n_j}(t), \quad j = 1, 2, \ldots, q, \tag{2.46}$$

$$N_{k_i} \dot{x}_{k_i}(t) = x_{k_i}(t) + u_{k_i}(t), \quad i = 1, 2, \ldots, l, \tag{2.47}$$

and

$$\dot{x}_h(t) = A_1 x_h(t) + u_h(t). \tag{2.48}$$

Thus, solving (2.40) is equivalent to solving the whole set of equations in (2.44)–(2.48).

2.3.2 Solution to the Basic Types of Equations

Now let us examine the solvability of each set of the equations in (2.44)–(2.48).

Solution to Equation (2.44) It is obvious that (2.44) can be solved only when $u_{\tilde{n}_o}(t) = 0$. In this case, (2.44) is an identical equation for any differentiable function $x_{n_o}(t)$. Therefore, (2.44) either does not have a solution or has an infinite number of solutions.

Solution to Equation (2.45) Equation (2.45) can be composed of a set of equations of the following form:

$$\begin{bmatrix} 1\ 0 \\ & 1\ 0 \\ & & \ddots\ \ddots \\ & & & 1\ 0 \end{bmatrix} \dot{z}(t) = \begin{bmatrix} 0\ 1 \\ & 0\ 1 \\ & & \ddots\ \ddots \\ & & & 0\ 1 \end{bmatrix} z(t) + u_z(t).$$

Suppose the dimension of the system is $(k-1) \times k$. Denote the elements of the state vector z and the input vector u_z by z_i and u_i, respectively, then by expanding this equation we obtain

$$\begin{cases} \dot{z}_1(t) = z_2(t) + u_1(t) \\ \dot{z}_2(t) = z_3(t) + u_2(t) \\ \quad \vdots \\ \dot{z}_{k-1}(t) = z_k(t) + u_{k-1}(t) \end{cases}$$

which can be converted into the following form:

$$\begin{cases} z_2(t) = \dot{z}_1(t) - u_1(t) \\ z_3(t) = \dot{z}_2(t) - u_2(t) \\ \quad \vdots \\ z_k(t) = \dot{z}_{k-1}(t) - u_{k-1}(t) \end{cases} \tag{2.49}$$

Clearly, for any given sufficiently differentiable function $z_1(t)$, the rest of the variables $z_2(t), \ldots, z_k(t)$ can be determined successively from (2.49). Therefore, such equations have an infinite number of solutions.

Solution to Equation (2.46) Every equation in (2.46) is of the following form:

$$\begin{bmatrix} 1 \\ 0\ 1 \\ & 0\ \ddots \\ & & \ddots\ 1 \\ & & & 0 \end{bmatrix} \dot{z}(t) = \begin{bmatrix} 0 \\ 1\ 0 \\ & 1\ \ddots \\ & & \ddots\ 0 \\ & & & 1 \end{bmatrix} z(t) + u_z(t).$$

Suppose the dimension of this system is $(k+1) \times k$. Again denote the elements of the state vector z and the input vector u_z by z_i and u_i, respectively, then the system is clearly equivalent to

$$z_k(t) + u_{k+1}(t) = 0, \tag{2.50}$$

and

$$\begin{cases} \dot{z}_1(t) = u_1(t) \\ \dot{z}_2(t) = z_1(t) + u_2(t) \\ \quad\vdots \\ \dot{z}_k(t) = z_{k-1}(t) + u_k(t) \end{cases} \qquad (2.51)$$

The set of equations in (2.51) can obviously be written into a standard normal linear system in state space form, and hence has a unique solution for arbitrarily given inputs $u_i(t)$ with sufficient order of differentials. However, the variable $z_k(t)$ solved from (2.51) generally does not satisfy the consistent condition (2.50) for arbitrarily given $u_{k+1}(t)$ with any order of differentials. Therefore, this type of equations generally do not have solutions, or more specifically, this type of equations may have solutions only for a special set of initial values and a special class of inputs satisfying the consistent condition (2.50).

Solution to Equation (2.47)   Equation (2.47) consists of equations of the following form:

$$\begin{bmatrix} 0 & 1 & & & \\ & 0 & 1 & & \\ & & \ddots & \ddots & \\ & & & 0 & 1 \\ & & & & 0 \end{bmatrix} \dot{z}(t) = z(t) + u_z(t),$$

which may be rewritten as

$$\begin{cases} \dot{z}_2(t) = z_1(t) + u_1(t) \\ \dot{z}_3(t) = z_2(t) + u_2(t) \\ \quad\vdots \\ \dot{z}_k(t) = z_{k-1}(t) + u_{k-1}(t) \\ z_k(t) + u_k(t) = 0 \end{cases}$$

This set of equations are clearly equivalent to

$$z_k(t) + u_k(t) = 0 \qquad (2.52)$$

and

$$\begin{cases} z_{k-1}(t) = \dot{z}_k(t) - u_{k-1}(t) \\ z_{k-2}(t) = \dot{z}_{k-1}(t) - u_{k-2}(t) \\ \quad\vdots \\ z_2(t) = \dot{z}_3(t) - u_2(t) \\ z_1(t) = \dot{z}_2(t) - u_1(t) \end{cases}$$

Therefore, when (2.52) has a unique solution $z_k(t)$, all the variables $z_1(t)$, $\ldots$, $z_{k-1}(t)$ can be successively determined for sufficiently differentiable function $u_k(t)$. Hence (2.47) has a unique solution for any sufficiently differentiable input function $u_k(t)$, and the initial value $z(0)$ with $z_k(0) = u_k(0)$ and $z_i(0), i = 1, 2, \ldots, k-1$, arbitrary.

Solution to Equation (2.48) Equation (2.48) is an ordinary differential equation, which has a unique solution for any piecewise continuous function $u(t)$.

It follows from the above analysis that the following facts hold:

- Equation (2.44) either does not have a solution or has an infinite number of solutions. Therefore, this type of equations should not exist if a solution to system (2.40) exists and is unique.
- The set of equations in (2.45) always has an infinite number of solutions for arbitrarily given sufficiently differentiable function $u(t)$. Therefore, this type of equation should not exist if the solution to system (2.40) exists and is unique.
- The set of equations in (2.46) has a unique solution when it has a solution. However, it has a solution only for a very special set of initial values and a very special set of system input $u_{n_j}(t)$, which satisfy certain strict consistent condition. Therefore, this type of equation should not exist if the solution to system (2.40) exists and is unique.
- All equations (2.47)–(2.48) have unique solutions for arbitrarily given sufficiently differentiable corresponding input functions.

It thus follows from the above points that the necessary and sufficient condition for the existence and uniqueness of the solution to (2.40) is the disappearance of (2.44)–(2.46). This fact immediately gives the following theorem for solution to the descriptor linear system (2.40).

Theorem 2.7. *The descriptor linear system (2.40) has a unique solution for any sufficiently differentiable $u(t)$ and some proper initial value $x(0) \in \mathbb{R}^n$ if and only if the matrices $\tilde{E}$ and $\tilde{A}$ in the Kronecker canonical equivalent form (2.42) of the system (2.40) take the following simple forms:*

$$\tilde{E} = QEP = \mathrm{diag}(I, N), \quad \tilde{A} = QAP = \mathrm{diag}(A_1, I). \tag{2.53}$$

Remark 2.1. Careful readers may have found a problem in the statement of the above theorem and may raise the question: why the above theorem contains the vague words "system (2.40) has a unique solution for sufficiently differentiable $u(t)$ and **some proper** initial values?" It can be seen from the above deduction that this is because the set of equations in (2.47) has a solution for only some special initial values. It is well known that a normal linear system has a unique solution for arbitrarily given initial values, while for descriptor linear systems this is no longer the case. Such a problem will be addressed more clearly and thoroughly in Sects. 3.4 and 3.5 in the next chapter, where classical and distributional solutions as well as the concept of consistent initial values of a descriptor linear system will be fully discussed.

2.4 Notes and References

This chapter first introduces the concept of restricted system equivalence between descriptor linear systems, which describes the lack of uniqueness of the state space representation of a descriptor linear system, and then gives three types of canonical equivalent forms for general descriptor linear systems, namely, the dynamics decomposition form, the Kronecker form, and the canonical equivalence forms for derivative feedback. Finally, based on the proposed Kronecker canonical equivalent form, conditions for existence and uniqueness of solutions to general descriptor linear systems are proposed.

The Kronecker canonical form for descriptor linear systems is of particular importance. It provides an efficient way of deriving the standard decomposition form of a descriptor linear system, which is introduced in Chap. 3 and frequently used in the sequential chapters. Algorithms for the computation of a Kronecker canonical form of a singular pencil are proposed in Beelen et al. (1986), Beelen and Van Dooren (1988), and Misra et al. (1994). The paper by Misra et al. (1994) presents a state space characterization of the transmission zeros of singular linear multi-variable systems, which is analogous to that of standard normal linear systems. This characterization is also a valid alternative to the algorithms given in Beelen et al. (1986) and Beelen and Van Dooren (1988) for computing the Kronecker structure of an arbitrary singular pencil. As a consequence, these have led to the development of a very useful Matlab Toolbox that has many analysis routines for linear descriptor systems (Varga 2000).

It should be noted that the restricted system equivalence relation introduced for descriptor linear systems involves, unlike that for the standard normal linear systems, two nonsingular transformation matrices. This actually gives more flexibility to the transformed canonical forms. Besides the ones stated in this chapter, there also exist many other canonical forms for descriptor linear systems. Among papers on restricted system equivalence, typical ones are Campbell (1982), Cullen (1984), Dai (1987), Fuhrmann (1977), Hayton et al. (1986), Pugh and Shelton (1978), Verghese (1981), Zhou et al. (1987). More recent work on canonical forms for descriptor linear systems includes Vafiadis and Karcanias (1997) and Zhu and Tian (2004).

Chapter 3
Regular Descriptor Linear Systems

We have pointed out in Sect. 1.3 that regularity is a very important property for descriptor linear systems. It guarantees the existence and uniqueness of solutions to descriptor linear systems. This chapter studies regular descriptor linear systems and starts in Sect. 3.1 with the definition of regularity of descriptor linear systems and its relation with solutions of descriptor linear systems. The restricted system equivalence for a regular descriptor linear system is introduced in Sect. 3.2. In the other sections of this chapter, some special features of regular descriptor linear systems are introduced. These include the eigenstructure, the state response, the stability as well as the transfer function of regular descriptor linear systems.

3.1 Regularity of Descriptor Linear Systems

Consider the following square descriptor linear system

$$E\dot{x}(t) = Ax(t) + Bu(t), \tag{3.1}$$

where $E, A \in \mathbb{R}^{n \times n}$, and $B \in \mathbb{R}^{n \times r}$ are the system coefficient matrices; $x \in \mathbb{R}^n$ and $u \in \mathbb{R}^r$ are, respectively, the state vector and the input vector of the system. Now let us first define the regularity of the system (3.1).

3.1.1 The Definition and Its Relation with Solutions

As a matter of fact, the regularity of the system (3.1) involves only the coefficient matrices E and A.

Definition 3.1. The system (3.1) is called regular if there exists a constant scalar $\gamma \in \mathbb{C}$ such that

$$\det(\gamma E - A) \neq 0, \tag{3.2}$$

or equivalently, the polynomial $\det(sE - A)$ is not identically zero. In this case, we also say that the matrix pair (E, A), or the matrix pencil $sE - A$, is regular.

G.-R. Duan, *Analysis and Design of Descriptor Linear Systems*, Advances
in Mechanics and Mathematics 23, DOI 10.1007/978-1-4419-6397-0_3,
© Springer Science+Business Media, LLC 2010

Example 3.1 (Duan 1992b, 1998a; Duan and Patton 1998a). Consider the matrix pair (E, A) with

$$E = \begin{bmatrix} 1 & 0 & 0 \\ 0 & 1 & 0 \\ 0 & 0 & 0 \end{bmatrix}, \quad A = \begin{bmatrix} 0 & 1 & 0 \\ 0 & 0 & 1 \\ 0 & 0 & -1 \end{bmatrix}.$$

Direct deduction shows that

$$\det(sE - A) = \det \begin{bmatrix} s & -1 & 0 \\ 0 & s & -1 \\ 0 & 0 & 1 \end{bmatrix} = s^2 \neq 0, \ \forall s \in \mathbb{C}, \ s \neq 0.$$

Thus, (E, A) is regular.

Example 3.2 (Kautsky et al. 1989; Chu et al. 1997; Duan and Patton 1998a). Consider the matrix pair (E, A) with

$$E = \begin{bmatrix} 0 & 0 & 0 & 1.72 & 0 \\ 0 & 0 & 0 & 0 & 0 \\ -0.82 & 0 & 0 & 0 & 0 \\ 0 & 0 & 0 & 0 & 0 \\ 0 & 0 & 0 & 0 & 1 \end{bmatrix}, \quad A = \begin{bmatrix} 0 & 1.1 & 0 & 0 & 0 \\ 0 & 0 & 1.56 & 0 & 0 \\ 1.23 & 0 & 0 & 1.98 & 0 \\ 0 & 0 & 0 & 0 & 0 \\ 0 & 0 & 1.01 & 0 & 0 \end{bmatrix}.$$

Note that the 4-th rows of $sE - A$ is composed of all zeros, we have

$$\det(sE - A) = 0, \ \forall s \in \mathbb{C}.$$

Thus, (E, A) is not regular.

In some references, for example, Bunse-Gerstner et al. (1992), the following definition for regularity is used.

Definition 3.2. The system (3.1), or the matrix pair (E, A), is called regular if there exist a pair of scalars $\alpha, \beta \in \mathbb{C}$ such that

$$\det(\alpha E - \beta A) \neq 0. \tag{3.3}$$

The above two definitions clearly suggest the following conclusion.

Proposition 3.1. *The following two conditions are equivalent:*

1. There exists a constant scalar $\gamma \in \mathbb{C}$ such that (3.2) holds.
2. There exist a pair of scalars $\alpha, \beta \in \mathbb{C}$ such that (3.3) holds.

Proof. Suppose that there exists a constant scalar $\gamma \in \mathbb{C}$ such that (3.2) holds. Then (3.3) obviously holds for $\alpha = \gamma$ and $\beta = 1$.

Now, let (3.3) holds for some $\alpha, \beta \in \mathbb{C}$. If $\beta \neq 0$, then it is easy to see that (3.2) holds for $\gamma = \alpha/\beta$. When $\beta = 0$, (3.3) reduces to

$$\det(\alpha E) \neq 0.$$

Note that $\det M$ is continuous with respect to the elements of the matrix M. Therefore, for sufficiently small but nonzero scalar $\varepsilon \in \mathbb{C}$, there holds

$$\det(\alpha E - \varepsilon A) \neq 0.$$

This indicates that (3.2) holds for $\gamma = \alpha/\varepsilon$.   □

Based on the above definitions, the following basic fact can be shown.

Proposition 3.2. *Regularity of descriptor linear systems is maintained under restricted system equivalence relations.*

Proof. Let (E, A) and $(\tilde{E}, \tilde{A})$ be two matrix pairs of the same dimension and satisfy the relations

$$\tilde{E} = QEP, \quad \tilde{A} = QAP \tag{3.4}$$

for some nonsingular matrices P and Q. Furthermore, suppose that the matrix pair (E, A) is regular, then according to the above Definition 3.2 that there exist a pair of scalars $\alpha, \beta \in \mathbb{C}$ such that (3.3) holds. Therefore, it follows from (3.4), (3.3) and the non-singularity of the matrices P and Q that

$$\begin{aligned}
\det(\alpha \tilde{E} - \beta \tilde{A}) \\
= \det(\alpha QEP - \beta QAP) \\
= \det Q \det(\alpha E - \beta A) \det P \\
\neq 0.
\end{aligned}$$

This states that the matrix pair $(\tilde{E}, \tilde{A})$ is also regular.   □

We pointed out in Sect. 1.3 that regularity guarantees the existence and uniqueness of solutions of descriptor linear systems. This important conclusion lies in the following fact and Theorem 2.7, which is about solutions to general descriptor linear systems.

Theorem 3.1. *The matrix pair (E, A), with $E, A \in \mathbb{R}^{n \times n}$, is regular if and only if there exist two nonsingular matrices Q, P such that*

$$QEP = \mathrm{diag}\left(I_{n_1}, N\right), \quad QAP = \mathrm{diag}(A_1, I_{n_2}), \tag{3.5}$$

where $n_1 + n_2 = n$, $A_1 \in \mathbb{R}^{n_1 \times n_1}$, $N \in \mathbb{R}^{n_2 \times n_2}$, and the matrix N is nilpotent.

Proof. Sufficiency: Suppose that the relations in (3.5) hold. Let $\gamma \notin \sigma(A_1)$, and without lost of generality, we restrict $\gamma \in \mathbb{R}$. Then

$$\det(\gamma I_{n_1} - A_1) \neq 0. \tag{3.6}$$

Further noting that
$$\det(\gamma N - I_{n_2}) = (-1)^{n_2}, \tag{3.7}$$

we have from (3.5) to (3.7) that

$$
\begin{aligned}
\det(\gamma E - A) \\
&= \det(Q^{-1}) \det(P^{-1}) \det(\gamma QEP - QAP) \\
&= \det(Q^{-1}) \det(P^{-1}) \det(\gamma I_{n_1} - A_1) \det(\gamma N - I_{n_2}) \\
&= (-1)^{n_2} \det(Q^{-1}) \det(P^{-1}) \det(\gamma I_{n_1} - A_1) \\
&\neq 0.
\end{aligned}
$$

This shows that the matrix pair (E, A) is regular.

Necessity: Suppose (E, A) is regular. Then it follows from Definition 3.1 that there exists a constant scalar $\gamma \in \mathbb{R}$ such that (3.2) holds. Therefore, $(\gamma E - A)^{-1}$ exists.

Consider the matrix pair $(\hat{E}, \hat{A})$ defined by

$$\hat{E} = (\gamma E - A)^{-1} E, \quad \hat{A} = (\gamma E - A)^{-1} A. \tag{3.8}$$

First, it is easy to obtain

$$
\begin{aligned}
\hat{A} &= (\gamma E - A)^{-1} (\gamma E + A - \gamma E) \\
&= \gamma (\gamma E - A)^{-1} E - I \\
&= \gamma \hat{E} - I.
\end{aligned} \tag{3.9}
$$

On the other hand, it follows from the Jordan canonical form decomposition in matrix theory that there exists a nonsingular matrix T such that

$$T \hat{E} T^{-1} = \mathrm{diag}(\hat{E}_1, \hat{E}_2), \tag{3.10}$$

where $T \in \mathbb{R}^{n \times n}$, $\hat{E}_1 \in \mathbb{R}^{n_1 \times n_1}$ is nonsingular, $\hat{E}_2 \in \mathbb{R}^{n_2 \times n_2}$ is nilpotent. This means that $\gamma \hat{E}_2 - I$ is nonsingular. Letting

$$Q = \mathrm{diag}\left(\hat{E}_1^{-1}, \ (\gamma \hat{E}_2 - I)^{-1}\right) T \, (\gamma E - A)^{-1}, \quad P = T^{-1}, \tag{3.11}$$

and using (3.8)–(3.10), we have

$$
\begin{aligned}
QEP &= \mathrm{diag}\left(\hat{E}_1^{-1}, \ (\gamma \hat{E}_2 - I)^{-1}\right) T \, (\gamma E - A)^{-1} E T^{-1} \\
&= \mathrm{diag}\left(\hat{E}_1^{-1}, \ (\gamma \hat{E}_2 - I)^{-1}\right) T \hat{E} T^{-1} \\
&= \mathrm{diag}\left(\hat{E}_1^{-1}, \ (\gamma \hat{E}_2 - I)^{-1}\right) \mathrm{diag}(\hat{E}_1, \hat{E}_2) \\
&= \mathrm{diag}(I_{n_1}, \ (\gamma \hat{E}_2 - I)^{-1} \hat{E}_2),
\end{aligned}
$$

and

$$
\begin{aligned}
QAP &= \operatorname{diag}\left(\hat{E}_1^{-1}, (\gamma\hat{E}_2 - I)^{-1}\right) T (\gamma E - A)^{-1} AT^{-1} \\
&= \operatorname{diag}\left(\hat{E}_1^{-1}, (\gamma\hat{E}_2 - I)^{-1}\right) T\hat{A}T^{-1} \\
&= \operatorname{diag}\left(\hat{E}_1^{-1}, (\gamma\hat{E}_2 - I)^{-1}\right) T(\gamma\hat{E} - I)T^{-1} \\
&= \operatorname{diag}\left(\hat{E}_1^{-1}, (\gamma\hat{E}_2 - I)^{-1}\right)(\gamma T\hat{E}T^{-1} - I) \\
&= \operatorname{diag}\left(\hat{E}_1^{-1}, (\gamma\hat{E}_2 - I)^{-1}\right)(\gamma\operatorname{diag}(\hat{E}_1, \hat{E}_2) - I) \\
&= \operatorname{diag}(\gamma I_{n_1}, \gamma(\gamma\hat{E}_2 - I)^{-1}\hat{E}_2) - \operatorname{diag}\left(\hat{E}_1^{-1}, (\gamma\hat{E}_2 - I)^{-1}\right) \\
&= \operatorname{diag}(\gamma I_{n_1} - \hat{E}_1^{-1}, I_{n_2}).
\end{aligned}
$$

Therefore, the relations in (3.5) hold with

$$
A_1 = \hat{E}_1^{-1}(\gamma\hat{E}_1 - I), \quad N = (\gamma\hat{E}_2 - I)^{-1}\hat{E}_2. \tag{3.12}
$$

It is easily verified that $(\gamma\hat{E}_2 - I)^{-1}\hat{E}_2 = \hat{E}_2(\gamma\hat{E}_2 - I)^{-1}$. This implies that the matrix N is nilpotent. The proof is then completed. □

Theorem 2.7 shows that if the system (3.1) has a unique solution for any sufficiently differentiable input function $u(t)$, the matrices E and A must be square and satisfy (2.53). This is equivalent to the regularity of (E, A) by Theorem 3.1.

3.1.2 Criteria for Regularity

The following theorem gives some criteria for regularity.

Theorem 3.2 (Yip and Sincovec 1981). *Let matrices $E, A \in \mathbb{R}^{n\times n}$. Then the following statements are equivalent:*

1. *The matrix pair (E, A) is regular.*
2. *Let $X_o = \ker A$, $X_i = \{x \mid Ax \in EX_{i-1}\}$, then $\ker E \cap X_i = \{0\}$ for $i = 0, 1, \dots$.*
3. *Let $Y_o = \ker A^{\mathrm{T}}$, $Y_i = \{x \mid A^{\mathrm{T}}x \in E^{\mathrm{T}}Y_{i-1}\}$, then $\ker E^{\mathrm{T}} \cap Y_i = \{0\}$ for $i = 0, 1, \dots$.*
4. *Let*

$$
G(k) = \begin{bmatrix}
E & & & & \\
A & E & & & \\
& A & \ddots & & \\
& & \ddots & E & \\
& & & A &
\end{bmatrix} \in \mathbb{R}^{(k+1)n\times nk}.
$$

Then $\operatorname{rank} G(k) = nk$ for $k = 1, 2, \dots$.

5. *Let*

$$F(k) = \begin{bmatrix} E & A & & & \\ & E & A & & \\ & & \ddots & \ddots & \\ & & & & A \\ & & & & E & A \end{bmatrix} \in \mathbb{R}^{nk \times n(k+1)}.$$

Then $\operatorname{rank} F(k) = nk$ *for* $k = 1, 2, \ldots$.

The conditions 2–5 in the above theorem are generally not convenient to verify. Moreover, for large k, the computation of $\operatorname{rank} G(k)$ and $\operatorname{rank} F(k)$ may subject to numerical problems. The following theorem gives a method for verifying the regularity of a given descriptor linear system based on its dynamics decomposition form introduced in Sect. 2.2.

Theorem 3.3. *Let* $E, A \in \mathbb{R}^{n \times n}$ *with* $\operatorname{rank} E = n_0$, *and* $P, Q \in \mathbb{R}^{n \times n}$ *be two nonsingular matrices satisfying*

$$QAP = \begin{bmatrix} E_0 & 0 \\ 0 & 0 \end{bmatrix}, \tag{3.13}$$

where $E_0 \in \mathbb{R}^{n_0 \times n_0}$ *is some nonsingular matrix. Further let*

$$QEP = \begin{bmatrix} A_{11} & A_{12} \\ A_{21} & A_{22} \end{bmatrix}. \tag{3.14}$$

Then the matrix pair (E, A) *is regular if and only if*

$$\det \left(A_{22} + A_{21} \left(\gamma E_0 - A_{11} \right)^{-1} A_{12} \right) \neq 0 \tag{3.15}$$

holds for arbitrary $\gamma \notin \sigma(E_0, A_{11})$.

Proof. Arbitrarily choosing $\gamma \notin \sigma(E_0, A_{11})$, then $(\gamma E_0 - A_{11})^{-1}$ exists. Using (3.13) and (3.14), we have

$$Q (\gamma E - A) P = \gamma QEP - QAP$$

$$= \gamma \begin{bmatrix} E_0 & 0 \\ 0 & 0 \end{bmatrix} - \begin{bmatrix} A_{11} & A_{12} \\ A_{21} & A_{22} \end{bmatrix}$$

$$= \begin{bmatrix} \gamma E_0 - A_{11} & -A_{12} \\ -A_{21} & -A_{22} \end{bmatrix}.$$

Since $\gamma E_0 - A_{11}$ is nonsingular, it follows from the above deduction and the Theorem A.2 in Appendix A about determinants of block matrices that $\gamma E - A$ is nonsingular if and only if

$$\det \left(-A_{22} - A_{21} \left(\gamma E_0 - A_{11}\right)^{-1} A_{12}\right) \neq 0,$$

which is equivalent to (3.15). The proof is then completed. $\qquad\qquad\Box$

In practical applications, the matrices P, Q and E_0 may be obtained via singular value decomposition. In this case, the matrices P and Q are orthogonal, and the matrix E_0 is diagonal.

The above criterion is convenient to use. However, since it involves a matrix inverse, it may be not numerically reliable in large dimension cases. The following theorem gives another criterion, which may be more preferable from the computational point of view.

Theorem 3.4. *The matrix pair* (E, A), *with* E, $A \in \mathbb{R}^{n \times n}$, *is regular if and only if*

$$\max \{\operatorname{rank}\left(\gamma_i E - A\right), \ i = 1, 2, \ldots, l + 1\} = n \tag{3.16}$$

holds for an arbitrarily selected set of distinct numbers $\gamma_i \in \mathbb{C}$, $i = 1, 2, \ldots, l + 1$, $l = \deg \det (sE - A)$.

Proof. The sufficiency is obvious. Let us only prove the necessity.

Suppose that the matrix pair (E, A) is regular, but (3.16) does not hold for a selected set of distinct numbers $\gamma_i \in \mathbb{C}$, $i = 1, 2, \ldots, l + 1, l = \deg \det (sE - A)$. We will derive a contradiction.

If (3.16) does not hold, we have

$$\det \left(\gamma_i E - A\right) = 0, \ i = 1, 2, \ldots, l + 1. \tag{3.17}$$

Denote

$$\det (sE - A) = a_l s^l + a_{l-1} s^{l-1} + \cdots + a_1 s + a_0,$$

then it follows from (3.17) that

$$a_l \gamma_i^l + a_{l-1} \gamma_i^{l-1} + \cdots + a_1 \gamma_i + a_0 = 0, \ i = 1, 2, \ldots, l + 1.$$

This can be arranged into the following linear equation

$$\begin{bmatrix} 1 & \gamma_1 & \cdots & \gamma_1^l \\ 1 & \gamma_2 & \cdots & \gamma_2^l \\ \vdots & \vdots & \cdots & \vdots \\ 1 & \gamma_{l+1} & \cdots & \gamma_{l+1}^l \end{bmatrix} \begin{bmatrix} a_0 \\ a_1 \\ \vdots \\ a_l \end{bmatrix} = 0.$$

Since the matrix of order $l + 1$ on the left-hand side of the above equation is a Vandermonde matrix and is nonsingular because $\gamma_i \in \mathbb{C}$, $i = 1, 2, \ldots, l + 1$, are distinct, the above equation obviously gives

$$a_i = 0, \ i = 1, 2, \ldots, l + 1.$$

This implies

$$\det(sE - A) \equiv 0,$$

which contradicts with the regularity assumption of the matrix pair (E, A).   □

Example 3.3. Consider the matrix pair (E, A) with

$$E = \begin{bmatrix} 0 & 0 \\ 1 & 0 \end{bmatrix}, \quad A = \begin{bmatrix} 1 & 0 \\ 1 & 0 \end{bmatrix}.$$

Choose $\gamma_1 = 0$, $\gamma_2 = 1$, $\gamma_3 = -1$, then it is easy to verify

$$\text{rank}(\gamma_1 E - A) = \text{rank} \begin{bmatrix} -1 & 0 \\ -1 & 0 \end{bmatrix} = 1 < 2,$$

$$\text{rank}(\gamma_2 E - A) = \text{rank} \begin{bmatrix} -1 & 0 \\ 0 & 0 \end{bmatrix} = 1 < 2,$$

$$\text{rank}(\gamma_3 E - A) = \text{rank} \begin{bmatrix} -1 & 0 \\ -2 & 0 \end{bmatrix} = 1 < 2,$$

it thus follows from the above theorem that this matrix pair (E, A) is not regular.

3.2 Equivalence of Regular Descriptor Linear Systems

In Sect. 2.2, we have introduced some canonical restricted equivalent forms for general descriptor linear systems. In this section, let us further consider the restricted system equivalence for regular descriptor linear systems.

Again, the system considered in this section is of the form

$$\begin{cases} E\dot{x} = Ax + Bu \\ y = Cx \end{cases}, \tag{3.18}$$

where $x \in \mathbb{R}^n$, $u \in \mathbb{R}^r$, $y \in \mathbb{R}^m$ are the state vector, the input vector and the output vector, respectively; $E, A \in \mathbb{R}^{n \times n}$, $B \in \mathbb{R}^{n \times r}$, and $C \in \mathbb{R}^{m \times n}$ are constant coefficient matrices.

3.2.1 Standard Decomposition Form

It has been seen from Sect. 2.2 that the Kronecker canonical equivalent form for a general descriptor linear system is very complicated. However, the following theorem shows that the Kronecker form for a regular descriptor system is very simple.

Theorem 3.5. *Given the descriptor system (3.18), or system (E, A, B, C), with $E, A \in \mathbb{R}^{n \times n}, B \in \mathbb{R}^{n \times r}, C \in \mathbb{R}^{m \times n}$, and (E, A) regular, there exist two nonsingular matrices Q and P such that*

$$(E, A, B, C) \overset{(P,Q)}{\Longleftrightarrow} (\tilde{E}, \tilde{A}, \tilde{B}, \tilde{C})$$

with

$$\begin{cases} \tilde{E} = QEP = \mathrm{diag}(I_{n_1}, N) \\ \tilde{A} = QAP = \mathrm{diag}(A_1, I_{n_2}) \\ \tilde{B} = QB = \begin{bmatrix} B_1 \\ B_2 \end{bmatrix} \\ \tilde{C} = CP = [C_1 \ C_2] \end{cases} \tag{3.19}$$

where $n_1 + n_2 = n$, and the involved partitions are compatible. Furthermore, the matrix $N \in \mathbb{R}^{n_2 \times n_2}$ is nilpotent.

Proof. It follows immediately from Theorems 2.7 and 3.1.   □

Based on the necessity part in the proof of Theorem 3.1, an algorithm for obtaining a standard decomposition of a descriptor linear system can be given as follows.

Algorithm 3.1. The standard decomposition. [Given regular matrix pair (E, A), finding the nonsingular matrices P and Q]

Step 1 Choose $\gamma \in \mathbb{R}$ such that $\det(\gamma E - A) \neq 0$.
Step 2 Let $\hat{E} = (\gamma E - A)^{-1} E$, find a matrix $T \in \mathbb{R}^{n \times n}$, and $\hat{E}_1, \hat{E}_2$ satisfying (3.10), where $\hat{E}_1 \in \mathbb{R}^{n_1 \times n_1}$ is nonsingular, $\hat{E}_2 \in \mathbb{R}^{n_2 \times n_2}$ is nilpotent. .
Step 3 Compute the matrices P and Q according to (3.11).
Step 4 Compute the matrices A_1 and N according to (3.12).

Let

$$\begin{bmatrix} x_1 \\ x_2 \end{bmatrix} = P^{-1}x, \quad x_1 \in \mathbb{R}^{n_1}, \quad x_2 \in \mathbb{R}^{n_2},$$

then according to the above theorem, system (3.18) is clearly r.s.e. to

$$\begin{cases} \dot{x}_1 = A_1 x_1 + B_1 u \\ y_1 = C_1 x_1 \end{cases} \tag{3.20}$$

and

$$\begin{cases} N\dot{x}_2 = x_2 + B_2 u \\ y_2 = C_2 x_2 \end{cases}, \tag{3.21}$$

with the joint measurement equation

$$y = C_1 x_1 + C_2 x_2 = y_1 + y_2. \tag{3.22}$$

The system represented by (3.20)–(3.22) is the Kronecker form for regular systems. For convenience, this form is called the standard decomposition form in the future. In this form, subsystems (3.20) and (3.21) are called the slow subsystem and the fast subsystem, respectively; x_1, x_2 are the slow and fast substates, respectively. Furthermore, if the nilpotent matrix N in the fast subsystem (3.21) has index h, then the original system (3.18) is called a system with index h. The meanings of the slow and fast subsystems will be revealed in the next section.

The above Theorem 3.5 clearly has the following corollary.

Corollary 3.1. *Let (E, A) be regular, and n_1 be the dimension in the decomposition (3.19), then*

$$\deg \det (sE - A) = n_1.$$

Proof. Noting that N is a nilpotent matrix, we have

$$\det \left(sN - I_{n_2} \right) = (-1)^{n_2}.$$

With this and the first two equations in (3.19), we obtain

$$\begin{aligned} \deg \det (sE - A) &= \deg \det [P (sE - A) Q] \\ &= \deg \det \left(\begin{bmatrix} sI_{n_1} & 0 \\ 0 & sN \end{bmatrix} - \begin{bmatrix} A_1 & 0 \\ 0 & I_{n_2} \end{bmatrix} \right) \\ &= \deg \det \left(sI_{n_1} - A_1 \right) + \deg \det \left(sN - I_{n_2} \right) \\ &= n_1. \end{aligned}$$

The proof is then completed.   □

The nonsingular matrices Q and P, which transfer a regular descriptor linear system into its standard decomposition form, are obviously not unique. This results in the lack of uniqueness of the standard decomposition form, i.e., the lack of uniqueness of the matrices A_1, B_1, B_2, C_1, C_2, and N. Assume that $\bar{Q}$ and $\bar{P}$ are another pair of nonsingular matrices and under the transformation $(\bar{P}, \bar{Q})$, the system (3.18) is transferred into another standard decomposition form given by

$$\begin{cases} \dot{\bar{x}}_1 = \bar{A}_1 \bar{x}_1 + \bar{B}_1 u \\ \bar{y}_1 = \bar{C}_1 \bar{x}_1 \end{cases}, \tag{3.23}$$

$$\begin{cases} N\dot{\bar{x}}_2 = \bar{x}_2 + \bar{B}_2 u \\ \bar{y}_2 = \bar{C}_2 \bar{x}_2 \end{cases}, \tag{3.24}$$

and

$$y = \bar{C}_1 \bar{x}_1 + \bar{C}_2 \bar{x}_2 = \bar{y}_1 + \bar{y}_2, \tag{3.25}$$

where $\bar{x}_1 \in \mathbb{R}^{\bar{n}_1}$, $\bar{x}_2 \in \mathbb{R}^{\bar{n}_2}$. Then, the relations between the two transformations (P, Q) and $(\bar{P}, \bar{Q})$ and their corresponding standard decomposition forms are revealed in the following theorem.

Theorem 3.6. *Suppose that (3.20)–(3.22) and (3.23)–(3.25) are both standard decomposition forms for system (3.18). Then $n_1 = \bar{n}_1$, $n_2 = \bar{n}_2$, and there exist nonsingular matrices $T_1 \in \mathbb{R}^{n_1 \times n_1}$ and $T_2 \in \mathbb{R}^{n_2 \times n_2}$ such that*

$$Q = \text{diag}(T_1, T_2)\bar{Q}, \quad P = \bar{P}\text{diag}\left(T_1^{-1}, T_2^{-1}\right), \tag{3.26}$$

$$A_1 = T_1 \bar{A}_1 T_1^{-1}, \quad N = T_2 \bar{N} T_2^{-1}, \tag{3.27}$$

and

$$B_i = T_i \bar{B}_i, \quad C_i = \bar{C}_i T_i^{-1}, \quad i = 1, 2. \tag{3.28}$$

Proof. Since

$$\det(sE - A)$$
$$= (-1)^{n_2} \det\left(Q^{-1}\right) \det\left(P^{-1}\right) \det\left(sI_{n_1} - A_1\right)$$
$$= (-1)^{\bar{n}_2} \det\left(\bar{Q}^{-1}\right) \det\left(\bar{P}^{-1}\right) \det\left(sI_{\bar{n}_1} - \bar{A}_1\right),$$

we know that $n_1 = \bar{n}_1$. Furthermore, note $n_1 + n_2 = n$, and $\bar{n}_1 + \bar{n}_2 = n$, we also have $n_2 = \bar{n}_2$.

By the definition of the standard decomposition and given conditions, we have

$$\begin{cases} Q^{-1}\text{diag}(I_{n_1}, N)P^{-1} = \bar{Q}^{-1}\text{diag}(I_{\bar{n}_1}, \bar{N})\bar{P}^{-1} = E \\ Q^{-1}\text{diag}(A_1, I_{n_2})P^{-1} = \bar{Q}^{-1}\text{diag}(\bar{A}_1, I_{\bar{n}_2})\bar{P}^{-1} = A \end{cases}.$$

Pre-multiplying by the matrix Q and post-multiplying by the matrix $\bar{P}$, both sides of the above equation, yield

$$\begin{cases} \text{diag}(I_{n_1}, N)P^{-1}\bar{P} = Q\bar{Q}^{-1}\text{diag}(I_{\bar{n}_1}, \bar{N}) \\ \text{diag}(A_1, I_{n_2})P^{-1}\bar{P} = Q\bar{Q}^{-1}\text{diag}(\bar{A}_1, I_{\bar{n}_2}) \end{cases}. \tag{3.29}$$

If we denote

$$Q\bar{Q}^{-1} = \begin{bmatrix} Q_{11} & Q_{12} \\ Q_{21} & Q_{22} \end{bmatrix}, \quad P^{-1}\bar{P} = \begin{bmatrix} P_{11} & P_{12} \\ P_{21} & P_{22} \end{bmatrix}, \tag{3.30}$$

substituting (3.30) into (3.29) yields

$$Q_{11} = P_{11}, \quad Q_{22} = P_{22}, \tag{3.31}$$

$$Q_{12}\bar{N} = P_{12}, \quad Q_{21}\bar{A}_1 = P_{21}, \tag{3.32}$$

$$Q_{21} = NP_{21}, \quad Q_{12} = A_1 P_{12}, \tag{3.33}$$

$$Q_{11}\bar{A}_1 = A_1 P_{11}, \quad Q_{22}\bar{N} = NP_{22}. \tag{3.34}$$

Furthermore, substituting (3.33) into (3.32), gives

$$P_{21} = NP_{21}\bar{A}_1, \quad P_{12} = A_1 P_{12}\bar{N}. \tag{3.35}$$

Noting that the matrices N and $\bar{N}$ are nilpotent, by repeatedly using (3.35) we obtain

$$\begin{cases} P_{21} = NP_{21}\bar{A}_1 = \cdots = N^n P_{21} \bar{A}_1^n = 0 \\ P_{12} = A_1 P_{12}\bar{N} = \cdots = A_1^n P_{12}\bar{N}^n = 0 \end{cases},$$

which, in conjunction with (3.33), yields

$$Q_{21} = 0, \quad Q_{12} = 0.$$

Denoting, in view of (3.31),

$$T_1 = Q_{11} = P_{11}, \quad T_2 = Q_{22} = P_{22},$$

then (3.30) becomes

$$Q\bar{Q}^{-1} = \begin{bmatrix} T_1 & 0 \\ 0 & T_2 \end{bmatrix}, \quad P^{-1}\bar{P} = \begin{bmatrix} T_1 & 0 \\ 0 & T_2 \end{bmatrix} \tag{3.36}$$

and (3.34) turns to be

$$T_1\bar{A}_1 = A_1 T_1, \quad T_2\bar{N} = NT_2. \tag{3.37}$$

The two relations in (3.36) and (3.37) clearly give (3.26) and (3.27), respectively.

Recalling that

$$QB = \begin{bmatrix} B_1 \\ B_2 \end{bmatrix}, \quad \bar{Q}B = \begin{bmatrix} \bar{B}_1 \\ \bar{B}_2 \end{bmatrix},$$

we can obtain

$$Q^{-1}\begin{bmatrix} B_1 \\ B_2 \end{bmatrix} = \bar{Q}^{-1}\begin{bmatrix} \bar{B}_1 \\ \bar{B}_2 \end{bmatrix},$$

which can be rearranged into

$$\begin{bmatrix} B_1 \\ B_2 \end{bmatrix} = Q\bar{Q}^{-1}\begin{bmatrix} \bar{B}_1 \\ \bar{B}_2 \end{bmatrix}.$$

This, together with the first relation in (3.36), gives the first equation in (3.28). The second equation in (3.28) can be shown similarly using the second relation in (3.36) and the following relations:

$$CP = [C_1 \ C_2], \quad C\bar{P} = [\bar{C}_1 \ \bar{C}_2].$$

The proof is then completed. □

The above theorem shows that, although different standard decomposition forms may be obtained under different transformations, these different forms obey a certain similarity relation. In other words, the standard decomposition form for a regular descriptor linear system is unique in the sense of similarity equivalence.

Example 3.4 (Dai 1989b). Consider the circuit network in Example 1.1 with $L=1$H, $C_0 = 1$F, $R = 1\Omega$. The state vector is chosen as

$$x(t) = \left[I(t) \ V_L(t) \ V_C(t) \ V_R(t) \right]^T,$$

and the measured variable is $y(t) = V_C(t)$. Then the system representation is

$$\begin{cases} \begin{bmatrix} 1 & 0 & 0 & 0 \\ 0 & 0 & 1 & 0 \\ 0 & 0 & 0 & 0 \\ 0 & 0 & 0 & 0 \end{bmatrix} \dot{x} = \begin{bmatrix} 0 & 1 & 0 & 0 \\ 1 & 0 & 0 & 0 \\ -1 & 0 & 0 & 1 \\ 0 & 1 & 1 & 1 \end{bmatrix} x + \begin{bmatrix} 0 \\ 0 \\ 0 \\ -1 \end{bmatrix} u(t) \\ y = \begin{bmatrix} 0 & 0 & 1 & 0 \end{bmatrix} x \end{cases}. \tag{3.38}$$

If we choose the following transformation matrices

$$Q = \begin{bmatrix} 1 & 0 & 1 & -1 \\ 0 & 1 & 0 & 0 \\ 0 & 0 & -1 & 1 \\ 0 & 0 & 1 & 0 \end{bmatrix}, \quad P = \begin{bmatrix} 1 & 0 & 0 & 0 \\ -1 & -1 & 1 & 0 \\ 0 & 1 & 0 & 0 \\ 1 & 0 & 0 & 1 \end{bmatrix},$$

and the state transformation

$$P^{-1}x = \begin{bmatrix} x_1 \\ x_2 \end{bmatrix}, \quad x_1 \in \mathbb{R}^2, x_2 \in \mathbb{R}^2,$$

then the standard decomposition form for the descriptor linear system (3.38) can be obtained as

$$\begin{cases} \dot{x}_1 = \begin{bmatrix} -1 & -1 \\ 1 & 0 \end{bmatrix} x_1 + \begin{bmatrix} 1 \\ 0 \end{bmatrix} u(t) \\ 0 = x_2 + \begin{bmatrix} -1 \\ 0 \end{bmatrix} u(t) \\ y = \begin{bmatrix} 0 & 1 \end{bmatrix} x_1 \end{cases}.$$

Example 3.5 (Fletcher 1988; Duan 1995, 1999). Consider the following descriptor linear system

$$\begin{cases} \begin{bmatrix} 1 & 0 & 0 & 0 \\ 0 & 1 & 0 & 0 \\ 0 & 0 & 1 & 0 \\ 0 & 0 & 0 & 0 \end{bmatrix} \dot{x} = \begin{bmatrix} 0 & 0 & 1 & 0 \\ 1 & 0 & 0 & 0 \\ 0 & 1 & 0 & 1 \\ 0 & 0 & 1 & 0 \end{bmatrix} x + \begin{bmatrix} 1 & 0 & 0 \\ 1 & -1 & 2 \\ 0 & 1 & 0 \\ 0 & 0 & 1 \end{bmatrix} u(t) \\ y = \begin{bmatrix} 0 & 1 & 0 & 0 \\ 0 & 0 & 0 & 1 \end{bmatrix} x \end{cases} \qquad (3.39)$$

If we choose the following transformation matrices

$$Q = \begin{bmatrix} 1 & 0 & 0 & -1 \\ 0 & 1 & 0 & 0 \\ 0 & 0 & 1 & 0 \\ 0 & 0 & 0 & 1 \end{bmatrix}, \quad P = \begin{bmatrix} 1 & 0 & 0 & 0 \\ 0 & 1 & 0 & 0 \\ 0 & 0 & 0 & 1 \\ 0 & -1 & 1 & 0 \end{bmatrix},$$

and the state transformation

$$P^{-1}x = \begin{bmatrix} x_1 \\ x_2 \end{bmatrix}, \quad x_1 \in \mathbb{R}^2, \quad x_2 \in \mathbb{R}^2,$$

then the standard decomposition form for the descriptor linear system (3.39) can be obtained as

$$\begin{cases} \dot{x}_1 = \begin{bmatrix} 0 & 0 \\ 1 & 0 \end{bmatrix} x_1 + \begin{bmatrix} 1 & 0 & -1 \\ 1 & -1 & 2 \end{bmatrix} u(t) \\ \begin{bmatrix} 0 & 1 \\ 0 & 0 \end{bmatrix} \dot{x}_2 = x_2 + \begin{bmatrix} 0 & 1 & 0 \\ 0 & 0 & 1 \end{bmatrix} u(t) \\ y = \begin{bmatrix} 0 & 1 \\ 0 & -1 \end{bmatrix} x_1 + \begin{bmatrix} 0 & 0 \\ 1 & 0 \end{bmatrix} x_2 \end{cases} \qquad (3.40)$$

3.2.2 The Inverse Form

Under the regularity assumption, another simple canonical equivalent form for a descriptor linear system can be easily obtained.

Theorem 3.7. *Consider the descriptor linear system (E, A, B, C) with $E, A \in \mathbb{R}^{n \times n}$, $B \in \mathbb{R}^{n \times r}$, $C \in \mathbb{R}^{m \times n}$ and (E, A) regular. Let γ be a scalar satisfying $\det(\gamma E - A) \neq 0$, and define*

$$Q = (\gamma E - A)^{-1}.$$

Then

$$(E, A, B, C) \overset{(I,Q)}{\Longleftrightarrow} (\tilde{E}, \tilde{A}, \tilde{B}, C)$$

with

$$\begin{cases} \tilde{E} = QE = (\gamma E - A)^{-1} E \\ \tilde{A} = QA = (\gamma E - A)^{-1} A \\ \tilde{B} = QB = (\gamma E - A)^{-1} B \end{cases}.$$

Furthermore, the following relation holds:

$$\tilde{A} = \gamma \tilde{E} - I.$$

Proof. The first conclusion clearly follows from the definition of restricted system equivalence, while the second one clearly follows from (3.8) to (3.9). □

The above theorem indicates that the system (3.18), when it is regular, has the following restricted equivalent form:

$$\begin{cases} \tilde{E}\dot{x} = (\gamma \tilde{E} - I)x + \tilde{B}u \\ y = Cx \end{cases}, \tag{3.41}$$

where

$$\tilde{E} = (\gamma E - A)^{-1} E, \quad \tilde{B} = (\gamma E - A)^{-1} B. \tag{3.42}$$

The system (3.41)–(3.42) is called the inverse form of the descriptor linear system (3.18). Obviously, for a fixed γ, this form is generally unique.

Like the standard decomposition form and the dynamics decomposition form, the inverse form can also provide some convenience in some analysis and design problems for descriptor linear systems. We will see in Chap. 4 that this form gives very simple and insightful conditions for controllability and observability analysis.

Example 3.6. Consider the system (3.39) in Example 3.5. Noting that $(E - A)^{-1}$ exists, we can choose $\gamma = 1$, and this leads to

$$Q = (E - A)^{-1} = \begin{bmatrix} 1 & 0 & 0 & -1 \\ 1 & 1 & 0 & -1 \\ 0 & 0 & 0 & -1 \\ -1 & -1 & -1 & 0 \end{bmatrix}.$$

Thus, the system (3.39) is r.s.e. to the system

$$\begin{cases} \tilde{E}\dot{x} = \tilde{A}x + \tilde{B}u(t) \\ y = Cx \end{cases},$$

where

$$\tilde{E} = QE = \begin{bmatrix} 1 & 0 & 0 & 0 \\ 1 & 1 & 0 & 0 \\ 0 & 0 & 0 & 0 \\ -1 & -1 & -1 & 0 \end{bmatrix},$$

$$\tilde{A} = QA = \begin{bmatrix} 0 & 0 & 0 & 0 \\ 1 & 0 & 0 & 0 \\ 0 & 0 & -1 & 0 \\ -1 & -1 & -1 & -1 \end{bmatrix},$$

$$\tilde{B} = QB = \begin{bmatrix} 1 & 0 & -1 \\ 2 & -1 & 1 \\ 0 & 0 & -1 \\ -2 & 0 & -2 \end{bmatrix}.$$

3.3   Transfer Function Matrices

Based on the state variable concept, the state space description method for descriptor
linear system analysis and synthesis uses a state space model of the following form:

$$\begin{cases} E\dot{x} = Ax + Bu \\ y = Cx \end{cases}, \tag{3.43}$$

where $x \in \mathbb{R}^n$, $u \in \mathbb{R}^r$ and $y \in \mathbb{R}^m$ are the state vector, the input vector, and the
output vector, respectively; $E, A \in \mathbb{R}^{n \times n}$, $B \in \mathbb{R}^{n \times r}$, and $C \in \mathbb{R}^{m \times n}$ are constant coefficient matrices. Such a model can be obtained from the physical sense of
the state variables and their relations, or through certain identification techniques.
One of the main advantages of the state space representation is that it allows us to
understand the inner structure of the system.

It is known from standard normal linear systems theory that transfer function
matrix is another representation of linear systems. Unlike the state space representation, the transfer function matrix reflects only the outer structure, that is, the
input–output relationship, or the transfer relationship from input to output. Although
the transfer function matrix representation is sometimes unable to characterize the
inner properties of the system, it provides a simple and succinct dependence relationship between input and output, and in certain cases is sufficient for certain
analysis and design problems.

3.3.1 The Definition

The main purpose of this subsection is to derive the transfer function matrix for the system (3.43) under the regularity assumption.
    Denote

$$\mathscr{L}[y(t)] = Y(s), \quad \mathscr{L}[u(t)] = U(s),$$

then the transfer function matrix for the system (3.43) is defined to be the rational function matrix $G(s)$ satisfying

$$Y(s) = G(s)U(s)$$

under the zero initial value condition $x(0) = 0$.

Theorem 3.8. *The regular descriptor linear system (3.43) has the following transfer function matrix*

$$G(s) = C(sE - A)^{-1} B.$$

Proof. Taking Laplace transform on both sides of the system equations, and using the properties for Laplace transforms introduced in Appendix A, we have

$$sEX(s) - Ex(0) = AX(s) + BU(s) \tag{3.44}$$

and

$$Y(s) = CX(s). \tag{3.45}$$

Under the assumption of regularity, $(sE - A)^{-1}$ exists. Therefore, (3.44) can be written, when taking $x(0) = 0$, as

$$X(s) = (sE - A)^{-1} BU(s).$$

Substituting this into (3.45) gives

$$Y(s) = C(sE - A)^{-1} BU(s).$$

The conclusion thus holds. □

Example 3.7. Again consider the system (3.39) in Example 3.5. Through some deductions, its transfer function can be obtained as follows:

$$\begin{aligned}
G(s) &= C(sE - A)^{-1} B \\
&= \frac{1}{s^2} \begin{bmatrix} 1+s & -s & 2s-1 \\ -1-s & s(1-s) & 1-2s-s^3 \end{bmatrix}.
\end{aligned}$$

3.3.2 Properties

We know from standard normal linear systems theory that the transfer function matrix of a normal linear system is always a proper fraction matrix. The following fact tells us that this is no longer true for descriptor linear systems.

Proposition 3.3. *The descriptor system with only a fast subsystem*

$$\begin{cases} N\dot{x} = x + Bu \\ y = Cx \end{cases}, \tag{3.46}$$

where N is a nilpotent matrix with nilpotent index h, has the transfer function matrix

$$G(s) = -CB - sCNB - \cdots - s^{h-1}CN^{h-1}B, \tag{3.47}$$

which is a polynomial matrix.

Proof. According to Theorem 3.8, the transfer function of the fast system (3.46) is

$$G(s) = C(sN - I)^{-1}B. \tag{3.48}$$

Noting that $N^h = 0$, we have

$$(sN - I)^{-1} = -(I - sN)^{-1}$$
$$= -(I + sN + \cdots + s^{h-1}N^{h-1}).$$

Substituting the above into (3.48) gives the expression (3.47).   □

Remark 3.1. In classical systems theory and normal linear systems theory, an improper transfer function is said not to be physically realizable. It is now clear that an improper transfer function cannot be realized in a normal linear system form, but can be realized in a descriptor linear system form.

The following result further shows that the transfer function matrix representation remains unchanged under restricted system equivalence.

Proposition 3.4. *Let matrix quadruple (E, A, B, C) be a regular descriptor linear system, and $(\tilde{E}, \tilde{A}, \tilde{B}, \tilde{C})$ be a restricted system equivalence form of (E, A, B, C). Then the two systems have the same transfer function matrix .*

Proof. Denote the transfer function matrices of the two systems (E, A, B, C) and $(\tilde{E}, \tilde{A}, \tilde{B}, \tilde{C})$ by $G(s)$ and $\tilde{G}(s)$, respectively, then

$$G(s) = C(sE - A)^{-1}B$$

and

$$\tilde{G}(s) = \tilde{C}(s\tilde{E} - \tilde{A})^{-1}\tilde{B}.$$

Let Q and P be two nonsingular matrices satisfying

$$\tilde{E} = QEP, \quad \tilde{A} = QAP, \quad \tilde{B} = QB, \quad \tilde{C} = CP,$$

then

$$\begin{aligned}
\tilde{G}(s) &= \tilde{C}(s\tilde{E} - \tilde{A})^{-1}\tilde{B} \\
&= CP(sQEP - QAP)^{-1}QB \\
&= C(sE - A)^{-1}B \\
&= G(s).
\end{aligned}$$

This gives the desired conclusion.   □

The above proposition reveals the fact that two restricted equivalent systems have the same input–output properties. Since restricted equivalent systems are often the different state space representations of the same practical system, this conclusion holds very naturally. Such a property provides a great advantage in system analysis and design. For systems with structures that are very complex, we may turn to study a relatively simpler r.s.e. form of the system.

3.4 State Responses of Regular Descriptor Linear Systems: Distributional Solutions

In Sect. 2.3, we have discussed the solution to a general descriptor system, and have pointed out in Sect. 3.1 the fact that a descriptor linear system has a unique solution for some proper initial value if and only if it is regular. In this section, let us further give the representation of the unique solution of a square regular descriptor linear system of the following form:

$$E\dot{x} = Ax + Bu, \tag{3.49}$$

where $x \in \mathbb{R}^n$ and $u \in \mathbb{R}^r$ are, respectively, the state and the input vectors. Also, it is assumed that the system has index h.

The development in this section is based on the standard decomposition form of the system and Laplace transforms. Eventually, readers will find that the solution obtained in this section contains generalized functions (or distributions), such a solution is called a distributional solution or a generalized solution. The reason for that lies in the following.

The frequency solution of (3.49) is clearly

$$X(s) = (sE - A)^{-1}[Ex(0) + BU(s)].$$

In the normal system case, $(sE - A)^{-1}$ is a rational fraction function since the matrix E is nonsingular, thus the inverse Laplace transform of $X(s)$ produces a classical function, which is continuously differentiable. While in the general descriptor system case, as we will see below that, $(sE - A)^{-1}$ may contain certain polynomial terms. The inverse Laplace transforms of these polynomial terms eventually produces in the time-domain a generalized function, or more precisely, the impulse function $\delta(t)$. This naturally results in the generalized solution or the distributional solution of descriptor linear systems. This point will be explained more clearly in the proof of the main theorem in the following subsection.

Distributional solution is an important and special issue in descriptor linear systems theory. For a more rigorous discussion, please refer to Yan and Duan (2005), and also some other earlier works, e.g., Verghese et al. (1981), Cobb (1983b), Zhou et al. (1987), Geerts (1993), and Campbell (1980).

3.4.1 Solutions of Slow and Fast Subsystems

Recall Theorem 3.1 that there exist two nonsingular matrices Q and P such that

$$(E, A, B) \overset{(P,Q)}{\Longleftrightarrow} (\tilde{E}, \tilde{A}, \tilde{B})$$

with

$$\begin{cases} \tilde{E} = QEP = \mathrm{diag}(I_{n_1}, N) \\ \tilde{A} = QAP = \mathrm{diag}(A_1, I_{n_2}) \\ \tilde{B} = QB = \begin{bmatrix} B_1 \\ B_2 \end{bmatrix} \end{cases}, \tag{3.50}$$

where $n_1 + n_2 = n$, and the partitions are compatible. Furthermore, the matrix $N \in \mathbb{R}^{n_2 \times n_2}$ is a nilpotent matrix with nilpotent index h.

Let

$$\begin{bmatrix} x_1 \\ x_2 \end{bmatrix} = P^{-1}x, \quad x_1 \in \mathbb{R}^{n_1}, \ x_2 \in \mathbb{R}^{n_2}, \tag{3.51}$$

then the system (3.49) is clearly r.s.e. to

$$\dot{x}_1 = A_1 x_1 + B_1 u \tag{3.52}$$

and

$$N\dot{x}_2 = x_2 + B_2 u, \tag{3.53}$$

which are the slow and fast subsystems, respectively.

It follows from the above that, in order to solve system (3.49), it suffices only to find the solutions to the slow and fast subsystems (3.52) and (3.53).

Solution of the Slow Subsystem

Noting that the slow subsystem (3.52) is nothing more than an ordinary differential equation, we have the following well-known result.

Proposition 3.5. *The slow subsystem (3.52) has a unique solution with any initial condition* $x_1(0) = x_{10}$ *(the initial time is supposed to be zero) for any piecewise continuous input function* $u(t)$*, and the solution is given by*

$$x_1(t, u, x_{10}) = x_{1i}(t, x_{10}) + x_{1u}(t, u),$$

where $x_{1i}(t, x_{10})$ *is the response due to the initial value* x_{10} *and is given by*

$$x_{1i}(t, x_{10}) = e^{A_1 t} x_{10},$$

and $x_{1u}(t, u)$ *is the response due to the input* $u(t)$ *and is given by*

$$x_{1u}(t, u) = \int_0^t e^{A_1(t-\tau)} B_1 u(\tau) \, d\tau.$$

It is seen from the above Proposition 3.5 that the response, $x_1(t)$, of the slow system (3.52) is completely determined by the initial value $x_1(0) = x_{10}$ and the control $u(\tau)$, $(0 \le \tau \le t)$.

Solution of the Fast Subsystem

For solution to the fast subsystem (3.53), we have the following theorem.

Theorem 3.9. *Given the fast system (3.53) with the matrix* N *being a nilpotent matrix with nilpotent index* h*, for any* h *times piecewise continuously differentiable input function* $u(t)$*, and initial value* $x_2(0) = x_{20}$*, the fast system (3.53) has a unique solution, which is given by*

$$x_2(t, u, x_{20}) = x_{2i}(t, x_{20}) + x_{2u}(t, u), \tag{3.54}$$

where $x_{2u}(t, u)$ *represents the response due to the input function* $u(t)$*, and is given by*

$$x_{2u}(t, u) = -\sum_{i=0}^{h-1} N^i B_2 \left(u^{(i)}(t) + \sum_{k=0}^{i-1} \delta^{(k)}(t) u^{(i-k-1)}(0) \right), \tag{3.55}$$

and $x_{2i}(t, x_{20})$ *represents the response due to the initial value condition* $x_2(0) = x_{20}$*, and is given by*

$$x_{2i}(t, x_{20}) = -\sum_{i=1}^{h-1} \delta^{(i-1)}(t) N^i x_{20}, \tag{3.56}$$

here $\delta(t)$ *is the Delta function, which has an infinite value at* $t = 0$ *and a zero value at any other point.*

Proof. Let
$$\mathscr{L}[x_2(t)] = X_2(s), \quad \mathscr{L}[u(t)] = U(s),$$

and take the Laplace transforms on both sides of (3.53), and use the properties of Laplace transforms provided in Appendix A, we have

$$(sN - I)X_2(s) = Nx_2(0) + B_2U(s).$$

From this equation, we obtain

$$X_2(s) = (sN - I)^{-1}(Nx_2(0) + B_2U(s)). \tag{3.57}$$

Furthermore, in view of the series expansion

$$(sN - I)^{-1} = -\sum_{i=0}^{h-1} N^i s^i,$$

we have, from (3.57),

$$X_2(s) = -\sum_{i=0}^{h-1} N^i s^i (Nx_2(0) + B_2U(s)). \tag{3.58}$$

Taking the inverse Laplace transform on both sides of (3.58) gives

$$x_2(t) = -\sum_{i=0}^{h-1} \mathscr{L}^{-1}\left[N^{i+1}s^i x_{20} + N^i s^i B_2 U(s)\right]$$

$$= -\sum_{i=0}^{h-1}\left[\delta^{(i)}(t)N^{i+1}x_{20} + N^i B_2 u^{(i)}(t) + \sum_{k=0}^{i-1}\delta^{(k)}(t)N^i B_2 u^{(i-k-1)}(0)\right]$$

$$= -\sum_{i=1}^{h-1}\delta^{(i-1)}(t)N^i x_{20} - \sum_{i=0}^{h-1}N^i B_2 u^{(i)}(t) - \sum_{i=0}^{h-1}\sum_{k=0}^{i-1}\delta^{(k)}(t)N^i B_2 u^{(i-k-1)}(0)$$

$$= -\sum_{i=1}^{h-1}\delta^{(i-1)}(t)N^i x_{20} - \sum_{i=0}^{h-1}N^i B_2\left(u^{(i)}(t) + \sum_{k=0}^{i-1}\delta^{(k)}(t)u^{(i-k-1)}(0)\right),$$

which can be arranged into the forms of (3.54)–(3.56).   □

Remark 3.2. The above Theorem 3.9 has improved some existing results in the literature, e.g., Cobb (1983a) and Dai (1989b). As a matter of fact, Cobb (1983a) and Dai (1989b) both have dropped the term

$$-\sum_{i=0}^{h-1}\sum_{k=0}^{i-1} N^i B_2\delta^{(k)}(t)u^{(i-k-1)}(0)$$

in their state response formulas for a fast descriptor linear system, while this term is truly important because it is an impulse term driven by the initial values of the control and the control derivatives, and has a great impact on the impulsive controllability to be examined in Chap. 4 (see, also, Yan and Duan 2005).

The solution (3.54)–(3.56) is a distributional solution since it contains the generalized function $\delta(t)$. It is clearly seen from the above proof that this is cased by $(sN - I)^{-1}$, which produces a frequency-domain solution (3.58) containing a polynomial term.

It follows from the above theorem that, the external input response of the fast system, $x_{2u}(t)$, is a linear combination of the derivatives of $u(t)$ at time t. For any scalar $\varepsilon > 0$, the properties of $u(\tau)$, $0 \leq \tau \leq t - \varepsilon$ do not have contribution to $x_{2u}(t)$. This shows an interesting phenomenon between the substates $x_{1u}(t)$ and $x_{2u}(t)$: $x_{1u}(t)$ represents a cumulative effect of $u(\tau)$, $0 \leq \tau \leq t$, with no relation to $u(t)$, while on the contrary, $x_{2u}(t)$ response so rapidly that it insistently reflects the properties of $u(t)$ at time t. This is why we call (3.52) and (3.53) the slow and fast subsystems, respectively.

Through some simple deductions, the following corollary of the above theorem can be immediately obtained.

Corollary 3.2. *Given the fast system (3.53) with the matrix N being a nilpotent matrix with nilpotent index h, for any h times piecewise continuously differentiable input function $u(t)$, and initial value $x_2(0) = x_{20}$, the fast system (3.53) has a unique solution, which is given by*

$$x_2(t, u, x_{20}) = x_{2\text{impulse}}(t) + x_{2\text{normal}}(t),$$

$$x_{2\text{impulse}}(t) = -\sum_{i=1}^{h-1} N^i \delta^{(i-1)}(t) (x_{20} + Q_c[N, B_2] U(0)),$$

$$x_{2\text{normal}} = -Q_c[N, B_2] U(t),$$

where

$$Q_c[N, B_2] = [\, B_2 \ B_2 N \ \cdots \ B_2 N^{h-1} \,], \quad U(t) = \begin{bmatrix} u(t) \\ \dot{u}(t) \\ \vdots \\ u^{(h-1)}(t) \end{bmatrix}. \qquad (3.59)$$

The above corollary further indicates that the state response of a fast subsystem contains certain impulse terms at $t = 0$, and these terms are resulted in by both the initial state value and the initial values of the input function as well as its derivatives.

3.4.2 The Distributional Solutions

Following from the relation (3.51), Proposition 3.5 and Theorem 3.9, we immediately have the following result for solution to the regular system (3.49).

Theorem 3.10. *Assume that the regular descriptor linear system (3.49) has index h, and its input function u(t) is h times piecewise continuously differentiable. Furthermore, let (3.52) and (3.53) be the slow and fast subsystems of system (3.49), respectively. Then, the distributional state response of the system (3.49), starting from the initial value $x(0) = x_0$, is given by*

$$x(t, u, x_0) = P \begin{bmatrix} x_{1i}(t, x_0) + x_{1u}(t, u) \\ x_{2i}(t, x_0) + x_{2u}(t, u) \end{bmatrix}, \tag{3.60}$$

with

- $x_{1i}(t, x_0)$ *being the initial value response of the slow system and being given by*

$$x_{1i}(t, x_0) = e^{A_1 t} \begin{bmatrix} I & 0 \end{bmatrix} P^{-1} x_0, \tag{3.61}$$

- $x_{1u}(t, u)$ *being the response of the slow system due to the control input u(t) and being given by*

$$x_{1u}(t, u) = \int_0^t e^{A_1(t-\tau)} B_1 u(\tau) \, d\tau, \tag{3.62}$$

- $x_{2i}(t, x_0)$ *being the initial value response of the fast system and being given by*

$$x_{2i}(t, x_0) = -\sum_{i=1}^{h-1} \delta^{(i-1)}(t) N^i \begin{bmatrix} 0 & I \end{bmatrix} P^{-1} x_0, \tag{3.63}$$

- $x_{2u}(t, u)$ *being the response of the fast system due to the control input u(t) and being given by*

$$x_{2u}(t, u) = -\sum_{i=0}^{h-1} N^i B_2 \left(u^{(i)}(t) + \sum_{k=0}^{i-1} \delta^{(k)}(t) u^{(i-k-1)}(0) \right). \tag{3.64}$$

It is seen from the above theorem that, unlike that in the normal systems theory, the solution of a descriptor system has a complicated form. It includes not only the normal exponential response part, which is created by the slow subsystem (a normal one), but also the impulse and input derivative portions (due to the fast subsystem). The complicated state response given by (3.60)–(3.64) characterizes the response of a regular descriptor linear system. On the other hand, like the response of normal linear systems, the response of a descriptor linear system is also the sum of two parts, the part of response stimulated by the initial condition and the part of response resulted in by the input signal. Such a fact can be easily seen from (3.60).

It is obviously observed from the above theorem that the response of a descriptor linear system may contain impulse terms. For convenience, we introduce the following definition.

Definition 3.3. If the state response of a descriptor linear system, starting from an arbitrary initial value, does not contain impulse terms, then the system is called impulse-free.

The above Theorem 3.10 obviously implies the following result.

Corollary 3.3. *Let (3.52) and (3.53) be the slow and fast subsystems of system (3.49), respectively. Then, system (3.49) is impulse-free if and only if $N = 0$.*

To finish this subsection, we mention another observation about the condition for the existence and uniqueness of solution to linear systems. A normal linear system always has a unique solution if the input function is piecewise continuous, or even weaker. However, for the descriptor system (3.49) to have a unique solution the input function $u(t)$ needs to be h times piecewise continuously differentiable. The latter is much stronger than that for the normal system case. Such characters stand for the special feature of descriptor systems.

3.4.3 Examples

Example 3.8. Consider the system (3.39) in Example 3.5 again. The standard decomposition form for this system is given in (3.40). Following from Proposition 3.5 and Theorem 3.9, the state responses of these slow and fast subsystems in (3.40) are given by

$$\begin{cases} x_1(t) = e^{A_1 t} x_1(0) + \int_0^t e^{A_1(t-\tau)} \begin{bmatrix} 1 & 0 & -1 \\ 1 & -1 & 2 \end{bmatrix} u(\tau) d\tau \\ x_2(t) = -\delta(t) N x_2(0) - B_2 u(t) - N B_2 \left(u^{(1)}(t) + \delta(t) u(0) \right) \end{cases}$$

with

$$N = \begin{bmatrix} 0 & 1 \\ 0 & 0 \end{bmatrix}.$$

In this example, $N \neq 0$, thus impulse terms appear in the state response, and so does the derivative of the input signal.

Example 3.9 (Dai 1989b). Consider the following system

$$\begin{bmatrix} 0 & 1 \\ 0 & 0 \end{bmatrix} \begin{bmatrix} \dot{x}_1(t) \\ \dot{x}_2(t) \end{bmatrix} = \begin{bmatrix} x_1(t) \\ x_2(t) \end{bmatrix} + \begin{bmatrix} -1 \\ -1 \end{bmatrix} u(t).$$

It is clear that the system is in a fast subsystem form, with

$$N = \begin{bmatrix} 0 & 1 \\ 0 & 0 \end{bmatrix}, \quad B = \begin{bmatrix} -1 \\ -1 \end{bmatrix}.$$

Assuming that $u(t) = 0$, $t \in [0, \alpha]$, $\alpha > 0$, according to Theorem 3.9, its solution is given by

$$\begin{bmatrix} x_1(t) \\ x_2(t) \end{bmatrix} = -\sum_{i=1}^{1} N^i \delta^{(i-1)}(t) \begin{bmatrix} x_1(0) \\ x_2(0) \end{bmatrix} - \sum_{i=0}^{1} N^i B u^{(i)}(t) - N B u(0) \delta(t), \quad (3.65)$$

which gives

$$\begin{cases} x_1(t) = -x_2(0) \, \delta(t) + u(t) + \dot{u}(t) + u(0) \delta(t) \\ x_2(t) = u(t) \end{cases} \quad (3.66)$$

This contains both the impulse and the derivative terms. To get an intuitive impression, let us plot the solution for the following two special cases.

Case 1. Let the input function be the unit step function

$$u(t) = f(t - \alpha) = \begin{cases} 1, & t > \alpha > 0 \\ 0, & t \leq \alpha, \quad \alpha \text{ is a constant} \end{cases}.$$

Then

$$\begin{cases} x_1(t) = -x_2(0) \, \delta(t) + f(t - \alpha) + \delta(t - \alpha) \\ x_2(t) = f(t - \alpha) \end{cases}.$$

This solution is illustrated in Fig. 3.1 for the case of $x_2(0) \neq 0$.

Case 2. Let the input function be

$$u(t) = g(t) = \begin{cases} 0, & t \leq \beta \\ t - \beta, & t > \beta > 0, \quad \beta \text{ is a constant} \end{cases},$$

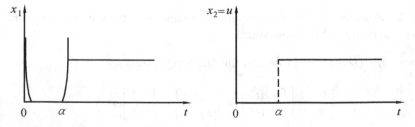

Fig. 3.1 State responses of the system for Case 1

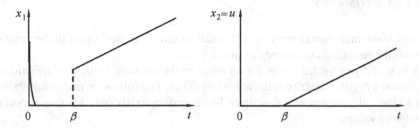

Fig. 3.2 State responses of the system for Case 2

then $u(t)$ is continuous. In this case, the solution is

$$\begin{cases} x_1(t) = -x_2(0)\delta(t) + g(t) + f(t - \beta) \\ x_2(t) = g(t) \end{cases},$$

which is shown in Fig. 3.2 for the case of $x_2(0) \neq 0$.

It follows from the normal linear system theory that the state response of a normal linear system is continuous provided that the input function is piecewise continuous (or weaker). However, this is no longer true in descriptor linear systems theory. Figures 3.1 and 3.2 show that impulse terms may be created in the state response by any jump behavior in the input due to the operation at the starting and closing switch actions (Fig. 3.1). Moreover, jump behavior may also appear in the state response, although the input function is continuous, as described in Fig. 3.2, which shows jump behavior appears at $t = \beta$ in $x_1(t)$ due to the discontinuous property in input derivatives.

Remark 3.3. Due to the fact mentioned in Remark 3.2, Dai (1989b) has dropped the terms $-NBu(0)\delta(t)$ and $u(0)\delta(t)$ in (3.65) and (3.66), respectively. However, since in both the two cases in the example the control $u(t)$ satisfies $u(0) = 0$, we have $u(0)\delta(t) = 0$, thus the final results in these two cases are still correct.

3.5 State Responses of Regular Descriptor Linear Systems: Classical Solutions

In the last section above, we have given the distributional solution of a regular descriptor linear system in the form of (3.49), which contains generalized functions (the impulse terms). In this section, we further investigate the classical solutions of the system, namely, solutions expressed by only classical functions which are continuously differentiable.

3.5.1 Consistency

We first show that the existence of a classical solution to a descriptor linear system requires the so-called consistency of initial values.

It is seen from Sect. 3.4 that for an arbitrary initial state value a distributional solution to a regular descriptor linear system exists. The following deduction shows that a classical solution of a descriptor linear system exists only for some special initial state values.

Consider a regular descriptor linear system which is of the following special form:

$$M\dot{x}(t) = x(t) + u(t), \tag{3.67}$$

where

$$M = \begin{bmatrix} 0 & 1 & & \\ & 0 & \ddots & \\ & & \ddots & 1 \\ & & & 0 \end{bmatrix} \in \mathbb{R}^n,$$

and the control $u(t)$ is sufficiently smooth. If we let

$$x(t) = \begin{bmatrix} x_1(t) & x_2(t) & \cdots & x_n(t) \end{bmatrix}^{\mathrm{T}}, u(t) = \begin{bmatrix} u_1(t) & u_2(t) & \cdots & u_n(t) \end{bmatrix}^{\mathrm{T}},$$

the system (3.67) can be rewritten as

$$\begin{cases} \dot{x}_2(t) = x_1(t) + u_1(t) \\ \dot{x}_3(t) = x_2(t) + u_2(t) \\ \quad\cdots \\ \dot{x}_n(t) = x_{n-1}(t) + u_{n-1}(t) \\ x_n(t) + u_n(t) = 0 \end{cases}$$

It is clear that this set of equations do not have solutions for arbitrary initial values $x_i(0), i = 1, 2, \ldots, n$. In fact, the last equation has a unique solution $x_n(t) = -u_n(t)$ only if the initial value of the state $x_n(t)$ satisfies $x_n(0) = -u_n(0)$. If the input $u_n(t)$ is sufficiently smooth, substituting the solution of $x_n(t)$ into the last second equation, we have

$$x_{n-1}(t) = \dot{u}_n(t) - u_{n-1}(t).$$

Obviously, this equation has a unique solution only if $x_{n-1}(0) = \dot{u}_n(0) - u_{n-1}(0)$. Along this line, we can obtain the following conclusion:

The descriptor linear system (3.67) has a solution only if the control input is sufficiently smooth and the initial value of the state satisfies

$$
\begin{bmatrix} x_1(0) \\ x_2(0) \\ \vdots \\ x_{n-1}(0) \\ x_n(0) \end{bmatrix} = \begin{bmatrix} -\sum_{i=1}^{n} u_i^{(i-1)}(0) \\ -\sum_{i=2}^{n} u_i^{(i-2)}(0) \\ \vdots \\ -\dot{u}_n(0) - u_{n-1}(0) \\ -u_n(0) \end{bmatrix}.
$$

In this case, the system (3.67) has the following corresponding unique solution

$$
\begin{bmatrix} x_1(t) \\ x_2(t) \\ \vdots \\ x_{n-1}(t) \\ x_n(t) \end{bmatrix} = \begin{bmatrix} -\sum_{i=1}^{n} u_i^{(i-1)}(t) \\ -\sum_{i=2}^{n} u_i^{(i-2)}(t) \\ \vdots \\ -\dot{u}_n(t) - u_{n-1}(t) \\ -u_n(t) \end{bmatrix}.
$$

This example shows that a regular descriptor linear system does not always have a solution for an arbitrary initial condition. This phenomenon arises because in Sect. 3.4 we allow generalized functions to exist in a solution of a regular linear system. While with the above treatment, the obtained solution is expressed only by classical functions when the control $u(t)$ is restricted to be a continuously differentiable function. Due to this fact, we introduce the following definition (Kunkel and Mehrmann 2006), which is special for descriptor linear systems.

Definition 3.4. A real finite vector x_0 is called a consistent initial value of the descriptor linear system (3.49) if the system has at least one classical response satisfying the initial condition $x(0) = x_0$.

3.5.2 The Classical Solutions

It is obvious that consistent initial conditions are in general different from each other for different control inputs. For convenience, for the regular descriptor linear system (3.49) we use $\mathscr{X}_0(u)$ to denote the set of all the consistent initial values corresponding to the input $u(t)$. The following theorem gives the characterization of $\mathscr{X}_0(u)$, and provides the classical solution corresponding to any initial value $x_0 \in \mathscr{X}_0(u)$ for the regular descriptor linear system (3.49).

Theorem 3.11. *Assume that the regular descriptor linear system (3.49) is of index h, and its input function $u(t)$ is h times piecewise continuously differentiable. Furthermore, let (3.52) and (3.53) be the slow and fast subsystems of system (3.49), respectively. Then the set of consistent initial conditions is given by*

$$
\mathscr{X}_0(u) = \left\{ \eta \ \middle| \ \begin{bmatrix} 0 & I \end{bmatrix} P^{-1} \eta = -\sum_{j=0}^{h-1} N^j B_2 u^{(j)}(0) \right\}, \tag{3.68}
$$

where P is defined by (3.50). For any $x_0 \in \mathcal{X}_0(u)$, the system has the classical response

$$x(t, u, x_0) = P \begin{bmatrix} x_{1i}(t, x_0) + x_{1u}(t, u) \\ -\sum_{i=0}^{h-1} N^i B_2 u^{(i)}(t) \end{bmatrix}, \tag{3.69}$$

where $x_{1i}(t, x_0)$ and $x_{1u}(t, u)$ are given by (3.61) and (3.62), respectively.

Proof. According to Theorem 3.10, the response of the system is given by (3.60). Since the response of the slow subsystem does not contain generalized functions, it suffices to consider the response of the fast subsystem.

Note the general relation

$$\sum_{i=0}^{h-1} \sum_{k=0}^{i-1} \alpha(i)\beta(k)\gamma(i, k) = \sum_{k=0}^{h-2} \sum_{i=k+1}^{h-1} \alpha(i)\beta(k)\gamma(i, k)$$

$$= \sum_{l=1}^{h-1} \beta(l - 1) \sum_{j=l}^{h-1} \alpha(j)\gamma(j, l - 1),$$

for the response of the fast system corresponding to the initial value $x(0) = x_0$, by letting $x_{20} = \begin{bmatrix} 0 & I \end{bmatrix} P^{-1} x_0$ we have

$$x_2(t, x_0) = x_{2i}(t, x_0) + x_{2u}(t, u)$$

$$= -\sum_{i=1}^{h-1} \delta^{(i-1)}(t) N^i x_{20} - \sum_{i=0}^{h-1} N^i B_2 \left(u^{(i)}(t) + \sum_{k=0}^{i-1} \delta^{(k)}(t) u^{(i-k-1)}(0) \right)$$

$$= -\sum_{i=1}^{h-1} \delta^{(i-1)}(t) N^i x_{20} - \sum_{i=0}^{h-1} N^i B_2 u^{(i)}(t) - \sum_{k=0}^{h-2} \sum_{i=k+1}^{h-1} \delta^{(k)}(t) N^i B_2 u^{(i-k-1)}(0)$$

$$= -\sum_{i=1}^{h-1} \delta^{(i-1)}(t) N^i x_{20} - \sum_{i=0}^{h-1} N^i B_2 u^{(i)}(t) - \sum_{l=1}^{h-1} \delta^{(l-1)}(t) \sum_{j=l}^{h-1} N^j B_2 u^{(j-l)}(0)$$

$$= -\sum_{i=1}^{h-1} \delta^{(i-1)}(t) \left(N^i x_{20} + \sum_{j=i}^{h-1} N^j B_2 u^{(j-i)}(0) \right) - \sum_{i=0}^{h-1} N^i B_2 u^{(i)}(t)$$

$$= -\sum_{i=1}^{h-1} N^i \delta^{(i-1)}(t) \left(x_{20} + \sum_{j=0}^{h-1} N^j B_2 u^{(j)}(0) \right) - \sum_{i=0}^{h-1} N^i B_2 u^{(i)}(t). \tag{3.70}$$

Letting $t = 0$ in the above equation gives

$$x_{20} = x_{2i}(0, x_0) + x_{2u}(0, u)$$

$$= -\sum_{i=1}^{h-1} N^i \delta^{(i-1)}(0) \left(x_{20} + \sum_{j=0}^{h-1} N^j B_2 u^{(j)}(0) \right) - \sum_{i=0}^{h-1} N^i B_2 u^{(i)}(0).$$

When $N \neq 0$, in view of the independency of functions $\delta^{(k)}(t)$ (see Theorem A.1 in Appendix A), it can be easily observed that for an arbitrary finite initial value x_{20}, the above equation holds if and only if

$$x_{20} = -\sum_{j=0}^{h-1} N^j B_2 u^{(j)}(0). \tag{3.71}$$

When $N = 0$, the above relation becomes

$$x_{20} = -B_2 u(0) = -\sum_{j=0}^{h-1} N^j B_2 u^{(j)}(0),$$

which is also in the form of (3.71). Therefore, the set $\mathscr{X}_0(u)$ of consistent initial conditions is given by (3.68).

Substituting (3.71) into (3.70) yields

$$x_2(t) = x_{2i}(t, x_0) + x_{2u}(t, u) = -\sum_{i=0}^{h-1} N^i B_2 u^{(i)}(t).$$

Thus, the classical response is obtained as in (3.69).   □

From the above theorem we know that the distributional solution of a regular descriptor system becomes the classical solution under the consistent initial conditions. Consequently, we can view the consistent initial conditions as the restriction imposed on the initial value $x(0)$ to guarantee that the distributional solution and the classical solution of the system are consistent.

For the fast system (3.53) with the matrix N being a nilpotent matrix with nilpotent index $h > 1$, and the input $u(t)$ being any h times piecewise continuously differentiable function, the set of consistent initial values and the corresponding classical solution for the system can be expressed as

$$\mathscr{X}_0(u) = \left\{ \eta \mid [0 \ I] P^{-1} \eta = -Q_c [N, B_2] U(0) \right\},$$

and

$$x_2(t, u, x_{20}) = -Q_c [N, B_2] U(t),$$

respectively, where $Q_c[N, B_2]$ and $U(t)$ are defined as in (3.59).

To finish this subsection, we finally give a corollary about the classical solution of the fast subsystem (3.53) with $N = 0$.

Corollary 3.4. *The fast subsystem (3.53) with $N = 0$ has a unique solution*

$$x_2(t) = -B_2 u(t),$$

with the unique consistent initial vale $x_{20} = -B_2 u(0)$.

3.5.3 The Example

Example 3.10. Consider the system

$$\begin{cases} E\dot{x}(t) = Ax(t) + Bu(t), \ 0 \le t \le \pi \\ y(t) = Cx(t) \end{cases},$$

with

$$E = \begin{bmatrix} 1 & 0 & 0 \\ 0 & 0 & 1 \\ 0 & 0 & 0 \end{bmatrix}, \ A = \begin{bmatrix} 2 & 0 & 0 \\ 0 & 1 & 0 \\ 0 & 0 & 1 \end{bmatrix}, \ B = \begin{bmatrix} 1 & 0 \\ 0 & 1 \\ 1 & 1 \end{bmatrix}, \ C = \begin{bmatrix} 1 & 2 & -1 \end{bmatrix},$$

and

$$u(t) = \begin{bmatrix} \sin t \\ \cos t \end{bmatrix}.$$

Obviously, this system is already in a standard decomposition form, so we have

$$P = I, \ A_1 = 2, \ B_1 = \begin{bmatrix} 1 & 0 \end{bmatrix}, \ N = \begin{bmatrix} 0 & 1 \\ 0 & 0 \end{bmatrix}, \ B_2 = \begin{bmatrix} 0 & 1 \\ 1 & 1 \end{bmatrix}.$$

For this system, the distributional solution is obtained as

$$x_1(t) = e^{A_1 t} x_{10} + \int_0^t e^{A_1(t-\tau)} B_1 u(\tau) d\tau$$

$$= e^{2t} x_{10} + \frac{1}{5} e^{2t} - \frac{2}{5} \sin t - \frac{1}{5} \cos t,$$

and

$$x_2(t) = -\delta(t) N x_{20} - B_2 u(t) - N B_2 \left(u^{(1)}(t) + \delta(t) u(0) \right)$$

$$= -\delta(t) \left(\begin{bmatrix} 0 & 1 \\ 0 & 0 \end{bmatrix} x_{20} + \begin{bmatrix} 1 \\ 0 \end{bmatrix} \right) - \begin{bmatrix} 2\cos t - \sin t \\ \cos t + \sin t \end{bmatrix}.$$

In the following we investigate the classical solution of this system.
Since

$$-\sum_{i=0}^{1} N^i B_2 u^i(0) = \begin{bmatrix} -2 \\ -1 \end{bmatrix},$$

according to Theorem 3.11, we can easily get the set of consistent initial conditions as

$$\mathscr{X}_0(u) = \left\{ \eta \ \middle| \ \begin{bmatrix} 0 & I_2 \end{bmatrix} \eta = \begin{bmatrix} -2 & -1 \end{bmatrix}^{\mathrm{T}} \right\}.$$

Thus, all the admissible initial value for the system takes the form

$$\eta = \begin{bmatrix} \alpha & -2 & -1 \end{bmatrix}^{\mathrm{T}}, \quad \alpha \in \mathbb{R} \text{ arbitrary.}$$

Particularly choosing $x_0 = \begin{bmatrix} 1 & -2 & -1 \end{bmatrix}^{\mathrm{T}} \in \mathcal{X}_0(u)$, we can further obtain the state response starting from this initial value as

$$x(t, u, x_0) = \begin{bmatrix} -\frac{2}{5}\sin t - \frac{1}{5}\cos t + \frac{6}{5}e^{2t} \\ \sin t - 2\cos t \\ -\sin t - \cos t \end{bmatrix}, \quad 0 \le t \le \pi,$$

and the corresponding output is given by

$$y(t) = \frac{13}{5}\sin t - \frac{16}{5}\cos t + \frac{6}{5}e^{2t}, \quad 0 \le t \le \pi.$$

Remark 3.4. Recall that in Remark 2.1 we have left a question: why Theorem 2.7 contains the imprecise words "system (2.40) has a unique solution for sufficiently differentiable $u(t)$ and **some proper** initial values?" We believe that by now the question has got its clear answer. As far as distributional solutions are concerned, the existence and uniqueness condition does not impose any requirement on the initial values, while as classical solutions are concerned, the initial values have to be consistent in order that the system has a solution.

3.6 Generalized Eigenvalues and Eigenvectors

In this section, we define generalized eigenvalues and eigenvectors for descriptor linear systems. Without loss of generality, we focus on the following descriptor linear system without control input:

$$E\dot{x} = Ax, \tag{3.72}$$

where $E, A \in \mathbb{R}^{n \times n}$. When the system (3.72) is not regular, we have

$$\det(sE - A) \equiv 0,$$

that is,

$$\det(sE - A) = 0, \quad \forall s \in \mathbb{C}.$$

In this case, the eigenvalues of the matrix pair (E, A) cannot be properly defined. Thus, in this whole section and the one that follows, we assume that the system (3.72) is regular.

3.6.1   Finite Eigenvalues and Eigenvectors

Finite Eigenvalues

Under the regularity assumption of the matrix pair (E, A), the following polynomial

$$\Delta(s) = \det (sE - A)$$

is not identically zero, and according to Corollary 3.1,

$$\deg \Delta(s) = \deg \det (sE - A) = n_1. \tag{3.73}$$

Therefore, $\Delta(s)$ is a polynomial with degree n_1, and can be written as

$$\Delta(s) = a_0 s^{n_1} + a_1 s^{n_1-1} + \cdots + a_{n_1-1}s + a_{n_1}, \quad a_0 \neq 0.$$

Recall from Sect. 1.1 that the above polynomial $\Delta(s)$ is the characteristic polynomial of the system (3.72), and the finite s's satisfying $\Delta(s) = \det(sE - A) = 0$ are called the finite poles or finite eigenvalues of the system, or the matrix pair (E, A). Thus, the set of finite poles of the system is

$$\sigma(E, A) = \{s \mid s \in \mathbb{C}, s \text{ finite}, \det(sE - A) = 0\}.$$

Since E is generally singular, the number of finite poles is always not greater than $n_0 = \text{rank}E \leq n$ for descriptor systems. Therefore, $\sigma(E, A)$ contains at most n_0 number of complex numbers. For convenience in the following $\sigma(I, A)$ is often specified as $\sigma(A)$.

When the system (3.72) is assumed to be regular, there exists the following standard decompositions

$$QEP = \begin{bmatrix} I_{n_1} & 0 \\ 0 & N \end{bmatrix} \text{ and } QAP = \begin{bmatrix} A_1 & 0 \\ 0 & I_{n_2} \end{bmatrix}, \tag{3.74}$$

where $Q, P \in \mathbb{R}^{n \times n}$ are nonsingular, $n_1 + n_2 = n$, $A_1 \in \mathbb{R}^{n_1 \times n_1}$, and $N \in \mathbb{R}^{n_2 \times n_2}$ is nilpotent.

Regarding the set of finite poles, $\sigma(E, A)$, we have the following lemma.

Lemma 3.1. *Assume that the descriptor linear system (3.72) is regular, and let Q and P be two nonsingular matrices satisfying (3.74). Then there holds*

$$\sigma(E, A) = \sigma(A_1).$$

Proof. Since N is nilpotent, we have

$$\det(sN - I_{n_2}) = (-1)^{n_2}.$$

Therefore,

$$\sigma\left(N, I_{n_2}\right) = \emptyset.$$

Using this relation and those in (3.74), we have

$$
\begin{aligned}
\sigma(E, A) &= \sigma(QEP, QAP) \\
&= \sigma\left(\text{diag}\left(I_{n_1}, N\right), \text{diag}\left(A_1, I_{n_2}\right)\right) \\
&= \sigma(I, A_1) \cup \sigma\left(N, I_{n_2}\right) \\
&= \sigma(A_1) \cup \emptyset \\
&= \sigma(A_1).
\end{aligned}
$$

The proof is then completed.   □

The above lemma clearly state that the set of finite poles of a regular descriptor linear system is identical with the set of poles of its slow subsystem.

The characteristic polynomial $\Delta(s)$ can be factored as

$$\Delta(s) = a_0 \left(s - s_1\right)^{m_1} \left(s - s_2\right)^{m_2} \cdots \left(s - s_p\right)^{m_p},$$

where $s_i \neq s_j$, $i, j = 1, 2, \ldots, p$, $i \neq j$, and $m_1 + m_2 + \cdots + m_p = n_1$. Obviously, s_i, $i = 1, 2, \ldots, p$, are the distinct eigenvalues of the matrix pair (E, A). We call the integer m_i the algebraic multiplicity of eigenvalue s_i.

Furthermore, define

$$q_i = n - \text{rank}\left(s_i E - A\right), \ i = 1, 2, \ldots, p, \tag{3.75}$$

then, similar to the standard normal system case q_i is called the geometric multiplicity of the eigenvalue s_i.

Finite Right Eigenvectors

Rewriting (3.75) in the following form:

$$\text{rank}\left(s_i E - A\right) = n - q_i, \ i = 1, 2, \ldots, p, \tag{3.76}$$

we can then easily see that q_i is actually the dimension of the kernel of $(s_i E - A)$, that is,

$$q_i = \dim \ker\left(s_i E - A\right), \ i = 1, 2, \ldots, p.$$

Therefore, there exist q_i linearly independent vectors v_{ij}^1, $j = 1, 2, \ldots, q_i$, satisfying

$$\left(A - s_i E\right) v_{ij}^1 = 0, \ j = 1, 2, \ldots, q_i. \tag{3.77}$$

These q_i vectors $v_{ij}^1, j = 1, 2, \ldots, q_i$, are called the finite right eigenvectors associated with the eigenvalues s_i.

Corresponding to each v_{ij}^1, there may exist a set of vectors $v_{ij}^2, v_{ij}^3, \ldots, v_{ij}^{p_{ij}}$ satisfying

$$\begin{cases} (A - s_i E) v_{ij}^2 = E v_{ij}^1 \\ (A - s_i E) v_{ij}^3 = E v_{ij}^2 \\ \quad \vdots \\ (A - s_i E) v_{ij}^{p_{ij}} = E v_{ij}^{p_{ij}-1} \end{cases} \qquad (3.78)$$

If, further, there does not exist a vector v satisfying

$$(A - s_i E) v = E v_{ij}^{p_{ij}},$$

then the set of vectors $v_{ij}^k, k = 2, 3, \ldots, p_{ij}$, are called a complete set of generalized eigenvectors associated with the eigenvalue s_i deduced from the eigenvector v_{ij}^1. For convenience, we also call $v_{ij}^k, k = 1, 2, \ldots, p_{ij}$, the j-th right eigenvector chain associated with the eigenvalue s_i, while p_{ij} is called the length of this eigenvector chain. Obviously, each eigenvalue s_i of the matrix pair (E, A) has q_i (the geometric multiplicity) right eigenvector chains, and

$$\sum_{i=1}^{p} \sum_{j=1}^{q_i} p_{ij} = n_1.$$

Combining (3.77) and (3.78) gives the following definition for the j-th eigenvector chain associated with eigenvalue s_i:

$$(A - s_i E) v_{ij}^k = E v_{ij}^{k-1}, \quad v_{ij}^0 = 0, \qquad (3.79)$$
$$k = 1, 2, \ldots, p_{ij}.$$

Denote

$$\begin{cases} J = \mathrm{diag}\left(J_1, J_2, \ldots, J_p\right) \\ J_i = \mathrm{diag}\left(J_{i1}, J_{i2}, \ldots, J_{iq_i}\right) \\ J_{ij} = \begin{bmatrix} s_i & 1 & & \\ & s_i & \ddots & \\ & & \ddots & 1 \\ & & & s_i \end{bmatrix}_{p_{ij} \times p_{ij}} \end{cases} \qquad (3.80)$$

Then J_{ij}, $j = 1, 2, \ldots, q_i$, are the q_i Jordan blocks associated with the eigenvalue s_i, and J is the Jordan matrix of the matrix pair (E, A) associated with all the finite eigenvalues of the matrix pair (E, A). Furthermore, define

$$\begin{cases} V = \begin{bmatrix} V_1 & V_2 & \cdots & V_p \end{bmatrix} \\ V_i = \begin{bmatrix} V_{i1} & V_{i2} & \cdots & V_{iq_i} \end{bmatrix} \\ V_{ij} = \begin{bmatrix} v_{ij}^1 & v_{ij}^2 & \cdots & v_{ij}^{p_{ij}} \end{bmatrix} \end{cases} \qquad (3.81)$$

Then the matrix V_i is a right eigenvector matrix associated with the eigenvalue s_i, and V is a right eigenvector matrix of the matrix pair (E, A) associated with all the finite eigenvalues of the matrix pair (E, A).

For convenience, we also call J the finite Jordan matrix of the matrix pair (E, A), and V a finite right eigenvector matrix of the matrix pair (E, A).

Rewrite (3.79) as

$$Av_{ij}^k = E\left(s_i v_{ij}^k + v_{ij}^{k-1}\right), \quad v_{ij}^0 = 0,$$
$$k = 1, 2, \ldots, p_{ij}, \quad j = 1, 2, \ldots, q_i, \quad i = 1, 2, \ldots, p. \quad (3.82)$$

Then, using the notations in (3.80) and (3.81), the equations in (3.82) can be compactly written as

$$AV_{ij} = EV_{ij}J_{ij}, \quad j = 1, 2, \ldots, q_i, \quad i = 1, 2, \ldots, p, \quad (3.83)$$

and the equations in (3.83) can also be further written as

$$AV_i = EV_i J_i, \quad i = 1, 2, \ldots, p, \quad (3.84)$$

while (3.84) can be compactly written as

$$AV = EVJ.$$

This is the equation that governs the relation of the finite Jordan matrix J and the finite right eigenvector matrix V of the matrix pair (E, A).

By using the above definition of V, the following result can be shown.

Proposition 3.6. *Let the matrix V be a finite right eigenvector matrix of the regular matrix pair (E, A). Then*

$$\text{rank}(V) = \text{rank}(EV) = n_1,$$

with n_1 given by (3.73).

As in the case for linear systems, we can also introduce the finite right eigenspace, defined and denoted by

$$\mathfrak{E}^R(E, A) = \text{span}(V).$$

Finite Left Eigenvectors

Again, it follows from (3.76) that there exist q_i linearly independent vectors $t_{ij}^{p_{ij}}$, $j = 1, 2, \ldots, q_i$, satisfying

$$\left(t_{ij}^{p_{ij}}\right)^{\text{T}} (A - s_i E) = 0, \quad j = 1, 2, \ldots, q_i. \quad (3.85)$$

These q_i vectors $t_{ij}^{p_{ij}}$, $j = 1, 2, \ldots, q_i$, are called the finite left eigenvectors associated with the eigenvalue s_i.

Corresponding to each $t_{ij}^{p_{ij}}$, there may exist a set of vectors $t_{ij}^{p_{ij}-1}, t_{ij}^{p_{ij}-2}, \ldots, t_{ij}^{1}$ satisfying

$$
\begin{cases}
\left(t_{ij}^{p_{ij}-1}\right)^{\mathrm{T}}(A - s_i E) = \left(t_{ij}^{p_{ij}}\right)^{\mathrm{T}} E \\
\left(t_{ij}^{p_{ij}-2}\right)^{\mathrm{T}}(A - s_i E) = \left(t_{ij}^{p_{ij}-1}\right)^{\mathrm{T}} E \\
\qquad\qquad \vdots \\
\left(t_{ij}^{1}\right)^{\mathrm{T}}(A - s_i E) = (t_{ij}^{2})^{\mathrm{T}} E
\end{cases}
\tag{3.86}
$$

If, further, there does not exist a vector t satisfying

$$
t^{\mathrm{T}}(A - s_i E) = (t_{ij}^{1})E,
$$

then the set of vectors $t_{ij}^{k}, k = 1, 2, \ldots, p_{ij} - 1$, are called a complete set of generalized eigenvectors associated with the eigenvalue s_i deduced from the eigenvector $t_{ij}^{p_{ij}}$. For convenience, we also call $t_{ij}^{k}, k = 1, 2, \ldots, p_{ij}$, the j-th left eigenvector chain associated with the eigenvalue s_i, while p_{ij} is called the length of this eigenvector chain. Obviously, each eigenvalue s_i of the matrix pair (E, A) has q_i (the geometric multiplicity) left eigenvector chains.

Combining (3.85) and (3.86) gives the following definition for the j-th left eigenvector chain associated with the eigenvalue s_i:

$$
(t_{ij}^{k})^{\mathrm{T}}(A - s_i E) = (t_{ij}^{k+1})^{\mathrm{T}} E, \ t_{ij}^{p_{ij}+1} = 0,
\tag{3.87}
$$
$$
k = 1, 2, \ldots, p_{ij}.
$$

Denote

$$
\begin{cases}
T = \begin{bmatrix} T_1 & T_2 & \cdots & T_P \end{bmatrix} \\
T_i = \begin{bmatrix} T_{i1} & T_{i2} & \cdots & T_{iq_i} \end{bmatrix}. \\
T_{ij} = \begin{bmatrix} t_{ij}^{1} & t_{ij}^{2} & \cdots & t_{ij}^{p_{ij}} \end{bmatrix}
\end{cases}
\tag{3.88}
$$

Then the matrix T_i is a left eigenvector matrix associated with the eigenvalue s_i, and T is a left eigenvector matrix of the matrix pair (E, A) associated with all the finite eigenvalues of (E, A). For convenience, we also call the matrix T a finite left eigenvector matrix of the matrix pair (E, A).

Rewrite (3.87) as

$$
(t_{ij}^{k})^{\mathrm{T}} A = \left(s_i t_{ij}^{k} + t_{ij}^{k+1}\right)^{\mathrm{T}} E, \ t_{ij}^{p_{ij}+1} = 0,
\tag{3.89}
$$
$$
k = 1, 2, \ldots, p_{ij}, \ j = 1, 2, \ldots, q_i, \ i = 1, 2, \ldots, p.
$$

Then, using the notations in (3.80) and (3.88), the equations in (3.89) can be compactly written as

$$
T_{ij}^{\mathrm{T}} A = J_{ij} T_{ij}^{\mathrm{T}} E, \ j = 1, 2, \ldots, q_i, \ i = 1, 2, \ldots, p,
\tag{3.90}
$$

and the equations in (3.90) can also be further written as

$$T_i^T A = J_i T_i^T E, \ i = 1, 2, \ldots, p, \tag{3.91}$$

while (3.91) can be compactly written as

$$T^T A = J T^T E.$$

This is the equation that governs the relation between the finite Jordan matrix J and the finite left eigenvector matrix T of the matrix pair (E, A).

By using the above definition of T, the following result can be shown.

Proposition 3.7. *Let the matrix T be a finite left eigenvector matrix of the regular matrix pair (E, A). Then*

$$\text{rank}(T) = \text{rank}(T^T E) = n_1,$$

with n_1 be given by (3.73).

As in the case for linear systems, we can also introduce the finite left eigenspace, defined and denoted by

$$\mathfrak{E}^L(E, A) = \text{span}(T).$$

3.6.2 Infinite Eigenvalues and Eigenvectors

The Infinite Eigenvalues

We have known that when the matrix pair (E, A) is regular, relation (3.73) holds. With this observation, we have the following proposition.

Proposition 3.8. *Let $E, A \in \mathbb{R}^{n \times n}$, and (E, A) be regular. Then the matrix pair (A, E) has a zero eigenvalue with algebraic multiplicity $n_2 = n - n_1$, or equivalently, the polynomial $\det(sA - E)$ has a multiple zero root of order n_2.*

Based on the above proposition, we can introduce the following definition.

Definition 3.5. *Let $E, A \in \mathbb{R}^{n \times n}$. If the matrix pair (A, E) has a zero eigenvalue with geometric multiplicity q and algebraic multiplicity m, then we say that the matrix pair (E, A) has an infinite eigenvalue, denoted by s_∞, with geometric multiplicity q and algebraic multiplicity m.*

Infinite Eigenvectors

When the matrix pair (E, A) has an infinite eigenvalue s_∞ with geometric multiplicity q, thus there are q right (left) eigenvector chains associated with the zero eigenvalue of the matrix pair (A, E). The definition of these q right eigenvector chains $\{v_{\infty j}^k, k = 1, 2, \ldots, p_{\infty j}, j = 1, 2, \ldots, q\}$ are clearly given by

$$E v_{\infty j}^k = A v_{\infty j}^{k-1}, \ v_{\infty j}^0 = 0, \tag{3.92}$$
$$k = 1, 2, \ldots, p_{\infty j}, \ j = 1, 2, \ldots, q,$$

and the definition of these q left eigenvector chains $\{t_{\infty j}^k, k = 1, 2, \ldots, p_{\infty j}, j = 1, 2, \ldots, q\}$ are clearly given by

$$\left(t_{\infty j}^k \right)^{\mathrm{T}} E = \left(t_{\infty j}^{k+1} \right)^{\mathrm{T}} A, \ t_{\infty j}^{p_{\infty j}+1} = 0, \tag{3.93}$$
$$k = 1, 2, \ldots, p_{\infty j}, \ j = 1, 2, \ldots, q,$$

where $p_{\infty j}, \ j = 1, 2, \ldots, q$, is called the length of the j-th chain, and they clearly satisfy, in view of Proposition 3.8,

$$\sum_{j=1}^{q} p_{\infty j} = n_2.$$

From this, we know that the algebraic multiplicity $m = n_2$.

Denote

$$\begin{cases} J_\infty = \mathrm{diag}\left(J_{\infty 1}, J_{\infty 2}, \ldots, J_{\infty q} \right) \\ J_{\infty j} = \begin{bmatrix} 0 & 1 & & & \\ & 0 & \ddots & & \\ & & \ddots & 1 \\ & & & 0 \end{bmatrix}_{p_{\infty j} \times p_{\infty j}} \end{cases}, \tag{3.94}$$

$$\begin{cases} V_\infty = \begin{bmatrix} V_{\infty 1} & V_{\infty 2} & \cdots & V_{\infty q} \end{bmatrix} \\ V_{\infty j} = \begin{bmatrix} v_{\infty j}^1 & v_{\infty j}^2 & \cdots & v_{\infty j}^{p_{\infty j}} \end{bmatrix} \end{cases}, \tag{3.95}$$

$$\begin{cases} T_\infty = \begin{bmatrix} T_{\infty 1} & T_{\infty 2} & \cdots & T_{\infty q} \end{bmatrix} \\ T_{\infty j} = \begin{bmatrix} t_{\infty j}^1 & t_{\infty j}^2 & \cdots & t_{\infty j}^{p_{\infty j}} \end{bmatrix} \end{cases}. \tag{3.96}$$

Then (3.92) and (3.93) can be, respectively, arranged into the forms of

$$E V_\infty = A V_\infty J_\infty, \ T_\infty^{\mathrm{T}} E = J_\infty T_\infty^{\mathrm{T}} A.$$

Corresponding to the finite case, the nilpotent matrix J_∞ is the Jordan form of system (3.72) associated with the infinite eigenvalue, the matrix V_∞ is a right eigenvector matrix of the matrix pair (E, A) associated with the infinite eigenvalue, and the matrix T_∞ is a left eigenvector matrix of the matrix pair (E, A) associated with the infinite eigenvalue. For convenience, the nilpotent matrix J_∞ is also called the infinite Jordan matrix of the matrix pair (E, A), and the matrices T_∞ and V_∞ are also called the infinite left and right eigenvector matrices of the matrix pair (E, A), respectively.

Directly using the above definitions of the infinite left and right eigenvector matrices, we can prove the following result.

Proposition 3.9. *Let matrices* $T_\infty \in \mathbb{R}^{n \times n_2}$ *and* $V_\infty \in \mathbb{R}^{n \times n_2}$ *be a pair of infinite left and right eigenvector matrices of the regular matrix pair* (E, A), *respectively. Then*

$$\mathrm{rank}(T_\infty) = \mathrm{rank}(T_\infty^\mathrm{T} A) = n_2,$$
$$\mathrm{rank}(V_\infty) = \mathrm{rank}(A V_\infty) = n_2.$$

As in the case for linear systems, we can also introduce the infinite left and right eigenspaces, defined and denoted by

$$\mathcal{C}_\infty^L(E, A) = \mathrm{span}(T_\infty),$$
$$\mathcal{C}_\infty^L(E, A) = \mathrm{span}(V_\infty).$$

A Special Case

In the special case of $p_{\infty j} = 1$, $j = 1, 2, \ldots, q (q = n_2)$, the matrix J_∞ becomes a zero matrix, and the above (3.92) and (3.93) reduce to

$$E v_{\infty j} = 0, \quad j = 1, 2, \ldots, n_2, \tag{3.97}$$

and

$$t_{\infty j}^\mathrm{T} E = 0, \quad j = 1, 2, \ldots, n_2, \tag{3.98}$$

respectively. These state that in this case the q right eigenvectors of the matrix pair (E, A) associated with the infinite eigenvalue s_∞ are in the kernel space of the matrix E, and the q left eigenvectors of the matrix pair (E, A) associated with the infinite eigenvalue s_∞ are in the kernel space of the matrix E^T. Furthermore, in this case the above (3.95) and (3.96) reduce to

$$V_\infty = \begin{bmatrix} v_{\infty 1} & v_{\infty 2} & \cdots & v_{\infty n_2} \end{bmatrix}$$

and

$$T_\infty = \begin{bmatrix} t_{\infty 1} & t_{\infty 2} & \cdots & t_{\infty n_2} \end{bmatrix},$$

respectively, and it follows from the above (3.97) and (3.98) that the infinite left and right eigenvector matrices T_∞ and V_∞ of the matrix pair (E, A) are now determined by

$$T_\infty^\mathrm{T} E = 0, \quad T_\infty \in \mathbb{R}_{n_2}^{n \times n_2},$$

and

$$EV_\infty = 0, \quad V_\infty \in \mathbb{R}_{n_2}^{n \times n_2},$$

respectively.

Remark 3.5. The problem of relative eigenvalues and eigenvectors of matrix pairs is much more complicated than that of a single matrix. Due to the length limit, the proofs of these propositions given in this section are omitted. Readers are recommended to refer to some related references, e.g. Gantmacher (1974), Liu and Patton (1996), Bavafa-Toosi et al. (2006), and Chu and Golub (2006).

3.7 Eigenstructure Decomposition with Relation to Standard Decomposition

In Sect. 3.6, we have defined the finite left and right eigenvector matrices T and V of the matrix pair (E, A), and the infinite left and right eigenvector matrices T_∞ and V_∞ of the matrix pair (E, A). In this section, we further discuss the so-called important eigenstructure decomposition.

3.7.1 Eigenstructure Decomposition

Define

$$\tilde{T} = \begin{bmatrix} T & T_\infty \end{bmatrix}, \quad \tilde{V} = \begin{bmatrix} V & V_\infty \end{bmatrix}, \tag{3.99}$$

where T and V are, respectively, the left and right finite eigenvector matrices defined by

$$T^\mathrm{T} A = J T^\mathrm{T} E, \quad AV = EVJ, \tag{3.100}$$

and T_∞ and V_∞ are, respectively, the left and right infinite eigenvector matrices defined by

$$T_\infty^\mathrm{T} E = J_\infty T_\infty^\mathrm{T} A, \quad EV_\infty = AV_\infty J_\infty. \tag{3.101}$$

Naturally, we call matrices $\tilde{T}$ and $\tilde{V}$, respectively, the entire left and right eigenvector matrices of the system (3.72), and the following proposition holds.

Proposition 3.10. *Given the matrix pair (E, A), then the entire left and right eigenvector matrices $\tilde{T}$ and $\tilde{V}$ defined by (3.99)–(3.101) are nonsingular.*

The following lemma gives some useful relations.

Lemma 3.2. *Let the matrix pair (E, A) be regular.*

1. *For any pair of finite left and right eigenvector matrices T and V, and any pair of infinite left and right eigenvector matrices T_∞ and V_∞, of the matrix pair (E, A), there hold*

$$T^T E V_\infty = T^T A V_\infty = 0, \tag{3.102}$$

$$T_\infty^T E V = T_\infty^T A V = 0. \tag{3.103}$$

2. *There exist a pair of finite left and right eigenvector matrices T and V, of the matrix pair (E, A), satisfying*

$$T^T E V = I_{n_1}, \tag{3.104}$$

$$T_\infty^T A V_\infty = I_{n_2}. \tag{3.105}$$

Proof. Using (3.100) and (3.101), we have

$$T^T E V_\infty = T^T A V_\infty J_\infty = J(T^T E V_\infty) J_\infty.$$

Since $\lambda(J)\lambda(J_\infty) = 0 \neq 1$, the above relation gives

$$T^T E V_\infty = 0.$$

With this and (3.100) again, we can obtain

$$T^T A V_\infty = J T^T E V_\infty = 0.$$

Thus, (3.102) holds.

Using (3.100) and (3.101) again, we have

$$T_\infty^T E V = J_\infty T_\infty^T A V = J_\infty (T_\infty^T E V) J.$$

Again because of $\lambda(J)\lambda(J_\infty) \neq 1$, the above equation gives

$$T_\infty^T E V = 0.$$

With this and (3.100), we further have

$$T_\infty^T A V = T_\infty^T E V J = 0.$$

This gives the proof of (3.103).

Noting

$$\begin{bmatrix} T^T \\ T_\infty^T \end{bmatrix} E V = \begin{bmatrix} T^T E V \\ T_\infty^T E V \end{bmatrix} = \begin{bmatrix} T^T E V \\ 0 \end{bmatrix},$$

and recalling

$$\det \begin{bmatrix} T^T \\ T_\infty^T \end{bmatrix} \neq 0, \ \mathrm{rank}(E V) = n_1,$$

we know that $\Pi = T^T E V$ is nonsingular. Let

$$\tilde{V} = V \Pi^{-1},$$

we have

$$JΠ = JT^T EV = T^T EV = T^T EVJ = ΠJ,$$

and hence $JΠ^{-1} = Π^{-1}J$. Thus,

$$A\tilde{V} = AVΠ^{-1} = EVJΠ^{-1} = EVΠ^{-1}J = E\tilde{V}J.$$

This states that $\tilde{V}$ is still a right eigenvector matrix. Furthermore, it satisfies

$$T^T E\tilde{V} = T^T EVΠ^{-1} = I_{n_1}.$$

This shows that there exist a pair of finite left and right eigenvector matrices T and V satisfying (3.104).

Finally, consider

$$T_∞^T A \begin{bmatrix} V & V_∞ \end{bmatrix} = \begin{bmatrix} T_∞^T AV & T_∞^T AV_∞ \end{bmatrix} = \begin{bmatrix} 0 & T_∞^T AV_∞ \end{bmatrix},$$

in view of

$$\det \begin{bmatrix} V & V_∞ \end{bmatrix} \neq 0, \ \mathrm{rank}(T_∞^T A) = n_1,$$

we know that $Π_∞ = T_∞^T AV_∞$ is nonsingular. Let

$$\tilde{V}_∞ = V_∞ Π_∞^{-1},$$

we have

$$J_∞ Π_∞ = J_∞ T_∞^T AV_∞ = T_∞^T EV_∞ = T^T AV_∞ J_∞ = Π_∞ J_∞,$$

hence $J_∞ Π_∞^{-1} = Π_∞^{-1} J_∞$. Thus,

$$E\tilde{V} = AV_∞ Π_∞^{-1} = AV_∞ J_∞ Π_∞^{-1} = AV_∞ Π_∞^{-1} J_∞ = A\tilde{V}_∞ J_∞.$$

This indicates that $\tilde{V}_∞$ is still a right eigenvector matrix. Furthermore, it satisfies

$$T_∞^T A\tilde{V}_∞ = T_∞^T AV_∞ Π_∞^{-1} = Π_∞ Π_∞^{-1} = I_{n_2}.$$

This shows that there exist a pair of infinite left and right eigenvector matrices $T_∞$ and $V_∞$ satisfying (3.105).   □

In view of the above lemma, we introduce the following definition.

Definition 3.6.
- The left eigenvector matrix T and the right eigenvector matrix V are said to be a normalized pair if they satisfy (3.104).
- The infinite left eigenvector matrix $T_∞$ and the infinite right eigenvector matrix $V_∞$ are said to be a normalized pair if they satisfy (3.105).

Using the above Lemma 3.2 and the basic relations in (3.100)–(3.101), we are now ready to present the following eigenstructure decomposition.

Theorem 3.12. [*Eigenstructure decomposition*] *Given a regular matrix pair* (E, A), *there exist nonsingular matrices* $T, V \in \mathbb{C}^{n \times n}$ *satisfying*

$$\tilde{T}^{\mathrm{T}} E \tilde{V} = \begin{bmatrix} I_{n_1} & 0 \\ 0 & J_\infty \end{bmatrix}, \quad \tilde{T}^{\mathrm{T}} A \tilde{V} = \begin{bmatrix} J & 0 \\ 0 & I_{n_2} \end{bmatrix}, \tag{3.106}$$

where J *and* J_∞ *are both in Jordan form, and* J_∞ *is also nilpotent.*

Proof. Post-multiplying the first equation in (3.101) by V_∞, and using (3.105), gives

$$T_\infty^{\mathrm{T}} E V_\infty = J_\infty T_\infty^{\mathrm{T}} A V_\infty = J_\infty.$$

Using the above, together with (3.102), (3.103), and (3.104), we have

$$\tilde{T}^{\mathrm{T}} E \tilde{V} = \begin{bmatrix} T^{\mathrm{T}} \\ T_\infty^{\mathrm{T}} \end{bmatrix} E \begin{bmatrix} V & V_\infty \end{bmatrix}$$

$$= \begin{bmatrix} T^{\mathrm{T}} E V & T^{\mathrm{T}} E V_\infty \\ T_\infty^{\mathrm{T}} E V & T_\infty^{\mathrm{T}} E V_\infty \end{bmatrix}$$

$$= \begin{bmatrix} I_{n_1} & 0 \\ 0 & J_\infty \end{bmatrix}.$$

Post-multiplying the first equation in (3.100) by V and using (3.104) gives

$$T^{\mathrm{T}} A V = T^{\mathrm{T}} E V J = J.$$

Using the above relations together with (3.102), (3.103), and (3.105), we can obtain

$$\tilde{T}^{\mathrm{T}} A \tilde{V} = \begin{bmatrix} T^{\mathrm{T}} \\ T_\infty^{\mathrm{T}} \end{bmatrix} A \begin{bmatrix} V & V_\infty \end{bmatrix}$$

$$= \begin{bmatrix} T^{\mathrm{T}} A V & T^{\mathrm{T}} A V_\infty \\ T_\infty^{\mathrm{T}} A V & T_\infty^{\mathrm{T}} A V_\infty \end{bmatrix}$$

$$= \begin{bmatrix} J & 0 \\ 0 & I_{n_2} \end{bmatrix}.$$

The proof is then completed. □

Due to the above lemma, we introduce the following definition.

Definition 3.7. A pair of the entire left and right eigenvector matrices $\tilde{T}$ and $\tilde{V}$ of a regular matrix pair (E, A) is said to be a corresponding pair if they satisfy the equations in (3.106).

3.7.2 Relation with Standard Decomposition

It is obvious that the above eigenstructure decomposition (3.106) is actually a special standard decomposition in view of the fact that the matrix J_∞ is nilpotent. The only problem with this decomposition is that the matrices T, V, and J may be complex. But this can be easily solved with very simple well-known techniques.

Conversely, we can also show that once a standard decomposition of a matrix pair is derived, an eigenstructure decomposition can also be easily obtained. This can also be viewed as alternative proof of the eigenstruture decomposition.

Theorem 3.13. *Let the system (3.72) be regular, and admit the standard decompositions in (3.74). Furthermore, let J be given by (3.80) and be the Jordan canonical form of matrix A_1, and let J_∞ be given by (3.94) and be the Jordan canonical form of N. Finally, let R_1 and R_2 be two nonsingular matrices satisfying*

$$R_1^{-1} A_1 R_1 = J, \quad R_2^{-1} N R_2 = J_\infty \tag{3.107}$$

with J and J_∞ be Jordan matrices of A_1 and N, respectively. Then the matrices $\tilde{T}$ and $\tilde{V}$ given by

$$\tilde{T}^{\mathrm{T}} = \begin{bmatrix} R_1^{-1} & 0 \\ 0 & R_2^{-1} \end{bmatrix} Q, \quad \tilde{V} = P \begin{bmatrix} R_1 & 0 \\ 0 & R_2 \end{bmatrix} \tag{3.108}$$

are a corresponding pair of entire left and right eigenvector matrices of system (3.72).

Proof. With the matrices $\tilde{T}$ and $\tilde{V}$ given by (3.108), it is easy to show, with the help of (3.74) and (3.107), the following relations:

$$
\begin{aligned}
\tilde{T}^{\mathrm{T}} E \tilde{V} &= \begin{bmatrix} R_1^{-1} & 0 \\ 0 & R_2^{-1} \end{bmatrix} QEP \begin{bmatrix} R_1 & 0 \\ 0 & R_2 \end{bmatrix} \\
&= \begin{bmatrix} R_1^{-1} & 0 \\ 0 & R_2^{-1} \end{bmatrix} \begin{bmatrix} I_{n_1} & 0 \\ 0 & N \end{bmatrix} \begin{bmatrix} R_1 & 0 \\ 0 & R_2 \end{bmatrix} \\
&= \operatorname{diag}\left(I_{n_1}, J_\infty\right),
\end{aligned}
$$

$$
\begin{aligned}
\tilde{T}^{\mathrm{T}} A \tilde{V} &= \begin{bmatrix} R_1^{-1} & 0 \\ 0 & R_2^{-1} \end{bmatrix} QAP \begin{bmatrix} R_1 & 0 \\ 0 & R_2 \end{bmatrix} \\
&= \begin{bmatrix} R_1^{-1} & 0 \\ 0 & R_2^{-1} \end{bmatrix} \begin{bmatrix} A_1 & 0 \\ 0 & I_{n_2} \end{bmatrix} \begin{bmatrix} R_1 & 0 \\ 0 & R_2 \end{bmatrix} \\
&= \operatorname{diag}\left(J, I_{n_2}\right).
\end{aligned}
$$

Therefore, the matrices $\tilde{T}$ and $\tilde{V}$ given by (3.108) are a corresponding pair of the entire left and right eigenvector matrices of system (3.72).   □

Remark 3.6. It can be seen from the above results and deduction that the eigenstructure of a matrix pair (E, A) can be obtained through the standard decomposition form of (E, A). Conversely, once the finite Jordan form, finite left and right eigenvector matrices as well as the infinite Jordan form, the infinite left and right eigenvector matrices associated with the matrix pair (E, A) are derived, a standard decomposition of the matrix pair (E, A) can also be immediately written out.

3.7.3 The Deflating Subspaces

Lemma 3.3. *Let the system (3.72) be regular, and admit the standard decompositions in (3.74). Partition the matrices P and Q as follows:*

$$Q^{\mathrm{T}} = [Q_1^{\mathrm{T}} \ Q_2^{\mathrm{T}}], \quad P = [P_1 \ P_2], \tag{3.109}$$

where $Q_1 \in \mathbb{R}^{n_1 \times n}_{n_1}, Q_2 \in \mathbb{R}^{n_2 \times n}_{n_2}, P_1 \in \mathbb{R}^{n \times n_1}_{n_1}, P_2 \in \mathbb{R}^{n \times n_2}_{n_2}$, then

$$AP_1 = EP_1A_1, \quad EP_2 = AP_2N, \tag{3.110}$$

and

$$Q_1A = A_1Q_1E, \quad Q_2E = NQ_2A. \tag{3.111}$$

Proof. We only show the equations in (3.110), and those in (3.111) can be shown similarly. Using (3.74) and (3.109), we clearly have

$$QEP_1 = \begin{bmatrix} I_{n_1} \\ 0 \end{bmatrix}, \quad QEP_2 = \begin{bmatrix} 0 \\ N \end{bmatrix} = \begin{bmatrix} 0 \\ I_{n_2} \end{bmatrix} N, \tag{3.112}$$

and

$$QAP_1 = \begin{bmatrix} A_1 \\ 0 \end{bmatrix} = \begin{bmatrix} I_{n_1} \\ 0 \end{bmatrix} A_1, \quad QAP_2 = \begin{bmatrix} 0 \\ I_{n_2} \end{bmatrix}. \tag{3.113}$$

Combining (3.112) and (3.113) gives

$$QAP_1 = QEP_1A_1, \quad QEP_2 = QAP_2N,$$

which is clearly equivalent to (3.110) in view of the nonsingularity of the matrix Q.
$\square$

Notice the similarity between (3.74) and (3.106), and that between the set of equations in (3.110)–(3.111) and those in (3.100)–(3.101), we introduce the following definition.

Definition 3.8. Let (E, A) be regular, and Q and P be two nonsingular matrices satisfying (3.74) and admit the decompositions in (3.109).

- Q_1 and P_1 are, respectively, called the finite left and right deflating matrices of the matrix pair (E, A);

- Q_2 and P_2 are, respectively, called the infinite left and right deflating matrices of the matrix pair (E, A).

The following lemma shows an interesting property of the deflating matrices.

Lemma 3.4. *Let (E, A) be regular. Suppose P and Q are defined by (3.74), and $\bar P$ and $\bar Q$ are another pair of transformation matrices satisfying*

$$\bar Q E \bar P = \begin{bmatrix} I_{n_1} & 0 \\ 0 & N \end{bmatrix} \text{ and } \bar Q A \bar P = \begin{bmatrix} A_1 & 0 \\ 0 & I_{n_2} \end{bmatrix}.$$

Furthermore, let

$$P = [\, P_1\ P_2 \,], \ \bar P = [\, \bar P_1\ \bar P_2 \,],$$
$$Q^{\mathrm T} = [\, Q_1\ Q_2 \,], \ \bar Q^{\mathrm T} = [\, \bar Q_1\ \bar Q_2 \,].$$

Then $n_1 = \bar n_1$, $n_2 = \bar n_2$, and the following relations hold:

$$\mathrm{span}(P_1) = \mathrm{span}(\bar P_1), \ \mathrm{span}(P_2) = \mathrm{span}(\bar P_2), \tag{3.114}$$
$$\mathrm{span}(Q_1) = \mathrm{span}(\bar Q_1), \ \mathrm{span}(Q_2) = \mathrm{span}(\bar Q_2). \tag{3.115}$$

Proof. It clearly follows from Theorem 3.6 that there exist nonsingular matrices $T_1 \in \mathbb R^{n_1 \times n_1}$ and $T_2 \in \mathbb R^{n_2 \times n_2}$ satisfying

$$[\, Q_1\ Q_2 \,] = [\, \bar Q_1\ \bar Q_2 \,] \begin{bmatrix} T_1^{\mathrm T} & 0 \\ 0 & T_2^{\mathrm T} \end{bmatrix} = [\, \bar Q_1 T_1^{\mathrm T}\ \bar Q_2 T_2^{\mathrm T} \,],$$

and

$$[\, P_1\ P_2 \,] = [\, \bar P_1\ \bar P_2 \,] \begin{bmatrix} T_1^{-1} & 0 \\ 0 & T_2^{-1} \end{bmatrix} = [\, \bar P_1 T_1^{-1}\ \bar P_2 T_2^{-1} \,].$$

These two relations clearly imply (3.114) and (3.115). $\qquad\Box$

Due to Lemma 3.4, we can now introduce the deflating subspaces of regular descriptor linear systems.

Definition 3.9. Let (E, A) be regular, and Q and P be two nonsingular matrices satisfying (3.74) and admit the decompositions in (3.109).

- $\mathrm{span}(Q_1)$ and $\mathrm{span}(P_1)$ are, respectively, called the finite left and right deflating subspace of the matrix pair (E, A), denoted by $\mathfrak D^L(E, A)$ and $\mathfrak D^R(E, A)$, respectively, i.e.

$$\mathfrak D^L(E, A) = \mathrm{span}(Q_1), \ \mathfrak D^R(E, A) = \mathrm{span}(P_1).$$

- span(Q_2) and span(P_2) are, respectively, called the infinite left and right deflating subspace of the matrix pair (E, A), denoted by $\mathfrak{D}_\infty^L(E, A)$ and $\mathfrak{D}_\infty^R(E, A)$, respectively, i.e.

$$\mathfrak{D}_\infty^L(E, A) = \text{span}(Q_2), \quad \mathfrak{D}_\infty^R(E, A) = \text{span}(P_2).$$

In view of Lemma 3.4 and Theorem 3.12, we have the following result.

Corollary 3.5. *For a given regular matrix pair (E, A), there hold*

$$\mathfrak{E}^L(E, A) = \mathfrak{D}^L(E, A), \quad \mathfrak{E}^R(E, A) = \mathfrak{D}^R(E, A),$$

$$\mathfrak{E}_\infty^L(E, A) = \mathfrak{D}_\infty^L(E, A), \quad \mathfrak{E}_\infty^R(E, A) = \mathfrak{D}_\infty^R(E, A).$$

Lemma 3.5. *Let the regular matrix pair (E, A) admit the standard decomposition given in (3.74). Define*

$$P_l = Q^{-1} \begin{bmatrix} I_{n_1} & 0 \\ 0 & 0 \end{bmatrix} Q, \quad P_r = P \begin{bmatrix} I_{n_1} & 0 \\ 0 & 0 \end{bmatrix} P^{-1}. \tag{3.116}$$

Then

$$\mathfrak{D}^R(E, A) = \text{Image}\, P_r, \quad \mathfrak{D}^L(E, A) = \text{Image}\, P_l. \tag{3.117}$$

Proof. We need only to prove the first relation in (3.117), the second one can be shown similarly.

First, let us show $\mathfrak{D}^R(E, A) \subset \text{Image}\, P_r$. In view of Corollary 3.5, we suffice to show

$$\mathfrak{E}^R(E, A) \subset \text{Image}\, P_r.$$

To achieve the above relation, let z be an arbitrary finite eigenvector of the matrix pair (E, A) corresponding to a finite eigenvalue λ, and we suffice to show $z \in \text{Image}\, P_r$.

Note that z satisfies

$$\lambda E z = A z,$$

which is equivalent to

$$\lambda (QEP) P^{-1} z = (QAP) P^{-1} z,$$

that is,

$$\lambda \begin{bmatrix} I_{n_1} & 0 \\ 0 & N \end{bmatrix} x = \begin{bmatrix} A_1 & 0 \\ 0 & I \end{bmatrix} x, \tag{3.118}$$

with

$$x = \begin{bmatrix} x_1 \\ x_2 \end{bmatrix} = P^{-1} z, \quad x_1 \in \mathbb{R}^{n_1}. \tag{3.119}$$

Equation (3.118) clearly gives $\lambda N x_2 = x_2$, which implies $x_2 = 0$. Furthermore, for an arbitrary $x_2' \in \mathbb{R}^{n_2}$, denote

$$x' = \begin{bmatrix} x_1 \\ x_2' \end{bmatrix}, \quad x'' = P x'.$$

Then we have from (3.119),

$$\begin{aligned} z = P x &= P \begin{bmatrix} x_1 \\ 0 \end{bmatrix} \\ &= P \begin{bmatrix} I_{n_1} & 0 \\ 0 & 0 \end{bmatrix} \begin{bmatrix} x_1 \\ x_2' \end{bmatrix}, \\ &= P \begin{bmatrix} I_{n_1} & 0 \\ 0 & 0 \end{bmatrix} P^{-1} P x', \\ &= P_r x'', \end{aligned}$$

which means $z \in \operatorname{Image} P_r$.

Second, let us show $\mathfrak{D}^R(E, A) \supset \operatorname{Image} P_r$.

$\forall z \in \operatorname{Image} P_r$, $\exists\, x$ such that

$$z = P_r x = P \begin{bmatrix} I_{n_1} & 0 \\ 0 & 0 \end{bmatrix} P^{-1} x. \tag{3.120}$$

Let

$$P^{-1} x = \begin{bmatrix} x_1 \\ x_2 \end{bmatrix} = x',$$

then (3.120) becomes

$$z = P \begin{bmatrix} x_1 \\ 0 \end{bmatrix} = P_1 x \in \operatorname{span}(P_1).$$

Hence, $\operatorname{Image} P_r \subset \operatorname{span}(P_1)$.

Combining the above two aspects gives the proof of the first conclusion. $\square$

The above lemma states that the matrices P_l and P_r are the spectral projections onto the left and right finite deflating subspaces of the matrix pair (E, A) along the left and right infinite deflating subspaces, respectively.

To end this section, we finally point out that the eigenstructure of a matrix pair is a very complicated problem. Due to length requirement, this section has only given the basic definitions and some of the most important facts. There are many aspects left unmentioned and unproven. Readers who are interested in this topic may refer to Wilkinson (1965) and Van Dooren (1981) and the references therein.

3.8 Stability

It is well known that a practical system should be stable. Otherwise, it may not work properly or may even be destroyed in practical use. Thus, stability is the most important property in a system. In control systems design, we are most concerned with the stability of the closed-loop systems. In this section and the following one, we investigate the stability of regular descriptor linear systems.

3.8.1 The Definition

Like the normal linear system case, when studying stability of descriptor linear systems, we need only to consider the following homogeneous equation:

$$E\dot{x} = Ax, \quad x(0) = x_0, \tag{3.121}$$

where $x \in \mathbb{R}^n$ is the state vector and $E, A \in \mathbb{R}^{n \times n}$ are the coefficient matrices. To make things easier, we assume in this section that the system (3.121) is regular since in this case the system has a unique solution.

Definition 3.10. The regular descriptor system (3.121) is said to be stable if there exist scalars α, $\beta > 0$ such that its state $x(t)$ satisfies

$$||x(t)||_2 \leq \alpha e^{-\beta t} ||x(0)||_2, \quad t > 0, \tag{3.122}$$

where the scalar β is called the decay rate.

Remark 3.7. It is clearly seen from the above definition that, when system (3.121) is stable, $\lim_{t \to \infty} x(t) = 0$. Thus, the stability defined for a descriptor linear system is actually the asymptotic stability in the Lyapunov sense. We are not calling it asymptotically stability because we want to maintain consistency between this concept with stability and stabilizability of the matrix pairs (E, A) to be introduced in the following context.

Since the stability of a descriptor linear system is described by the state response of the system, recalling that the state response of the two restricted equivalent descriptor systems is linked by a constant nonsingular linear transform, we immediately have the following conclusion.

Proposition 3.11. *Two restricted equivalent descriptor linear systems in the form of (3.121) possess the same stability property.*

3.8.2 The Direct Criterion

Recall from Sects. 1.1 and 3.4.1 that

$$\sigma(E, A) = \{s \mid s \in \mathbb{C}, s \text{ finite}, \det(sE - A) = 0\}$$

is the set of finite poles of the system, it contains at most $n_0 = \operatorname{rank} E \leq n$ number of complex numbers. For convenience, $\sigma(I, A)$ is often simply specified as $\sigma(A)$.

The following theorem gives a direct criterion for the stability of regular descriptor linear systems. It indicates that the stability of the system (3.121) is totally determined by $\sigma(E, A)$.

Theorem 3.14. *The regular descriptor linear system (3.121) is stable if and only if*

$$\sigma(E, A) \subset \mathbb{C}^- = \{s \mid s \in \mathbb{C}, \operatorname{Re}(s) < 0\}. \tag{3.123}$$

Proof. The proof adopts the standard decomposition of regular descriptor linear systems. According to Theorem 3.5, there exist two nonsingular matrices Q and P such that (3.74) holds. Therefore, by letting

$$x(t) = P \begin{bmatrix} x_1(t) \\ x_2(t) \end{bmatrix}, \quad x_0 = P \begin{bmatrix} x_{10} \\ x_{20} \end{bmatrix}, \tag{3.124}$$

the system (3.121) is restricted equivalent to the following systems

$$\dot{x}_1(t) = A_1 x_1(t), \quad x_1(0) = x_{10}, \tag{3.125}$$

and

$$N\dot{x}_2(t) = x_2(t), \quad x_2(0) = x_{20}, \tag{3.126}$$

where N is a nilpotent matrix.

It follows from Proposition 3.5 that the system (3.125) has the solution

$$x_1(t) = e^{A_1 t} x_{10}, \ t > 0,$$

and it follows from Theorem 3.9 that the system (3.126) has the solution

$$x_2(t) = 0, \ t > 0. \tag{3.127}$$

Now let inequality (3.122) hold. Then we have, using (3.124) and (3.127)

$$\begin{aligned} \|x_1(t)\|_2 &= \|P^{-1}x(t)\|_2 \\ &\leq \|P^{-1}\|_2 \|x(t)\|_2 \\ &= \|P^{-1}\|_2 \alpha e^{-\beta t} \|x(0)\|_2, \ t > 0. \end{aligned}$$

Therefore, system (3.125) is asymptotically stable, hence

$$\sigma(A_1) \subset \mathbb{C}^-. \qquad (3.128)$$

Conversely, let (3.128) hold, we can find positive scalars α and β such that

$$\|x_1(t)\|_2 \leq \|P\|_2^{-1} \alpha e^{-\beta t} \|x_1(0)\|_2, \quad t > 0,$$

with this, (3.124) and (3.127), we can easily obtain

$$\|x(t)\|_2 \leq \|P\|_2 \left\| \begin{bmatrix} x_1(t) \\ x_2(t) \end{bmatrix} \right\|_2$$

$$= \|P\|_2 \|x_1(t)\|_2$$

$$\leq \alpha e^{-\beta t} \|x_1(0)\|_2$$

$$\leq \alpha e^{-\beta t} \|x(0)\|_2, \quad t > 0,$$

which is identical with (3.122).

It follows from the above reasoning that the system is stable if and only if (3.128) holds. Further, in view of Lemma 3.1 we know that the system is stable if and only if (3.123) holds. Thus the second conclusion of the theorem is true. □

Due to the above theorem, we introduce the following definition.

Definition 3.11. Consider the system (3.121) or the matrix pair (E, A).

1. A pole of the system, or a finite eigenvalue of the matrix pair (E, A), is called stable if it has negative real part, is called critically stable if it has zero real part, and is called unstable if it has positive real part.
2. The matrix pair (E, A) is called stable if it has only stable finite eigenvalues, is called critically stable if it has critical finite eigenvalues but not unstable ones, and is called unstable if it has unstable finite eigenvalues.

The above definition clearly clarifies the equivalence between the stability of system (3.121) and that of the matrix pair (E, A).

Theorem 3.14 provides a direct criterion for the stability of descriptor linear systems in the sense that it verifies the stability of a system by directly checking the positions of the finite poles of the system. This is clearly a direct generalization of a stability criterion for normal linear systems.

There are basically two ways of finding the finite poles of a descriptor linear system computationally. One is to first convert the system into its standard decomposition form, and then find the eigenvalues of the matrix A_1 and the other is to use the command eig in the Matrix Functions Toolbox of Matlab.

3.8.3   Criterion via Lyapunov Equation

In this subsection, we further discuss the stability of regular descriptor linear system (E, A) and present criterions via Lyapunov matrix equations. The main purpose of this subsection is to present a general relation between the stability of the system (E, A) and the solutions of a type of generalized Lyapunov matrix equations in the form of

$$E^T X A + A^T X E = -P_r^T Y P_r, \tag{3.129}$$

where Y is positive definite, and P_r, defined by (3.116), is the spectral projection onto the right finite deflating subspace of the pair (E, A) along the right infinite deflating subspace.

The result (Stykel 2002b,c) is stated as follows.

Theorem 3.15. *Let P_r be the spectral projection onto the right finite deflating subspace of a regular pair (E, A) given by (3.74) and (3.116).*

1. *If there exist a positive definite matrix Y and a positive semidefinite matrix X satisfying (3.129), then the matrix pair (E, A) is stable.*
2. *If the matrix pair (E, A) is stable, then for every positive definite matrix Y, (3.129) has a positive semidefinite solution X.*

Proof. Let

$$Q^{-T} X Q^{-1} = \begin{bmatrix} X_{11} & X_{12} \\ X_{12}^T & X_{22} \end{bmatrix}, \quad P^T Y P = \begin{bmatrix} Y_{11} & Y_{12} \\ Y_{12}^T & Y_{22} \end{bmatrix},$$

and use (3.74), we have

$$\begin{aligned} E^T X A &= P^{-T} \begin{bmatrix} I_{n_1} & 0 \\ 0 & N^T \end{bmatrix} Q^{-T} X Q^{-1} \begin{bmatrix} I_{n_1} & 0 \\ 0 & N^T \end{bmatrix} P^{-1} \\ &= P^{-T} \begin{bmatrix} I_{n_1} & 0 \\ 0 & N^T \end{bmatrix} \begin{bmatrix} X_{11} & X_{12} \\ X_{12}^T & X_{22} \end{bmatrix} \begin{bmatrix} A_1 & 0 \\ 0 & I_{n_2} \end{bmatrix} P^{-1} \\ &= P^{-T} \begin{bmatrix} X_{11} A_1 & X_{12} \\ N^T X_{12}^T A_1 & N^T X_{22} \end{bmatrix} P^{-1}, \end{aligned}$$

and

$$\begin{aligned} P_r^T Y P_r &= P^{-T} \begin{bmatrix} I_{n_1} & 0 \\ 0 & 0 \end{bmatrix} P^{-1} Y P \begin{bmatrix} I_{n_1} & 0 \\ 0 & 0 \end{bmatrix} P^{-1} \\ &= P^{-T} \begin{bmatrix} I_{n_1} & 0 \\ 0 & 0 \end{bmatrix} \begin{bmatrix} Y_{11} & Y_{12} \\ Y_{12}^T & Y_{22} \end{bmatrix} \begin{bmatrix} I_{n_1} & 0 \\ 0 & 0 \end{bmatrix} P^{-1} \\ &= P^{-T} \begin{bmatrix} Y_{11} & 0 \\ 0 & 0 \end{bmatrix} P^{-1}. \end{aligned}$$

Hence,

$$E^T X A + A^T X E = P^{-T} \begin{bmatrix} X_{11}A_1 + A_1^T X_{11} & X_{12} + A_1^T X_{12}N \\ X_{12}^T + N^T X_{12}^T A_1 & N^T X_{22} + X_{22}^T N \end{bmatrix} P^{-1},$$

and the Lyapunov equation (3.129) is equivalent to the following set of equations:

$$X_{11}A_1 + A_1^T X_{11} = -Y_{11}, \tag{3.130}$$

$$X_{12} + A_1^T X_{12}N = 0, \tag{3.131}$$

$$N^T X_{22} + X_{22}N = 0. \tag{3.132}$$

Proof of conclusion 1. Suppose that there exist $X \geq 0$, and $Y > 0$ satisfying (3.129), then we know that $X_{11} \geq 0$, $Y_{11} > 0$. We can further show $X_{11} > 0$. In fact, if there exists $z \neq 0$ such that $X_{11}z = 0$, by pre-multiplying by z^T and post-multiplying by z both sides of (3.130), we have

$$x^T Y_{11} x = 0, \; x \neq 0.$$

This contradicts with the fact that $Y_{11} > 0$. Therefore, for any given $Y_{11} > 0$, there exists a $X_{11} > 0$ satisfying the Lyapunov equation (3.129). It thus follows from the Lyapunov theory for normal linear systems that A_1 is stable. Therefore, (E, A) is stable following from Lemma 3.1 and Theorem 3.14. □

Proof of conclusion 2. Now let the regular system (E, A) be assumed to be stable. Then according to Lemma 3.1 and Theorem 3.14, we know that A_1 is stable. In this case, it follows again from Lyapunov stability theory that for each $Y_{11} > 0$, we can find a unique $X_{11} > 0$ satisfying the Lyapunov equation (3.130). Therefore, it is easy to verify that, for $X_{12} = 0$, and $X_{22} = 0$, (3.131) and (3.132) are satisfied. Therefore, for any $Y > 0$, we have found the following

$$X = \begin{bmatrix} X_{11} & 0 \\ 0 & 0 \end{bmatrix} \geq 0$$

satisfying the Lyapunov equation (3.129). □

The above theorem is clearly a generalization of the Lyapunov stability result for normal linear systems.

3.8.4 Examples

Example 3.11. Consider the three-link planar manipulator system (1.29) in Sect. 1.2.4 again. The characteristic polynomial of this system is

$$\begin{aligned} f(s) &= \det(sE - A) \\ &= 31.81820s^2 + 3.28747s + 1.68624 \\ &= 31.81820(s + 0.05162 - 0.22435i)(s + 0.05162 + 0.22435i). \end{aligned}$$

Thus, the finite pole set is

$$\sigma(E, A) = \{-0.05162 + 0.22435i, -0.05162 - 0.22435i\},$$

which is formed by a pair of stable eigenvalues. Therefore, the system is stable by Theorem 3.14.

Example 3.12. Consider the system in Example 3.1 again. Since

$$\det(sE - A) = s^2,$$

we have

$$\sigma(E, A) = \{0, 0\}.$$

The system has a pair of zero poles, it thus follows from Theorem 3.14 that the system is not stable.

Example 3.13. Consider the following system

$$E = \begin{bmatrix} 0 & -1 & 3 \\ 0 & 0 & -1 \\ 0 & 0 & -1 \end{bmatrix}, \ A = \begin{bmatrix} 2 & 2 & -2 \\ -1 & 0 & 0 \\ -1 & 0 & 1 \end{bmatrix}.$$

It can be easily verified that the following transformation matrices

$$Q = \begin{bmatrix} 1 & 2 & 0 \\ 0 & 0 & -1 \\ 0 & -1 & 1 \end{bmatrix} \ \text{and} \ P = \begin{bmatrix} 0 & 1 & 1 \\ -1 & 0 & 1 \\ 0 & 0 & 1 \end{bmatrix},$$

transform the system into a standard decomposition form with

$$\tilde{E} = QEP = \begin{bmatrix} 1 & 0 & 0 \\ 0 & 0 & 1 \\ 0 & 0 & 0 \end{bmatrix}, \ \tilde{A} = QAP = \begin{bmatrix} -2 & 0 & 0 \\ 0 & 1 & 0 \\ 0 & 0 & 1 \end{bmatrix}.$$

Now we show the stability of the system using Theorem 3.15. Note

$$
\begin{aligned}
P_r &= P \begin{bmatrix} 1 & 0 & 0 \\ 0 & 0 & 0 \\ 0 & 0 & 0 \end{bmatrix} P^{-1} \\
&= \begin{bmatrix} 0 & 0 & 0 \\ 0 & 1 & -1 \\ 0 & 0 & 0 \end{bmatrix}.
\end{aligned}
$$

Furthermore, let

$$X = \begin{bmatrix} x_{11} & x_{12} & x_{13} \\ x_{12} & x_{22} & x_{23} \\ x_{13} & x_{23} & x_{33} \end{bmatrix}, \quad Y = \begin{bmatrix} y_{11} & y_{12} & y_{13} \\ y_{12} & y_{22} & y_{23} \\ y_{13} & y_{23} & y_{33} \end{bmatrix},$$

then the Lyapunov matrix equation expands as follows:

$$\begin{cases} -2x_{11} + x_{12} + x_{13} = 0 \\ 6x_{11} - 5x_{12} - 5x_{13} + x_{22} + 2x_{23} + x_{33} = 0 \\ -4x_{11} = -y_{22} \\ 8x_{11} - 2x_{12} - 3x_{13} = y_{22} \\ -12x_{11} + 4x_{12} + 10x_{13} - 2x_{23} - 2x_{33} = -y_{22} \end{cases}.$$

Through examining this set of equations, we can find a pair of matrices Y and X satisfying this Lyapunov matrix equation as follows:

$$Y = \begin{bmatrix} 2 & 1 & 0 \\ -1 & 4 & 1 \\ 0 & 1 & 2 \end{bmatrix} \quad \text{and} \quad X = \begin{bmatrix} 1 & 2 & 0 \\ 2 & 8 & -4 \\ 0 & -4 & 4 \end{bmatrix}.$$

Furthermore, it can be easily verified that $Y > 0$, and $X \geq 0$. Therefore, the system is stable by Theorem 3.15.

3.9 Admissibility: Stability plus Impulse-Freeness

In the above subsection, we have given a criterion for the stability of a general descriptor linear systems in terms of a type of generalized Lyapunov matrix equations. This criterion is more general in the sense that it may be applicable to systems whose responses contain impulse terms.

3.9.1 The Definition

In most situations, impulse terms are not desirable since they may saturate the state response or even destroy the system. So, in this subsection we are interested in giving a criterion for the system to be stable and impulse-free. This raises the concept of admissibility.

Definition 3.12. The system (3.121), or the matrices pair (E, A), is called admissible if it is stable and impulse-free.

Following Corollary 3.3, Lemma 3.1, and Theorem 3.14, we immediately have the following result on admissibility of descriptor linear systems.

Proposition 3.12. *Let the system (3.121) be regular, and admit the standard decomposition (3.74). Then the system is admissible if and only if $N = 0$ and one of the following conditions holds:*

- $\sigma(E, A) = \sigma(A_1) \subset \mathbb{C}^-$;
- $\forall Q_s > 0, \exists P_s > 0,\ s.t.\ A_1^T P_s + P_s^T A_1 = -Q_s$.

The conditions in the above proposition clearly depend on the standard decomposition of the system. In the following, we present a criterion of the Lyapunov type which require only the original system parameters E and A.

3.9.2 The Criterion

Like the case of stability, the admissibility of the descriptor linear system (3.121) is also related to a type of generalized Lyapunov matrix equations, but in the following form:

$$E^T X A + A^T X E = -E^T Y E, \tag{3.133}$$

where $Y = Y^T > 0$.

Theorem 3.16. *Let (E, A) be regular. Then the following hold:*

1. *If there exist $X = X^T \geq 0$ and $Y = Y^T > 0$ satisfying (3.133), then the system (E, A) is admissible.*
2. *If (E, A) is admissible, then for each $Y > 0$ there exists $X > 0$ satisfying (3.133).*

Proof. Denote

$$\tilde{X} = Q^{-T} X Q^{-1} = \begin{bmatrix} X_{11} & X_{12} \\ X_{12}^T & X_{22} \end{bmatrix}, \quad \tilde{Y} = Q^{-T} Y Q^{-1} = \begin{bmatrix} Y_{11} & Y_{12} \\ Y_{12}^T & Y_{22} \end{bmatrix}. \tag{3.134}$$

Then, pre-multiplying by P^T and post-multiplying by P the Lyapunov equation (3.133), and using (3.74), (3.133)–(3.134), gives

$$\tilde{E}^T \tilde{X} \tilde{A} + \tilde{A}^T \tilde{X} \tilde{E} = -\tilde{E}^T \tilde{Y} \tilde{E}. \tag{3.135}$$

Note, in view of (3.74), (3.133)–(3.134),

$$\tilde{E}^T \tilde{X} \tilde{A} = (P^T E^T Q^T)(Q^{-T} X Q^{-1})(Q A P)$$

$$= \begin{bmatrix} I_{n_1} & 0 \\ 0 & N^T \end{bmatrix} \begin{bmatrix} X_{11} & X_{12} \\ X_{12}^T & X_{22} \end{bmatrix} \begin{bmatrix} A_1 & 0 \\ 0 & I_{n_2} \end{bmatrix}$$

$$= \begin{bmatrix} X_{11}A_1 & X_{12} \\ N^T X_{12}^T A_1 & N^T X_{22} \end{bmatrix},$$

$$\tilde{E}^T \tilde{Y} \tilde{E} = P^T E^T Q^T Q^{-T} Y Q^{-1} Q E P$$

$$= \begin{bmatrix} I_{n_1} & 0 \\ 0 & N^T \end{bmatrix} \begin{bmatrix} Y_{11} & Y_{12} \\ Y_{12}^T & Y_{22} \end{bmatrix} \begin{bmatrix} I_{n_1} & 0 \\ 0 & N \end{bmatrix}$$

$$= \begin{bmatrix} Y_{11} & Y_{12}N \\ N^T Y_{12}^T & N^T Y_{22} N \end{bmatrix},$$

the Lyapunov equation (3.135) can be equivalently arranged into

$$\begin{bmatrix} X_{11}A_1 + A_1^T X_{11} + Y_{11} & X_{12} + A_1^T X_{12}N + Y_{12}N \\ N^T X_{12}^T A_1 + X_{12}^T + N^T Y_{12}^T & N^T X_{22} + X_{22}N + N^T Y_{22}N \end{bmatrix} = 0, \qquad (3.136)$$

which is equivalent to the following set of equations:

$$X_{11}A_1 + A_1^T X_{11} + Y_{11} = 0, \qquad (3.137)$$

$$X_{12} + A_1^T X_{12}N + Y_{12}N = 0, \qquad (3.138)$$

$$N^T X_{12}^T A_1 + X_{12}^T + N^T Y_{12}^T = 0, \qquad (3.139)$$

$$N^T X_{22} + X_{22}N + N^T Y_{22}N = 0. \qquad (3.140)$$

Proof of conclusion 1. Suppose there exist $X \geq 0$ and $Y > 0$ satisfying (3.133), then it follows from the above deduction that (3.137)–(3.140) hold. Let the index of the nilpotent matrix N be h, that is,

$$h = \min \left\{ k > 0 \mid N^k = 0 \right\},$$

and suppose $N \neq 0$, then $h > 1$. Pre-multiplying (3.140) by $(N^T)^{h-1}$, gives

$$(N^T)^{h-1} X_{22} N = 0 \implies (N^T)^{h-1} X_{22} N^{h-1} = 0$$

$$\implies (N^T)^{h-1} \left(X_{22}^T \right)^{\frac{1}{2}} (X_{22})^{\frac{1}{2}} N^{h-1} = 0$$

$$\implies \left((X_{22})^{\frac{1}{2}} N^{h-1} \right)^T \left((X_{22})^{\frac{1}{2}} N^{h-1} \right) = 0,$$

which yields

$$X_{22} N^{h-1} = 0. \qquad (3.141)$$

Pre-multiplying by $(N^T)^{h-2}$ and post-multiplying by N^{h-2} (3.140) produces

$$(N^T)^{h-1} X_{22} N^{h-2} + (N^T)^{h-2} X_{22} N^{h-1} + (N^T)^{h-1} Y_{22} N^{h-1} = 0,$$

which is equivalent to, in view of (3.141) and $Y_{22} > 0$,

$$(N^T)^{h-1} Y_{22} N^{h-1} = 0 \Longrightarrow \left((Y_{22})^{\frac{1}{2}} N^{h-1} \right)^T (Y_{22})^{\frac{1}{2}} N^{h-1} = 0$$

$$\Longrightarrow (Y_{22})^{\frac{1}{2}} N^{h-1} = 0$$

$$\Longrightarrow N^{h-1} = 0.$$

This contradicts with the minimality of h. Therefore, there must hold $h = 1$, that is, $N = 0$. Therefore, the system is impulse-free according to Corollary 3.3.

Now let us show the stability of system (E, A). Recall that $Y_{11} > 0$ and $X \geq 0$, we know $X_{11} \geq 0$. Since (3.137) holds, we can further show $X_{11} > 0$. In fact, if there exists $x \neq 0$ such that $X_{11} x = 0$, by pre-multiplying by x^T and post-multiplying by x both sides of (3.137), we have

$$x^T Y_{11} x = 0, \ x \neq 0.$$

This contradicts with the fact that $Y_{11} > 0$. Therefore, for any given $Y_{11} > 0$, there exists a $X_{11} > 0$ satisfying the Lyapunov equation (3.137). It thus follows from the Lyapunov theory for normal linear systems that A_1 is stable. Therefore, (E, A) is stable following from Lemma 3.1 and Theorem 3.14.   □

Proof of conclusion 2. Now let the regular system (E, A) be impulse-free and stable. Then according to Corollary 3.3, Lemma 3.1, and Theorem 3.14, we know that $N = 0$ and A_1 is stable. In this case, (3.136) can be written as

$$\begin{bmatrix} X_{11} A_1 + A_1^T X_{11} + Y_{11} & X_{12} \\ X_{12}^T & 0 \end{bmatrix} = 0. \tag{3.142}$$

Furthermore, it follows again from Lyapunov stability theory that for each $Y > 0$, we can find a unique $X_{11} > 0$ satisfying the Lyapunov equation (3.137). Therefore, it is easy to verify that, for any $X_{22} > 0$,

$$\tilde{X} = \begin{bmatrix} X_{11} & 0 \\ 0 & X_{22} \end{bmatrix} > 0$$

is a solution to (3.142). Therefore,

$$X = Q^T \begin{bmatrix} X_{11} & 0 \\ 0 & X_{22} \end{bmatrix} Q > 0$$

is a solution to the original Lyapunov equation (3.133).   □

It is obvious that Theorem 3.16 is also a direct generalization of the Lyapunov Theorem for normal linear systems. The advantage of the criterion is that it does not require the standard decomposition of the system.

3.9.3 The Example

Example 3.14. Consider again the system (3.38) in Example 3.4. Denote

$$
X = \begin{bmatrix} x_{11} & x_{12} & x_{13} & x_{14} \\ x_{12} & x_{22} & x_{23} & x_{24} \\ x_{13} & x_{23} & x_{33} & x_{34} \\ x_{14} & x_{24} & x_{34} & x_{44} \end{bmatrix}, \quad Y = \begin{bmatrix} y_{11} & y_{12} & y_{13} & y_{14} \\ y_{12} & y_{22} & y_{23} & y_{24} \\ y_{13} & y_{23} & y_{33} & y_{34} \\ y_{14} & y_{24} & y_{34} & y_{44} \end{bmatrix},
$$

we have

$$
E^T X A + A^T X E
$$
$$
= \begin{bmatrix} 2(x_{12} - x_{13}) & x_{11} + x_{14} & x_{14} + x_{22} - x_{23} & x_{13} + x_{14} \\ x_{11} + x_{14} & 0 & x_{12} + x_{24} & 0 \\ x_{14} + x_{22} - x_{23} & x_{12} + x_{24} & 2x_{24} & x_{23} + x_{24} \\ x_{13} + x_{14} & 0 & x_{23} + x_{24} & 0 \end{bmatrix},
$$

and

$$
E^T Y E = \begin{bmatrix} y_{11} & 0 & y_{12} & 0 \\ 0 & 0 & 0 & 0 \\ y_{12} & 0 & y_{22} & 0 \\ 0 & 0 & 0 & 0 \end{bmatrix}.
$$

If we choose

$$
X = \begin{bmatrix} 2 & 1 & 2 & -2 \\ 1 & 3 & 1 & -1 \\ 2 & 1 & 3 & 0 \\ -2 & -1 & 0 & 8 \end{bmatrix}, \quad Y = \begin{bmatrix} 2 & 0 & 0 & 0 \\ 0 & 2 & 0 & 0 \\ 0 & 0 & 1 & 0 \\ 0 & 0 & 0 & 1 \end{bmatrix},
$$

it is easily verified that $X > 0$, $Y > 0$ and they satisfy (3.133). Therefore, the system is admissible by Theorem 3.16.

3.10 Notes and References

This chapter examines some of the main properties of regular descriptor linear systems, and also presents the system responses as well as the eigenstructures of descriptor linear systems.

Regularity is a special feature for descriptor linear systems. All normal linear systems are regular. It is explored in this chapter that regularity guarantees the existence and uniqueness of solutions to descriptor linear systems. Typical earlier papers concerning regularity are Yip and Sincovec (1981) and Fletcher (1986).

It has been shown in this chapter that a descriptor linear system is regular if and only if its Kronecker canonical equivalence form reduces to the so-called standard decomposition form, which contains a slow subsystem and a fast subsystem. This standard decomposition form performs a very fundamental role in this book, as is demonstrated in this chapter with response analysis for regular descriptor linear systems, it serves as a basic technique for many analysis and design problems. For deriving computationally a standard decomposition form of a given descriptor linear system, the Matlab Toolbox for linear descriptor systems introduced in Varga (2000) can be readily used, which was developed based on the basic techniques proposed in Misra et al. (1994).

Based on the standard decomposition form, the response analysis of a descriptor linear system is converted into response analysis of the slow and the fast subsystems. The response of the slow subsystem is nothing but that of a normal linear system, while that of the fast subsystem is quite special: it contains the derivatives of the control input, and is instantaneous with respect to time. What is more, different from that of a normal linear system, the "normal sense" response of a descriptor linear system may not exist for an arbitrarily given initial state value. Such a phenomenon raises two concepts. One is the so-called distributional solution of a regular descriptor linear system, which always exists for an arbitrarily given initial value, but the payoff is to allow it to contain generalized functions, namely, the delta function. Thus these types of solutions are also called generalized solutions. The other is the so-called consistent initial values, which clarifies the allowable initial values for the existence of a classical solution, that is, a solution containing only classical functions which are continuously differentiable.

Regarding the solution of a regular descriptor linear system, in the literature there are mainly two opinions. One argues that the initial condition should be restricted to the consistent ones; while the other says that any possible initial condition should be acceptable. In this chapter, both solutions are discussed. In fact, they are the different representations due to different treatments, while both are effective and reasonable. In essence, the problem can be understood in the following way: If generalized functions are allowed to exist in the solution of a descriptor linear system, the solution always exists for arbitrarily chosen initial values; the classical response exists only when the initial value is consistent. In other words, the consistent initial conditions can be viewed as the restriction imposed on the initial value $x(0)$ to guarantee that the distributional solution and the classical solution of the system are consistent. Concerning the consistent initial values and distributional solutions, one may further refer to Cobb (1983b, 1984), Verghese et al. (1981), Cullen (1984), Geerts (1993), Yan and Duan (2005), and Kunkel and Mehrmann (2006). Particularly, Yan and Duan (2005) have modified an existing distributional solution of a descriptor linear system and has also given satisfactory explanations of some "doubtful points" about solutions to descriptor linear systems. Furthermore, Yan and Duan (2006) have examined the impulsive parts in the solutions to the initial value problem of linear time-varying descriptor systems.

Eigenstructures of descriptor linear systems are also introduced in this chapter. This concept and related results are closely related with the design problem treated in Chap. 9 – eigenstructure assignment. Compared with normal systems theory, infinite eigenvalues and infinite eigenvectors are special features of descriptor linear systems, which do not exist in normal linear systems. It is interesting to note that the eigenstructure of a descriptor linear system has a very close relation with the standard decomposition of the system.

Like the case for any type of dynamical systems, stability is also a very basic issue for descriptor linear systems. In this chapter, we have given the basic definition and also provided some basic necessary and sufficient conditions for stability of regular descriptor linear systems. Besides the direct criterion, two criterions using generalized Lyapunov equations have also been provided. It is worth pointing out that, for stability of descriptor systems, there are numerous results in the literature. For linear descriptor systems, Lewis (1985b) has established stability condition in terms of a type of generalized Lyapunov matrix equations in the form of

$$A^{T}XE + E^{T}XA + E^{T}C^{T}CE = 0,$$

where C is the measurement matrix in the output equation, while Takaba et al. (1995) have given stability conditions in terms of two Lyapunov equations. Furthermore, Stykel (2002a,b,c) and Muller (2006) have also given stability criteria in terms of certain types of generalized Lyapunov matrix equations, and the inertia theorem was also provided for the types generalized Lyapunov equations. For linear time-invariant descriptor systems with structured perturbations, a robust stability problem is considered in Zhai et al. (2002). In addition, Chaabanc ct al. (2006) also dealt with the problems of robust stability of uncertain continuous descriptor systems, and moreover, the robust stability of linear interval descriptor systems has attracted much attention (Lin et al. 1997, 2001; Su et al. 2002). Regarding stability of time-delay descriptor systems, Fridman (2001) provided a Lyapunov-based approach, Yang et al. (2005a, 2006) introduced the concepts of practical stability in terms of two measurements and give some results of practical stability parallel to Lyapunov stability theorems. Also, the robust exponential stability of descriptor systems with time-varying delays and time-varying parameter uncertainties is dealt with in Yue et al. (2005). Besides the above, many other results about stability exist, these include the finite time stability of linear discrete descriptor systems (Debeljkovic et al. 1998), generalized Lyapunov theorems for rectangular descriptor systems (Ishihara and Terra 2001), and absolute stability of nonlinear descriptor systems (Yang et al. 2007a,b).

Chapter 4
Controllability and Observability

We know from standard normal linear systems theory that controllability and observability are a pair of very important concepts for linear systems. They provide us with certain structural features of the system as well as some internal properties of the system, and are essential for certain system design problems studied in the later chapters.

This chapter focuses on topics related with controllability and observability. As a preliminary, the first section introduces the concept of reachable state and characterizes the state reachable subspaces. In Sects. 4.2 and 4.3, definitions of controllability and observability are proposed, respectively. As we will find out, unlike the normal linear system case, there are several types of controllabilities and observabilities for descriptor linear systems. Conditions for these types of controllabilities and observabilities in terms of the fast and slow subsystems are given. Based on the results introduced in Sects. 4.2 and 4.3, the so-called dual principle is proposed in Sect. 4.4, which reveals the close relation between the controllabilities and observabilities of a regular descriptor linear system and its dual system. This principle gives us great convenience in the sequential sections in deriving criteria for various types of controllability and observability. Sections 4.5 and 4.6 further give some criteria for these types of controllabilities and observabilities of descriptor systems. Section 4.5 stresses on those using the original system coefficient matrices, while Sect. 4.6 gives criteria based on certain restricted equivalent forms. Afterwards, two problems which are closely related with the concepts of controllability and observability, namely, system structural decomposition and minimal realization, are investigated in Sects. 4.7 and 4.8. Notes and references follow in Sect. 4.9.

4.1 State Reachable Subsets

Consider the regular descriptor system:

$$E\dot{x} = Ax + Bu, \tag{4.1}$$

G.-R. Duan, *Analysis and Design of Descriptor Linear Systems*, Advances in Mechanics and Mathematics 23, DOI 10.1007/978-1-4419-6397-0_4, © Springer Science+Business Media, LLC 2010

where $x \in \mathbb{R}^n$ and $u \in \mathbb{R}^r$ are the state vector and the control input vector, respectively; and E, $A \in \mathbb{R}^{n \times n}$, $B \in \mathbb{R}^{n \times r}$ are constant matrices. The dynamical order of the system is $n_0 = \mathrm{rank}\, E \leq n$.

It follows from the standard decomposition of descriptor linear systems that there exist two nonsingular matrices Q and P such that under the transformation (P, Q) the system (4.1) is transformed into the following equivalent standard decomposition form:

$$\begin{cases} \dot{x}_1 = A_1 x_1 + B_1 u \\ N \dot{x}_2 = x_2 + B_2 u \end{cases}, \tag{4.2}$$

where $x_1 \in \mathbb{R}^{n_1}$, $x_2 \in \mathbb{R}^{n_2}$, $n_1 + n_2 = n$, the matrix $N \in \mathbb{R}^{n_2 \times n_2}$ is nilpotent, and the nilpotent index is denoted by h. The relations between the states and the coefficients of the two equivalent systems are given below:

$$P^{-1} x = \begin{bmatrix} x_1 \\ x_2 \end{bmatrix},$$

$$QEP = \begin{bmatrix} I & 0 \\ 0 & N \end{bmatrix}, \quad QAP = \begin{bmatrix} A_1 & 0 \\ 0 & I \end{bmatrix}, \quad QB = \begin{bmatrix} B_1 \\ B_2 \end{bmatrix}.$$

4.1.1 The Definition

For the sake of simplicity, in this section we discuss the descriptor linear system in the standard decomposition form (4.2). The obtained results are easily applicable to systems in the general form of (4.1).

Let us start from the concept of the reachable states.

Definition 4.1. Given the regular descriptor linear system (4.2), a vector $w \in \mathbb{R}^n$ is said to be a reachable vector of system (4.2), if there exists an initial condition $x_1(0) = x_{10}$, an admissible control input $u(t) \in \mathbb{C}_p^{h-1}$, and some $t_1 > 0$ such that the response of the system (4.2) satisfies

$$x(t_1, u, x_{10}) = \begin{bmatrix} x_1(t_1) \\ x_2(t_1) \end{bmatrix} = w.$$

Please note that in the above definition the admissible control input is confined to $u(t) \in \mathbb{C}_p^{h-1}$, which represents the set of $h - 1$ times piecewise continuously differentiable functions.

Let

$$\mathscr{R}_t[x_{10}] = \{ w \mid \exists u(t) \in \mathbb{C}_p^{h-1} \text{ s.t. } x(t, u, x_{10}) = w \in \mathbb{R}^n \},$$

then $\mathscr{R}_t[x_{10}]$ is the set of reachable states at time t from the initial condition $x_1(0) = x_{10}$. It clearly reduces to $\mathscr{R}_t[0]$ when $x_{10} = 0$. Note that the response of the system (4.2) is given by

$$\begin{cases} x_1(t) = e^{A_1 t} x_1(0) + \int_0^t e^{A_1(t-\tau)} B_1 u(\tau) \, d\tau \\ x_2(t) = -\sum_{i=0}^{h-1} N^i B_2 u^{(i)}(t) \end{cases} , \quad t > 0,$$

we clearly have

$$\mathscr{R}_t[x_{10}] = \left\{ x \, \middle| \, x = \begin{bmatrix} e^{A_1 t} x_{10} + \int_0^t e^{A_1(t-\tau)} B_1 u(\tau) \, d\tau \\ -\sum_{i=0}^{h-1} N^i B_2 u^{(i)}(t) \end{bmatrix}, \ u(t) \in \mathbb{C}_p^{h-1} \right\},$$

(4.3)

and

$$\mathscr{R}_t[0] = \left\{ x \, \middle| \, x = \begin{bmatrix} \int_0^t e^{A_1(t-\tau)} B_1 u(\tau) \, d\tau \\ -\sum_{i=0}^{h-1} N^i B_2 u^{(i)}(t) \end{bmatrix}, \ u(t) \in \mathbb{C}_p^{h-1} \right\}. \quad (4.4)$$

Furthermore, denote

$$\mathscr{R}_t = \bigcup_{x_{10} \in \mathbb{R}^{n_1}} \mathscr{R}_t[x_{10}], \quad (4.5)$$

and

$$\mathscr{H}_t = \{ x \mid x = x_I(t, x_{10}), \ x_{10} \in \mathbb{R}^{n_1} \}, \quad (4.6)$$

with

$$x_I(t, x_{10}) = \begin{bmatrix} e^{A_1 t} x_{10} \\ 0 \end{bmatrix}, \quad (4.7)$$

then $\mathscr{R}_t$ is the state reachable set at time t for system (4.2) from all possible initial condition $x_1(0) = x_{10} \in \mathbb{R}^{n_1}$, and $\mathscr{H}_t$ is the set of free reachable state at time t starting from all possible initial condition x_{10}. Further denote by "$\oplus$" the external direct sum in a vector space, we then have the following basic facts.

Proposition 4.1. *Given the descriptor linear system (4.2). Then for any $t > 0$, the following hold:*

1. *$\mathscr{R}_t[0]$ is a subspace in $\mathbb{R}^n$.*
2. *$\mathscr{R}_t[x_{10}] = \mathscr{R}_t[0] + x_I(t, x_{10})$ is a manifold in $\mathbb{R}^n$.*
3. *$\mathscr{H}_t = \mathbb{R}^{n_1} \oplus \{0_{n_2}\}$.*
4. *$\mathscr{R}_t = \mathscr{R}_t[0] + \mathscr{H}_t$ is a subspace in $\mathbb{R}^n$.*

Proof. Since both $\int_0^t e^{A_1(t-\tau)} B_1 u(\tau) d\tau$ and $\sum_{i=0}^{h-1} N^i B_2 u^{(i)}(t)$ are linear operators in $u(t)$, the first conclusion holds obviously. The second conclusion clearly follows from the definition of $\mathscr{R}_t[x_{10}]$ and $\mathscr{R}_t[0]$. Note that $e^{A_1 t}$ is a nonsingular matrix, we have

$$\left\{ x \mid x = e^{A_1 t} x_{10}, \ x_{10} \in \mathbb{R}^{n_1} \right\} = \mathbb{R}^{n_1}.$$

With this fact and the definition of $\mathscr{H}_t$, the third conclusion is clearly seen to be true. The fourth conclusion is obvious. □

The above proposition provides us with some deep insight about the reachable sets $\mathscr{R}_t[0]$, $\mathscr{R}_t[x_{10}]$, $\mathscr{H}_t$ and $\mathscr{R}_t$. A direct phenomenon may be seen from the above proposition is that, instead of filling up the whole space as in the normal system case, the state response for a descriptor linear system lies on a manifold in the space. This is a special feature that is possessed by descriptor linear systems.

4.1.2 Characterization of $\mathscr{R}_t[0]$ and $\mathscr{R}_t$

We have known from Proposition 4.1 that some general structural properties of the reachable subsets $\mathscr{R}_t[0]$, $\mathscr{R}_t[x_{10}]$, $\mathscr{H}_t$ and $\mathscr{R}_t$. In this subsection, we further characterize the reachable subspaces $\mathscr{R}_t[0]$ and $\mathscr{R}_t$.

Introduce for any matrix pair (A, B), with $A \in \mathbb{R}^{n \times n}$ and $B \in \mathbb{R}^{n \times r}$, the controllability matrix

$$Q_c[A, B] = [B \; AB \cdots A^{n-1} B]. \tag{4.8}$$

Then, the main result of this subsection is given as follows.

Theorem 4.1. *Let $\mathscr{R}_t[0]$ be the state reachable subspace of the regular descriptor linear system (4.2) at time t. Then*

$$\mathscr{R}_t[0] = \mathrm{Image}Q_c[A_1, B_1] \oplus \mathrm{Image}Q_c[N, B_2], \, \forall t > 0, \tag{4.9}$$

where "$\oplus$" is the external direct sum in the vector space $\mathbb{R}^n$.

To prove the above theorem, we need first to make some preparations.

Lemma 4.1. *Let $A \in \mathbb{R}^{n \times n}$, $B \in \mathbb{R}^{n \times r}$. Then the following hold:*

1. $\ker Q_c^{\mathrm{T}}[A, B] = \bigcap\limits_{i=0}^{n-1} \ker \left(B^{\mathrm{T}}(A^{\mathrm{T}})^i \right).$
2. $\int_0^t \mathrm{e}^{A(t-\tau)} B u(\tau) \, \mathrm{d}\tau \in \mathrm{Image}Q_c[A, B], \; \forall u(\tau) \in \mathbb{R}^r.$
3. $x \in \ker(B^{\mathrm{T}} \mathrm{e}^{A^{\mathrm{T}} t}), \; t_1 < t < t_2 \iff x \in \bigcap\limits_{i=0}^{n-1} \ker \left(B^{\mathrm{T}}(A^{\mathrm{T}})^i \right).$

Proof. The first conclusion directly follows from the definition of $Q_c[A, B]$ and that of kernel spaces. It follows from Cayley–Hamilton Theory that there exist $\beta_i(t)$, $i = 0, 1, \cdots, n-1$, such that

$$\mathrm{e}^{At} = \beta_0(t) I + \beta_1(t) A + \cdots + \beta_{n-1}(t) A^{n-1}, \tag{4.10}$$

we have

$$\int_0^t \mathrm{e}^{A(t-\tau)} B u(\tau) \, \mathrm{d}\tau = \sum_{i=0}^{n-1} A^i B \int_0^t \beta_i(t-\tau) u(\tau) \, \mathrm{d}\tau$$
$$= Q_c[A, B]z$$

with

$$z^T = \int_0^t \left[\beta_0 (t - \tau) u^T (\tau) \quad \beta_1 (t - \tau) u^T (\tau) \quad \cdots \quad \beta_{n-1} (t - \tau) u^T (\tau) \right] d\tau.$$

Therefore, the second conclusion holds.

Again using (4.10), we have

$$B^T e^{A^T t} = \beta_0 (t) B^T + \beta_1 (t) B^T A^T + \cdots + \beta_{n-1} (t) B^T (A^T)^{n-1}.$$

Thus, $x \in \ker(B^T e^{A^T t})$, $t_1 < t < t_2$ is equivalent to

$$\left[\beta_0 (t) B^T + \beta_1 (t) B^T A^T + \cdots + \beta_{n-1} (t) B^T (A^T)^{n-1} \right] x = 0, \ t_1 < t < t_2.$$

Furthermore, due to the arbitrariness of t, the above relation holds if and only if

$$B^T (A^T)^i x = 0, \ i = 0, 1, \cdots, n - 1,$$

which is obviously equivalent to

$$x \in \bigcap_{i=0}^{n-1} \ker \left(B^T (A^T)^i \right).$$

Therefore, the third conclusion holds.   □

Lemma 4.2. *Let $A \in \mathbb{R}^{n \times n}$, $B \in \mathbb{R}^{n \times r}$, and $f(t)$ be an arbitrary polynomial which is nonidentically zero. Define*

$$W (f, t) = \int_0^t f (\tau) e^{A\tau} B B^T e^{A^T \tau} f (\tau) d\tau.$$

Then

$$\text{Image} W(f, t) = \text{Image} Q_c [A, B], \ \forall t > 0.$$

Proof. In view of the definition of $Q_c [A, B]$ in (4.8) and the first conclusion in Lemma 4.1, it suffices only to prove

$$\ker W(f, t) = \bigcap_{i=0}^{n-1} \ker \left(B^T (A^T)^i \right).$$

When $x \in \ker W(f, t)$, we have

$$x^T W (f, t) x = x^T 0 = 0,$$

i.e.,

$$x^{\mathrm{T}} W\,(f,t)\,x = \int_0^t x^{\mathrm{T}} f\,(\tau)\,\mathrm{e}^{A\tau} B B^{\mathrm{T}} \mathrm{e}^{A^{\mathrm{T}}\tau} f\,(\tau)\,x\,\mathrm{d}\tau$$
$$= \int_0^t \left\| B^{\mathrm{T}} \mathrm{e}^{A^{\mathrm{T}}\tau} f\,(\tau)\,x \right\|_2^2 \mathrm{d}\tau$$
$$= 0. \tag{4.11}$$

Since

$$\left\| B^{\mathrm{T}} \mathrm{e}^{A^{\mathrm{T}}\tau} f\,(\tau)\,x \right\|_2^2 \geq 0,\ 0 \leq \iota \leq l$$

and $f\,(\tau)$ is continuous, we know from (4.11) that

$$B^{\mathrm{T}} \mathrm{e}^{A^{\mathrm{T}}\tau} f\,(\tau)\,x = 0, \quad 0 \leq \tau \leq t. \tag{4.12}$$

Since polynomial $f\,(\tau)$ is not identically zero, it has only a finite number of zeros on $0 \leq \tau \leq t$. Thus, (4.12) is equivalent to

$$B^{\mathrm{T}} \mathrm{e}^{A^{\mathrm{T}}\tau} x = 0, \quad 0 \leq \tau \leq t.$$

Applying the third conclusion of Lemma 4.1 further yields

$$x \in \bigcap_{i=0}^{n-1} \ker \left(B^{\mathrm{T}} (A^{\mathrm{T}})^i \right).$$

Thus,

$$\ker W(f,t) \subset \bigcap_{i=0}^{n-1} \ker \left(B^{\mathrm{T}} (A^{\mathrm{T}})^i \right).$$

If

$$x \in \bigcap_{i=0}^{n-1} \ker \left(B^{\mathrm{T}} (A^{\mathrm{T}})^i \right),$$

reversing the above process gives $x \in \ker W\,(f,t)$, i.e.,

$$\bigcap_{i=0}^{n-1} \ker \left(B^{\mathrm{T}} (A^{\mathrm{T}})^i \right) \subset \ker W(f,t).$$

Hence, the proof is done. $\square$

Lemma 4.3. *For any h vectors $x_i \in \mathbb{R}^n$, $i = 0, 1, 2, \cdots, h-1$, and an arbitrary $t_1 > 0$, there always exists a vector polynomial $f(t) \in \mathbb{R}^n$ whose order is $h-1$, such that*

$$f^{(i)}\,(t_1) = x_i, \quad i = 0, 1, 2, \cdots, h-1. \tag{4.13}$$

Proof. Set

$$f(t) = x_0 + x_1 (t - t_1) + \cdots + \frac{1}{(h-1)!} x_{h-1} (t - t_1)^{h-1}.$$

Then $f(t)$ is clearly a polynomial satisfying (4.13). □

Proof of Theorem 4.1

First, let us show (4.9).

When $x_1(0) = 0$, the response of system (4.2) is

$$\begin{cases} x_1(t) = \int_0^t e^{A_1(t-\tau)} B_1 u(\tau) d\tau \\ x_2(t) = -\sum_{i=0}^{h-1} N^i B_2 u^{(i)}(t) \end{cases}, \quad t > 0.$$

Thus, $x_2(t) \in \text{Image} Q_c[N, B_2]$ obviously follows. Furthermore, it follows the second conclusion of Lemma 4.1 that $x_1(t) \in \text{Image} Q_c[A_1, B_1]$. Therefore,

$$x(t) \in \text{Image} Q_c[A_1, B_1] \oplus \text{Image} Q_c[N, B_2].$$

This gives, in view of (4.4),

$$\mathcal{R}_t[0] \subset \text{Image} Q_c[A_1, B_1] \oplus \text{Image} Q_c[N, B_2]. \tag{4.14}$$

On the other hand, let

$$x = \begin{bmatrix} x_1 \\ x_2 \end{bmatrix} \in \text{Image} Q_c[A_1, B_1] \oplus \text{Image} Q_c[N, B_2],$$

with

$$x_1 \in \text{Image} Q_c[A_1, B_1], \quad x_2 \in \text{Image} Q_c[N, B_2].$$

Since $x_2 \in \text{Image} Q_c[N, B_2]$, there exist $x_{2i} \in \mathbb{R}^r$, $i = 0, 1, 2, \cdots, h - 1$, such that

$$x_2 = -\sum_{i=0}^{h-1} N^i B_2 x_{2i}.$$

It follows from Lemma 4.3 that, for any fixed $t > 0$, there exists a polynomial $f_2(\tau)$ of order $h - 1$ such that $f_2^{(i)}(t) = x_{2i}$. Thus

$$-\sum_{i=0}^{h-1} N^i B_2 f_2^{(i)}(t) = x_2.$$

If we impose for system (4.2) the input control

$$u(t) = u_1(t) + f_2(t),$$

then there holds, for $t > 0$,

$$x_2(t) = -\sum_{i=0}^{h-1} N^i B_2 u^{(i)}(t)$$

$$= -\sum_{i=0}^{h-1} N^i B_2 u_1^{(i)}(t) - \sum_{i=0}^{h-1} N^i B_2 f_2^{(i)}(t)$$

$$= x_2 - \sum_{i=0}^{h-1} N^i B_2 u_1^{(i)}(t), \tag{4.15}$$

and

$$x_1(t, u, 0) = \int_0^t e^{A_1(t-\tau)} B_1 u_1(\tau) \, d\tau + \int_0^t e^{A_1(t-\tau)} B_1 f_2(\tau) \, d\tau. \tag{4.16}$$

Since $x_1 \in \text{Image} Q_c[A_1, B_1]$ and, according to the second conclusion of Lemma 4.1,

$$\int_0^t e^{A_1(t-\tau)} B_1 f_2(\tau) \, d\tau \in \text{Image} Q_c[A_1, B_1],$$

we have

$$\tilde{x}_1 = x_1 - \int_0^t e^{A_1(t-\tau)} B_1 f_2(\tau) \, d\tau \in \text{Image} Q_c[A_1, B_1]. \tag{4.17}$$

For any fixed $t > 0$, let

$$f_1(\tau) = \tau^h (\tau - t)^h,$$

then $f_1(\tau)$ is not identically zero and Lemma 4.2 indicates that a vector $z \in \mathbb{R}^{n_1}$ can be chosen such that

$$W(f_1, t) z = \tilde{x}_1. \tag{4.18}$$

For this chosen vector z, we construct

$$u_1(\tau) = f_1^2(\tau) B_1^T e^{A_1^T(t-\tau)} z, \quad 0 \le \tau \le t, \tag{4.19}$$

which can be verified to satisfy

$$u_1^{(i)}(t) = 0, \quad i = 0, 1, \cdots, h-1. \tag{4.20}$$

Therefore, it follows from (4.15) and (4.20) that $x_2(t) = x_2$ holds. Furthermore, it follows from (4.16) to (4.19) that

$$
\begin{aligned}
x_1(t, u, 0) &= \int_0^t e^{A_1(t-\tau)} B_1 u_1(\tau)\, d\tau + \int_0^t e^{A_1(t-\tau)} B_1 f_2(\tau)\, d\tau \\
&= x_1 - x_1 + \int_0^t e^{A_1(t-\tau)} B_1 f_2(\tau)\, d\tau + \int_0^t e^{A_1(t-\tau)} B_1 u_1(\tau)\, d\tau \\
&= x_1 - \tilde{x}_1 + \int_0^t e^{A_1(t-\tau)} B_1 u_1(\tau)\, d\tau \\
&= x_1 - \tilde{x}_1 + W(f_1, t)\, z \\
&= x_1.
\end{aligned}
$$

Thus,

$$
\mathscr{R}_t[0] \supset \mathrm{Image}\, Q_c[A_1, B_1] \oplus \mathrm{Image}\, Q_c[N, B_2]. \tag{4.21}
$$

The combination of (4.14) with (4.21) obviously results in (4.9). With this, the proof of the theorem is completed. $\qquad\square$

The above proof process also shows that, with a suitably chosen control $u(t)$, reachable vectors may be reached in any short period.

Corollary 4.1. *Let*

$$
\mathscr{R}_t^s[0] = \left\{ x \,\middle|\, x = \int_0^t e^{A_1(t-\tau)} B_1 u(\tau)\, d\tau,\ u(t) \in \mathbb{C}_p^{h-1} \right\},
$$

and

$$
\mathscr{R}_t^f[0] = \left\{ x \,\middle|\, x = -\sum_{i=0}^{h-1} N^i B_2 u^{(i)}(t),\ u(t) \in \mathbb{C}_p^{h-1} \right\}.
$$

Then the following hold:

1. $\mathscr{R}_t^s[0] = \mathrm{Image}\, Q_c[A_1, B_1];\ \mathscr{R}_t^f[0] = \mathrm{Image}\, Q_c[N, B_2],\ \forall t > 0;$
2. $\mathscr{R}_t = \mathbb{R}^{n_1} \oplus \mathrm{Image}\, Q_c[N, B_2], \forall t > 0.$

Proof. The first conclusion follows immediately from Theorem 4.1 and the definitions of $\mathscr{R}_t^s[0]$, $\mathscr{R}_t^f[0]$ and $\mathscr{R}_t[0]$.

It follows from the third and the fourth conclusions of Proposition 4.1 that

$$
\begin{aligned}
\mathscr{R}_t &= \mathscr{H}_t + \mathscr{R}_t[0] \\
&= \left(\mathbb{R}^{n_1} \oplus \{0_{n_2}\} \right) + \mathscr{R}_t[0] \\
&= \left(\mathbb{R}^{n_1} \oplus \{0_{n_2}\} \right) + \left(\mathscr{R}_t^s[0] \oplus \mathscr{R}_t^f[0] \right) \\
&= \left(\mathbb{R}^{n_1} + \mathscr{R}_t^s[0] \right) \oplus \left(\{0_{n_2}\} + \mathscr{R}_t^f[0] \right) \\
&= \left(\mathbb{R}^{n_1} + \mathscr{R}_t^s[0] \right) \oplus \mathscr{R}_t^f[0].
\end{aligned}
$$

Since

$$\mathbb{R}^{n_1} + \mathscr{R}_t^s[0] = \mathbb{R}^{n_1}, \ \mathscr{R}_t^f[0] = \text{Image}\, Q_c[N, B_2],$$

the second conclusion clearly follows.   □

4.1.3 Two Examples

Example 4.1. Consider the system

$$\begin{cases} \dot{x}_1 = \begin{bmatrix} 0 & 0 \\ 1 & 0 \end{bmatrix} x_1 + \begin{bmatrix} 1 & 0 & -1 \\ 1 & -1 & 2 \end{bmatrix} u(t) \\[2mm] \begin{bmatrix} 0 & 1 \\ 0 & 0 \end{bmatrix} \dot{x}_2 = x_2 + \begin{bmatrix} 0 & 1 & 0 \\ 0 & 0 & 1 \end{bmatrix} u(t) \\[2mm] y = \begin{bmatrix} 0 & 1 \\ 0 & -1 \end{bmatrix} x_1 + \begin{bmatrix} 0 & 0 \\ 1 & 0 \end{bmatrix} x_2 \end{cases} \qquad (4.22)$$

which is actually a standard decomposition form for the system (3.39) in Example 3.5. For this system,

$$A_1 = \begin{bmatrix} 0 & 0 \\ 1 & 0 \end{bmatrix}, \ B_1 = \begin{bmatrix} 1 & 0 & -1 \\ 1 & -1 & 2 \end{bmatrix},$$

$$N = \begin{bmatrix} 0 & 1 \\ 0 & 0 \end{bmatrix}, \ B_2 = \begin{bmatrix} 0 & 1 & 0 \\ 0 & 0 & 1 \end{bmatrix},$$

$$C_1 = \begin{bmatrix} 0 & 1 \\ 0 & -1 \end{bmatrix}, \ C_2 = \begin{bmatrix} 0 & 0 \\ 1 & 0 \end{bmatrix}.$$

It can be easily verified that

$$\text{Image}\, Q_c[A_1, B_1] = \text{Image}\, [B_1 \ A_1 B_1] = \mathbb{R}^2,$$

and

$$\text{Image}\, Q_c[N, B_2] = \text{Image}\, [B_2 \ N B_2] = \mathbb{R}^2.$$

Thus,

$$\mathscr{R}_t[0] = \text{Image}\, Q_c[A_1, B_1] \oplus \text{Image}\, Q_c[N, B_2] = \mathbb{R}^4,$$

$$\mathscr{R}_t = \mathbb{R}^2 \oplus \text{Image}\, Q_c[N, B_2] = \mathbb{R}^4.$$

In this example, $\mathscr{R}_t = \mathscr{R}_t[0]$, or in other words, the reachable set for the whole system is identical to the reachable set from the zero initial condition.

Example 4.2. Consider the following system

$$\begin{bmatrix} 0 & 1 & 0 \\ 0 & 0 & 1 \\ 0 & 0 & 0 \end{bmatrix} \dot{x}(t) = \begin{bmatrix} 0 & 1 & 0 \\ 0 & 0 & 1 \\ 0 & 0 & -1 \end{bmatrix} x(t) + \begin{bmatrix} 1 \\ -1 \\ 1 \end{bmatrix} u(t).$$

For this system, which is composed of only a fast one, it is clearly seen that

$$\mathcal{R}_t = \mathcal{R}_t[0] = \text{Image}\, Q_c[N, B_2] = \mathbb{R}^3,$$

that is, both $\mathcal{R}_t$ and $\mathcal{R}_t[0]$ are the whole vector space. Thus for arbitrary $w \in \mathbb{R}^3$, a time t_1 and a permissible control $u(t)$ can theoretically be found such that $x(t_1) = 0$.

The state response of the system is given by

$$x(t) = -\sum_{i=0}^{h-1} N^i B_2 u^{(i)}(t) = \begin{bmatrix} -u(t) + \dot{u}(t) - \ddot{u}(t) \\ u(t) - \dot{u}(t) \\ -u(t) \end{bmatrix}, \quad t > 0.$$

Thus, for any

$$w = \begin{bmatrix} w_1 \\ w_2 \\ w_3 \end{bmatrix} \in \mathbb{R}^3 \text{ and } t_1 > 0,$$

the equality $x(t_1) = w$ implies

$$u(t_1) = -w_3, \ \dot{u}(t_1) = -w_3 - w_2, \ \ddot{u}(t_1) = -w_2 - w_1. \tag{4.23}$$

Therefore, the following control law

$$u(t) = (-w_2 - w_1)(t - t_1)^2 + (-w_3 - w_2)(t - t_1) - w_3$$

can be constructed, which drives the state of the system to w at time $t = t_1$. However, when $w_1 + w_2$ and $w_2 + w_3$ are large, it follows from the second and the third equations in (4.23) that the variation rate and acceleration of the control law $u(t)$ at t_1 is large. In this case, the control law $u(t)$ will increase or decrease greatly at t_1 and may thus be difficult to realize practically.

4.2 Controllability

In this section, we consider the controllability of descriptor linear systems. Different from the normal system case, for descriptor linear systems, there are four types of controllabilities, namely, C-controllability, R-controllability, I-controllability, and S-controllability.

As in the normal systems case, all the four types of controllabilities are described by the trajectory of the system. Since the trajectories of two restricted equivalent systems are related by a constant nonsingular linear transformation, all the four types of controllabilities are invariant under restricted system equivalence. Note that every regular descriptor linear system is r.s.e. to a standard decomposition form of (4.2), without loss of generality, in this section we again study the descriptor linear system in the form of (4.2), which has the following slow and fast subsystems:

$$\dot{x}_1 = A_1 x_1 + B_1 u, \tag{4.24}$$

$$N\dot{x}_2 = x_2 + B_2 u. \tag{4.25}$$

In this section, the definitions of the various controllabilities of the descriptor linear system (4.2) are given, and their relations with the controllability of the slow and fast subsystems (4.24) and (4.25) are presented.

4.2.1 C-Controllability

First let us investigate the complete controllability.

Definition 4.2. The regular system (4.2) is called completely controllable (or C-controllable) if, for any $t_1 > 0$, $x_0 \in \mathbb{R}^n$ and $w \in \mathbb{R}^n$, there exists a control input $u(t) \in \mathbb{C}_p^{h-1}$ such that the response of the system (4.2) starting from the initial condition $x(0) = x_0$ satisfies $x(t_1, u, x_0) = w$.

This definition states that under the C-controllability assumption, for any initial condition $x(0) = x_0$, we can always find a control input for the system such that the state response of the system is driven from $x(0)$ to any prescribed position in $\mathbb{R}^n$ in any specified time period. It is easy to see that the definition is a natural generalization of the controllability for normal linear systems.

Let $A \in \mathbb{R}^{n \times n}$, $B \in \mathbb{R}^{n \times r}$. As in the normal systems case, we define the controllability matrix pair (A, B) as

$$Q_c[A, B] = \begin{bmatrix} B & AB & \cdots & A^{n-1}B \end{bmatrix}.$$

For the matrix pair (N, B_2), since

$$\begin{bmatrix} B_2 & NB_2 & \cdots & N^{n_2-1}B_2 \end{bmatrix} = \begin{bmatrix} B_2 & NB_2 & \cdots & N^{h-1}B_2 & 0 \end{bmatrix},$$

we convert

$$Q_c[N, B_2] = \begin{bmatrix} B_2 & NB_2 & \cdots & N^{h-1}B_2 \end{bmatrix}.$$

Based on the above definition and notation, we can derive the following lemma.

Lemma 4.4. *The regular descriptor linear system (4.2) is C-controllable if and only if*

$$\mathscr{R}_t[0] = \mathbb{R}^n, \ \forall t > 0, \tag{4.26}$$

or equivalently,

$$\text{Image} Q_c[A_1, B_1] = \mathbb{R}^{n_1}, \quad \text{Image} Q_c[N, B_2] = \mathbb{R}^{n_2}. \tag{4.27}$$

Proof. It follows directly from the above definition that system (4.2) is C-controllable if and only if

$$\mathscr{R}_t[x_{10}] = \mathbb{R}^n, \ \forall x_{10} \in \mathbb{R}^{n_1}, \ t > 0. \tag{4.28}$$

Furthermore, note the second conclusion of Proposition 4.1, it is obvious that (4.26) is equivalent to (4.28).

It follows from Theorem 4.1 that

$$\mathscr{R}_t[0] = \text{Image} Q_c[A_1, B_1] \oplus \text{Image} Q_c[N, B_2] = \mathbb{R}^n, \ \forall t > 0.$$

Therefore, (4.27) is clearly equivalent to (4.26). The proof is then completed. □

Regarding the complete controllability of the system (4.2) as well as its slow and fast subsystems (4.24) and (4.25), we have the following theorem.

Theorem 4.2. *Consider the regular descriptor linear system (4.2) and its slow subsystem (4.24) and fast subsystem (4.25).*

1. *The slow subsystem (4.24) is C-controllable if and only if*

$$\text{rank} Q_c[A_1, B_1] = n_1, \tag{4.29}$$

   or equivalently,

$$\text{rank}\begin{bmatrix} sI - A_1 & B_1 \end{bmatrix} = n_1, \ \forall s \in \mathbb{C}, \ s \ \text{finite}. \tag{4.30}$$

2. *The fast subsystem (4.25) is C-controllable if and only if*

$$\text{rank} Q_c[N, B_2] = n_2, \tag{4.31}$$

   or equivalently,

$$\text{rank}[N \ B_2] = n_2. \tag{4.32}$$

3. *The system (4.2) is C-controllable if and only if both its slow and fast subsystems (4.24) and (4.25) are C-controllable.*

Proof. Proof of conclusion 1. The slow subsystem (4.24) is a normal one, for which Definition 4.2 becomes the controllability definition in normal linear system theory. Therefore, the conclusion follows immediately from well-known necessary and sufficient conditions for controllability of standard normal linear systems.

Proof of conclusion 2. According to Lemma 4.4 that the fast subsystem (4.25) is C-controllable if and only if

$$\text{Image} \, Q_c[N, B_2] = \mathbb{R}^{n_2},$$

or, equivalently, condition (4.31) holds, which indicates that the matrix pair (N, B_2) is controllable. It therefore follows from the PBH criterion for controllability of standard normal linear systems that condition (4.31) is equivalent to

$$\text{rank} \, [sI - N \ \ B_2] = n_2, \quad \forall s \in \sigma \, (N). \tag{4.33}$$

Since N is nilpotent, $\sigma \, (N) = \{0\}$. Equation (4.33) is true if and only if

$$\text{rank} \, [-N \ \ B_2] = \text{rank} \, [N \ \ B_2] = n_2.$$

This is equivalent to the condition (4.32).

Proof of conclusion 3. According to Lemma 4.4, and the first two conclusions of this theorem, we have,

$$
\begin{aligned}
&\text{the system (4.2) is C-controllable} \\
&\Leftrightarrow \text{Image} \, Q_c[A_1, B_1] = \mathbb{R}^{n_1}, \quad \text{Image} \, Q_c[N, B_2] = \mathbb{R}^{n_2} \\
&\Leftrightarrow \text{rank} \, Q_c[A_1, B_1] = n_1, \quad \text{rank} \, Q_c[N, B_2] = n_2 \\
&\Leftrightarrow \text{both the fast and slow subsystems are C-controllable.}
\end{aligned}
$$

$\square$

Example 4.3 (Dai 1989b). Consider the system

$$
\begin{cases}
\dot{x}_1 = \begin{bmatrix} -1 & -1 \\ 1 & 0 \end{bmatrix} x_1 + \begin{bmatrix} 1 \\ 0 \end{bmatrix} u \, (t) \\[2mm]
0 = x_2 + \begin{bmatrix} -1 \\ 0 \end{bmatrix} u \, (t) \\[2mm]
y = \begin{bmatrix} 0 & 1 \end{bmatrix} x_1
\end{cases}
\tag{4.34}
$$

which is actually a standard decomposition form for the circuit network system in Example 1.1 with $L = 1\text{H}$, $C_0 = 1\text{F}$, $R = 1\Omega$ (see also Example 3.4). Direct computation shows that

$$\text{rank} \, [B_1 \ \ A_1 B_1] = \text{rank} \begin{bmatrix} 1 & -1 \\ 0 & 1 \end{bmatrix} = 2,$$

$$\text{rank} \, [B_2 \ \ N B_2] = \text{rank} \begin{bmatrix} -1 & 0 \\ 0 & 0 \end{bmatrix} = 1 < 2.$$

From Theorem 4.2 we know that the system is not C-controllable, while its slow subsystem is C-controllable. In fact, the descriptor system (1.13) contains

$$V_R(t) = RI(t),$$

which is an identical equation. Thus,

$$V_R(t) - RI(t) = 0,$$

and cannot be affected by the control input.

Remark 4.1. Since C-controllability of a normal linear system is identical with its controllability defined in normal systems theory, in the following we will simply call a normal linear system controllable when it is C-controllable.

4.2.2 R-Controllability

Based on the reachable subspace $\mathscr{R}_t$ defined in (4.5), the R-controllability of the regular descriptor linear system (4.2) can be given as follows.

Definition 4.3. The regular descriptor system (4.2) is called R-controllable, if it is controllable in the reachable subspace $\mathscr{R}_t$, or more precisely, for any prescribed $t_1 > 0$, $x_{10} \in \mathbb{R}^{n_1}$ and $w \in \mathscr{R}_{t_1}$, there always exists an admissible control input $u(t) \in \mathbb{C}_p^{h-1}$ such that the response of system (4.2) starting from the initial value $x_1(0) = x_{10}$ satisfies $x(t_1, u, x_{10}) = w$.

The R-controllability guarantees the controllability for the system from any admissible initial condition $x_1(0)$ to any reachable state and this process can be finished in any given time period if the control $u(t)$ is suitably chosen.

Directly following the above definition, we have the following lemma.

Lemma 4.5. *The regular descriptor linear system (4.2) is R-controllable if and only if*

$$\mathscr{R}_t[0] = \mathscr{R}_t, \quad \forall t > 0, \tag{4.35}$$

or equivalently,

$$\text{Image} Q_c[A_1, B_1] = \mathbb{R}^{n_1}. \tag{4.36}$$

Proof. It directly follows from the above definition that system (4.2) is R-controllable if and only if

$$\mathscr{R}_t[x_{10}] = \mathscr{R}_t, \quad \forall x_{10} \in \mathbb{R}^{n_1}, \ t > 0. \tag{4.37}$$

Thus, it suffices only to show the equivalence between (4.35) and (4.37). Obviously, (4.37) implies (4.35). On the other hand, let (4.35) holds. Then it follows from the second conclusion of Proposition 4.1 that

$$\mathscr{R}_t[x_{10}] = \mathscr{R}_t[0] + x_I(t, x_{10}) = \mathscr{R}_t + x_I(t, x_{10}), \quad \forall x_{10} \in \mathbb{R}^{n_1}, \ t > 0.$$

Furthermore, since $x_I(t, x_{10}) \in \mathscr{R}_t$, the above relation clearly gives (4.37).

It follows from Corollary 4.1 that

$$\mathscr{R}_t = \mathbb{R}^{n_1} \oplus \mathrm{Image}\, Q_c[N, B_2].$$

On the other hand, it follows from Theorem 4.1 that

$$\mathscr{R}_t[0] = \mathrm{Image}\, Q_c[A_1, B_1] \oplus \mathrm{Image}\, Q_c[N, B_2].$$

Therefore, condition (4.35) holds if and only if (4.36) is valid. The proof is completed.   □

The following is an immediately corollary of the above lemma.

Lemma 4.6. *The following descriptor linear system with only a fast subsystem:*

$$N\dot{x} = x + Bu,$$

where N is nilpotent, is always R-controllable.

It follows from the above lemma that the fast subsystem (4.25) does not affect the R-controllability of the system (4.2), the following theorem then holds naturally.

Theorem 4.3. *The regular descriptor linear system (4.2), with slow subsystem (4.24) and fast subsystem (4.25), is R-controllable if and only if the slow subsystem (4.24) is C-controllable, or equivalently, the condition (4.29) or (4.30) holds.*

Proof. It follows from Lemma 4.5 and Theorem 4.2 that

> the system (4.2) is R-controllable
> $\Leftrightarrow \mathrm{Image}\, Q_c[A_1, B_1] = \mathbb{R}^{n_1}$
> $\Leftrightarrow \mathrm{rank}\, Q_c[A_1, B_1] = n_1$
> $\Leftrightarrow$ the slow subsystem is C-controllable
> $\Leftrightarrow$ condition (4.30).

                                                                                                     □

This theorem, combined with Theorem 4.2, shows that system (4.2) is R-controllable if it is C-controllable, but the inverse is not true.

Example 4.4. According to Theorem 4.3, system (4.34) is R-controllable, but is not C-controllable.

4.2.3 I-Controllability and S-Controllability

Clearly, the concepts of C-controllability and R-controllability are concerned with only the terminal behavior of the system. We have noted that in the state response of a descriptor linear system, there may exist impulse terms at $t = 0$. For descriptor linear systems, it is necessary to analyze the control effect on the impulse terms in the state response.

Recalling the state response of a regular descriptor linear system, there are no impulse terms in the substate $x_1(t)$ when $u \in \mathbb{C}_p^{h-1}$ and it follows from Corollary 3.2 that the impulse part in $x_2(t)$ is determined by

$$x_{2\text{impulse}}(t) = -\sum_{i=1}^{h-1} N^i \delta^{(i-1)}(t) \left(x_{20} + \sum_{j=0}^{h-1} N^j B_2 u^{(j)}(0) \right), \qquad (4.38)$$

which is resulted in by the initial values of the state vector and the control vector as well as its derivatives. To eliminate the impulse in the state response by control means to select an appropriate control law $u \in \mathbb{C}_p^{h-1}$ such that the state response of the system at $t = 0$ is a limited number. In view of this fact, we can introduce the following definition for I-controllability.

Definition 4.4. The regular descriptor system (4.2) is called I-controllable (impulse controllable) if for any vector $x_{20} \in \mathbb{R}^{n_2}$, there exists an admissible control input $u \in \mathbb{C}_p^{h-1}$ such that impulse part in the response of the fast subsystem given by (4.38) is identically zero, that is,

$$x_{2\text{impulse}}(t) = 0, \ \forall t \geq 0.$$

This definition characterizes the ability to eliminate by admissible control the impulse terms in descriptor linear systems. It directly follows from the definition and (4.38) that the following basic fact holds.

Proposition 4.2. *The regular descriptor linear system (4.2) is I-controllable if and only if for any vector* $x_{20} \in \mathbb{R}^{n_2}$, *there exists an admissible control input* $u \in \mathbb{C}_p^{h-1}$ *such that*

$$N x_{20} + \begin{bmatrix} NB_2 & N^2 B_2 & \cdots & N^{h-1} B_2 \end{bmatrix} \begin{bmatrix} u(0) \\ \dot{u}(0) \\ \vdots \\ u^{(h-2)}(0) \end{bmatrix} = 0. \qquad (4.39)$$

Proof. It follows from Definition 4.4 that the regular descriptor linear system (4.2) is I-controllable if and only if for any vector $x_{20} \in \mathbb{R}^{n_2}$, there exists an admissible control input $u \in \mathbb{C}_p^{h-1}$ such that

$$\left[\sum_{i=1}^{h-1} \delta^{(i-1)}(t) \left(N^i x_2(0) + \sum_{j=i}^{h-1} N^j B_2 u^{(j-i)}(0) \right) \right]_{t=0} = 0. \qquad (4.40)$$

It follows from Theorem A.1 in Appendix A that the group of functions $\delta^{(i)}(t)$, $i = 0, 1, 2, \cdots, h - 2$, is linearly independent, the above equation is equivalent to

$$N^i x_2(0) + \sum_{j=i}^{h-1} N^j B_2 u^{(j-i)}(0) = 0, \ i = 1, 2, \ldots, h - 1,$$

which can be written as

$$N^i x_2(0) + \begin{bmatrix} N^i B_2 & N^{i+1} B_2 & \cdots & N^{h-1} B_2 \end{bmatrix} \begin{bmatrix} u(0) \\ \dot{u}(0) \\ \vdots \\ u^{(h-i-1)}(0) \end{bmatrix} = 0, \qquad (4.41)$$

$$i = 1, 2, \ldots, h - 1.$$

Furthermore, note the nilpotent property of the matrix N, it is easy to show that the above set of equations in (4.41) holds if and only if (4.39) holds. Therefore, (4.39) is equivalent to (4.40).

To complete the proof, we now need only to show that given the values $u^{(i)}(0) = u_0^{(i)}$, $i = 0, 1, 2, \cdots, h - 1$, satisfying (4.39), there exists an admissible control input $\hat{u}(t) \in \mathbb{C}_p^{h-1}$ such that

$$\hat{u}^{(i)}(0) = u_0^{(i)}, \ i = 0, 1, 2, \cdots, h - 1.$$

This is guaranteed by Lemma 4.3. Thus, the proof is done. $\qquad\qquad\qquad$ $\square$

I-controllability is necessary to eliminate the impulse portions in a system in which impulse terms are generally not expected to appear. Otherwise, strong impulse behavior may stop the system from working or even destroy the system.

Directly following from the definition, we have the following lemma.

Lemma 4.7. *The descriptor system with only the slow subsystem (or normal system)*

$$\dot{x} = Ax + Bu$$

is always I-controllable.

Since the slow subsystem (4.24) is always I-controllable, the I-controllability of the system (4.2) must then be completely determined by the I-controllability of its fast subsystem (4.25). This is confirmed by the following theorem.

Theorem 4.4. *Consider the regular descriptor linear system (4.2) with its slow subsystem (4.24) and fast subsystem (4.25).*

1. *System (4.2) is I-controllable if and only if its fast subsystem (4.25) is I-controllable.*

2. *The fast subsystem (4.25) is I-controllable if and only if one of the following conditions holds:*

   (a) $\text{Image} N = \text{Image} \begin{bmatrix} NB_2 & N^2 B_2 & \cdots & N^{h-1} B_2 \end{bmatrix}$.
   (b) $\ker N + \text{Image} Q_c [N, B_2] = \mathbb{R}^{n_2}$.
   (c) $\text{Image} N + \ker N + \text{Image} B_2 = \mathbb{R}^{n_2}$.

Proof. The first conclusion is obvious. Here, we only prove the second one.

Proof of (a). It follows from Proposition 4.2 that the fast subsystem (4.25) is I-controllable if and only if for any vector $x_{20} \in \mathbb{R}^{n_2}$, there exists an admissible control input $u \in \mathbb{C}_p^{h-1}$ such that (4.39) holds. This is actually equivalent to condition (a).

Proof of $(a) \Leftrightarrow (b)$. Noting that

$$\begin{bmatrix} NB_2 & N^2 B_2 & \cdots & N^{h-1} B_2 & 0 \end{bmatrix}$$
$$= \begin{bmatrix} NB_2 & N^2 B_2 & \cdots & N^{h-1} B_2 & N^h B_2 \end{bmatrix}$$
$$= N \begin{bmatrix} B_2 & N^1 B_2 & \cdots & N^{h-2} B_2 & N^{h-1} B_2 \end{bmatrix}$$
$$= N Q_c [N, B_2],$$

and

$$\text{Image} \begin{bmatrix} NB_2 & N^2 B_2 & \cdots & N^{h-1} B_2 \end{bmatrix} = \text{Image} \begin{bmatrix} NB_2 & N^2 B_2 & \cdots & N^{h-1} B_2 & 0 \end{bmatrix},$$

we have

$$\text{Image} \begin{bmatrix} NB_2 & N^2 B_2 & \cdots & N^{h-1} B_2 \end{bmatrix} = \text{Image} (N Q_c [N, B_2]).$$

Therefore, it follows from the above relation and Proposition A.6 that

$$(a) \Leftrightarrow \text{Image} (N Q_c [N, B_2]) = \text{Image} N$$
$$\Leftrightarrow \text{Image} Q_c [N, B_2] + \ker N = \mathbb{R}^{n_2}$$
$$\Leftrightarrow (b).$$

Proof of $(b) \Leftrightarrow (c)$. Noting that

$$\text{Image} Q_c [N, B_2] = \text{Image} B_2 + \text{Image} \begin{bmatrix} NB_2 & N^2 B_2 & \cdots & N^{h-1} B_2 \end{bmatrix},$$

we can write the condition in (b) as

$$\ker N + \text{Image} B_2 + \text{Image} \begin{bmatrix} NB_2 & N^2 B_2 & \cdots & N^{h-1} B_2 \end{bmatrix} = \mathbb{R}^{n_2}.$$

Furthermore, by using condition (a), the above equation can be converted into the condition in (c). □

It clearly follows from Theorems 4.2 and 4.4 that a system is I-controllable if it is C-controllable. However, the inverse is not true.

Using the condition (a) in Theorem 4.4, we can further obtain the following result about I-controllability of the regular system (4.2).

Theorem 4.5. *Consider the regular descriptor linear system (4.2) with the fast subsystem (4.25). Let the controllability decomposition of (N, B_2) be*

$$\left(\begin{bmatrix} N_{11} & N_{12} \\ 0 & N_{22} \end{bmatrix}, \begin{bmatrix} B_{21} \\ 0 \end{bmatrix} \right). \tag{4.42}$$

Then the system (4.2) is I-controllable if and only if

$$N_{22} = 0 \text{ and } N_{12} = N_{11} M \tag{4.43}$$

hold for some matrix M.

Proof. Since I-controllability is invariant under restricted system equivalence, without loss of generality we assume that the matrix pair (N, B_2) directly possesses the form of (4.42).

It follows from the condition (a) in Theorem 4.4 that the system (4.2) is I-controllable if and only if

$$\text{Image} N = \text{Image} \begin{bmatrix} NB_2 & N^2 B_2 & \cdots & N^{h-1} B_2 \end{bmatrix}. \tag{4.44}$$

Assume that $N_{11} \in \mathbb{R}^{\tilde{n}_1 \times \tilde{n}_1}$, $N_{22} \in \mathbb{R}^{\tilde{n}_2 \times \tilde{n}_2}$. Since (N_{11}, B_{21}) is controllable, we have

$$\text{Image} \begin{bmatrix} B_{21} & N_{11} B_{21} & \cdots & N_{11}^{h-1} B_{21} \end{bmatrix} = \mathbb{R}^{\tilde{n}_1}, \tag{4.45}$$

Using the decomposed forms of matrices N and B_2, we have

$$\text{Image} N = \text{Image} \begin{bmatrix} N_{11} & N_{12} \\ 0 & N_{22} \end{bmatrix}, \tag{4.46}$$

and

$$\text{Image} \begin{bmatrix} NB_2 & N^2 B_2 & \cdots & N^{h-1} B_2 \end{bmatrix}$$

$$= \text{Image} \begin{bmatrix} NB_2 & N^2 B_2 & \cdots & N^{h-1} B_2 & 0 \end{bmatrix}$$

$$= \text{Image} \begin{bmatrix} N_{11} B_{21} & N_{11}^2 B_{21} & \cdots & N_{11}^{h-1} B_{21} & 0 \\ 0 & 0 & \cdots & 0 & 0 \end{bmatrix}$$

$$= \text{Image} \begin{bmatrix} N_{11} B_{21} & N_{11}^2 B_{21} & \cdots & N_{11}^{h-1} B_{21} & N_{11}^h B_{21} \\ 0 & 0 & \cdots & 0 & 0 \end{bmatrix}$$

$$= \text{Image} \left\{ \begin{bmatrix} N_{11} \\ 0 \end{bmatrix} \begin{bmatrix} B_{21} & N_{11}B_{21} & \cdots & N_{11}^{h-1}B_{21} \end{bmatrix} \right\}$$

$$\subset \text{Image} \begin{bmatrix} N_{11} \\ 0 \end{bmatrix}. \tag{4.47}$$

On the other hand, arbitrarily choose

$$x = \begin{bmatrix} x_1 \\ 0 \end{bmatrix} \in \text{Image} \begin{bmatrix} N_{11} \\ 0 \end{bmatrix},$$

where $x_1 \in \text{Image} N_{11} \subset \mathbb{R}^{\tilde{n}_1}$, then there exists a vector y_1 satisfying

$$x_1 = N_{11}y_1. \tag{4.48}$$

In view of (4.45), there exist a series of vectors α_i, $i = 0, 1, \cdots, h - 1$, such that

$$y_1 = B_{21}\alpha_0 + N_{11}B_{21}\alpha_1 + \cdots + N_{11}^{h-1}B_{21}\alpha_{h-1}. \tag{4.49}$$

Combining (4.48) and (4.49) yields

$$x_1 = N_{11}(B_{21}\alpha_0 + N_{11}B_{21}\alpha_1 + \cdots + N_{11}^{h-1}B_{21}\alpha_{h-1}).$$

Letting

$$\tilde{y}_1 = [\alpha_0 \ \alpha_1 \ \cdots \ \alpha_{h-1}]^{\mathrm{T}},$$

we have

$$x_1 = N_{11} \begin{bmatrix} B_{21} & N_{11}B_{21} & \cdots & N_{11}^{h-1}B_{21} \end{bmatrix} \tilde{y}_1$$

$$\in \text{Image} \left\{ N_{11} \begin{bmatrix} B_{21} & N_{11}B_{21} & \cdots & N_{11}^{h-1}B_{21} \end{bmatrix} \right\}.$$

Therefore,

$$x = \begin{bmatrix} x_1 \\ 0 \end{bmatrix}$$

$$\in \text{Image} \left\{ \begin{bmatrix} N_{11} \\ 0 \end{bmatrix} \begin{bmatrix} B_{21} & N_{11}B_{21} & \cdots & N_{11}^{h-1}B_{21} \end{bmatrix} \right\}$$

$$= \text{Image} \begin{bmatrix} NB_2 & N^2B_2 & \cdots & N^{h-1}B_2 \end{bmatrix}.$$

Thus,

$$\text{Image} \begin{bmatrix} N_{11} \\ 0 \end{bmatrix} \subset \text{Image} \begin{bmatrix} NB_2 & N^2B_2 & \cdots & N^{h-1}B_2 \end{bmatrix}. \tag{4.50}$$

Combining (4.47) and (4.50) gives

$$\text{Image} \begin{bmatrix} NB_2 & N^2B_2 & \cdots & N^{h-1}B_2 \end{bmatrix} = \text{Image} \begin{bmatrix} N_{11} \\ 0 \end{bmatrix}. \tag{4.51}$$

Substituting (4.46) and (4.51) into (4.44) produces

$$\text{Image} \begin{bmatrix} N_{11} & N_{12} \\ 0 & N_{22} \end{bmatrix} = \text{Image} \begin{bmatrix} N_{11} \\ 0 \end{bmatrix}.$$

This equation is true if and only if there exists an M such that (4.43) holds.   □

Besides the above three types of controllabilities, there is another type of controllability for regular descriptor linear systems, which is defined below.

Definition 4.5. The descriptor linear system (4.2) is said to be strongly controllable, or S-controllable, if it is both R-controllable and I-controllable.

The relations among these four types of controllability concepts can be described as follows:

C-controllability

⇓

S-controllability ⟺ R-controllability + I-controllability

4.3 Observability

Having discussed the various types of controllabilities of descriptor linear systems, in this section we introduce the dual concepts–C-observability, R-observability, I-observability, and S-observability. These types of observabilities are concerned with the ability to reconstruct the state from the system inputs and the measured outputs. Therefore, different from the preceding two sections, in this section the system to be considered is of the following form:

$$\begin{cases} E\dot{x} = Ax + Bu \\ \tilde{y} = Cx \end{cases}, \tag{4.52}$$

where $x \in \mathbb{R}^n$, $u \in \mathbb{R}^r$, and $\tilde{y} \in \mathbb{R}^m$ are the state vector, the control input vector, and the measured output vector, respectively; and E, $A \in \mathbb{R}^{n \times n}$, $B \in \mathbb{R}^{n \times r}$, $C \in \mathbb{R}^{m \times n}$ are constant matrices, and (E, A) is regular. The dynamical order of the system is $n_0 = \text{rank} E \leq n$.

It follows from Sect. 3.4 that the response of system (4.52) can be expressed as follows:

$$x(t, u, x_0) = x_i(t, x_0) + x_u(t, u),$$

where $x_i(t, x_0)$ is determined by the initial condition $x(0) = x_0$, while $x_u(t, u)$ the control input $u(t)$. Therefore, the output of the system is

$$\tilde{y}(t) = C x_i(t, x_0) + C x_u(t, u).$$

Introduce

$$y(t) = \tilde{y}(t) - Cx_u(t, u) = Cx_i(t, x_0),$$

then $y(t)$ is determined by the system input and output data of system (4.52). Furthermore, since $x_i(t, x_0)$ and $y(t)$ are, respectively, the state response and the output of the following system

$$\begin{cases} E\dot{x} = Ax \\ y = Cx \end{cases}, \tag{4.53}$$

the problem of reconstructing the state $x(t, u, x_0)$ of the system (4.52) from its input and output data $u(t)$ and $\tilde{y}(t)$ is equivalent to the problem of reconstructing the state $x(t, x_0)$ of the system (4.53) by its output data $y(t)$. Therefore, in this whole section, which is concerned with observability only, we need only to consider the system (4.53) which, does not have the input term.

As in Sect. 4.2, standard decomposition also performs an important role in the development. The slow and fast subsystems of system (4.53) are assumed to be

$$\begin{cases} \dot{x}_1 = A_1 x_1 \\ y_1 = C_1 x_1 \end{cases}, \tag{4.54}$$

and

$$\begin{cases} N\dot{x}_2 = x_2 \\ y_2 = C_2 x_2 \end{cases}, \tag{4.55}$$

respectively, and the original system output is given by

$$y = C_1 x_1 + C_2 x_2 = y_1 + y_2.$$

The pair of transformation matrices is assumed to be (P, Q), thus,

$$P^{-1}x = \begin{bmatrix} x_1 \\ x_2 \end{bmatrix},$$

$$QEP = \begin{bmatrix} I & 0 \\ 0 & N \end{bmatrix}, \quad QAP = \begin{bmatrix} A_1 & 0 \\ 0 & I \end{bmatrix}, \quad CP = [C_1 \ C_2].$$

4.3.1 C-Observability

First, let us define the C-observability of system (4.53).

Definition 4.6. The regular system (4.53) is called completely observable, or C-observable, if the initial condition $x(0)$ of the system can be uniquely determined from the output data $y(t)$, $0 \le t \le \infty$.

The definition of C-observability given above is concerned with observing the initial condition $x(0)$ from the output data. Recall that, for a regular descriptor linear system, once the initial condition $x(0)$ is determined, the system state response at any time t can be obtained. Thus, the state at any time point of a C-observable system can be uniquely determined from the system output data.

Clearly, Definition 4.6 reduces to the observability in normal linear system theory when the system (4.53) is normal.

Given a pair of matrices $A \in \mathbb{R}^{n \times n}$ and $C \in \mathbb{R}^{m \times n}$, denote the observability matrix for the matrix pair (A, C) by

$$Q_o[A, C] = \begin{bmatrix} C \\ CA \\ \vdots \\ CA^{n-1} \end{bmatrix},$$

then $Q_o[A_1, C_1]$ and $Q_o[N, C_2]$ are the observability matrices of the slow and fast subsystems (4.24) and (4.25), respectively. Noting that $N^i = 0$, $i \geq h$, we have

$$\begin{bmatrix} C_2 \\ C_2 N \\ \vdots \\ C_2 N^{n_2 - 1} \end{bmatrix} = \begin{bmatrix} C_2 \\ C_2 N \\ \vdots \\ C_2 N^{h-1} \\ 0 \end{bmatrix}.$$

Therefore, in the following we convert that

$$Q_o[N, C_2] = \begin{bmatrix} C_2 \\ C_2 N \\ \vdots \\ C_2 N^{h-1} \end{bmatrix}.$$

In terms of the two matrices $Q_o[A_1, C_1]$ and $Q_o[N, C_2]$, we can state the following lemma.

Lemma 4.8. *Let (4.54) and (4.55) be the slow subsystem and the fast subsystem of the regular descriptor linear system (4.53), respectively. Then $y(t) \equiv 0$, $t \geq 0$, if and only if*

$$x(0) \in \ker Q_o[A_1, C_1] \oplus \ker Q_o[N, C_2]. \tag{4.56}$$

Proof. The state response of system (4.53) is given by

$$
\begin{cases}
x(t) = P \begin{bmatrix} x_1(t) \\ x_2(t) \end{bmatrix} \\
x_1(t) = e^{A_1 t} x_{10}, \quad y_1 = C_1 x_1 \\
x_2(t) = -\sum_{i=1}^{h-1} \delta^{(i-1)}(t) N^i x_{20}, \quad y_2 = C_2 x_2 \\
y = y_1 + y_2 = Cx
\end{cases}
$$

Noting that $x_2(t) = 0$, $t > 0$ and $x_2(t)|_{t=0}$ is infinity when $N x_{20} \neq 0$, further in view of the special forms of y_1 and y_2, we know $y(t) \equiv 0$ if and only if $y_1(t) \equiv 0$ and $y_2(t) \equiv 0$.

If

$$
y_1(t) = C_1 x_1 = C_1 e^{A_1 t} x_{10} \equiv 0,
$$

taking differentials on both sides at $t = 0$ gives

$$
\begin{bmatrix}
C_1 \\
C_1 A_1 \\
\vdots \\
C_1 A_1^{n_1 - 1}
\end{bmatrix} x_{10} = 0,
$$

which is equivalent to

$$
Q_o[A_1, C_1] x_{10} = 0,
$$

i.e.,

$$
x_{10} \in \ker Q_o[A_1, C_1]. \tag{4.57}
$$

On the other hand, noting that

$$
\begin{cases}
y_2(t) = C_2 x_2(t) = -\sum_{i=1}^{h-1} \delta^{(i-1)}(t) C_2 N^i x_{20} \\
y_2(0) = C_2 x_{20}.
\end{cases}
$$

and combining this with the Theorem A.1 in Appendix A, we know that $y_2(t) \equiv 0$ if and only if

$$
C_2 N^i x_2(0) = 0, \quad i = 0, 1, 2, \cdots, h - 1.
$$

This set of equations can be arranged into

$$
Q_o[N, C_2] x_{20} = 0.
$$

Hence,

$$
x_{20} \in \ker Q_o[N, C_2]. \tag{4.58}
$$

Combining (4.57) and (4.58) gives (4.56). Furthermore, since the above process is reversal, the proof is then completed.   □

Based on the above lemma, we can now prove the following main result about C-observability of descriptor linear systems.

Theorem 4.6. *Let (4.54) and (4.55) be the slow subsystem and the fast subsystem of the regular descriptor linear system (4.53), respectively.*

1. *The slow subsystem (4.54) is C-observable if and only if*

$$\mathrm{rank}\, Q_o[A_1, C_1] = n_1, \tag{4.59}$$

   or equivalently,

$$\mathrm{rank} \begin{bmatrix} sI - A_1 \\ C_1 \end{bmatrix} = n_1, \ \forall s \in \mathbb{C}, \ s \ \text{finite}. \tag{4.60}$$

2. *The fast subsystem (4.55) is C-observable if and only if*

$$\mathrm{rank}\, Q_o[N, C_2] = n_2, \tag{4.61}$$

   or equivalently,

$$\mathrm{rank} \begin{bmatrix} N \\ C_2 \end{bmatrix} = n_2. \tag{4.62}$$

3. *The system (4.53) is C-observable if and only if both its slow and fast subsystems (4.54) and (4.55) are C-observable.*

Proof. Proof of conclusion 1. The slow subsystem (4.54) is normal. For normal linear systems, Definition 4.6 reduces to the one for observability of normal linear systems. Thus, the slow subsystem (4.54) is observable if and only if $(A_1, \ C_1)$ is C-observable. It thus follows from the well-known criteria for observability of normal linear systems that the observability of the matrix pair $(A_1, \ C_1)$ is equivalent to (4.59) or (4.60).

Proof of conclusion 2. By definition, C-observability for the fast subsystem (4.55) means that $x_2(0) = 0$ if $y_2(t) \equiv 0$, $t > 0$. As indicated by Lemma 4.8, this is equivalent to

$$\ker Q_o[N, C_2] = \{0\},$$

which is clearly equivalent to (4.61). Therefore, the fast subsystem (4.55) is C-observable if and only if condition (4.61) holds.

Condition (4.61) is clearly equivalent to

$$\mathrm{rank} \begin{bmatrix} C_2^{\mathrm{T}} & N^{\mathrm{T}} C_2^{\mathrm{T}} & \cdots & (N^{\mathrm{T}})^{h-1} C_2^{\mathrm{T}} \end{bmatrix} = n_2,$$

which indicates that the matrix pair $\left(N^{\mathrm{T}}, C_2^{\mathrm{T}}\right)$ is controllable. By the second conclusion of Theorem 4.2, we have rank $\left[N^{\mathrm{T}} \ C_2^{\mathrm{T}}\right] = n_2$, which is equivalent to (4.62).

Proof of conclusion 3. It follows from Lemma 4.8 that the regular descriptor linear system (4.53) is C-observable if and only if

$$\ker Q_o[A_1, C_1] \oplus \ker Q_o[N, C_2] = \{0\}.$$

This is equivalent to

$$\ker Q_o[A_1, C_1] = \{0\}, \ \ker Q_o[N, C_2] = \{0\}. \tag{4.63}$$

These two relations in (4.63) are equivalent to the conditions (4.59) and (4.61), respectively. Thus, the third conclusion holds in view of the first two conclusions of the theorem. □

Example 4.5. Consider the standard decomposition (4.22) for the system (3.39) in Example 3.5. Since $n_1 = n_2 = 2$ and

$$\mathrm{rank}\begin{bmatrix} C_1 \\ C_1 A_1 \end{bmatrix} = 2, \ \mathrm{rank}\begin{bmatrix} C_2 \\ C_2 N \end{bmatrix} = 2,$$

its slow subsystem and fast subsystem are both C-observable. Therefore, the system (4.22) and (3.39) are also C-observable.

Remark 4.2. Since C-observability of a normal linear system is identical with its observability defined in normal systems theory, in the following we will simply call a normal linear system observable when it is C-observable.

4.3.2 R-Observability

It follows from (4.3) to (4.7) that for system (4.53), the reachable subspace $\mathscr{R}_t$ reduces to $\mathscr{H}_t$.

Corresponding to the concept of R-controllability, we have the following R-observability for descriptor linear systems.

Definition 4.7. The regular system (4.53) is R-observable if for arbitrarily given $x(t) \in \mathscr{H}_t$, there exists a time point $t_f \geq t$, such that the state $x(t)$ can be uniquely determined by the system output data $y(\tau)$, $t \leq \tau \leq t_f$.

It follows from the above definition that the regular system (4.53) is R-observable if and only if it is C-observable in the reachable subspace $\mathscr{H}_t$. While C-observability reflects the reconstruction ability of the whole state (impulse terms are included) from the measured output, R-observability characterizes the ability to reconstruct only the reachable state (impulse portions are not included) from the output data. Obviously, from the definition, system (4.53) is R-observable if it is C-observable.

Lemma 4.9. *The fast system*

$$\begin{cases} N\dot{x} = x \\ y = Cx \end{cases},$$

where N is nilpotent, is always R-observable.

Proof. The system responses for this system are $x(t) = 0$, $y(t) = 0$, $\forall t > 0$, the conclusion clearly follows from the definition. □

Since the fast subsystem (4.55) of system (4.53) is always R-observable, the R-observability of the system (4.53) must be determined completely by that of its slow subsystem (4.54). This is confirmed by the following theorem.

Theorem 4.7. *The regular descriptor linear system (4.53) with slow subsystem (4.54) and fast subsystem (4.55) is R-observable if and only if its slow subsystem (4.54) is observable.*

Proof. According to the state space representation, any reachable state has the form

$$\begin{cases} x(t) = P \begin{bmatrix} x_1(t) \\ x_2(t) \end{bmatrix} \\ \dot{x}_1(t) = A_1 x_1(t) \\ x_2(t) = 0, \quad t > 0 \\ y = y_1 + y_2 = C_1 x_1 \end{cases}.$$

Thus, reconstructing the reachable state $x(t)$ from $y(t)$ is equivalent to recovering $x_1(t)$ from $y(t)$ only. This is equivalent to the observability of the matrix pair (A_1, C_1). Therefore, system (4.53) is R-observable if and only if its slow subsystem (4.54) is observable. □

Example 4.6. Consider again the system (4.34). Since

$$A_1 = \begin{bmatrix} -1 & -1 \\ 1 & 0 \end{bmatrix}, \quad C_1 = \begin{bmatrix} 0 & 1 \end{bmatrix},$$

and (A_1, C_1) is observable, the system is R-observable.

4.3.3 I-Observability and S-Observability

The I-observability of system (4.53) is concerned with observing the impulse terms in the system state response from the output data of the system. Since the state response of the fast subsystem (4.55) satisfies

$$x_2(0) = \begin{cases} \infty, & N x_{20} \neq 0 \\ 0, & \text{otherwise} \end{cases},$$

the output at $t = 0$, that is, $y_2(0)$, is either zero or infinity. Further noting that

$$x_2(t) = 0, \ t > 0,$$

we have

$$y_2(t) = Cx_2(t) \equiv 0 \Leftrightarrow y_2(0) = 0.$$

With these observations, we can give the following definition for I-observability of system (4.53).

Definition 4.8. The regular descriptor linear system (4.53) with the slow subsystem (4.54) and the fast subsystem (4.55) is called impulse observable, or simply, I-observable, if $y_2(0) = 0$ implies $x_2(t)|_{t=0} = 0$.

Impulse observability guarantees the ability to uniquely determine the impulse behavior in $x(t)$ from information of the impulse behavior in output. C-observability and R-observability are on the finite value terms in the state response, while impulse observability focuses on the impulse terms that take infinite values.

Noting that a normal linear system does not have impulse behavior, by the above definition we immediately have the following lemma.

Lemma 4.10. *The slow system (or normal system)*

$$\begin{cases} \dot{x} = Ax \\ y = Cx \end{cases}$$

is always I-observable.

Noting that

$$Q_o[N, C_2]N = \begin{bmatrix} C_2 N \\ C_2 N^2 \\ \vdots \\ C_2 N^{h-1} \\ 0 \end{bmatrix},$$

we have

$$\ker(Q_o[N, C_2]N) = \ker \begin{bmatrix} C_2 N \\ C_2 N^2 \\ \vdots \\ C_2 N^{h-1} \end{bmatrix}.$$

To present the main result in this section about I-observability, we need the following lemma.

Lemma 4.11. *For the fast system (4.55), $y_2(0) = 0$ if and only if*

$$x_2(0) \in \ker(Q_o[N, C_2]N). \tag{4.64}$$

Proof. Since

$$x_2(t) = -\sum_{i=1}^{h-1} \delta^{(i-1)}(t) N^i x_{20},$$

we have

$$y_2(t) = -\sum_{i=1}^{h-1} \delta^{(i-1)}(t) C_2 N^i x_{20}.$$

Thus, by using Theorem A.1 in Appendix A, $y(t)|_{t=0} = 0$ if and only if

$$C_2 N^i x_{20} = 0, \ i = 1, 2, \ldots, h-1,$$

which can be equivalently arranged into the form of ,

$$\begin{bmatrix} C_2 N \\ C_2 N^2 \\ \vdots \\ C_2 N^{h-1} \end{bmatrix} x_{20} = 0.$$

This states that (4.64) holds. □

Based on the above lemma, we can now show the following main result about I-observability.

Theorem 4.8. *Let (4.54) and (4.55) be the slow subsystem and the fast subsystem of the regular descriptor linear system (4.53), respectively.*

1. *System (4.53) is I-observable if and only if its fast subsystem (4.55) is I-observable.*
2. *The fast subsystem (4.55) is I-observable if and only if one of the following conditions holds:*

   (a) $\ker(Q_o[N, C_2]N) = \ker N$;
   (b) $\ker Q_o[N, C_2] \cap \text{Image} N = \{0\}$;
   (c) $\ker N \cap \text{Image} N \cap \ker C_2 = \{0\}$.

Proof. Note that, when $x_1(0) = x_{10} = 0$,

$$\begin{cases} x(t) = P\begin{bmatrix} x_1(t) \\ x_2(t) \end{bmatrix} = P\begin{bmatrix} 0 \\ x_2(t) \end{bmatrix} \\ y(t) = C_1 x_1(t) + C_2 x_2(t) = C_2 x_2(t) = y_2(t) \end{cases}.$$

From this, we know that the first conclusion holds. In the following, let us show the second conclusion.

Proof of (a). Again recalling that

$$x_2(t) = -\sum_{i=1}^{h-1} \delta^{(i-1)}(t) N^i x_{20},$$

we know that $x_2(t)|_{t=0} = 0$ if and only if

$$N x_2(0) = 0.$$

This is equivalent to

$$x_2(0) \in \ker N.$$

Combining the above relation with Lemma 4.11 immediately gives the equivalent condition (a).

Proof of $(a) \Leftrightarrow (b)$. Assume that (a) is true. Then for any

$$\alpha \in \ker Q_o[N, C_2] \cap \text{Image} N,$$

there exists a vector β such that $\alpha = N\beta$. Moreover,

$$\alpha = N\beta \in \ker Q_o[N, C_2].$$

This indicates that

$$\beta \in \ker(Q_o[N, C_2]N) = \ker N,$$

i.e., $N\beta = 0$. Thus, $\alpha = N\beta = 0$. Therefore, condition (b) holds, and $(a) \Rightarrow (b)$ is shown. Now let us show $(b) \Rightarrow (a)$.

For any $\alpha \in \ker(Q_o[N, C_2]N)$, there holds

$$N\alpha \in \ker Q_o[N, C_2].$$

Furthermore, noting that $N\alpha \in \text{Image} N$, we have, when condition (b) is true,

$$N\alpha \in \ker Q_o[N, C_2] \cap \text{Image} N = \{0\}.$$

Thus $N\alpha = 0$, which gives $\alpha \in \ker N$. Therefore,

$$\ker(Q_o[N, C_2]N) \subset \ker N.$$

Furthermore, since

$$\ker N \subset \ker(Q_o[N, C_2]N)$$

holds due to the definition of $\ker(Q_o[N, C_2]N)$, condition (a) is valid.

Proof of $(b) \Leftrightarrow (c)$. Let condition (b) hold. Taking an arbitrary

$$\alpha \in \ker N \cap \ker C_2 \cap \text{Image} N.$$

Since $\alpha \in \ker N \cap \ker C_2$, there hold

$$N\alpha = 0,$$
$$C_2\alpha = 0,$$
$$C_2 N\alpha = 0,$$
$$\vdots$$
$$C_2 N^{h-1}\alpha = 0.$$

These imply

$$\alpha \in \ker Q_o[N, C_2].$$

It further follows from $\alpha \in \text{Image} N$ that there exists a vector β such that $\alpha = N\beta$. Therefore,

$$\alpha = N\beta \in \ker Q_o[N, C_2] \cap \text{Image} N = \{0\}.$$

This shows $(b) \Rightarrow (c)$.

Conversely, let (c) hold. For any

$$\alpha \in \ker Q_o[N, C_2] \cap \text{Image} N,$$

there exists a vector β such that $\alpha = N\beta$ and

$$C_2 N^i \beta = 0, \quad i = 1, 2, \ldots, h-1.$$

Therefore, we can get

$$N^{h-1}\beta \in \ker N \cap \text{Image} N \cap \ker C_2 = \{0\},$$

which shows that $N^{h-1}\beta = 0$. Then, there exist

$$N^{h-2}\beta \in \ker N \cap \text{Image} N \cap \ker C_2 = \{0\}.$$

Thus, $N^{h-2}\beta = 0$. Following this procedure, we can derive in the end that

$$N\beta \in \ker N \cap \text{Image} N \cap \ker C_2 = \{0\},$$

and $\alpha = N\beta = 0$. This implies

$$\ker Q_o[N, C_2] \cap \text{Image} N = \{0\},$$

and condition (b) holds. This shows $(c) \Rightarrow (b)$. $\square$

It follows from Theorems 4.6 and 4.8 that

> the descriptor linear system (4.53) is C-observable
> $\Rightarrow \operatorname{rank} Q_o[N, C_2] = n_2$
> $\Leftrightarrow \ker Q_o[N, C_2] = \{0\}$
> $\Rightarrow \ker Q_o[N, C_2] \cap \operatorname{Image} N = \{0\}$
> $\Rightarrow$ the descriptor linear system (4.53) is I-observable,

that is, a descriptor linear system is I-observable if it is C-observable, but the inverse is not true.

By now we have introduced three types of observabilities, namely, C-observability, R-observability, and I-observability. Corresponding to the concept of S-controllability, we also have the following concept of S-observability.

Definition 4.9. The regular descriptor linear system (4.53) is called S-observable if it is both R-observable and I-observable.

The relations among the different types of observabilities for descriptor linear systems are clearly seen to be as follows:

$$\text{C-observability}$$
$$\Downarrow$$
$$\text{S-observability} \Longleftrightarrow \text{R-observability} + \text{I-observability}$$

4.4 The Dual Principle

As in the case for standard normal linear systems, a so-called dual principle holds, which reveals the close relation between controllabilities and observabilities.

4.4.1 The Dual System

To introduce the dual principle for descriptor linear systems, like the case for normal linear systems, let us first introduce the dual system of a general descriptor linear system in the form of

$$\begin{cases} E\dot{x} = Ax + Bu \\ y = Cx \end{cases}, \tag{4.65}$$

where the variables are as stated before. A pair of slow and fast subsystems of the system (4.65) are

$$\begin{cases} \dot{x}_1 = A_1 x_1 + B_1 u \\ y_1 = C_1 x_1 \end{cases} \tag{4.66}$$

and

$$\begin{cases} N\dot{x}_2 = x_2 + B_2 u \\ y_2 = C_2 x_2 \end{cases}.$$ (4.67)

Definition 4.10. The following system

$$\begin{cases} E^T \dot{z} = A^T z + C^T v \\ w = B^T z \end{cases}$$ (4.68)

is called the dual system of the descriptor linear system (4.65).

According to the above definition, the dual system of system (E, A, B, C) is (E^T, A^T, C^T, B^T).

4.4.2 The Dual Principle

The following dual principle reveals the relation between the controllabilities (observabilities) of system (4.65) and the observabilities (controllabilities) of its dual system (4.68).

Theorem 4.9. *Let (4.68) be the dual system of the regular descriptor linear system (4.65).*

1. *System (4.65) is C-controllable (C-observable) if and only if its dual system (4.68) is C-observable (C-controllable).*
2. *System (4.65) is R-controllable (R-observable) if and only if its dual system (4.68) is R-observable (R-controllable).*
3. *System (4.65) is I-controllable (I-observable) if and only if its dual system (4.68) is I-observable (I-controllable).*
4. *System (4.65) is S-controllable (S-observable) if and only if its dual system (4.68) is S-observable (S-controllable).*

The dual principle plays an important role in system theory. Because of this principle, there is no need to study the controllabilities and observabilities separately. For example, once we obtain certain criterion for a type of controllability, we can then apply it to the dual system, and as a result, the corresponding criterion for the corresponding type of observability can be easily obtained. Such an idea is adopted in the following two sections.

To prove the above dual principle, let us first state the following simple fact without a proof.

Lemma 4.12. *Consider the regular descriptor linear system (4.65) and its dual system (4.68). If (4.66) and (4.67) are slow and fast subsystems of system (4.65), respectively, then the following systems*

$$\begin{cases} \dot{z}_1 = A_1^T z_1 + C_1^T v \\ w_1 = B_1^T x_1 \end{cases} \tag{4.69}$$

and

$$\begin{cases} N_2^T \dot{z}_2 = z_2 + C_2^T v \\ w_2 = B_2^T z_2 \end{cases} \tag{4.70}$$

are slow and fast subsystems of the dual system (4.68), respectively.

Proof of Theorem 4.9

It follows from Theorems 4.2 and 4.6 that the following equivalence relations hold:

The system (4.65) is C-controllable

$\Leftrightarrow \begin{cases} \text{rank} Q_c [A_1, \ B_1] = n_1 \\ \text{rank} Q_c [N, \ B_2] = n_2 \end{cases}$

$\Leftrightarrow \begin{cases} \text{rank} \left[B_1 \ A_1 B_1 \ \cdots \ A_1^{n_1-1} B_1 \right] = n_1 \\ \text{rank} \left[B_2 \ N B_2 \ \cdots \ N^{h-1} B_2 \right] = n_2 \end{cases}$

$\Leftrightarrow \text{rank} \begin{bmatrix} B_1^T \\ B_1^T A_1^T \\ \vdots \\ B_1^T \left(A_1^T \right)^{n_1-1} \end{bmatrix} = n_1 \text{ and rank} \begin{bmatrix} B_2^T \\ B_2^T N^T \\ \vdots \\ B_2^T \left(N^T \right)^{h-1} \end{bmatrix} = n_2$

$\Leftrightarrow \begin{cases} \text{rank} Q_o \left[A_1^T, \ B_1^T \right] = n_1 \\ \text{rank} Q_o \left[N^T, \ B_2^T \right] = n_2 \end{cases}$

$\Leftrightarrow$ the slow and fast subsystems (4.69) and (4.70) are C-observable

$\Leftrightarrow$ The system (4.68) is C-observable.

Thus, the first conclusion is shown. The second one is an obvious corollary of the first one.

It follows from Theorems 4.4 and 4.8 that the following equivalent relations hold:

The system (4.65) is I-controllable

$\Leftrightarrow \ker N + \text{Image} Q_c [N, B_2] = \mathbb{R}^{n_2}$

$\Leftrightarrow \ker N + \text{Image} \left[B_2, N B_2, \cdots, N^{h-1} B_2 \right] = \mathbb{R}^{n_2}$

$\Leftrightarrow \text{Image} N^T \cap \ker \begin{bmatrix} B_2^T \\ B_2^T N^T \\ \vdots \\ B_2^T \left(N^T \right)^{h-1} \end{bmatrix} = \{0\}$

$\Leftrightarrow \text{Image}N^{\mathrm{T}} \cap \ker Q_o \left[N^{\mathrm{T}}, B_2^{\mathrm{T}} \right] = \{0\}$

$\Leftrightarrow$ The fast subsystem (4.70) is I-observable

$\Leftrightarrow$ The system (4.68) is I-observable.

This proves the third conclusion. The fourth one is an immediate corollary of the second and the third conclusions.   □

4.5 Direct Criteria

In the preceding Sects. 4.2 and 4.3, we introduced the various types of controllabilities and observabilities of descriptor linear systems, and have given some basic criteria based on the standard decomposition of descriptor linear systems. In this section, we further provide some direct criteria for controllabilities and observabilities, which are based on the original system coefficient matrices only.

In this section, we stress on the case of controllabilities. Results for observabilities are easily obtained by using the dual principle introduced in the preceding section.

Again, we still study the descriptor linear system in the form of

$$\begin{cases} E\dot{x} = Ax + Bu \\ y = Cx \end{cases}, \tag{4.71}$$

where $x \in \mathbb{R}^n$, $u \in \mathbb{R}^r$, and $y \in \mathbb{R}^m$ are the state vector, the control input vector, and the measured output vector, respectively; and E, $A \in \mathbb{R}^{n \times n}$, $B \in \mathbb{R}^{n \times r}$, $C \in \mathbb{R}^{m \times n}$ are constant matrices. It is assumed that the system has the following slow and fast subsystems

$$\begin{cases} \dot{x}_1 = A_1 x_1 + B_1 u \\ y_1 = C_1 x_1 \end{cases} \tag{4.72}$$

and

$$\begin{cases} N\dot{x}_2 = x_2 + B_2 u \\ y_2 = C_2 x_2 \end{cases}. \tag{4.73}$$

The relations between the coefficient matrices of the two systems are given by

$$\begin{cases} QEP = \text{diag}(I_{n_1}, N) \\ QAP = \text{diag}(A_1, I_{n_2}) \\ QB = \begin{bmatrix} B_1 \\ B_2 \end{bmatrix} \\ CP = [C_1 \ C_2] \end{cases}, \tag{4.74}$$

where the matrices Q and P are the left and right transformation matrices, respectively.

4.5.1 C-Controllability and C-Observability

Regarding the C-controllability of the regular descriptor linear system (4.71) as well as its slow subsystem (4.72) and fast subsystem (4.73), we have the following theorem.

Theorem 4.10. *Consider the regular descriptor linear system (4.71) with its slow subsystem (4.72) and fast subsystem (4.73).*

1. The slow subsystem (4.72) is C-controllable if and only if

$$\text{rank}[sE - A\ B] = n, \quad \forall s \in \mathbb{C}, \ s \text{ finite.} \tag{4.75}$$

2. The fast subsystem (4.73) is C-controllable if and only if

$$\text{rank}\begin{bmatrix} E & B \end{bmatrix} = n. \tag{4.76}$$

3. System (4.71) is C-controllable if and only if (4.75) and (4.76) hold, or

$$\text{rank}[\alpha E - \beta A\ B] = n, \quad \forall\, (\alpha, \beta) \in \mathbb{C}^2 \setminus \{(0,0)\}. \tag{4.77}$$

Proof. Proof of conclusion 1. It follows from Theorem 4.2 that the slow subsystem (4.72) is C-controllable if and only if

$$\text{rank}\,[sI - A_1\ B_1] = n_1, \quad \forall s \in \mathbb{C}, \ s \text{ finite.} \tag{4.78}$$

Noticing that

$$\text{rank}[sE - A\ B] = \text{rank}[sQEP - QAP\ QB]$$

$$= \text{rank}\begin{bmatrix} sI - A_1 & 0 & B_1 \\ 0 & sN - I & B_2 \end{bmatrix},$$

and $sN - I$ is nonsingular for any finite $s \in \mathbb{C}$, we have

$$\text{rank}[sE - A\ B] = n_2 + \text{rank}[sI - A_1\ B_1],$$

which indicates that (4.78) holds if and only if (4.75) is true.
Proof of conclusion 2. It again follows from Theorem 4.2 that the fast subsystem (4.73) is C-controllable if and only if

$$\text{rank}\,[N\ B_2] = n_2. \tag{4.79}$$

Furthermore, note that

$$\text{rank}\begin{bmatrix} E & B \end{bmatrix} = \text{rank}\begin{bmatrix} QEP & QB \end{bmatrix}$$

$$= \text{rank}\begin{bmatrix} I & 0 & B_1 \\ 0 & N & B_2 \end{bmatrix}$$

$$= n_1 + \text{rank}\begin{bmatrix} N & B_2 \end{bmatrix}.$$

This clearly indicates that (4.79) holds if and only if (4.76) is true.

Proof of conclusion 3. Suppose (4.77) holds. Letting $(\alpha, \beta) = (s, 1)$ and $(\alpha, \beta) = (1, 0)$ gives (4.75) and (4.76), respectively. Therefore, the two subsystems (4.72) and (4.73) are C-controllable, and so is the original system (4.71) following from Theorem 4.2.

Conversely, suppose that system (4.71) is C-controllable, then conditions (4.75) and (4.76) hold. It follows from (4.76) that for any $\alpha \neq 0$, there holds

$$\text{rank}\begin{bmatrix} \alpha E & B \end{bmatrix} = n.$$

This is condition (4.77) for the case of $\beta = 0$.

When $\beta \neq 0$, taking $s = \alpha/\beta$, in (4.75), yields

$$\text{rank}\begin{bmatrix} \dfrac{\alpha}{\beta}E - A & B \end{bmatrix} = n, \quad \forall \alpha, \beta \in \mathbb{C}, \ \beta \neq 0,$$

which is clearly equivalent to (4.77) when $\beta \neq 0$. Combining the above two aspects completes the proof of conclusion 3. $\square$

Example 4.7. Consider again the system in Example 3.1,

$$\begin{bmatrix} 1 & 0 & 0 \\ 0 & 1 & 0 \\ 0 & 0 & 0 \end{bmatrix} \dot{x} = \begin{bmatrix} 0 & 1 & 0 \\ 0 & 0 & 1 \\ 0 & 0 & -1 \end{bmatrix} x + \begin{bmatrix} 0 & 0 \\ 1 & 0 \\ 0 & 1 \end{bmatrix} u. \tag{4.80}$$

For this system, we have

$$\text{rank}\begin{bmatrix} sE - A & B \end{bmatrix} = \text{rank}\begin{bmatrix} s & -1 & 0 & 0 & 0 \\ 0 & s & -1 & 1 & 0 \\ 0 & 0 & 1 & 0 & 1 \end{bmatrix}$$

$$= 3, \ \forall s \in \mathbb{C}, \ s \text{ finite},$$

and

$$\text{rank}\begin{bmatrix} E & B \end{bmatrix} = \text{rank}\begin{bmatrix} 1 & 0 & 0 & 0 & 0 \\ 0 & 1 & 0 & 1 & 0 \\ 0 & 0 & 0 & 0 & 1 \end{bmatrix} = 3.$$

Therefore, by Theorem 4.10 the above system (4.80) is C-controllable.

Using the above Theorem 4.10 and the dual principle, the following result about C-observability can be easily derived.

Theorem 4.11. *Consider the regular descriptor linear system (4.71) with slow subsystem (4.72) and fast subsystem (4.73).*

1. *The slow subsystem (4.72) is C-observable if and only if*

$$\text{rank}\begin{bmatrix} sE - A \\ C \end{bmatrix} = n, \ \forall s \in \mathbb{C}, \ s \ \text{finite}. \tag{4.81}$$

2. *The fast subsystem (4.73) is C-observable if and only if*

$$\text{rank}\begin{bmatrix} E \\ C \end{bmatrix} = n. \tag{4.82}$$

3. *System (4.71) is C-observable if and only if conditions (4.81) and (4.82) hold, or*

$$\text{rank}\begin{bmatrix} \alpha E - \beta A \\ C \end{bmatrix} = n, \ \forall \, (\alpha, \beta) \in \mathbb{C}^2 \backslash \{(0,0)\}.$$

Example 4.8. Consider the rolling drive system in Example 1.5. The system model is

$$\begin{cases} \begin{bmatrix} I_4 & 0 & 0 \\ 0 & I_4 & 0 \\ 0 & 0 & 0 \end{bmatrix} \dot{x} = \begin{bmatrix} 0 & I_4 & 0 \\ -K & -D & J \\ H & G & 0 \end{bmatrix} x + Bu \\ y = Cx \end{cases} \tag{4.83}$$

where

$$D = K = \begin{bmatrix} 0 & 0 & 0 & 0 \\ 0 & 1 & -1 & 0 \\ 0 & -1 & 2 & -1 \\ 0 & 0 & -1 & 1 \end{bmatrix}, \ J = \begin{bmatrix} 0 & 0 \\ 0 & 1 \\ 0 & -1 \\ 1 & 0 \end{bmatrix},$$

$$G = \begin{bmatrix} 0 & 0 & 0 & 0 \\ 0 & 0 & 0 & 1 \end{bmatrix}, \ H = \begin{bmatrix} 0 & -1 & 1 & 0 \\ -1 & 0 & 0 & 0 \end{bmatrix},$$

$$B^{\text{T}} = \begin{bmatrix} 0 & 0 & 0 & 0 \vdots 0 & 1 & 0 & 0 \vdots 0 & 0 \end{bmatrix}.$$

In the following, we treat two different choices of the observation matrix C.

Case I:

$$C = \begin{bmatrix} 0 & 0 & 1 & 0 \vdots 0 & 0 & 0 & 0 \vdots 0 & 0 \\ 0 & 0 & 0 & 1 \vdots 0 & 0 & 0 & 0 \vdots 0 & 0 \end{bmatrix}.$$

In this case, it can be easily checked that

$$\text{rank} \begin{bmatrix} sE - A \\ C \end{bmatrix} = 10, \ \forall s \in \mathbb{C}, \ s \text{ finite,}$$

and

$$\text{rank} \begin{bmatrix} E \\ C \end{bmatrix} = 8 < 10.$$

This indicates that its slow subsystem is C-observable while its fast one is not. Hence, the rolling ring drive system (4.83) is not C-observable.

Case II:

$$C = \begin{bmatrix} 0 & 0 & 1 & 0 & \vdots & 0 & 0 & 0 & 0 & \vdots & 0 & 1 \\ 0 & 0 & 0 & 1 & \vdots & 0 & 0 & 0 & 0 & \vdots & 1 & 0 \end{bmatrix}. \tag{4.84}$$

In this case, we have,

$$\text{rank} \begin{bmatrix} sE - A \\ C \end{bmatrix} = 10, \ \forall s \in \mathbb{C}, \ s \text{ finite,}$$

and

$$\text{rank} \begin{bmatrix} E \\ C \end{bmatrix} = 10.$$

Therefore, this system with the observer matrix (4.84) is C-observable by Theorem 4.11.

4.5.2 R-Controllability and R-Observability

For R-controllability of descriptor linear systems, we have the following result.

Theorem 4.12. *The regular descriptor system (4.71) is R-controllable if and only if*

$$\text{rank}[sE - A \ B] = n, \ \forall s \in \mathbb{C}, \ s \text{ finite.} \tag{4.85}$$

Proof. The descriptor system (4.71) is R-controllable if and only if its slow subsystem (4.72) is controllable, while the subsystem (4.72) is controllable if and only if (4.85) holds according to Theorem 4.10. Thus, the conclusion holds. □

Example 4.9. Consider the following system

$$\begin{bmatrix} 1 & 1 & 0 & 0 \\ 0 & 0 & 0 & 0 \\ 0 & 0 & 0 & 0 \\ 1 & 0 & 0 & 0 \end{bmatrix} \dot{x} = \begin{bmatrix} 0 & -1 & 1 & 1 \\ 0 & 0 & 0 & 1 \\ 0 & 0 & 1 & 1 \\ -1 & -1 & 0 & 0 \end{bmatrix} Ax + \begin{bmatrix} 0 \\ 0 \\ -1 \\ 1 \end{bmatrix} Bu. \tag{4.86}$$

It can be easily verified that

$$\text{rank} \, [sE - A \; B] = \begin{bmatrix} s-1 & 2 & -1 & -1 \\ 0 & 0 & 0 & -1 \\ 0 & 0 & -1 & -1 \\ 2 & 1 & 0 & 0 \end{bmatrix} = 4, \; \forall s \in \mathbb{C}, \; s \text{ finite}.$$

Thus, this system is R-controllable. However,

$$\text{rank} \, [E \; B] = \begin{bmatrix} 1 & 1 & 0 & 0 & 0 \\ 0 & 0 & 0 & 0 & 0 \\ 0 & 0 & 0 & 0 & -1 \\ 1 & 0 & 0 & 0 & 1 \end{bmatrix} = 3 < 4.$$

The system is not C-controllable.

Based on Theorem 4.12 and the dual principle, the following result about R-observability can be easily shown.

Theorem 4.13. *The regular descriptor system (4.71) is R-observable if and only if*

$$\text{rank} \begin{bmatrix} sE - A \\ C \end{bmatrix} = n, \; \forall s \in \mathbb{C}, \; s \text{ finite}. \tag{4.87}$$

Combining the above Theorems 4.12 and 4.13 and Definition 2.3 immediately gives the following corollary.

Corollary 4.2. *The regular descriptor system (4.71) is R-controllable (R-observable) if and only if it does not have input (output) transmission zeros.*

Example 4.10. It is seen from Examples 4.6 and 4.8 that systems (4.34) and (4.83) are both R-observable.

4.5.3 I-Controllability and I-Observability

First let us prove the following theorem about I-controllability.

Theorem 4.14. *The regular descriptor linear system (4.71) is I-controllable if and only if*

$$\text{rank} \, [E \; AV_\infty \; B] = n \tag{4.88}$$

holds for an arbitrary matrix $V_\infty \in \mathbb{R}^{n \times (n - n_0)}$ *whose columns span* $\ker E$, *or equivalently,*

$$\text{rank} \begin{bmatrix} E & 0 & 0 \\ A & E & B \end{bmatrix} = n + \text{rank} E. \tag{4.89}$$

Proof. Let

$$\tilde{V}_\infty = \begin{bmatrix} \tilde{V}^1_\infty \\ \tilde{V}^2_\infty \end{bmatrix} = P^{-1} V_\infty,$$

then from $E V_\infty = 0$ we can obtain

$$(QEP)\,\tilde{V}_\infty = \begin{bmatrix} I & 0 \\ 0 & N \end{bmatrix} \begin{bmatrix} \tilde{V}^1_\infty \\ \tilde{V}^2_\infty \end{bmatrix} = 0,$$

that is,

$$\tilde{V}^1_\infty = 0, \quad N \tilde{V}^2_\infty = 0. \tag{4.90}$$

Using (4.90) and the relations in (4.74), we have

$$\mathrm{rank}\,[E \quad AV_\infty \quad B] = \mathrm{rank}\,[QEP \quad QAV_\infty \quad QB]$$

$$= \mathrm{rank}\,[QEP \quad QAP\tilde{V}_\infty \quad QB]$$

$$= \mathrm{rank}\begin{bmatrix} I & 0 & 0 & B_1 \\ 0 & N & \tilde{V}^2_\infty & B_2 \end{bmatrix}$$

$$= \mathrm{rank}\begin{bmatrix} I & 0 & 0 & 0 \\ 0 & N & \tilde{V}^2_\infty & B_2 \end{bmatrix}$$

$$= n_1 + \mathrm{rank}\begin{bmatrix} N & \tilde{V}^2_\infty & B_2 \end{bmatrix}.$$

Using the above relation, the third condition in Theorem 4.4, and the fact that the columns of $\tilde{V}^2_\infty$ span $\ker N$, we have

$$\text{system (4.71) is I-controllable}$$
$$\Leftrightarrow \mathrm{Image}\,N + \ker N + \mathrm{Image}\,B_2 = \mathbb{R}^{n_2}$$
$$\Leftrightarrow \mathrm{rank}\begin{bmatrix} N & \tilde{V}^2_\infty & B_2 \end{bmatrix} = n_2$$
$$\Leftrightarrow \mathrm{rank}\,[E \quad AV_\infty \quad B] = n.$$

This shows that (4.88) is a necessary and sufficient condition for the I-controllability of the regular descriptor linear system (4.71).

Now let us show the equivalence of condition (4.88) and (4.89).

Let V be such a matrix that $[V_\infty \ V]$ is nonsingular. Noting that

$$E\,[V_\infty \ V] = [0 \ EV],$$

we have

$$\mathrm{rank}\,(EV) = \mathrm{rank}\,E.$$

Therefore,

$$\text{rank} \begin{bmatrix} E & 0 & 0 \\ A & E & B \end{bmatrix}$$

$$= \text{rank} \left\{ \begin{bmatrix} E & 0 & 0 \\ A & E & B \end{bmatrix} \text{diag}\left([V_\infty \ V], I, I\right) \right\}$$

$$= \text{rank} \begin{bmatrix} EV_\infty & EV & 0 & 0 \\ AV_\infty & AV & E & B \end{bmatrix}$$

$$= \text{rank} \begin{bmatrix} EV & 0 & 0 & 0 \\ AV & E & AV_\infty & B \end{bmatrix}$$

$$= \text{rank}\, E + \text{rank} \begin{bmatrix} E & AV_\infty & B \end{bmatrix}.$$

With this, we complete the proof of the theorem. □

Example 4.11. Consider the system (4.80) in Example 4.7. We have

$$\text{rank} \begin{bmatrix} E & 0 & 0 \\ A & E & B \end{bmatrix}$$

$$= \text{rank} \begin{bmatrix} 1 & 0 & 0 & 0 & 0 & 0 & 0 & 0 \\ 0 & 1 & 0 & 0 & 0 & 0 & 0 & 0 \\ 0 & 0 & 0 & 0 & 0 & 0 & 0 & 0 \\ 0 & 1 & 0 & 1 & 0 & 0 & 0 & 0 \\ 0 & 0 & 1 & 0 & 1 & 0 & 1 & 0 \\ 0 & 0 & -1 & 0 & 0 & 0 & 0 & 1 \end{bmatrix}$$

$$= 5 = n + \text{rank}\, E.$$

Thus, the system is I-controllable. Moreover, the C-controllability of the system has been proven in Example 4.7.

Example 4.12. Consider the system (4.86). It has been shown in Example 4.9 that the system is not C-controllable. However since

$$\text{rank} \begin{bmatrix} E & 0 & 0 \\ A & E & B \end{bmatrix} = 6 = n + \text{rank}\, E,$$

the system is I-controllable.

Based on the above Theorem 4.14 and the dual principle, the following result for I-observability holds.

Theorem 4.15. *The regular descriptor linear system (4.71) is I-observable if and only if*

$$\text{rank} \begin{bmatrix} E & A \\ 0 & E \\ 0 & C \end{bmatrix} = n + \text{rank} E, \tag{4.91}$$

or equivalently

$$\text{rank} \begin{bmatrix} E \\ T_\infty^T A \\ C \end{bmatrix} = n, \tag{4.92}$$

where $T_\infty \in \mathbb{R}^{n \times (n - n_0)}$ is a matrix whose columns span $\ker E^T$.

Example 4.13. Consider a descriptor linear system in the form of (4.71) with the following parameters

$$E = \begin{bmatrix} 2 & 1 & 0 & 0 \\ 2 & 0 & -1 & 0 \\ 0 & 0 & 0 & 0 \\ 2 & 1 & -1 & 0 \end{bmatrix}, \quad A = \begin{bmatrix} 0 & 0 & 1 & 1 \\ -1 & 1 & 0 & 1 \\ 1 & 0 & 0 & 2 \\ 0 & 1 & 1 & 3 \end{bmatrix},$$

$$B^T = \begin{bmatrix} 0 & 0 & -1 & -1 \end{bmatrix}, \quad C = \begin{bmatrix} 0 & 1 & 0 & 0 \end{bmatrix}.$$

Since

$$\text{rank} \begin{bmatrix} E \\ C \end{bmatrix} = 3 < 4, \quad \text{rank} \begin{bmatrix} sE - A \\ C \end{bmatrix} = 4, \ \forall s \in \mathbb{C}, \ s \text{ finite},$$

this system is R-observable but not C-observable. However, since

$$\text{rank} \begin{bmatrix} E & A \\ 0 & E \\ 0 & C \end{bmatrix} = \text{rank} \begin{bmatrix} 2 & 1 & 0 & 0 & 0 & 0 & 1 & 1 \\ 2 & 0 & -1 & 0 & -1 & 1 & 0 & 1 \\ 0 & 0 & 0 & 0 & 1 & 0 & 0 & 2 \\ 2 & 1 & -1 & 0 & 0 & 1 & 1 & 3 \\ 0 & 0 & 0 & 0 & 2 & 1 & 0 & 0 \\ 0 & 0 & 0 & 0 & 2 & 0 & -1 & 0 \\ 0 & 0 & 0 & 0 & 0 & 0 & 0 & 0 \\ 0 & 0 & 0 & 0 & 2 & 1 & -1 & 0 \\ 0 & 0 & 0 & 0 & 0 & 1 & 0 & 0 \end{bmatrix}$$

$$= 7 = n + \text{rank} E,$$

the system is I-observable according to Theorem 4.15.

4.5.4 S-Controllability and S-Observability

It follows from the definitions of S-controllability and S-observability and the previous results for R- and I-controllability and observability that the following results clearly hold.

Theorem 4.16. *The regular descriptor linear system (4.71) is S-controllable if and only if one of the following statements is true:*

1. *Conditions (4.75) and (4.89) hold.*
2. *Conditions (4.75) and (4.88) hold.*

Theorem 4.17. *The regular descriptor linear system (4.71) is S-observable if and only if one of the following statements is true:*

1. *Conditions (4.87) and (4.91) hold.*
2. *Conditions (4.87) and (4.92) hold.*

4.6 Criteria Based on Equivalent Forms

In Sects. 4.2 and 4.3, we have given criteria for controllabilities and observabilities of descriptor linear systems based on the standard decomposition form, and in Sect. 4.5 some criteria based on directly the original system data. In this section, we further look at some other criteria for controllabilities and observabilities of descriptor linear systems based on the equivalent canonical dynamics decomposition form, the inverse form and the form for derivative feedback.

It follows from the definitions that the various types of controllabilities and observabilities are concerned with the state trajectory and its observation, and thus are invariant under nonsingular and constant coordinate transformations. Therefore, all these types of controllabilities and observabilities remain the same under restricted system equivalence.

The descriptor linear system to be studied in this section still takes the form of (4.71)

4.6.1 Criteria Based on the Dynamics Decomposition Form

Suppose $\text{rank} E = n_0$, then it follows from Theorem 2.2 that there exist nonsingular matrices Q and P such that

$$\tilde{E} = QEP = \begin{bmatrix} I_{n_0} & 0 \\ 0 & 0 \end{bmatrix}.$$

Decompose the state vector of the system as follows:

$$\tilde{x} = P^{-1}x = \begin{bmatrix} x_1 \\ x_2 \end{bmatrix}, \ x_1 \in \mathbb{R}^{n_0}, \ x_2 \in \mathbb{R}^{n-n_0},$$

and let

$$\begin{cases} \tilde{A} = QAP = \begin{bmatrix} A_{11} & A_{12} \\ A_{21} & A_{22} \end{bmatrix} \\ \tilde{B} = QB = \begin{bmatrix} B_1 \\ B_2 \end{bmatrix} \\ \tilde{C} = CP = \begin{bmatrix} C_1 & C_2 \end{bmatrix} \end{cases},$$

where the partitions are compatible, then the system (4.71) can be written in the following dynamics decomposition form:

$$\begin{cases} \dot{x}_1 = A_{11}x_1 + A_{12}x_2 + B_1 u \\ 0 = A_{21}x_1 + A_{22}x_2 + B_2 u \\ y = C_1 x_1 + C_2 x_2 \end{cases}. \tag{4.93}$$

Theorem 4.18. *Let (4.93) be a dynamics decomposition form of the regular descriptor linear system (4.71) with* $\operatorname{rank} E = n_0$.

1. *System (4.71) (or (4.93)) is R-controllable if and only if*

$$\operatorname{rank} \begin{bmatrix} sI_{n_0} - A_{11} & A_{12} & B_1 \\ -A_{21} & A_{22} & B_2 \end{bmatrix} = n, \ \forall s \in \mathbb{C}. \tag{4.94}$$

2. *System (4.71) (or (4.93)) is C-controllable if and only if (4.94) holds and*

$$\operatorname{rank} B_2 = n - n_0. \tag{4.95}$$

3. *System (4.71) is I-controllable if and only if*

$$\operatorname{rank} \begin{bmatrix} A_{22} & B_2 \end{bmatrix} = n - n_0. \tag{4.96}$$

Proof. The first conclusion is obvious because condition (4.94) is clearly equivalent to (4.75).

It follows from Theorem 4.10 that a necessary and sufficient condition for C-controllability of system (4.71) is (4.75) and $\operatorname{rank} \begin{bmatrix} E & B \end{bmatrix} = n$. Noting that

$$\operatorname{rank} \begin{bmatrix} E & B \end{bmatrix} = \operatorname{rank} \begin{bmatrix} QEP & QB \end{bmatrix}$$
$$= \operatorname{rank} \begin{bmatrix} I_{n_0} & 0 & B_1 \\ 0 & 0 & B_2 \end{bmatrix}$$
$$= n_0 + \operatorname{rank} B_2,$$

we see that rank $\begin{bmatrix} E & B \end{bmatrix} = n$ is equivalent to (4.95). Thus, the second conclusion holds true.

By Theorem 4.14, system (4.71) is I-controllable if and only if

$$\text{rank} \begin{bmatrix} E & 0 & 0 \\ A & E & B \end{bmatrix} = n + n_0. \tag{4.97}$$

Since

$$\text{rank} \begin{bmatrix} E & 0 & 0 \\ A & E & B \end{bmatrix} = \text{rank} \begin{bmatrix} QEP & 0 & 0 \\ QAP & QEP & QB \end{bmatrix}$$

$$= \text{rank} \begin{bmatrix} I_{n_0} & 0 & 0 & 0 & 0 \\ 0 & 0 & 0 & 0 & 0 \\ A_{11} & A_{12} & I_{n_0} & 0 & B_1 \\ A_{21} & A_{22} & 0 & 0 & B_2 \end{bmatrix}$$

$$= \text{rank} \begin{bmatrix} I_{n_0} & 0 & 0 & 0 & 0 \\ 0 & 0 & 0 & 0 & 0 \\ 0 & 0 & I_{n_0} & 0 & 0 \\ 0 & A_{22} & 0 & 0 & B_2 \end{bmatrix}$$

$$= 2n_0 + \text{rank} \begin{bmatrix} A_{22} & B_2 \end{bmatrix}.$$

Equation (4.97) holds if and only if (4.96) is true. The third conclusion is proven. □

From the dual principle, we immediately have the following theorem.

Theorem 4.19. *Let (4.93) be a dynamics decomposition form of the regular descriptor linear system (4.71) with* $\text{rank} E = n_0$.

1. *System (4.71) (or (4.93)) is R-observable if and only if*

$$\text{rank} \begin{bmatrix} sI_{n_0} - A_{11} & -A_{12} \\ A_{21} & A_{22} \\ C_1 & C_2 \end{bmatrix} = n, \quad \forall s \in \mathbb{C}. \tag{4.98}$$

2. *System (4.71) (or (4.93)) is C-observable if and only if (4.98) holds and*

$$\text{rank} C_2 = n - n_0.$$

3. *System (4.71) is I-observable if and only if*

$$\text{rank} \begin{bmatrix} A_{22} \\ C_2 \end{bmatrix} = n - n_0.$$

Example 4.14. Consider the system (3.38) with the standard decomposition form given in Example 3.4, whose R-controllability has been proven in Example 4.3 (see also, Example 4.4), i.e.,

$$\text{rank}\,[sE - A \ B] = n, \quad \forall s \in \mathbb{C}, \ s \ \text{finite}.$$

Since

$$\text{rank}\,B_2 = \text{rank} \begin{bmatrix} 0 \\ -1 \end{bmatrix} = 1 < 2 = n - \text{rank}E,$$

and

$$\text{rank}\,[A_{22} \ B_2] = 2 = n - \text{rank}E,$$

this system is I-controllable, but not C-controllable. This coincides with the conclusion in Examples 4.3 and 4.4.

4.6.2 Criteria Based on the Inverse Form

Since the system (4.71) is assumed regular, there exists a scalar γ satisfying $\det(\gamma E - A) \neq 0$. Define

$$Q = (\gamma E - A)^{-1}.$$

Then it follows from Theorem 3.7 that under the transformation (I, Q) the system (4.71) can be converted into the following restricted equivalent inverse form

$$\begin{cases} \hat{E}\dot{x} = (\gamma\hat{E} - I)x + \hat{B}u \\ y = Cx \end{cases}, \tag{4.99}$$

where

$$\hat{E} = (\gamma E - A)^{-1}E, \quad \hat{B} = (\gamma E - A)^{-1}B. \tag{4.100}$$

Theorem 4.20. *Let (4.99) be an inverse form of the regular system (4.71).*

1. *System (4.71) is C-controllable if and only if*

$$\text{rank}[sI - \hat{E} \ \hat{B}] = n, \ \forall s \in \mathbb{C}, \ s \ \text{finite}, \tag{4.101}$$

or in other words, the matrix pair $(\hat{E}, \hat{B})$ is controllable and hence does not have uncontrollable modes.

2. *System (4.71) is R-controllable if and only if*

$$\text{rank}[sI - \hat{E} \ \hat{B}] = n, \ \forall s \in \mathbb{C}, s \neq 0, s \ \text{finite}, \tag{4.102}$$

or in other words, all the uncontrollable poles of $(\hat{E}, \hat{B})$, if they exist, must be zero.

3. The system (4.71) is I-controllable if and only if

$$\text{rank}\left(\hat{E}\begin{bmatrix}\hat{E} & \hat{B}\end{bmatrix}\right) = \text{rank}\hat{E}, \tag{4.103}$$

or, there exists a nonsingular matrix T such that

$$T^{-1}\hat{E}T = \begin{bmatrix}\hat{E}_1 & 0 \\ 0 & 0\end{bmatrix}, \quad T^{-1}B = \begin{bmatrix}\hat{B}_1 \\ 0\end{bmatrix},$$

where the partitions are consistent in dimensions, and $(\hat{E}_1, \hat{B}_1)$ has no zero uncontrollable poles.

Proof. Proof of conclusion 1. Using the definitions of $\hat{E}$, $\hat{B}$, we have

$$\begin{aligned}
&\text{rank}[sI - \hat{E} \quad \hat{B}] \\
&= \text{rank}\left[s\left(\gamma E - A\right) - E \quad B\right] \\
&= \begin{cases} \text{rank}\begin{bmatrix}E & B\end{bmatrix}, \text{ when } s = 0 \\ \text{rank}\left[\left(\gamma - \frac{1}{s}\right)E - A \quad B\right], \text{ when } s \neq 0 \end{cases}.
\end{aligned} \tag{4.104}$$

Recall that a necessary and sufficient condition for C-controllability of system (4.71) is $\text{rank}\,[E\ B] = n$ and $\text{rank}\,[\ sE - A\quad B\ ] = n$, $\forall s \in \mathbb{C}$, s finite, (4.104) indicates that system (4.71) is C-controllable if and only if (4.101) holds.

Proof of conclusion 2. It follows from Theorem 4.12 that system (4.71) is R-controllable if and only if

$$\begin{aligned}
\text{rank}\begin{bmatrix}sE - A & B\end{bmatrix} &= \text{rank}[(s - \gamma)\hat{E} + I \quad \hat{B}] \\
&= n, \ \forall s \in \mathbb{C}, \ s \text{ finite.}
\end{aligned} \tag{4.105}$$

If $s - \gamma = 0$, the second equation in (4.105) automatically holds. When $s - \gamma \neq 0$, we can define

$$\lambda = -\frac{1}{s - \gamma},$$

and (4.105) is equivalent to

$$\text{rank}[(s - \gamma)\hat{E} + I \quad \hat{B}] = \text{rank}[\lambda I - \hat{E} \quad \hat{B}] = n, \ \forall \lambda \in \mathbb{C}, \lambda \neq 0, \lambda \text{ finite.}$$

Therefore, (4.102) is a necessary and sufficient condition for R-controllability of system (4.71).

Proof of conclusion 3. System (4.71) is I-controllable if and only if

$$\text{rank}\begin{bmatrix}E & 0 & 0 \\ A & E & B\end{bmatrix} = n + \text{rank}E.$$

Noting that

$$\text{rank}\begin{bmatrix} E & 0 & 0 \\ A & E & B \end{bmatrix} = \text{rank}\begin{bmatrix} \hat{E} & 0 & 0 \\ \gamma\hat{E} - I_n & \hat{E} & \hat{B} \end{bmatrix}$$

$$= \text{rank}\begin{bmatrix} \hat{E} & 0 & 0 \\ I_n & \hat{E} & \hat{B} \end{bmatrix}$$

$$= \text{rank}\begin{bmatrix} 0 & -\hat{E}\begin{bmatrix} \hat{E} & \hat{B} \end{bmatrix} \\ I_n & \begin{bmatrix} \hat{E} & \hat{B} \end{bmatrix} \end{bmatrix}$$

$$= \text{rank}\begin{bmatrix} 0 & -\hat{E}\begin{bmatrix} \hat{E} & \hat{B} \end{bmatrix} \\ I_n & 0 \end{bmatrix},$$

we know immediately that system (4.71) is I-controllable if and only if (4.103) holds.

To prove the other condition, we make use of a decomposition result. Linear system theory assures the existence of a nonsingular matrix T such that

$$T\hat{E}T^{-1} = \begin{bmatrix} \hat{E}_{11} & \hat{E}_{12} & \hat{E}_{13} \\ 0 & \hat{E}_{22} & 0 \\ 0 & 0 & \hat{E}_{33} \end{bmatrix}, \quad T\hat{B} = \begin{bmatrix} \hat{B}_{11} \\ 0 \\ 0 \end{bmatrix},$$

where $(\hat{E}_{11}, \hat{B}_{11})$ is controllable, $\hat{E}_{22}$ is nonsingular and $\hat{E}_{33}$ is nilpotent.

Using the controllability of $(\hat{E}_{11}, \hat{B}_{11})$ and the nonsingularity of $\hat{E}_{22}$, we can obtain

$$\text{rank}\left(\hat{E}\begin{bmatrix} \hat{E} & \hat{B} \end{bmatrix}\right)$$

$$= \text{rank}\left(T\hat{E}T^{-1}\begin{bmatrix} T\hat{E}T^{-1} & T\hat{B} \end{bmatrix}\right)$$

$$= \text{rank}\begin{bmatrix} \hat{E}_{11}^2 & \hat{E}_{11}\hat{E}_{12} + \hat{E}_{12}\hat{E}_{22} & E_{11}\hat{E}_{13} + \hat{E}_{13}\hat{E}_{33} & \hat{E}_{11}\hat{B}_{11} \\ 0 & \hat{E}_{22}^2 & 0 & 0 \\ 0 & 0 & \hat{E}_{33}^2 & 0 \end{bmatrix}$$

$$= \text{rank}\begin{bmatrix} \hat{E}_{11} & \hat{E}_{11}\hat{E}_{12} + \hat{E}_{12}\hat{E}_{22} & \hat{E}_{11}\hat{E}_{13} + \hat{E}_{13}\hat{E}_{33} & 0 \\ 0 & \hat{E}_{22}^2 & 0 & 0 \\ 0 & 0 & \hat{E}_{33}^2 & 0 \end{bmatrix}$$

$$= \text{rank}\begin{bmatrix} \hat{E}_{11} & \hat{E}_{12}\hat{E}_{22} & \hat{E}_{13}\hat{E}_{33} \\ 0 & \hat{E}_{22}^2 & 0 \\ 0 & 0 & \hat{E}_{33}^2 \end{bmatrix}$$

$$= \text{rank}\begin{bmatrix} \hat{E}_{11} & 0 & \hat{E}_{13}\hat{E}_{33} \\ 0 & \hat{E}_{22} & 0 \\ 0 & 0 & \hat{E}_{33}^2 \end{bmatrix}.$$

Therefore, (4.103) is equivalent to

$$
\text{rank} \begin{bmatrix} \hat{E}_{11} & 0 & \hat{E}_{13}\hat{E}_{33} \\ 0 & \hat{E}_{22} & 0 \\ 0 & 0 & \hat{E}_{33}^2 \end{bmatrix} = \text{rank} \begin{bmatrix} \hat{E}_{11} & \hat{E}_{12} & \hat{E}_{13} \\ 0 & \hat{E}_{22} & 0 \\ 0 & 0 & \hat{E}_{33} \end{bmatrix}. \tag{4.106}
$$

Noticing the nilpotent property of $\hat{E}_{33}$, (4.106) holds if and only if

$$
\hat{E}_{33} = 0 \text{ and } \text{rank}\,\hat{E}_{11} = \text{rank}[\hat{E}_{11}\ \hat{E}_{13}].
$$

Thus, there exists a matrix M such that $\hat{E}_{13} = \hat{E}_{11}M$. Let

$$
\hat{T} = \begin{bmatrix} I & 0 & M \\ 0 & I & 0 \\ 0 & 0 & I \end{bmatrix}.
$$

Then $\hat{T}$ is nonsingular and

$$
(\hat{T}T)\hat{E}(\hat{T}T)^{-1} = \text{diag}(\hat{E}_1, 0), \ (\hat{T}T)\hat{B} = \begin{bmatrix} \hat{B}_1 \\ 0 \end{bmatrix},
$$

where

$$
\hat{E}_1 = \begin{bmatrix} \hat{E}_{11} & \hat{E}_{12} \\ 0 & \hat{E}_{22} \end{bmatrix}, \quad \hat{B}_1 = \begin{bmatrix} \hat{B}_{11} \\ 0 \end{bmatrix}.
$$

With this, we complete the proof.   □

The following result immediately follows from the above theorem and the dual principle.

Theorem 4.21. *Let (4.99) be an inverse form of the regular system (4.71).*

1. *System (4.71) is C-observable if and only if*

$$
\text{rank} \begin{bmatrix} sI - \hat{E} \\ C \end{bmatrix} = n, \ \forall s \in \mathbb{C}, \ s \neq 0, \ s \text{ finite.}
$$

In other words, the matrix pair $(\hat{E}, C)$ is observable and hence does not have unobservable modes.

2. *System (4.71) is R-observable if and only if*

$$
\text{rank} \begin{bmatrix} sI - \hat{E} \\ C \end{bmatrix} = n, \ \forall s \in \mathbb{C}, \ s \neq 0, \ s \text{ finite,}
$$

i.e., all the unobservable modes of $(\hat{E}, C)$, if they exist, must be zero.

3. *System (4.71) is I-observable if and only if*

$$\text{rank}\left(\begin{bmatrix} \hat{E} \\ C \end{bmatrix} \hat{E}\right) = \text{rank}\hat{E},$$

or, there exists a nonsingular matrix T such that

$$T^{-1}\hat{E}T = \text{diag}(\hat{E}_1, 0), \; T^{-1}CT = [\hat{C}_1 \; 0]$$

with $(\hat{E}_1, \hat{C}_1)$ having no zero unobservable poles.

The above two theorems obviously give the following corollary, which clearly reveals the difference between C-controllability (C-observability) and R-controllability (R-observability).

Corollary 4.3. *Let (4.99) be an inverse form of the regular system (4.71). Then the following hold:*

1. *The regular descriptor linear system (4.71) is C-controllable (or C-observable) if and only if the matrix pair $(\hat{E}, \hat{B})$ (or $(\hat{E}, C)$) does not have uncontrollable (unobservable) modes.*
2. *The regular descriptor linear system (4.71) is R-controllable (or R-observable) if and only if all the uncontrollable (unobservable) modes of the matrix pair $(\hat{E}, \hat{B})$ (or $(\hat{E}, C)$), if they exist, are zero ones.*

Theorems 4.18, 4.19, 4.20, and 4.21 establish the various controllability and observability criteria under different equivalent forms. Particularly, Theorem 4.20 shows that for the inverse form, the zero uncontrollable poles of $(\hat{E}, \hat{B})$ are the only difference between C-controllability and R-controllability.

Example 4.15. Consider the system (4.86) in Example 4.9 again. Since the matrix A is nonsingular, we can choose $\gamma = 0$. Then we have the transformation matrix

$$Q = -A^{-1} = \begin{bmatrix} -1 & 0 & 1 & 1 \\ 1 & 0 & -1 & 0 \\ 0 & 1 & -1 & 0 \\ 0 & -1 & 0 & 0 \end{bmatrix},$$

and the corresponding inverse form for the system (4.86) is obtained as

$$\tilde{E} = QE = \begin{bmatrix} 0 & -1 & 0 & 0 \\ 1 & 1 & 0 & 0 \\ 0 & 0 & 0 & 0 \\ 0 & 0 & 0 & 0 \end{bmatrix}, \; \tilde{B} = QB = \begin{bmatrix} 0 \\ 1 \\ 1 \\ 0 \end{bmatrix}.$$

It can be easily verified that $(\hat{E}, \hat{B})$ is not controllable, thus the system is not C-controllable. However, since

$$
\text{rank}[sI - \hat{E}\ \hat{B}] = \text{rank}
\begin{bmatrix}
s & 1 & 0 & 0 & 0 \\
-1 & s-1 & 0 & 0 & 1 \\
0 & 0 & s & 0 & 1 \\
0 & 0 & 0 & s & 0
\end{bmatrix}
$$

$$
= 4, \ \forall s \in \mathbb{C}, \ s \neq 0, \ s \text{ finite,}
$$

and

$$
\text{rank}\left(\hat{E}\left[\hat{E}\ \hat{B}\right]\right) = \text{rank}
\begin{bmatrix}
-1 & 0 & -1 & 0 & -1 \\
0 & 0 & 1 & 0 & 0 \\
1 & 0 & 0 & 0 & 1 \\
-1 & 0 & -1 & 0 & -1
\end{bmatrix}
= 2 = \text{rank}\hat{E},
$$

the system (4.86) is R-controllable and I-controllable, and hence S-controllable. This coincides with the already proven results in Example 4.9.

4.6.3 Criteria Based on Equivalent Form for Derivative Feedback

It follows from Theorem 2.6 that, for any descriptor linear system (E, A, B) with $E, A \in \mathbb{R}^{n \times n}$, $B \in \mathbb{R}^{n \times r}$, and $\text{rank}E = n_0$, $\text{rank}B = r > 0$, there exist a nonsingular matrix $Q_3 \in \mathbb{R}^{n \times n}$ and two orthogonal matrices $P_3 \in \mathbb{R}^{n \times n}$ and $U_3 \in \mathbb{R}^{r \times r}$ such that

$$
(E, A, B) \overset{(P_3, Q_3)}{\Longleftrightarrow} (\tilde{E}, \tilde{A}, \tilde{B}),
$$

where $(\tilde{E}, \tilde{A}, \tilde{B})$ is the third canonical equivalent form for derivative feedback for the system (E, A, B), and its coefficient matrices are given by

$$
\tilde{E} = Q_3 E P_3 =
\begin{bmatrix}
I_l & 0_{l \times (n-l)} \\
0_{r \times l} & M \\
0_{(n-n_2) \times l} & 0_{(n-n_2) \times (n-l)}
\end{bmatrix},
\tag{4.107}
$$

$$
\tilde{A} = Q_3 A P_3 =
\begin{bmatrix}
A_{11} & A_{12} \\
A_{21} & A_{22} \\
A_{31} & A_{32}
\end{bmatrix}, \ A_{32} \in \mathbb{R}^{(n-n_2) \times (n-l)},
\tag{4.108}
$$

and

$$
\tilde{B} = Q_3 B =
\begin{bmatrix}
0_{l \times r} \\
I_r \\
0_{(n-n_2) \times r}
\end{bmatrix} U_3^T,
\tag{4.109}
$$

with M given in (2.32), and

$$\begin{cases} n_2 = \text{rank}\begin{bmatrix} E & B \end{bmatrix} \\ l = n_2 - r \end{cases}. \tag{4.110}$$

Based on the above fact, we can derive the following theorem about the controllability of the regular descriptor linear system (4.71).

Theorem 4.22. *Let* $\text{rank} E = n_0$, *and* n_2 *and* l *be defined in (4.110). Furthermore, let* $\tilde{E}, \tilde{A}$ *and* $\tilde{B}$ *be given by (4.107)–(4.109), with the nonsingular matrix* $Q_3 \in \mathbb{R}^{n \times n}$ *and the two orthogonal matrices* $P_3 \in \mathbb{R}^{n \times n}$ *and* $U_3 \in \mathbb{R}^{r \times r}$ *obtained by Algorithm 2.2.*

1. System (4.71) is R-controllable if and only if

$$\text{rank}\begin{bmatrix} sI_l - A_{11} & A_{12} \\ A_{31} & A_{32} \end{bmatrix} = n - r, \ \forall s \in \mathbb{C}. \tag{4.111}$$

2. System (4.71) is C-controllable if and only if $n_2 = l + r = n$, *and*

$$\text{rank}\begin{bmatrix} sI_l - A_{11} & A_{12} \end{bmatrix} = l, \ \forall s \in \mathbb{C}. \tag{4.112}$$

3. System (4.71) is I-controllable if and only if the matrix A_{32}^1 *has full-row rank, that is,*

$$\text{rank} A_{32}^1 = n - n_2, \tag{4.113}$$

where $A_{32}^1 \in \mathbb{R}^{(n-n_2) \times (n-n_0)}$ *is determined by the partition*

$$A_{32} = \begin{bmatrix} A_{32}^0 & A_{32}^1 \end{bmatrix}. \tag{4.114}$$

Proof. In view of (4.107)–(4.109), we have

$$\text{rank}\begin{bmatrix} sE - A & B \end{bmatrix} = \text{rank}\begin{bmatrix} sQ_3 E P_3 - Q_3 A P_3 & Q_3 B U_3 \end{bmatrix}$$

$$= \text{rank}\begin{bmatrix} sI_l - A_{11} & -A_{12} & 0_{l \times r} \\ -A_{21} & sM - A_{22} & I_r \\ -A_{31} & -A_{32} & 0 \end{bmatrix}$$

$$= r + \text{rank}\begin{bmatrix} sI_l - A_{11} & -A_{12} \\ -A_{31} & -A_{32} \end{bmatrix}.$$

Therefore, it follows from Theorem 4.12 that the system (4.71) is R-controllable if and only if (4.111) holds.

It follows from Theorem 4.10 that system (4.71) is C-controllable if and only if (4.111) and $\text{rank}\begin{bmatrix} E & B \end{bmatrix} = n$ hold. Note that, by definition,

$$\text{rank}\begin{bmatrix} E & B \end{bmatrix} = n_2 = r + l,$$

we clearly see that $\text{rank}\begin{bmatrix} E & B \end{bmatrix} = n$ if and only if $r + l = n_2 = n$. Further, since $A_{32} \in \mathbb{R}^{(n-n_2) \times (n-l)}$, the matrices A_{31} and A_{32} vanish when $r + l = n_2 = n$. Therefore, in this case (4.111) becomes (4.112). The second conclusion is also shown.

Finally, using (4.107)–(4.109) again, and in view of the structure of matrix M and the partition (4.114), we have

$$\text{rank}\begin{bmatrix} E & 0 & 0 \\ A & E & B \end{bmatrix} = \text{rank}\begin{bmatrix} Q_3 E P_3 & 0 & 0 \\ Q_3 A P_3 & Q_3 E P_3 & Q_3 B U_3 \end{bmatrix}$$

$$= \text{rank}\begin{bmatrix} I_l & 0 & 0 & 0 & 0 \\ 0 & M & 0 & 0 & 0 \\ 0 & 0 & 0 & 0 & 0 \\ A_{11} & A_{12} & I_l & 0 & 0 \\ A_{21} & A_{22} & 0 & M & I_r \\ A_{31} & A_{32} & 0 & 0 & 0 \end{bmatrix}$$

$$= \text{rank}\begin{bmatrix} I_l & 0 & 0 & 0 & 0 \\ 0 & M & 0 & 0 & 0 \\ 0 & 0 & I_l & 0 & 0 \\ 0 & 0 & 0 & 0 & I_r \\ 0 & A_{32} & 0 & 0 & 0 \end{bmatrix}$$

$$= r + 2l + \text{rank}\begin{bmatrix} M \\ A_{32} \end{bmatrix}$$

$$= r + 2l + \text{rank}\begin{bmatrix} I_{n_0-l} & 0 \\ 0 & 0 \\ A_{32}^0 & A_{32}^1 \end{bmatrix}$$

$$= n_0 + n_2 + \text{rank} A_{32}^1.$$

Thus, by using Theorem 4.14 we obtain the condition (4.113) for I-controllability of system (4.71). □

Due to the above Theorem 4.22, we have the following corollary.

Corollary 4.4. *Let* $\text{rank} E = n_0$, *and* n_2 *and* l *be defined in (4.110). Then the system (4.71) is I-controllable if and only if there exist a nonsingular matrix* $Q \in \mathbb{R}^{n \times n}$ *and two orthogonal matrices* $P \in \mathbb{R}^{n \times n}$ *and* $U_3 \in \mathbb{R}^{r \times r}$ *such that*

$$(E, A, B) \overset{(P, Q)}{\Longleftrightarrow} (\tilde{E}, \tilde{A}, \tilde{B}),$$

where the matrices $\tilde{E}, \tilde{A}, \tilde{B}$ *possess the following special forms:*

$$\tilde{E} = Q E P = \text{diag}(\Pi_1, 0_{(n-n_0) \times (n-n_0)}), \tag{4.115}$$

$$\tilde{A} = QAP = \begin{bmatrix} A_{11} & A_{12} & A_{13} \\ A_{21} & A_{22} & A_{23} \\ A_{31} & \Delta & 0_{(n-n_2)\times(n_2-n_0)} \end{bmatrix}, \qquad (4.116)$$

$$\tilde{B} = Q_3 B = \begin{bmatrix} \Pi_2 & 0_{n_0\times(n_2-n_0)} \\ 0 & I_{n_2-n_0} \\ 0_{(n-n_2)\times(n_0-l)} & 0 \end{bmatrix} U_3^{\mathrm{T}}, \qquad (4.117)$$

with $A_{23} \in \mathbb{R}^{(n_2-n_0)\times(n_2-n_0)}$, and $\Delta \in \mathbb{R}^{(n-n_2)\times(n-n_2)}$ being a diagonal matrix with positive diagonal elements and

$$\Pi_1 = \mathrm{diag}(I_l,\ d_1,\ d_2,\dots,\ d_{n_0-l}) > 0,$$

$$\Pi_2 = \begin{bmatrix} 0_{l\times(n_0-l)} \\ I_{n_0-l} \end{bmatrix}.$$

This corollary can be easily proven. Here we give, with the help of Algorithm 2.2, an algorithm for deriving the above equivalent form stated in the above corollary.

Algorithm 4.1. The equivalent form for derivative feedback.

Step 1 Obtain the orthogonal matrices $Q_3 \in \mathbb{R}^{\tilde{n}\times\tilde{n}}$, $P_3 \in \mathbb{R}^{n\times n}$ and $U_3 \in \mathbb{R}^{r\times n}$, and the matrix $M \in \mathbb{R}^{r\times l}$ using Algorithm 2.2.

Step 2 Compute $\bar{A} = Q_3 A P_3$ and partition the matrix $\bar{A}$ as

$$\bar{A} = \begin{bmatrix} \bar{A}_{11} & \bar{A}_{12} \\ \bar{A}_{21} & \bar{A}_{22} \\ \bar{A}_{31} & \bar{A}_{32} \end{bmatrix}, \quad \bar{A}_{32} \in \mathbb{R}^{(n-n_2)\times(n-l)}.$$

Step 3 Obtain based on singular value decomposition two orthogonal matrices $Q_1 \in \mathbb{R}^{(n-n_2)\times(n-n_2)}$ and $Q_\mathrm{r} \in \mathbb{R}^{(n-n_0)\times(n-n_0)}$ satisfying

$$Q_1 \bar{A}_{32} Q_\mathrm{r} = \begin{bmatrix} \Delta & 0_{(n-n_2)\times(n_2-n_0)} \end{bmatrix}.$$

Step 4 Compute the matrix Q and P by, respectively,

$$Q = \begin{bmatrix} I_{n_2} & 0 \\ 0 & Q_1 \end{bmatrix}, \quad P = \begin{bmatrix} I_{n_0} & 0 \\ 0 & Q_1 \end{bmatrix}.$$

By the dual principle, to verify the observability of the regular descriptor linear system (4.71), we can use the above theorem to check the controllability of the dual system $\left(E^{\mathrm{T}}, A^{\mathrm{T}}, C^{\mathrm{T}}, B^{\mathrm{T}}\right)$.

Example 4.16. Consider a descriptor linear system in the form of (4.71) with parameters

$$E = \begin{bmatrix} 0 & 1 & 0 \\ 0 & 0 & 1 \\ 0 & 0 & 0 \end{bmatrix}, \ A = \begin{bmatrix} 1 & 0 & 0 \\ 0 & 1 & 0 \\ 0 & 0 & 1 \end{bmatrix}, \ B = \begin{bmatrix} 0 \\ 1 \\ 0 \end{bmatrix}.$$

For this system, $n_0 = 2$, $n_2 = \text{rank} \begin{bmatrix} E & B \end{bmatrix} = 2$, $n_1 = l = 1$. We now check the controllability of the system using Theorem 4.22.

Letting

$$Q_3 = I, \ P_3 = \begin{bmatrix} 0 & 0 & 1 \\ 1 & 0 & 0 \\ 0 & 1 & 0 \end{bmatrix}, U_3 = 1,$$

one can obtain

$$\tilde{E} = Q_3 E P_3 = \begin{bmatrix} I_l & 0 \\ 0 & M \\ 0 & 0 \end{bmatrix} = \begin{bmatrix} 1 & 0 & 0 \\ 0 & 1 & 0 \\ 0 & 0 & 0 \end{bmatrix},$$

$$\tilde{A} = Q_3 A P_3 = \begin{bmatrix} 0 & 0 & 1 \\ 1 & 0 & 0 \\ 0 & 1 & 0 \end{bmatrix}, \ \tilde{B} = Q_3 B = \begin{bmatrix} 0 \\ 1 \\ 0 \end{bmatrix}.$$

Since

$$\text{rank} \begin{bmatrix} sI_l - A_{11} & A_{12} \\ A_{31} & A_{32} \end{bmatrix} = \text{rank} \begin{bmatrix} s & 0 & 1 \\ 0 & 1 & 0 \end{bmatrix}$$
$$= 2$$
$$= n - r, \quad \forall s \in \mathbb{C},$$

the system is R-controllable. However, since $n_2 = 2 < n = 3$, the system is not C-controllable. Also, since $A_{32}^1 = 0$, the system is not I-controllable.

4.7 System Decomposition

It is well known from normal linear systems theory that a system can be decomposed into a canonical form according to controllability and observability. In this section, we investigate this topic for regular descriptor linear systems of the following form:

$$\begin{cases} E\dot{x} = Ax + Bu \\ y = Cx \end{cases}, \tag{4.118}$$

where $x \in \mathbb{R}^n$, $u \in \mathbb{R}^r$, and $y \in \mathbb{R}^m$ are the state vector, the control input vector, and the measured output vector, respectively; and $E, \ A \in \mathbb{R}^{n \times n}$, $B \in \mathbb{R}^{n \times r}$, $C \in \mathbb{R}^{m \times n}$ are constant coefficient matrices.

4.7.1 The General Structural Decomposition

Here, let us introduce the general system decomposition by controllability and observability for system (4.118) based on its standard decomposition. The process involves four steps.

Step 1. *Deriving the standard decomposition form*

Under regularity assumption, we can find two nonsingular matrices Q and P such that under transformation (P, Q) the system (4.118) is r.s.e. to the standard decomposition form

$$\begin{cases} \dot{x}_1 = A_1 x_1 + B_1 u \\ N\dot{x}_2 = x_2 + B_2 u , \\ y = C_1 x_1 + C_2 x_2 \end{cases}$$

where the coefficient matrices are determined by (4.74), with $x_1 \in \mathbb{R}^{n_1}$, $x_2 \in \mathbb{R}^{n_2}$, and $N \in \mathbb{R}^{n_2 \times n_2}$ being a nilpotent matrix.

Step 2. *Realizing the structural decomposition of* (A_1, B_1, C_1)

According to the well-known results in linear system theory, for the matrix triple (A_1, B_1, C_1), a nonsingular matrix T_1 can be found such that

$$T_1 A_1 T_1^{-1} = \begin{bmatrix} A_{11} & 0 & A_{13} & 0 \\ A_{21} & A_{22} & A_{23} & A_{24} \\ 0 & 0 & A_{33} & 0 \\ 0 & 0 & A_{43} & A_{44} \end{bmatrix}, \quad T_1 B_1 = \begin{bmatrix} B_{11} \\ B_{12} \\ 0 \\ 0 \end{bmatrix},$$

$$C_1 T_1^{-1} = \begin{bmatrix} C_{11} & 0 & C_{13} & 0 \end{bmatrix},$$

where

$$A_{ij} \in \mathbb{R}^{n_{1i} \times n_{1j}}, \ i, \ j = 1, 2, 3, 4, \ \sum_{i=1}^{4} n_{1i} = n_1,$$

and the following are true:

- the subsystem (A_{11}, B_{11}, C_{11}) is both controllable and observable;
- the subsystem

$$\left(\begin{bmatrix} A_{11} & 0 \\ A_{21} & A_{22} \end{bmatrix}, \begin{bmatrix} B_{11} \\ B_{12} \end{bmatrix}, \begin{bmatrix} C_{11} & 0 \end{bmatrix} \right)$$

  is controllable, but unobservable;
- the subsystem

$$\left(\begin{bmatrix} A_{11} & A_{13} \\ 0 & A_{33} \end{bmatrix}, \begin{bmatrix} B_{11} \\ 0 \end{bmatrix}, \begin{bmatrix} C_{11} & C_{13} \end{bmatrix} \right)$$

  is observable, but uncontrollable;
- the subsystem $(A_{44}, 0, 0)$ is neither controllable nor observable.

Step 3. *Realizing the structural decomposition of* (N, B_2, C_2)

Similarly, we can find a nonsingular matrix T_2 such that (N, B_2, C_2) may be decomposed into the controllability and observability form:

$$
T_2 N T_2^{-1} = \begin{bmatrix} N_{11} & 0 & N_{13} & 0 \\ N_{21} & N_{22} & N_{23} & N_{24} \\ 0 & 0 & N_{33} & 0 \\ 0 & 0 & N_{43} & N_{44} \end{bmatrix}, \quad T_2 B_2 = \begin{bmatrix} B_{21} \\ B_{22} \\ 0 \\ 0 \end{bmatrix},
$$

$$
C_2 T_2^{-1} = \begin{bmatrix} C_{21} & 0 & C_{23} & 0 \end{bmatrix},
$$

where $N_{ij} \in \mathbb{R}^{n_{2i} \times n_{2j}}$, $i, j = 1, 2, 3, 4$, $\sum_{i=1}^{4} n_{2i} = n_2$, particularly, $N_{ii} \in \mathbb{R}^{n_{2i} \times n_{2i}}$, $i = 1, 2, 3, 4$, are nilpotent, and the following are true:

- The subsystem (N_{11}, B_{21}, C_{21}) is both controllable and observable.
- The subsystem

$$
\left(\begin{bmatrix} N_{11} & 0 \\ N_{21} & N_{22} \end{bmatrix}, \begin{bmatrix} B_{21} \\ B_{22} \end{bmatrix}, \begin{bmatrix} C_{21} & 0 \end{bmatrix} \right)
$$

is controllable, but unobservable.
- The subsystem

$$
\left(\begin{bmatrix} N_{11} & N_{13} \\ 0 & N_{33} \end{bmatrix}, \begin{bmatrix} B_{21} \\ 0 \end{bmatrix}, \begin{bmatrix} C_{21} & C_{23} \end{bmatrix} \right)
$$

is observable, but uncontrollable.
- The subsystem $(N_{44}, 0, 0)$ is neither controllable nor observable.

Step 4. *Obtaining the canonical decomposition form*

Denote

$$
Q_1 = \mathrm{diag}\,(T_1, T_2)\, Q, \quad P_1 = P \mathrm{diag}\,\left(T_1^{-1}, \ T_2^{-1}\right).
$$

Obviously, under the transformation (P_1, Q_1), system (4.118) is r.s.e. to

$$
\begin{cases}
\begin{bmatrix} \dot{x}_{11} \\ \dot{x}_{12} \\ \dot{x}_{13} \\ \dot{x}_{14} \end{bmatrix} = \begin{bmatrix} A_{11} & 0 & A_{13} & 0 \\ A_{21} & A_{22} & A_{23} & A_{24} \\ 0 & 0 & A_{33} & 0 \\ 0 & 0 & A_{43} & A_{44} \end{bmatrix} \begin{bmatrix} x_{11} \\ x_{12} \\ x_{13} \\ x_{14} \end{bmatrix} + \begin{bmatrix} B_{11} \\ B_{12} \\ 0 \\ 0 \end{bmatrix} u \\[4em]
\begin{bmatrix} N_{11} & 0 & N_{13} & 0 \\ N_{21} & N_{22} & N_{23} & N_{24} \\ 0 & 0 & N_{33} & 0 \\ 0 & 0 & N_{43} & N_{44} \end{bmatrix} \begin{bmatrix} \dot{x}_{21} \\ \dot{x}_{22} \\ \dot{x}_{23} \\ \dot{x}_{24} \end{bmatrix} = \begin{bmatrix} x_{21} \\ x_{22} \\ x_{23} \\ x_{24} \end{bmatrix} + \begin{bmatrix} B_{21} \\ B_{22} \\ 0 \\ 0 \end{bmatrix} u
\end{cases}, \quad (4.119)
$$

$$
y = \begin{bmatrix} C_{11} & 0 & C_{13} & 0 & C_{21} & 0 & C_{23} & 0 \end{bmatrix} \begin{bmatrix} T_1 x_1 \\ T_2 x_2 \end{bmatrix}, \quad (4.120)
$$

where

$$T_i x_i = \begin{bmatrix} x_{i1} \\ x_{i2} \\ x_{i3} \\ x_{i4} \end{bmatrix}, \quad x_{ij} \in \mathbb{R}^{n_{ij}}, \quad j = 1, 2, 3, 4, \quad i = 1, 2,$$

$$\sum_{j=1}^{4} n_{ij} = n_i, \quad i = 1, 2, \quad n_1 + n_2 = n,$$

$N_{ii}, i = 1, 2, 3, 4$ are nilpotent.

Rewrite the coordinate vector as

$$\tilde{x}_i = \begin{bmatrix} x_{1i} \\ x_{2i} \end{bmatrix}, \quad i = 1, 2, 3, 4.$$

This means the combination of corresponding state variables in the slow and fast substates. Direct computation verifies that, under such a coordinate transformation, system (4.118) is r.s.e. to the following canonical structural decomposition form

$$\begin{cases} \tilde{E}\dot{\tilde{x}} = \tilde{A}\tilde{x} + \tilde{B}u \\ y = \tilde{C}\tilde{x} \end{cases} \tag{4.121}$$

with

$$\tilde{E} = \begin{bmatrix} \tilde{E}_{11} & 0 & \tilde{E}_{13} & 0 \\ \tilde{E}_{21} & \tilde{E}_{22} & \tilde{E}_{23} & \tilde{E}_{24} \\ 0 & 0 & \tilde{E}_{33} & 0 \\ 0 & 0 & \tilde{E}_{43} & \tilde{E}_{44} \end{bmatrix}, \quad \tilde{A} = \begin{bmatrix} \tilde{A}_{11} & 0 & \tilde{A}_{13} & 0 \\ \tilde{A}_{21} & \tilde{A}_{22} & \tilde{A}_{23} & \tilde{A}_{24} \\ 0 & 0 & \tilde{A}_{33} & 0 \\ 0 & 0 & \tilde{A}_{43} & \tilde{A}_{44} \end{bmatrix},$$

$$\tilde{B}^{\mathrm{T}} = \begin{bmatrix} \tilde{B}_1^{\mathrm{T}} & \tilde{B}_2^{\mathrm{T}} & 0 & 0 \end{bmatrix}, \quad \tilde{C} = \begin{bmatrix} \tilde{C}_1 & 0 & \tilde{C}_3 & 0 \end{bmatrix},$$

where

$$\begin{cases} \tilde{E}_{ii} = \mathrm{diag}\,(I, \, N_{ii}); \\ \tilde{A}_{ii} = \mathrm{diag}\,(A_{ii}, \, I), \, i = 1, 2, 3, 4; \\ \tilde{E}_{ij} = \mathrm{diag}\,(0, \, N_{ij}); \\ \tilde{A}_{ij} = \mathrm{diag}\,(A_{ij}, \, 0), \, i \neq j, \, i, j = 1, 2, 3, 4; \\ \tilde{B}_i = \begin{bmatrix} B_{1i} \\ B_{2i} \end{bmatrix}, \, i = 1, 2; \\ \tilde{C}_j = \begin{bmatrix} C_{1j} & C_{2j} \end{bmatrix}, \, j = 1, 3, \end{cases} \tag{4.122}$$

with $\tilde{x}_i \in \mathbb{R}^{\tilde{n}_i}$, $\tilde{n}_i = n_{1i} + n_{2i}$, $i = 1, 2, 3, 4$, $\sum_{i=1}^{4} \tilde{n}_i = n$.

From the controllability and observability criteria (Theorems 4.2 and 4.11), and the decomposition of (4.121)–(4.122), the following result clearly holds.

Theorem 4.23. *The structural canonical decomposition form (4.121)–(4.122) of the regular descriptor system (4.118) has the following properties.*

1. *The subsystem* $(\tilde{E}_{11}, \tilde{A}_{11}, \tilde{B}_1, \tilde{C}_1)$ *is both C-controllable and C-observable.*
2. *The subsystem*

$$\left(\begin{bmatrix} \tilde{E}_{11} & 0 \\ \tilde{E}_{21} & \tilde{E}_{22} \end{bmatrix}, \begin{bmatrix} \tilde{A}_{11} & 0 \\ \tilde{A}_{21} & \tilde{A}_{22} \end{bmatrix}, \begin{bmatrix} \tilde{B}_1 \\ \tilde{B}_2 \end{bmatrix}, [\tilde{C}_1 \ 0] \right)$$

   is C-controllable, but not C-observable.
3. *The subsystem*

$$\left(\begin{bmatrix} \tilde{E}_{11} & \tilde{E}_{13} \\ 0 & \tilde{E}_{33} \end{bmatrix}, \begin{bmatrix} \tilde{A}_{11} & \tilde{A}_{13} \\ 0 & \tilde{A}_{33} \end{bmatrix}, \begin{bmatrix} \tilde{B}_1 \\ 0 \end{bmatrix}, [\tilde{C}_1 \ \tilde{C}_3] \right)$$

   is C-observable, but not C-controllable.
4. *The subsystem* $(\tilde{E}_{44}, \tilde{A}_{44}, 0, 0)$ *is neither C-controllable nor C-observable.*

4.7.2 Special Cases

By properly selecting the state vectors, some special canonical forms can be obtained.

The Controllability Canonical Form

Particularly, let

$$\bar{x}_1 = \begin{bmatrix} \tilde{x}_1 \\ \tilde{x}_2 \end{bmatrix} \text{ and } \bar{x}_2 = \begin{bmatrix} \tilde{x}_3 \\ \tilde{x}_4 \end{bmatrix}.$$

Then, (4.121)–(4.122) become

$$\begin{cases} \begin{bmatrix} \bar{E}_{11} & \bar{E}_{12} \\ 0 & \bar{E}_{22} \end{bmatrix} \begin{bmatrix} \dot{\bar{x}}_1 \\ \dot{\bar{x}}_2 \end{bmatrix} = \begin{bmatrix} \bar{A}_{11} & \bar{A}_{12} \\ 0 & \bar{A}_{22} \end{bmatrix} \begin{bmatrix} \bar{x}_1 \\ \bar{x}_2 \end{bmatrix} + \begin{bmatrix} \bar{B}_1 \\ 0 \end{bmatrix} u \\ y = [\bar{C}_1 \ \bar{C}_2] \begin{bmatrix} \bar{x}_1 \\ \bar{x}_2 \end{bmatrix} \end{cases}, \quad (4.123)$$

where $\bar{E}_{ij}, \bar{A}_{ij}, \bar{B}_i, \bar{C}_i, i, j = 1, 2$, are certain constant matrices, and $(\bar{E}_{11}, \bar{A}_{11}, \bar{B}_1)$ is C-controllable. System (4.123) is called the controllability canonical form for system (4.118).

The Observability Canonical Form

If we set

$$\hat{x}_1 = \begin{bmatrix} \tilde{x}_1 \\ \tilde{x}_3 \end{bmatrix} \text{ and } \hat{x}_2 = \begin{bmatrix} \tilde{x}_2 \\ \tilde{x}_4 \end{bmatrix},$$

system (4.121)–(4.122) can be rewritten as

$$
\begin{cases}
\begin{bmatrix} \hat{E}_{11} & 0 \\ \hat{E}_{12} & \hat{E}_{13} \end{bmatrix} \begin{bmatrix} \dot{\hat{x}}_1 \\ \dot{\hat{x}}_2 \end{bmatrix} = \begin{bmatrix} \hat{A}_{11} & 0 \\ \hat{A}_{21} & \hat{A}_{22} \end{bmatrix} \begin{bmatrix} \hat{x}_1 \\ \hat{x}_2 \end{bmatrix} + \begin{bmatrix} \hat{B}_1 \\ \hat{B}_2 \end{bmatrix} u \\
y = \begin{bmatrix} \hat{C}_1 & 0 \end{bmatrix} \begin{bmatrix} \hat{x}_1 \\ \hat{x}_2 \end{bmatrix}
\end{cases}
$$

where $\hat{E}_{ij}, \hat{A}_{ij}, \hat{B}_i, \hat{C}_i, i, j = 1, 2$, are certain constant matrices, and $(\hat{E}_{11}, \hat{A}_{11}, \hat{C}_1)$ is C-observable. This form is called the observability canonical form of system (4.118).

The Normalizability Canonical Form

If we set

$$
\overline{\overline{x}}_1 = \begin{bmatrix} x_1 \\ x_{21} \\ x_{22} \end{bmatrix} \text{ and } \overline{\overline{x}}_2 = \begin{bmatrix} x_{23} \\ x_{24} \end{bmatrix},
$$

system (4.119)–(4.120) become

$$
\begin{cases}
\begin{bmatrix} \overline{\overline{E}}_{11} & \overline{\overline{E}}_{12} \\ 0 & \overline{\overline{E}}_{22} \end{bmatrix} \begin{bmatrix} \dot{\overline{\overline{x}}}_1 \\ \dot{\overline{\overline{x}}}_2 \end{bmatrix} = \begin{bmatrix} \overline{\overline{A}}_{11} & \overline{\overline{A}}_{12} \\ 0 & \overline{\overline{A}}_{22} \end{bmatrix} \begin{bmatrix} \overline{\overline{x}}_1 \\ \overline{\overline{x}}_2 \end{bmatrix} + \begin{bmatrix} \overline{\overline{B}}_1 \\ 0 \end{bmatrix} u \\
y = \begin{bmatrix} \overline{\overline{C}}_1 & \overline{\overline{C}}_2 \end{bmatrix} \begin{bmatrix} \overline{\overline{x}}_1 \\ \overline{\overline{x}}_2 \end{bmatrix}
\end{cases}
$$

where, $\begin{bmatrix} \overline{\overline{E}}_{11} & \overline{\overline{B}}_1 \end{bmatrix}$ is of full-row rank and $\overline{\overline{E}}_{22}$ is nilpotent. This decomposition is called the normalizability decomposition.

Example 4.17. Consider the two connected one-mass oscillators system in Example 1.4. Under the transformation matrix (P, Q),

$$
P = \begin{bmatrix} \frac{1}{4} & 0 & 0 & 0 & \frac{1}{2} \\ -\frac{1}{4} & 0 & 0 & 0 & \frac{1}{2} \\ 0 & -\frac{\sqrt{6}}{8} & 0 & \frac{1}{2} & 0 \\ 0 & \frac{\sqrt{6}}{8} & 0 & \frac{1}{2} & 0 \\ \frac{1}{8} & 0 & \frac{1}{2} & 1 & -\frac{1}{4} \end{bmatrix}, \quad Q = \begin{bmatrix} 2 & -2 & 0 & 0 & 0 \\ 0 & 0 & -\frac{2\sqrt{6}}{3} & \frac{2\sqrt{6}}{3} & -\frac{\sqrt{6}}{3} \\ 0 & 0 & 1 & 1 & 2 \\ 1 & 1 & 0 & 0 & 0 \\ 0 & 0 & 0 & 0 & 1 \end{bmatrix},
$$

we can obtain the standard decomposition for the system as

$$
\begin{cases}
\dot{x}_1 = \begin{bmatrix} 0 & -\frac{\sqrt{6}}{2} \\ \frac{\sqrt{6}}{2} & 0 \end{bmatrix} x_1 + \begin{bmatrix} 0 \\ \frac{4\sqrt{6}}{3} \end{bmatrix} u(t) \\[4mm]
\begin{bmatrix} 0 & 1 & 0 \\ 0 & 0 & 1 \\ 0 & 0 & 0 \end{bmatrix} \dot{x}_2 = x_2 \\[6mm]
y = \begin{bmatrix} \frac{1}{4} & 0 \end{bmatrix} x_1 + \begin{bmatrix} 0 & 0 & \frac{1}{2} \end{bmatrix} x_2
\end{cases}
\tag{4.124}
$$

Since

$$
\left(\begin{bmatrix} 0 & -\frac{\sqrt{6}}{2} \\ \frac{\sqrt{6}}{2} & 0 \end{bmatrix}, \begin{bmatrix} 0 \\ \frac{4\sqrt{6}}{3} \end{bmatrix}, \begin{bmatrix} \frac{1}{4} & 0 \end{bmatrix} \right)
$$

is both C-controllable and C-observable, we need to make decomposition on only the fast subsystem. If we denote $x_2 = \begin{bmatrix} x_{21} & x_{22} & x_{23} \end{bmatrix}^{\mathrm{T}}$ in (4.124), then the subsystem having state x_{23} is C-observable but not C-controllable, and the subsystem having state x_{21} and x_{22} is neither C-controllable nor C-observable. Hence, by letting

$$
\bar{x}^{\mathrm{T}} = \begin{bmatrix} x_1^{\mathrm{T}} & x_{23} & x_{21} & x_{22} \end{bmatrix},
$$

the canonical form of system (4.124) can be obtained as follows:

$$
\begin{cases}
\bar{E}\dot{\bar{x}} = \bar{A}\bar{x} + \bar{B}u \\
y = \bar{C}\bar{x}
\end{cases}
$$

with

$$
\bar{E} = \begin{bmatrix} 1 & 0 & 0 & 0 & 0 \\ 0 & 1 & 0 & 0 & 0 \\ \hline 0 & 0 & 0 & 0 & 0 \\ 0 & 0 & 0 & 1 & 0 \\ 0 & 0 & 0 & 0 & 1 \end{bmatrix}, \quad
\bar{A} = \begin{bmatrix} 0 & -\frac{\sqrt{6}}{2} & 0 & 0 & 0 \\ \frac{\sqrt{6}}{2} & 0 & 0 & 0 & 0 \\ \hline 0 & 0 & 0 & 0 & 1 \\ 0 & 0 & 1 & 0 & 0 \\ 0 & 0 & 0 & 1 & 0 \end{bmatrix},
$$

$$
\bar{B}^{\mathrm{T}} = \begin{bmatrix} 0 & \frac{4\sqrt{6}}{3} & 0 & 0 & 0 \end{bmatrix}, \quad
\bar{C} = \begin{bmatrix} \frac{1}{4} & 0 & \frac{1}{2} & 0 & 0 \end{bmatrix}.
$$

Example 4.18. Consider a descriptor linear system with the following parameters:

$$
E = \begin{bmatrix} 1 & 0 & 0 & 0 \\ 0 & 1 & 0 & 0 \\ 0 & 0 & -1 & 0 \\ 0 & 0 & 0 & 0 \end{bmatrix}, \quad
A = \begin{bmatrix} 0 & 0 & 0 & 1 \\ 0 & 0 & 1 & 0 \\ -1 & 1 & 0 & 0 \\ 1 & 0 & 1 & 1 \end{bmatrix},
$$

$$
B^{\mathrm{T}} = \begin{bmatrix} 0 & 0 & 0 & -1 \end{bmatrix}, \quad C = \begin{bmatrix} 0 & 1 & 0 & 0 \end{bmatrix}.
$$

It can be easily verified that this system is C-controllable, but not C-observable. To obtain a canonical decomposition, we begin by determining its standard decomposition. In this system, matrix A is nonsingular. Let $Q_1 = A^{-1}$, $P_1 = I_4$. Then

$$
E_1 = Q_1 E P_1 = A^{-1} E =
\begin{bmatrix}
-1 & -1 & 0 & 0 \\
-1 & -1 & -1 & 0 \\
0 & 1 & 0 & 0 \\
1 & 0 & 0 & 0
\end{bmatrix},
\tag{4.125}
$$

and

$$
Q_1 A P_1 = I_4.
$$

Choosing

$$
T =
\begin{bmatrix}
1 & 0 & 0 & 0 \\
0 & 1 & 0 & 0 \\
0 & 0 & 1 & 0 \\
1 & 0 & 0 & 1
\end{bmatrix},
$$

we have

$$
T E_1 T^{-1} =
\begin{bmatrix}
-1 & -1 & 0 & 0 \\
-1 & -1 & -1 & 0 \\
0 & 1 & 0 & 0 \\
0 & -1 & 0 & 0
\end{bmatrix}.
\tag{4.126}
$$

Denoting $Q = T Q_1$ and $P = P_1 T^{-1}$, from (4.125) and (4.126), it is easy to obtain

$$
QEP = T E_1 T^{-1}, \quad QAP = I_4.
$$

Thus, the coordinate transformation

$$
\begin{bmatrix} x_1 \\ x_2 \end{bmatrix} = P^{-1} x = T x, \quad x_1 \in \mathbb{R}^3, \quad x_2 \in \mathbb{R}
$$

transfers the above system into the following canonical form

$$
\begin{cases}
\begin{bmatrix}
-1 & -1 & 0 & 0 \\
-1 & -1 & -1 & 0 \\
0 & 1 & 0 & 0 \\
\hline
0 & -1 & 0 & 0
\end{bmatrix}
\begin{bmatrix} \dot{x}_1 \\ \dot{x}_2 \end{bmatrix}
=
\begin{bmatrix} x_1 \\ x_2 \end{bmatrix}
+
\begin{bmatrix}
-1 \\ -1 \\ 0 \\ \hline -1
\end{bmatrix} u \\
\\
y = \begin{bmatrix} 0 & 1 & 0 & 0 \end{bmatrix}
\begin{bmatrix} x_1 \\ x_2 \end{bmatrix}
\end{cases}.
$$

The above decomposition is not only the canonical form but also the C-observable canonical form. In this decomposition, the substate x_1 is both C-controllable and C-observable; but substate x_2 is only C-controllable.

Remark 4.3. By using the inverse form for regular descriptor linear systems, we may obtain the canonical decomposition more directly. First take the coordinate transformation to obtain the inverse canonical form, then perform the controllability and observability decomposition on $(\hat{E}, \hat{B}, C)$, and finally the canonical form can be obtained. This approach will not be discussed here.

4.8 Transfer Function Matrix and Minimal Realization

In this section, we study the transfer function matrix and minimal realization of descriptor linear systems based on the canonical structural decomposition introduced in the preceding section.

4.8.1 Transfer Function Matrix

As previously proven, for a given system (4.118), there exist two nonsingular matrices Q and P such that (4.118) is r.s.e. to the canonical form (4.121)–(4.122). Since the transfer function matrices of restricted equivalent systems are the same, to study the transfer function matrix of system (4.118), we need only to investigate that of the canonical form (4.121)–(4.122). Based on this idea, through some direct but tedious induction, the following result can be shown.

Theorem 4.24. *Let the structural canonical decomposition form of the regular descriptor system (4.118) be given by (4.121). Then*

$$G(s) = C(sE - A)^{-1} B = \tilde{C}_1(s\tilde{E}_{11} - \tilde{A}_{11})^{-1}\tilde{B}_1. \qquad (4.127)$$

The above result tells us that the transfer function matrix for a regular descriptor linear system is determined by the C-controllable and C-observable subsystem, and has no relation with the other uncontrollable or unobservable parts. Accounting for this, the transfer function matrix reflects only the most important part of a system: the C-controllable and C-observable part, which is the control effect of the input on the measured outputs, without any knowledge of control on the state.

Using (4.127) and the relations (4.122), we can further obtain the following result.

Theorem 4.25. *Let the structural canonical decomposition form of the regular descriptor system (4.118) be given by (4.121)–(4.122). Then*

$$G(s) = C(sE - A)^{-1} B = G_1(s) + G_2(s)$$

with

$$G_1(s) = C_{11}(sI - A_{11})^{-1} B_{11}$$

and

$$G_2(s) = C_{21}(sN_{11} - I)^{-1}B_{21}.$$

Another distinguishing feature may be seen from the above theorem: Different from the case of normal systems, whose transfer function matrix is strictly proper. The transfer function matrix of a descriptor system generally has two parts. One is $G_1(s)$, which is determined by the slow subsystem and is strictly proper; the other is a polynomial $G_2(s)$, which is determined by the fast subsystem (refer to Proposition 3.3).In the general case, $G_2(s) \neq 0$. Different from the strict properness in the normal case, $G_2(s)$ for a descriptor system is rational, but not necessarily proper. This is a special feature of descriptor systems.

Example 4.19. Consider again the system (3.39) in Example 3.5, whose standard decomposition is given in (4.22). According to Theorem 4.25, the transfer function matrix for the system is

$$
\begin{aligned}
G(s) &= C_1(sI - A_1)^{-1}B_1 + C_2(sN - I)^{-1}B_2 \\
&= \frac{1}{s^2}\begin{bmatrix} 1+s & -s & 2s-1 \\ -1-s & s & 1-2s \end{bmatrix} + \begin{bmatrix} 0 & 0 & 0 \\ 0 & -1 & -s \end{bmatrix} \\
&= \frac{1}{s^2}\begin{bmatrix} 1+s & -s & 2s-1 \\ -1-s & s(1-s) & 1-2s-s^3 \end{bmatrix},
\end{aligned}
$$

which coincides with the results in Example 3.7.

For a single input single output system, its transfer function

$$G(s) = \frac{b_{n-1}s^{n-1} + \cdots + b_1 s + b_0}{s^{n_1} + a_{n_1-1}s^{n_1-1} + \cdots + a_1 s + a_0}$$

is a rational transfer function.

In general, when $n_1 < n$, $G(s)$ is not proper if $b_{n-1} \neq 0$. Thus, viewing from the point of transfer function, the only difference between normal and descriptor systems is the strict properness of transfer function. If it is strictly proper, the input–output relationship may be described by a normal system, otherwise, the state space representation must be in a descriptor system form. As for multi-input multi-output systems, the difference lies in the properness of the transfer function matrices.

As pointed out earlier, two r.s.e. systems have the same transfer function matrix. Furthermore, we can prove the following theorem.

Theorem 4.26. *Two C-controllable and C-observable systems* (E, A, B, C) *and* $(\tilde{E}, \tilde{A}, \tilde{B}, \tilde{C})$ *have the same transfer function matrix if and only if they are r.s.e.*

Proof. This is an immediate corollary of Theorems 4.24 and 3.6.   □

This theorem assures that the transfer function matrix is uniquely determined by the C-controllable and C-observable subsystem. If controllability and observability are not assumed, this theorem may be not true.

Example 4.20. For the following descriptor linear system

$$
\left\{
\begin{aligned}
\begin{bmatrix} 1 & 2 & 8 & 2 & 4 \\ 0 & 1 & 4 & 0 & 2 \\ 2 & 2 & 0 & 2 & 0 \\ 4 & 0 & 0 & 8 & 0 \\ 0 & 4 & 16 & 0 & 8 \end{bmatrix} \dot{x} =
\begin{bmatrix} 2 & 0 & 2 & 4 & 1 \\ 5 & 4 & 0 & 6 & 0 \\ 0 & 0 & 5 & 0 & 6 \\ 10 & 10 & 0 & 10 & 0 \\ 20 & 16 & 2 & 24 & 1 \end{bmatrix} x +
\begin{bmatrix} 3 & -2 & 3 \\ 1 & -1 & 4 \\ 0 & 1 & 0 \\ 4 & 0 & 1 \\ 4 & -4 & 16 \end{bmatrix} u \\
y = \begin{bmatrix} 0 & 1 & 4 & 0 & 2 \\ 0 & -1 & 1 & 0 & 4 \end{bmatrix} x
\end{aligned}
\right.
$$

$$(4.128)$$

we can derive the transfer function as follows:

$$
G(s) = \frac{1}{s^2} \begin{bmatrix} 1+s & -s & 2s-1 \\ -1-s & s(1-s) & 1-2s-s^3 \end{bmatrix}.
$$

It is shown in Examples 3.7 and 4.19 that system (3.39) has the same transfer function. But systems (3.39) and (4.128) are not r.s.e. to each other since they have different orders, and hence do not have the same controllability and observability properties.

4.8.2 Minimal Realization

As mentioned earlier, for a given descriptor system, we may obtain its transfer function matrix. Conversely, for a given transfer function matrix $G(s)$, a certain descriptor system may be found whose transfer function matrix coincides with $G(s)$. This is the so-called realization theory.

Basic Theory

To be precise, let us give the following definition.

Definition 4.11. Assume that $G(s) \in \mathbb{R}^{m \times r}(s)$ is a rational matrix. If there exist constant matrices $E, A \in \mathbb{R}^{n \times n}$, $B \in \mathbb{R}^{n \times r}$ and $C \in \mathbb{R}^{m \times n}$ such that

$$
G(s) = C(sE - A)^{-1}B,
$$

then the system

$$
\begin{cases} E\dot{x} = Ax + Bu \\ y = Cx \end{cases}
$$

$$(4.129)$$

is called a descriptor system realization of $G(s)$, or simply, a realization of $G(s)$. Furthermore, if any other realization has an order greater than n, system (4.129) is called the minimal order realization or, simply, minimal realization.

Example 4.21. It is seen from Examples 4.19 and 4.20 that both the systems in (3.39) and (4.128) are the realizations of the transfer function

$$G(s) = \frac{1}{s^2} \begin{bmatrix} 1+s & -s & 2s-1 \\ -1-s & s(1-s) & 1-2s-s^3 \end{bmatrix}.$$

However, system (4.128) is of higher order than the system (3.39).

Lemma 4.13. *Let $g(s)$ be a rational function*

$$g(s) = \frac{b(s)}{a(s)}.$$

Then there always exists a strictly proper rational function $g_1(s)$ and a polynomial $g_2(s)$ such that

$$g(s) = g_1(s) + g_2(s). \tag{4.130}$$

Proof. Let $d = \deg b(s)$, $l = \deg a(s)$. Here, deg represents the degree of a polynomial. If $d < l$, $g(s)$ is strictly proper. In this case, the lemma is true for $g_1(s) = g(s)$, $g_2(s) = 0$; otherwise, $d \geq l$, using the polynomial division theorem, we know that there exist polynomials $q(s)$ and $r(s)$ such that

$$b(s) = q(s)a(s) + r(s)$$

and

$$\deg(r(s)) < \deg(a(s)).$$

Therefore,

$$g(s) = \frac{b(s)}{a(s)} = \frac{r(s)}{a(s)} + q(s).$$

Let

$$g_1(s) = \frac{r(s)}{a(s)} \text{ and } g_2(s) = q(s),$$

then $g_1(s)$ is strictly proper, $g_2(s)$ is a polynomial, and they satisfy (4.130). The proof is then done. $\qquad\square$

Theorem 4.27. *Any rational matrix $G(s) \in \mathbb{R}^{m \times r}(s)$ may be represented as*

$$G(s) = G_1(s) + G_2(s), \tag{4.131}$$

where $G_1(s)$ is a strictly proper rational matrix and $G_2(s)$ is a polynomial matrix.

Proof. Let $G(s) = (g_{ij}(s))_{m \times r}$, where $g_{ij}(s)$ is the element at the i-th row and j-th column, which is rational. From Lemma 4.13, there exist strictly proper function $g_{ij}^1(s)$ and polynomial $g_{ij}^2(s)$ such that

$$g_{ij}(s) = g_{ij}^1(s) + g_{ij}^2(s), \; i = 1, 2, \ldots, m, \; j = 1, 2, \ldots, r.$$

Setting

$$G_1(s) = \left[g_{ij}^1(s) \right]_{m \times r}, \; G_2(s) = \left[g_{ij}^2(s) \right]_{m \times r},$$

it is easy to verify that (4.131) holds.   □

Example 4.22. Consider the transfer function matrix

$$G(s) = \frac{1}{s^2 + 2s + 1} \begin{bmatrix} s+1 \\ s^4 + 2s^3 - s \end{bmatrix}.$$

Since

$$\frac{s^4 + 2s^3 - s}{s^2 + 2s + 1} = s^2 - 1 + \frac{1}{s+1},$$

$G(s)$ may be rewritten in the form of

$$G(s) = G_1(s) + G_2(s),$$

with

$$G_1(s) = \frac{1}{s+1} \begin{bmatrix} 1 \\ 1 \end{bmatrix}, \; G_2(s) = \begin{bmatrix} 0 \\ s^2 - 1 \end{bmatrix},$$

where $G_1(s)$ is clearly strictly proper rational, and $G_2(s)$ is a polynomial.

Lemma 4.14. *For any polynomial matrix $P(s)$, there always exist matrices N, B_2 and C_2, where N is nilpotent, such that*

$$P(s) = C_2(sN - I)^{-1} B_2.$$

Proof. Consider the strictly proper matrix $\bar{G}(s) = -\frac{1}{s} P\left(\frac{1}{s}\right)$, which may be viewed as the transfer function matrix of a normal system. Thus, according to linear system theory, there exist matrices N, B_2 and C_2, such that

$$\bar{G}(s) = -\frac{1}{s} P\left(\frac{1}{s}\right) = C_2(sI - N)^{-1} B_2.$$

Since $\bar{G}(s)$ has only zero poles, N has only zero eigenvalues. Thus, N is nilpotent. Direct computation results in

$$P(s) = C_2 (sN - I)^{-1} B_2.$$

Hence, system

$$N\dot{x} = x + B_2 u, \; y = C_2 x$$

is a descriptor system realization for $P(s)$.   □

Lemma 4.14 shows that any polynomial matrix $P(s)$ has a descriptor system realization with only a fast subsystem.

Theorem 4.28. *Any rational matrix may have a realization of (4.129), which satisfies*

$$G(s) \doteq C(sE - A)^{-1}B.$$

Furthermore, the realization is minimal if and only if the system is both C-controllable and C-observable.

Proof. According to Theorem 4.27, any rational matrix $G(s)$ may have a decomposition

$$G(s) = G_1(s) + G_2(s),$$

in which $G_1(s)$ is strictly proper and $G_2(s)$ is polynomial. For the strict properness of $G_1(s)$, from linear system theory, it always has a realization A_1, B_1, C_1, satisfying

$$G_1(s) = C_1(sI - A_1)^{-1}B_1.$$

Also, Lemma 4.14 guarantees the existence of the matrices N, B_2 and C_2, with N being nilpotent, such that

$$G_2(s) = C_2(sN - I)^{-1}B_2.$$

Let

$$E = \text{diag}(I, N), \; A = \text{diag}(A_1, I), \; B = \begin{bmatrix} B_1 \\ B_2 \end{bmatrix}, \; C = [C_1 \; C_2]. \qquad (4.132)$$

It is easy to verify that

$$G(s) = G_1(s) + G_2(s) = C(sE - A)^{-1}B.$$

The system determined by (4.132) is thus a realization of $G(s)$.

As for the second conclusion, we note that any transfer function matrix $G(s)$ may be decomposed into the sum of a strictly proper rational matrix $G_1(s)$ and a polynomial matrix $G_2(s)$, and the order of its realization is the sum of those of $G_1(s)$ and $G_2(s)$. Thus, system (4.129) is a minimal realization if and only if both (A_1, B_1, C_1) and (N, B_2, C_2) are minimal, or equivalently, if and only if (A_1, B_1, C_1) and (N, B_2, C_2) are minimal realizations of the strictly proper matrix $G_1(s)$ and the polynomial $G_2(s)$, respectively, indicating both (A_1, B_1, C_1) and (N, B_2, C_2) are C-controllable and C-observable. This in turn implies the minimal realization (4.132) is C-controllable and C-observable. □

A direct result of the combination of Theorems 4.26 and 4.28 is the following corollary.

Corollary 4.5. *Any two minimal realizations for a rational matrix are r.s.e.*

Example 4.23. Consider again the systems (3.39) and (4.128). It follows from Examples 4.19 and 4.20 that they are both realizations of

$$G(s) = \frac{1}{s^2} \begin{bmatrix} 1+s & -s & 2s-1 \\ -1-s & s(1-s) & 1-2s-s^3 \end{bmatrix}.$$

Since the system (3.39) is C-controllable and C-observable, it is minimal, but (4.128) is not.

Realization Algorithm

Theorem 4.28 shows the existence of realizations and minimal realizations. Meanwhile, the proof of Theorem 4.28 also provides us with a method to find a realization of a general $m \times r$ rational matrix $G(s)$.

Algorithm 4.2. Minimal realization of rational matrices.

Step 1 Decompose the rational matrix $G(s)$ into two parts $G(s) = G_1(s) + G_2(s)$, with G_1 strictly proper and G_2 a polynomial.

Step 2 Find the minimal realization of the strictly proper matrix $G_1(s)$ using any effective algorithm in normal linear systems theory for realization of transfer functions.

Step 3 Find the minimal realization of the polynomial matrix $G_2(s)$.

Step 4 Obtain the minimal realization (4.129) using the matrix in the way of (4.132).

Regarding the usage of the above algorithm, a natural question is how Step 3 is to be carried out. In the following, let us provide an algorithm to solve the Step 3 in the above algorithm.

For a given polynomial matrix,

$$P(s) = P_0 + P_1 s + \cdots + P_{k-1} s^{k-1} \in \mathbb{R}^{m \times r} [s], \qquad (4.133)$$

Lemma 4.14 assures the existence of $B_2 \in \mathbb{R}^{n_2 \times r}$, $C_2 \in \mathbb{R}^{m \times n_2}$ and the nilpotent matrix $N \in \mathbb{R}^{n_2 \times n_2}$ such that

$$P(s) = C_2(sN - I)^{-1} B_2. \qquad (4.134)$$

Let us now introduce the Silverman-Ho algorithm to search the minimal realization.

Algorithm 4.3. Minimal realization of polynomial matrices.

Step 1 Form the matrices

$$M_0 = \begin{bmatrix} -P_0 & -P_1 & \cdots & -P_{k-2} & -P_{k-1} \\ -P_1 & -P_2 & \cdots & -P_{k-1} & 0 \\ \cdots & \cdots & \cdots & \cdots & \cdots \\ -P_{k-2} & -P_{k-1} & \cdots & & 0 \\ -P_{k-1} & 0 & \cdots & & 0 \end{bmatrix} \in \mathbb{R}^{km \times kr}$$

and

$$M_1 = \begin{bmatrix} -P_1 & -P_2 & \cdots & -P_{k-1} & 0 \\ -P_2 & -P_3 & \cdots & 0 & 0 \\ \cdots & \cdots & \cdots & \cdots & 0 \\ -P_{k-1} & 0 & \cdots & \cdots & 0 \\ 0 & 0 & \cdots & \cdots & 0 \end{bmatrix} \in \mathbb{R}^{km \times kr}.$$

Step 2 Compute $\tilde{n}_2 = \text{rank} M_0$. Then take the full-rank decomposition

$$M_0 = L_1 L_2, \tag{4.135}$$

where $L_1 \in \mathbb{R}^{km \times \tilde{n}_2}$, $L_2 \in \mathbb{R}^{\tilde{n}_2 \times kr}$ are of full-column and full-row rank, respectively.

Step 3 Let $\tilde{B}_2$ and $\tilde{C}_2$ be matrices formed by the first r columns of L_2 and the first m rows of L_1, respectively, and compute

$$\tilde{N} = \left(L_1^{\mathrm{T}} L_1\right)^{-1} L_1^{\mathrm{T}} M_1 L_2^{\mathrm{T}} \left(L_2 L_2^{\mathrm{T}}\right)^{-1}. \tag{4.136}$$

Proposition 4.3. *For the polynomial matrix (4.133), the matrices $\tilde{N}$, $\tilde{B}_2$ and $\tilde{C}_2$ obtained by the above algorithm form a minimal realization for $P(s)$.*

Proof. The existence of minimal realization is proven in Lemma 4.14 and Theorem 4.28. There exists a C-controllable and C-observable triple (N, B_2, C_2), with N nilpotent, such that (4.134) holds, i.e.,

$$P_0 + P_1 s + \cdots + P_{k-1} s^{k-1} = -C_2 B_2 - C_2 N B_2 s - \cdots - C_2 N^{k-1} B_2 s^{k-1}.$$

Equaling the coefficients of the same order on both sides of this equation gives

$$-P_i = C_2 N^i B_2, \quad i = 0, 1, 2, \cdots, k - 1.$$

(If k is less than the nilpotent index h of N, the coefficients of the higher order terms are considered to be zero). Thus,

$$M_0 = H_1 H_2, \quad M_1 = H_1 N H_2, \tag{4.137}$$

where

$$H_1 = \begin{bmatrix} C_2 \\ C_2 N \\ \vdots \\ C_2 N^{k-1} \end{bmatrix}, \quad H_2 = \begin{bmatrix} B_2 & N B_2 & \cdots & N^{k-1} B_2 \end{bmatrix}.$$

Recalling our assumption that (N, B_2, C_2) is minimal, we see that H_1 and H_2 are of full-column and full-row rank, respectively.

Noticing $n_2 = \text{rank} H_1 = \text{rank} H_2$, we have $\tilde{n}_2 = \text{rank} M_0 = \text{rank} H_1 = n_2$. Furthermore, from (4.135) and (4.137) we can easily see

$$L_1 L_2 = H_1 H_2. \tag{4.138}$$

Pre-multiplying both sides of (4.138) by L_1^T and paying attention to the nonsingularity of $L_1^T L_1$, we obtain

$$L_2 = \left(L_1^T L_1\right)^{-1} L_1^T H_1 H_2.$$

By selection, $n_2 = \text{rank} L_2 = \text{rank} H_2$, and if

$$T = \left(L_1^T L_1\right)^{-1} L_1^T H_1 \in \mathbb{R}^{n_2 \times n_2},$$

we know that T is nonsingular and

$$L_2 = T H_2 = \begin{bmatrix} TB_2 & TNB_2 & \cdots & TN^{k-1}B_2 \end{bmatrix}. \tag{4.139}$$

According to the selection of $\tilde{B}_2$, from (4.139) we have

$$\tilde{B}_2 = TB_2. \tag{4.140}$$

The substitution of (4.139) into (4.138) results in $L_1 T H_2 = H_1 H_2$. Post-multiplying this equation by H_2^T and noticing that $H_2 H_2^T$ is nonsingular, we have

$$L_1 = H_1 T^{-1} = \begin{bmatrix} C_2 T^{-1} \\ C_2 N T^{-1} \\ \vdots \\ C_2 N^{k-1} T^{-1} \end{bmatrix}.$$

Thus,

$$\tilde{C}_2 = C_2 T^{-1}. \tag{4.141}$$

Furthermore, the selection method of $\tilde{N}$ in (4.136), together with (4.137) and

$$L_2 = T H_2, \quad L_1 = H_1 T^{-1},$$

yields

$$\tilde{N} = \left(L_1^T L_1\right)^{-1} L_1^T H_1 N H_2 L_2^T \left(L_2 L_2^T\right)^{-1} = TNT^{-1}. \tag{4.142}$$

Equations (4.140)–(4.142) show that the realization has the same order with the minimal one (N, B_2, C_2), and the two realizations are similar. Thus, $\left(\tilde{N}, \tilde{B}_2, \tilde{C}_2\right)$ is a minimal realization for $P(s)$.   □

The above method for finding the minimal realization for a polynomial matrix is relatively simple.

For finding computationally a minimal realization of a transfer function matrix of a descriptor linear system, one can also use directly the command tm2dss in the Matlab Toolbox for linear descriptor systems introduced in Varga (2000).

Example 4.24 (Dai 1989b). Consider the following polynomial matrix

$$G_2(s) = \begin{bmatrix} 0 \\ s^2 - 1 \end{bmatrix}.$$

Clearly, its minimal realization is ready when the minimal realization of polynomial $P(s) = s^2 - 1$ is obtained. Next, let us find its minimal realization using the above Algorithm 4.3.

For $P(s)$, we have

$$M_0 = \begin{bmatrix} 1 & 0 & -1 \\ 0 & -1 & 0 \\ -1 & 0 & 0 \end{bmatrix}, \ M_1 = \begin{bmatrix} 0 & 1 & 0 \\ -1 & 0 & 0 \\ 0 & 0 & 0 \end{bmatrix}.$$

Note that M_0 is nonsingular, thus we may choose $L_1 = M_0$ and $L_2 = I_3$. Therefore,

$$\tilde{B}_2 = \begin{bmatrix} 1 & 0 & 0 \end{bmatrix}^{\mathrm{T}}, \ \tilde{C}_2 = \begin{bmatrix} 1 & 0 & -1 \end{bmatrix},$$

and according to (4.136),

$$\tilde{N} = \left(L_1^{\mathrm{T}} L_1\right)^{-1} L_1^{\mathrm{T}} M_1 L_2^{\mathrm{T}} \left(L_2 L_2^{\mathrm{T}}\right)^{-1} = M_0^{-1} M_1 = \begin{bmatrix} 0 & 0 & 0 \\ 1 & 0 & 0 \\ 0 & 1 & 0 \end{bmatrix}.$$

Hence, $(\tilde{N}, \tilde{B}_2, \tilde{C}_2)$ is a minimal realization of $P(s) = s^2 - 1$, and the descriptor system

$$\begin{cases} \begin{bmatrix} 0 & 0 & 0 \\ 1 & 0 & 0 \\ 0 & 1 & 0 \end{bmatrix} \dot{x} = x + \begin{bmatrix} 1 \\ 0 \\ 0 \end{bmatrix} u \\ y = \begin{bmatrix} 0 & 0 & 0 \\ 1 & 0 & -1 \end{bmatrix} x \end{cases}$$

is a minimal realization of $G_2(s)$.

4.9   Notes and References

This chapter introduces the various types of controllabilities and observabilities as well as their criteria. Decomposition of descriptor linear systems by controllability and/or observability is also investigated. Furthermore, transfer function matrices of descriptor linear systems are discussed and minimal realizations of transfer function matrices are also explored.

Relations among the different types of controllabilities and observabilities have been shown at the end of Sects. 4.2 and 4.3. Table 4.1 gives criteria for the different types of Controllabilities and observabilities of a descriptor linear system in terms of its slow and fast subsystems. To use these criteria, one needs to carry out the standard decomposition form of the system. Again, to derive computationally a standard decomposition form of a given descriptor linear system, the Matlab Toolbox for linear descriptor systems introduced in Varga (2000) can be readily used, which was developed based on the basic techniques proposed in Misra et al. (1994). As a matter of fact, Misra et al. (1994) has developed an algorithm for computing the zeros of a generalized state space model. This algorithm can be viewed as an (almost) universal analysis tool for linear time-invariant systems. Properties such as stability, controllability, observability, stabilizability, or detectability, etc., can be easily obtained by computing the zeros of appropriate system matrices.

Regarding criteria for controllability and observability, both direct and indirect ones are presented in the chapter. The direct ones employ the original system coefficient matrices and are thus more convenient to use. Tables 4.2 and 4.3 list the direct criteria for controllabilities and observabilities.

As in the case for normal linear systems, a dual principle holds for the corresponding controllability and observability. This principle allows us to derive, based on a result about controllability, the corresponding result about observability very easily, and vice versa.

Table 4.1 Criteria based on standard decomposition

Subsystem	Controllability			Observability		
	C-	R-	I-	C-	R-	I-
slow-subsystem	✓	✓		✓	✓	
fast-subsystem	✓		✓	✓		✓

Table 4.2 Direct criteria for controllabilities

Item	Criterion
C-controllability	$\operatorname{rank}[\alpha E - \beta A\ B] = n, \ \forall\, (\alpha, \beta) \in \mathbb{C}^2 \backslash \{(0,0)\}$
R-controllability	$\operatorname{rank}[sE - A\ B] = n, \ \forall s \in \mathbb{C}, \ s \text{ finite}$
I-controllability	$\operatorname{rank} \begin{bmatrix} E & 0 & 0 \\ A & E & B \end{bmatrix} = n + \operatorname{rank} E$

Table 4.3 Direct criteria for observabilities

Item	Criterion
C-observability	$\operatorname{rank} \begin{bmatrix} \alpha E - \beta A \\ C \end{bmatrix} = n, \ \forall\, (\alpha, \beta) \in \mathbb{C}^2 \backslash \{(0,0)\}$
R-observability	$\operatorname{rank} \begin{bmatrix} sE - A \\ C \end{bmatrix} = n, \ \forall s \in \mathbb{C}, \ s \text{ finite}$
I-observability	$\operatorname{rank} \begin{bmatrix} E & A \\ 0 & E \\ 0 & C \end{bmatrix} = n + \operatorname{rank} E$

It is well known that the transfer function matrix of a normal linear system is always physically realizable or proper. While for descriptor linear systems, as we have seen in this chapter, the transfer function matrices may be improper. This is a big difference between normal linear systems and descriptor linear systems.

Regarding the concepts introduced in Sect. 4.1, a close reference is Yip and Sincovec (1981), while concerning the controllability and observability theory for descriptor linear systems, there are numerous reported results. For theories of various controllabilities, observabilities and duality relations, please refer to Cobb (1984), Bender and Laub (1985), Campbell (1982), Christodoulou and Paraskevopoulos (1985), Cullen (1986a), Kalyuthnaya (1978), Lewis and Ozcaldiran (1984), Lewis (1985a), Zhou et al. (1987), Gontian and Tarn (1982), Verghese et al. (1981), Chen and Duan (2006), Yan and Duan (2006), and Yan (2007). For indirect criteria which are based on the three equivalent forms, one can refer to Dai (1987) and Zhou et al. (1987). The main contents in Sects. 4.7 and 4.8 are closely related to Verghese et al. (1981) and Cullen (1986b).

Besides the concepts introduced in this chapter, there are also other types of controllabilities and observabilities for descriptor linear systems, for example, the structural controllability (Aoki et al. 1983; Murota 1983; Yamada and Luenberger 1985a,b), D_i–controllability (Pandolfi 1980), and robust controllability (Chou et al. 2006).

It needs to be pointed out that under the strong restricted system equivalence (Verghese et al. 1981), the I-controllability, controllability at infinity (see, e.g., Verghese et al. 1981), and the controllability of the fast subsystem turns to be the same.

More recently, many researchers have paid attention to the controllability and/or observability of more complicated descriptor linear systems. Su et al. (2004) studied impulsive controllability of linear periodic descriptor systems, Chou et al. (2001) gave sufficient conditions for the controllability of linear descriptor systems with both time-varying structured and unstructured parameter uncertainties, Byers and Mehrmann (1997) investigated the structure that can be achieved by feedback in descriptor systems that lack controllability at infinity; while Xie and Wang (2002) and Xie and Wang (2003) treated the controllability of linear descriptor systems with multiple time delays in control. Also, causal observability of descriptor systems has also been studied by Hou and Muller (1999) using a matrix pencil approach.

This chapter has also considered realization of transfer function matrices of descriptor linear systems. Again, we point out that besides the algorithms presented in Sect. 4.8, to find computationally a minimal realization of a transfer function matrix of a descriptor linear system, one can also use directly the command tm2dss in the Matlab Toolbox for linear descriptor systems introduced in Varga (2000). Regarding an extension of the realization problem, one can refer to Benner and Sokolov (2006), which provides a general solution to the partial realization problem for linear descriptor systems using the Markov parameters of the system defined via its Laurent series.

Part II
Descriptor Linear Systems Design

Chapter 5
Regularization of Descriptor Linear Systems

It has been seen from Chapter 3 that regularity is an important property for
descriptor linear systems and it offers many nice features, including the existence
and uniqueness of solutions to descriptor linear systems. Therefore, for descriptor
linear systems which are not regular, finding some feedback controllers such that
the closed-loop system is regular is a very basic and important issue in descriptor
linear system theory. This issue is termed as regularization of descriptor systems
via feedback and is investigated in this chapter.

Sections 5.1 and 5.2 consider regularizability of a linear descriptor system via
state proportional plus/or derivative feedback. Simple and convenient necessary and
sufficient conditions are obtained, which are only dependent upon the open-loop
coefficient matrices. In Sect. 5.3, the problem of finding a regularizing feedback
controller for a regularizable descriptor linear system is examined. It is shown that
once the system is regularizable via a certain type of feedback controllers, then "almost all" of this type of feedback controllers can regularize the system. Section 5.4
provides a proof of Theorem 5.7 about the regularizability conditions, and some
notes and comments are given in Sect. 5.5.

The main results in this chapter are taken from Duan and Zhang (2002c, 2003).

5.1 Regularizability under P- (D-) Feedback

In this section, we investigate regularizability of a descriptor linear system via
proportional or derivative feedback. Let us start with the case of proportional
feedback.

5.1.1 Proportional Feedback

First, let us give the formulation of the problem of regularizability under proportional feedback.

G.-R. Duan, *Analysis and Design of Descriptor Linear Systems*, Advances
in Mechanics and Mathematics 23, DOI 10.1007/978-1-4419-6397-0_5,
© Springer Science+Business Media, LLC 2010

Problem Formulation

Consider the following linear, time-invariant descriptor system

$$\begin{cases} E\dot{x} = Ax + Bu \\ y_p = C_p x \end{cases}, \tag{5.1}$$

where $x \in \mathbb{R}^n$, $u \in \mathbb{R}^r$ and $y_p \in \mathbb{R}^m$ are, respectively, the state-variable vector, the input vector, and the proportional output vector, and A, $E \in \mathbb{R}^{n \times n}$, $B \in \mathbb{R}^{n \times r}$ and $C_p \in \mathbb{R}^{m \times n}$ are known matrices. Without loss of generality, we make the following assumption.

Assumption 5.1. $\operatorname{rank} B = r$ and $\operatorname{rank} C_p = m$.

When an *output feedback controller* of the form

$$u = K_p y_p = K_p C_p x \tag{5.2}$$

is applied to system (5.1), the closed-loop system is obtained as

$$E\dot{x} = (A + BK_p C_p)x. \tag{5.3}$$

In the special case of $C_p = I$, we have $y_p = x$. In this case, the above controller (5.2) reduces to the following *state feedback controller*

$$u = K_p x, \tag{5.4}$$

and the corresponding closed-loop system (5.3) becomes

$$E\dot{x} = (A + BK_p)x.$$

Recall that system (5.1) is said to be regular if the matrix pair (E, A) is regular, i.e.,

$$\det(\alpha E - \beta A) \neq 0, \quad \text{for some } (\alpha, \beta) \in \mathbb{C}^2,$$

we have the following definition.

Definition 5.1. The system (5.1) is said to be regularizable via the *output feedback* controller (5.2) if the resulted closed-loop system (5.3) or the matrix pair $\left(E, \ A + BK_p C_p\right)$ is regular for some matrix $K_p \in \mathbb{R}^{r \times m}$. The system (5.1) is simply said to be regularizable if $C_p = I$. Furthermore, if the dual system of system (5.1) is regularizable, the system (5.1) is called dual regularizable.

With the above preparation, the problem to be solved in this subsection can be stated as follows.

Problem 5.1. Determine the necessary and sufficient conditions for regularizability of the descriptor linear system (5.1) via the output feedback controller (5.2), that is,

to establish necessary and sufficient conditions for the existence of the feedback
gain matrix $K_p \in \mathbb{R}^{r \times m}$ such that the matrix pair $(E, A + BK_pC_p)$ is regular.

Regularizability Conditions

Now let us represent the condition for the regularizability problem. Let us start with
the special case of $C_p = I$, the state feedback case.

Theorem 5.1. *Suppose* $\operatorname{rank} B = r$. *Then the following statements are equivalent:*

1. *The system (5.1) is regularizable via the state feedback controller (5.4).*
2. *There exist some* α, $\beta \in \mathbb{C}$ *such that*

$$\operatorname{rank} \begin{bmatrix} \alpha E - \beta A & B \end{bmatrix} = n.$$

3. *There exist some* $\gamma \in \mathbb{C}$ *such that*

$$\operatorname{rank} \begin{bmatrix} \gamma E - A & B \end{bmatrix} = n,$$

or

$$\operatorname{rank} \begin{bmatrix} E - \gamma A & B \end{bmatrix} = n.$$

4. *For any group of mutually distinct scalars* $\gamma_i \in \mathbb{C}, i = 1, \ldots, n - r + 1$, *there holds*

$$\max \left\{ \operatorname{rank} \begin{bmatrix} \gamma_i E - A & B \end{bmatrix}, i = 1, \ldots, n - r + 1 \right\} = n.$$

The above theorem is a special case of the following Theorem 5.2. It offers a
simple test method to determine whether the open-loop system (5.1) can be regularized via the state feedback controller (5.4). It clearly follows from the second and
the third conditions of the theorem that system (5.1) is regularizable via the state
feedback controller (5.4) if the system (5.1) is R-controllable.

Specially letting $(\alpha, \beta) = (1, 0)$ and $(\alpha, \beta) = (0, -1)$ in the second condition
of the above Theorem 5.1 immediately gives the following corollary.

Corollary 5.1. *Subject to* $\operatorname{rank} B = r$, *the system (5.1) is regularizable via the state
feedback controller (5.4) if*

$$\operatorname{rank} \begin{bmatrix} A & B \end{bmatrix} = n, \tag{5.5}$$

or

$$\operatorname{rank} \begin{bmatrix} E & B \end{bmatrix} = n. \tag{5.6}$$

Example 5.1. Consider a system in the form of (5.1) with the following parameters
(Dai, 1989b, p. 160):

$$E = \begin{bmatrix} 1 & 0 & 0 & 0 \\ 0 & 1 & 1 & 0 \\ 0 & 0 & 0 & 0 \\ 0 & 0 & 0 & 1 \end{bmatrix}, \quad B = \begin{bmatrix} 0 & 0 \\ 0 & 1 \\ 1 & 0 \\ 0 & 0 \end{bmatrix},$$

$$A = \begin{bmatrix} 1 & -1 & 0 & 0 \\ 0 & 1 & -1 & 1 \\ 1 & -2 & 0 & -1 \\ 0 & 0 & 0 & 0 \end{bmatrix}.$$

For this system we have $n = 4$, $r = 2$, and $n_0 = \text{rank} E = 3$. Since

$$\text{rank}\begin{bmatrix} E & -A & B \end{bmatrix} = 4,$$

it thus follows from Theorem 5.1 that the above system is regularizable via state feedback.

For solution to Problem 5.1 in the general case, we have the following theorem.

Theorem 5.2. *Suppose Assumption 5.1 holds. Then the following statements are equivalent:*

1. The system (5.1) is regularizable via the output feedback controller (5.2).
2. There exist some α, $\beta \in \mathbb{C}$ such that

$$\text{rank}\begin{bmatrix} \alpha E - \beta A & B \end{bmatrix} = \text{rank}\begin{bmatrix} \alpha E - \beta A \\ C_{\text{p}} \end{bmatrix} = n.$$

3. There exists some $\gamma \in \mathbb{C}$ such that

$$\text{rank}\begin{bmatrix} \gamma E - A & B \end{bmatrix} = \text{rank}\begin{bmatrix} \gamma E - A \\ C_{\text{p}} \end{bmatrix} = n,$$

or

$$\text{rank}\begin{bmatrix} E - \gamma A & B \end{bmatrix} = \text{rank}\begin{bmatrix} E - \gamma A \\ C_{\text{p}} \end{bmatrix} = n.$$

4. For any two groups of mutually distinct scalars $\gamma_i^c \in \mathbb{C}$, $i = 1,\ldots,n-r+1$, and $\gamma_j^o \in \mathbb{C}$, $j = 1,\ldots,n-m+1$, there hold

$$\max \left\{ \text{rank}\begin{bmatrix} \gamma_i^c E - A & B \end{bmatrix}, \ i = 1,\ldots,n-r+1 \right\} = n$$

and

$$\max \left\{ \text{rank}\begin{bmatrix} \gamma_j^o E - A \\ C_{\text{p}} \end{bmatrix}, \ j = 1,\ldots,n-m+1 \right\} = n.$$

The above theorem is a special case of the Theorem 5.7 presented in the next section, whose proof is provided in Sect. 5.4. It follows from Theorem 5.2 that the descriptor linear system (5.1) is regularizable by proportional output feedback if the system (E, A, B, C_{p}) is both R-controllable and R-observable.

Specially letting $\gamma = 0$ in the third condition of the above Theorem 5.2 immediately gives the following corollary.

Corollary 5.2. *Subject to Assumption 5.1, the system (5.1) is regularizable via the output feedback controller (5.2) if*

$$\text{rank} \begin{bmatrix} A & B \end{bmatrix} = \text{rank} \begin{bmatrix} A \\ C_p \end{bmatrix} = n,$$

or

$$\text{rank} \begin{bmatrix} E & B \end{bmatrix} = \text{rank} \begin{bmatrix} E \\ C_p \end{bmatrix} = n.$$

Example 5.2. Consider a system in the form of (5.1) with the following parameters (Fletcher 1988; Duan 1995; Duan et al. 1999b):

$$E = \begin{bmatrix} 1 & 0 & 0 & 0 \\ 0 & 1 & 0 & 0 \\ 0 & 0 & 1 & 0 \\ 0 & 0 & 0 & 0 \end{bmatrix}, \quad A = \begin{bmatrix} 0 & 0 & 1 & 0 \\ 1 & 0 & 0 & 0 \\ 0 & 1 & 0 & 1 \\ 0 & 0 & 1 & 0 \end{bmatrix},$$

$$B = \begin{bmatrix} 1 & 0 & 0 \\ 1 & -1 & 2 \\ 0 & 1 & 0 \\ 0 & 0 & 1 \end{bmatrix}, \quad C_p = \begin{bmatrix} 0 & 1 & 0 & 0 \\ 0 & 0 & 0 & 1 \end{bmatrix}.$$

For this system, it is easy to check that

$$\text{rank} \begin{bmatrix} A & B \end{bmatrix} = \text{rank} \begin{bmatrix} A \\ C_p \end{bmatrix} = 4.$$

Therefore, it follows from the above Corollary 5.2 that the system is regularizable via output feedback.

5.1.2 Derivative Feedback

Problem Formulation

Consider the following linear, time-invariant descriptor system

$$\begin{cases} E\dot{x} = Ax + Bu \\ y_d = C_d \dot{x} \end{cases}, \tag{5.7}$$

where $x \in \mathbb{R}^n$, $u \in \mathbb{R}^r$, and $y_d \in \mathbb{R}^q$ are the state-variable vector, the input vector, and the measured derivative output vector, respectively, and A, $E \in \mathbb{R}^{n \times n}$, $B \in \mathbb{R}^{n \times r}$, and $C_d \in \mathbb{R}^{q \times n}$ are known matrices. Without loss of generality, we assume the following:

Assumption 5.2. $\text{rank} B = r$ and $\text{rank} C_d = q$.

When a *partial-state derivative feedback controller* of the form

$$u = K_d y_d = K_d C_d \dot{x} \tag{5.8}$$

is applied to system (5.7), the closed-loop system is obtained as

$$(E - BK_d C_d)\dot{x} = Ax. \tag{5.9}$$

In the special case of $C_d = I$, we have $y_d = \dot{x}$. In this case, the above controller (5.8) reduces to the following *full-state derivative feedback controller*

$$u = K_d \dot{x}, \tag{5.10}$$

and the corresponding closed-loop system (5.9) becomes

$$(E - BK_d)\dot{x} = Ax.$$

Definition 5.2. The system (5.7) is said to be regularizable via the *partial-state derivative feedback* controller (5.8) if the resulted closed-loop system (5.9) or the matrix pair $(E - BK_d C_d,\ A)$ is regular for some matrix $K_d \in \mathbb{R}^{r \times q}$.

Based on the above preparation, the problem to be solved in this subsection can be stated as follows.

Problem 5.2. Determine the necessary and sufficient conditions for regularizability of the descriptor linear system (5.7) via the partial-state derivative feedback controller (5.8), that is, to establish necessary and sufficient conditions for the existence of the feedback gain matrix $K_d \in \mathbb{R}^{r \times q}$ such that the matrix pair $(E - BK_d C_d,\ A)$ is regular.

Regularizability Conditions

Regarding solution to the above regularizability problem, we first have the following result for the special case of $C_d = I$.

Theorem 5.3. *Suppose* $\text{rank} B = r$. *Then the following statements are equivalent:*

1. *The system (5.7) is regularizable via the full-state derivative feedback controller (5.10).*
2. *The system (5.7) is regularizable via the full-state feedback controller* $u = K_p x$.
3. *Any of the conditions 2~4 in Theorem 5.1 holds.*

Proof. We need only to show the equivalence between the first two conclusions. This clearly follows from the following relation

$$\det\left(\gamma\,(E - BK_{\mathrm{d}}) - A\right) = \det\left(\gamma E - A - B(\gamma K_{\mathrm{d}})\right)$$
$$= \det\left(\gamma E - (A + BK_{\mathrm{p}})\right),$$

where

$$K_{\mathrm{p}} = \gamma K_{\mathrm{d}}.$$

$\square$

Combining the above theorem with Corollary 5.1 immediately gives the following corollary.

Corollary 5.3. *Subject to* rank $B = r$, *the system (5.7) is regularizable via the full-state derivative feedback controller (5.10) if (5.5) or (5.6) holds.*

For solution to Problem 5.2 in the general case, we have the following result.

Theorem 5.4. *Suppose Assumption 5.2 holds. Then the following statements are equivalent:*

1. *The system (5.7) is regularizable via the partial-state derivative feedback controller (5.8).*
2. *There exist some* α, $\beta \in \mathbb{C}$ *such that*

$$\operatorname{rank}\left[\alpha E - \beta A \;\; B\right] = \operatorname{rank}\left[\begin{matrix} \alpha E - \beta A \\ C_{\mathrm{d}} \end{matrix}\right] = n.$$

3. *There exist some* $\gamma \in \mathbb{C}$ *such that*

$$\operatorname{rank}\left[\gamma E - A \;\; B\right] = \operatorname{rank}\left[\begin{matrix} \gamma E - A \\ C_{\mathrm{d}} \end{matrix}\right] = n,$$

or

$$\operatorname{rank}\left[E - \gamma A \;\; B\right] = \operatorname{rank}\left[\begin{matrix} E - \gamma A \\ C_{\mathrm{d}} \end{matrix}\right] = n.$$

4. *For any two groups of mutually distinct scalars* $\gamma_i^c \in \mathbb{C}$, $i = 1, \ldots, n - r + 1$, *and* $\gamma_j^o \in \mathbb{C}$, $j = 1, \ldots, n - q + 1$, *there hold*

$$\max\left\{\operatorname{rank}\left[\gamma_i^c E - A \;\; B\right], \; i = 1, \ldots, n - r + 1\right\} = n$$

and

$$\max\left\{\operatorname{rank}\left[\begin{matrix} \gamma_j^o E - A \\ C_{\mathrm{d}} \end{matrix}\right], \; j = 1, \ldots, n - q + 1\right\} = n.$$

Proof. Note

$$\det\left(\gamma\left(E - BK_dC_d\right) - A\right) = \det\left(\gamma E - A - B\gamma K_d C_d\right)$$
$$= \det\left(\gamma E - (A + BK_pC_d)\right),$$

where

$$K_p = \gamma K_d.$$

Therefore, applying Theorem 5.2 to the system (5.7) with the feedback controller

$$u = K_p C_d x$$

immediately gives the results.   □

It follows from Theorem 5.4 that the descriptor linear system (5.7) is regu-larizable by partial-state derivative feedback if the system (E, A, B, C_d) is both R-controllable and R-observable.

Specially letting $\gamma = 0$ in the third condition of the above Theorem 5.4 immedi-ately gives the following corollary.

Corollary 5.4. *Subject to Assumption 5.2, the system (5.7) is regularizable via the partial-state derivative feedback controller (5.8) if*

$$\text{rank}\begin{bmatrix} A & B \end{bmatrix} = \text{rank}\begin{bmatrix} A \\ C_d \end{bmatrix} = n,$$

or

$$\text{rank}\begin{bmatrix} E & B \end{bmatrix} = \text{rank}\begin{bmatrix} E \\ C_d \end{bmatrix} = n.$$

Example 5.3. Consider a system in the form of (5.7) with the matrices E, A, B, and C_d given as

$$E = \begin{bmatrix} 1 & 0 & 0 & 0 \\ 0 & 1 & 1 & 0 \\ 0 & 0 & 0 & 0 \\ 0 & 0 & 0 & 1 \end{bmatrix}, \quad A = \begin{bmatrix} 1 & -1 & 0 & 0 \\ 0 & 1 & -1 & 1 \\ 1 & -2 & 0 & -1 \\ 0 & 0 & 0 & 0 \end{bmatrix},$$

$$B = \begin{bmatrix} 0 & 0 \\ 0 & 1 \\ 1 & 0 \\ 0 & 0 \end{bmatrix}, \quad C_d = \begin{bmatrix} 1 & 0 & 0 & 0 \\ 0 & 0 & 0 & -1 \end{bmatrix},$$

where the matrices E, A, and B are taken from Example 5.1. For this system, we have $n = 4$, $r = 2$, $q = 2$ and $n_0 = 3$. Since

$$\text{rank}\begin{bmatrix} E - A & B \end{bmatrix} = 4$$

and

$$\text{rank} \begin{bmatrix} E - A \\ C_d \end{bmatrix} = 4,$$

it follows from the above Corollary 5.4 that this system is regularizable via partial-state derivative feedback.

5.2 Regularizability under P-D Feedback

5.2.1 Problem Formulation

Consider the following linear, time-invariant descriptor system

$$\begin{cases} E\dot{x} = Ax + Bu \\ y_p = C_p x \\ y_d = C_d \dot{x} \end{cases}, \tag{5.11}$$

where $x \in \mathbb{R}^n$, $u \in \mathbb{R}^r$, $y_p \in \mathbb{R}^m$, and $y_d \in \mathbb{R}^q$ are the state-variable vector, the input vector, the proportional output vector, and the measured derivative output vector, respectively, and A, $E \in \mathbb{R}^{n \times n}$, $B \in \mathbb{R}^{n \times r}$, $C_p \in \mathbb{R}^{m \times n}$ and $C_d \in \mathbb{R}^{q \times n}$ are known matrices. Without loss of generality, we assume the following:

Assumption 5.3. $\text{rank} B = r$, $\text{rank} C_p = m$ and $\text{rank} C_d = q$.

When an *output plus partial-state derivative feedback controller* of the form

$$\begin{aligned} u &= K_p y_p + K_d y_d \\ &= K_p C_p x + K_d C_d \dot{x} \end{aligned} \tag{5.12}$$

is applied to system (5.11), the closed-loop system is obtained as

$$(E - BK_d C_d)\dot{x} = (A + BK_p C_p)x. \tag{5.13}$$

The above controller (5.12) has the following special cases.

Case 1. $C_p = I$. In this case, we have $y_p = x$, and the above controller (5.12) reduces to the following *full-state plus partial-state derivative feedback controller*

$$\begin{aligned} u &= K_p x + K_d y_d \\ &= K_p x + K_d C_d \dot{x}, \end{aligned} \tag{5.14}$$

and the corresponding closed-loop system (5.13) becomes

$$(E - BK_d C_d)\dot{x} = (A + BK_p)x. \tag{5.15}$$

Case 2. $C_d = I$. In this case, we have $y_d = \dot{x}$. The above controller (5.12) reduces to the following *output plus full-state derivative feedback controller*

$$u = K_p y_p + K_d \dot{x}$$
$$= K_p C_p x + K_d \dot{x}, \qquad (5.16)$$

and the corresponding closed-loop system (5.13) becomes

$$(E - BK_d)\dot{x} = (A + BK_p C_p)x.$$

Case 3. $C_p = I$ and $C_d = I$. In this case, we have $y_p - x$ and $y_d = \dot{x}$. The above controller (5.14) reduces to the following *full-state plus full-state derivative feedback controller*

$$u = K_p x + K_d \dot{x},$$

and the corresponding closed-loop system (5.15) becomes

$$(E - BK_d)\dot{x} = (A + BK_p)x.$$

Case 4. $C_d = C_p$. In this case, we have $y_d = \dot{y}_p$. The above controller (5.12) reduces to the following *output plus output derivative feedback controller*

$$u = K_p y_p + K_d \dot{y}_p$$
$$= K_p C_p x + K_d C_p \dot{x}, \qquad (5.17)$$

and the corresponding closed-loop system (5.13) becomes

$$(E - BK_d C_p)\dot{x} = (A + BK_p C_p)x.$$

Definition 5.3. The system (5.11) is said to be regularizable via the *output plus partial-state derivative feedback* controller (5.12) if the resulted closed-loop system (5.13) or the matrix pair $\left(E - BK_d C_d, \ A + BK_p C_p\right)$ is regular for some matrices $K_d \in \mathbb{R}^{r \times q}$ and $K_p \in \mathbb{R}^{r \times m}$.

Based on the above preparation, the problem to be solved in this section can be stated as follows.

Problem 5.3. Determine the necessary and sufficient conditions for regularizability of the descriptor linear system (5.11) via the output plus partial-state derivative feedback controller (5.12), that is, to establish necessary and sufficient conditions for the existence of the feedback gain matrices $K_d \in \mathbb{R}^{r \times q}$ and $K_p \in \mathbb{R}^{r \times m}$ such that the matrix pair $(E - BK_d C_d, A + BK_p C_p)$ is regular.

5.2.2 Regularizability Conditions

For solution to Problem 5.3 in the special case of $C_p = I$ and/or $C_d = I$, we have the following theorem.

Theorem 5.5. *Suppose Assumption 5.3 holds. Then the following statements are equivalent:*

1. *The system (5.11) is regularizable via the full-state plus partial-state derivative feedback controller (5.14).*
2. *The system (5.11) is regularizable via the output plus full-state derivative feedback controller (5.16).*
3. *The system (5.11) is regularizable via the full-state derivative feedback controller $u = K_d \dot{x}$.*
4. *The system (5.11) is regularizable via the full-state feedback controller $u = K_p x$.*
5. *Any of the conditions 2~4 in Theorem 5.1 holds.*

Proof. Note

$$\det \left(\gamma \left(E - BK_dC_d \right) - (A + BK_p) \right)$$
$$= \det \left(\gamma E - A - B(\gamma K_dC_d + K_p) \right)$$
$$= \det \left(\gamma E - (A + BK_p') \right),$$

with

$$K_p' = \gamma K_dC_d + K_p.$$

This shows the equivalence between the first and the fourth conclusions.

Similarly, note

$$\det \left(\gamma \left(E - BK_d \right) - (A + BK_pC_p) \right)$$
$$= \det \left(\gamma E - A - B(\gamma K_d + K_pC_p) \right)$$
$$= \det \left(\gamma E - (A + BK_p') \right),$$

where

$$K_p' = \gamma K_d + K_pC_p.$$

This shows the equivalence between the second and the fourth conclusions.

The third conclusion is equivalent to the fourth one according to Theorem 5.3, and according to Theorem 5.1, the fifth conclusion is also equivalent to the fourth one. The proof is then completed. □

Combining the above theorem with Corollary 5.1 immediately gives the following.

Corollary 5.5. *Subject to Assumption 5.3, the system (5.11) is regularizable via the output plus full-state derivative feedback controller (5.12) (or the full-state plus partial-state derivative feedback controller (5.14)) if (5.5) or (5.6) holds.*

For the special case of $C_d = C_p$, we have the following result.

Theorem 5.6. *Suppose Assumption 5.3 holds and $C_d = C_p$. Then the following statements are equivalent:*

1. *The system (5.11) is regularizable via the output plus output derivative feedback controller (5.17).*
2. *The system (5.11) is regularizable via the output feedback controller $u = K_p y_p$.*
3. *The system (5.11) is regularizable via the partial-state derivative feedback controller $u = K_d y_d$.*
4. *Any of the conditions 2~4 in Theorem 5.2 (or Theorem 5.4) holds.*

Proof. Note

$$
\begin{aligned}
\det\left(\gamma\left(E - BK_d C_p\right) - (A + BK_p C_p)\right) \\
= \det\left(\gamma E - A - B(\gamma K_d + K_p)C_p\right) \\
= \det\left(\gamma E - (A + BK_p' C_p)\right),
\end{aligned}
$$

where

$$
K_p' = \gamma K_d + K_p.
$$

This shows the equivalence between the first two conclusions.

The equivalence between the second and the third conclusions clearly holds when $C_d = C_p$ in view of Theorems 5.2 and 5.4, the proof is then completed.   □

Combining the above theorem with Corollaries 5.2 and 5.4 immediately gives the following corollary.

Corollary 5.6. *Subject to Assumption 5.3, the system (5.11) is regularizable via the output plus output derivative feedback controller (5.17) if the conditions in Corollary 5.2 or Corollary 5.4 hold.*

For solution to Problem 5.3 in the general case, we have the following result.

Theorem 5.7. *Suppose Assumption 5.3 holds, and let*

$$
C = \begin{bmatrix} C_d \\ C_p \end{bmatrix}.
$$

Then the following statements are equivalent:

1. *The system (5.11) is regularizable via the output plus partial-state derivative feedback controller (5.12).*
2. *There exist some α, $\beta \in \mathbb{C}$ such that*

$$
\mathrm{rank}\begin{bmatrix} \alpha E - \beta A & B \end{bmatrix} = \mathrm{rank}\begin{bmatrix} \alpha E - \beta A \\ C \end{bmatrix} = n. \tag{5.18}
$$

3. *There exist some $\gamma \in \mathbb{C}$ such that*

$$\text{rank}\left[\gamma E - A \ B\right] = \text{rank}\left[\begin{matrix} \gamma E - A \\ C \end{matrix}\right] = n. \tag{5.19}$$

4. *There exist some $\gamma \in \mathbb{C}$ such that*

$$\text{rank}\left[E - \gamma A \ B\right] = \text{rank}\left[\begin{matrix} E - \gamma A \\ C \end{matrix}\right] = n.$$

5. *For any two groups of mutually distinct scalars $\gamma_i^c \in \mathbb{C}$, $i = 1, \ldots, n - r + 1$, and $\gamma_j^o \in \mathbb{C}$, $j = 1, \ldots, n - l + 1$, $l = \text{rank}C$, there hold*

$$\max\left\{\text{rank}\left[\gamma_i^c E - A \ B\right], \ i = 1, \ldots, n - r + 1\right\} = n \tag{5.20}$$

and

$$\max\left\{\text{rank}\left[\begin{matrix} \gamma_j^o E - A \\ C \end{matrix}\right], \ j = 1, \ldots, n - l + 1\right\} = n. \tag{5.21}$$

The proof of Theorem 5.7 is provided in Sect. 5.4.

Specially letting $\gamma = 0$ in the third and fourth conditions of the above Theorem 5.7 immediately gives the following corollary.

Corollary 5.7. *Subject to Assumption 5.3, the system (5.11) is regularizable via the output plus partial-state derivative feedback controller (5.12) if*

$$\text{rank}\left[A \ B\right] = \text{rank}\left[\begin{matrix} A \\ C \end{matrix}\right] = n, \tag{5.22}$$

or

$$\text{rank}\left[E \ B\right] = \text{rank}\left[\begin{matrix} E \\ C \end{matrix}\right] = n. \tag{5.23}$$

Example 5.4. Consider a system in the form of (5.11) with the following parameters

$$E = \begin{bmatrix} 1 & 0 & 0 & 0 \\ 0 & 1 & 1 & 0 \\ 0 & 0 & 0 & 0 \\ 0 & 0 & 0 & 1 \end{bmatrix}, \quad A = \begin{bmatrix} 1 & -1 & 0 & 0 \\ 0 & 1 & -1 & 1 \\ 1 & -2 & 0 & -1 \\ 0 & 0 & 0 & 0 \end{bmatrix},$$

$$B = \begin{bmatrix} 0 & 0 \\ 0 & 1 \\ 1 & 0 \\ 0 & 0 \end{bmatrix}, \quad C_p = \begin{bmatrix} 0 & 1 & 0 & 0 \\ 1 & 0 & 0 & -1 \\ 0 & 0 & 1 & 1 \end{bmatrix},$$

$$C_d = \begin{bmatrix} 1 & 0 & 0 & 0 \\ 0 & 0 & 0 & -1 \end{bmatrix},$$

where the matrices E, A, and B are taken from Example 5.1. It is easy to see that

$$\text{rank}\begin{bmatrix} E & B \end{bmatrix} = \text{rank} C = 4,$$

that is, (5.23) holds and the system is thus regularizable via output plus partial-state derivative feedback according to Corollary 5.7.

5.3 Regularizing Controllers

In the above two sections, we have given necessary and sufficient conditions for regularizability of descriptor linear systems via various feedback controllers. In this section, let us consider, in addition, the problem related to the solution of the regularizing controllers when the considered descriptor linear system is regularizable.

5.3.1 Problem Formulation

Consider, again, the following linear time-invariant descriptor system

$$\begin{cases} E\dot{x} = Ax + Bu \\ y_p = C_p x \\ y_d = C_d \dot{x} \end{cases}, \tag{5.24}$$

where $x \in \mathbb{R}^n$, $u \in \mathbb{R}^r$, $y_p \in \mathbb{R}^m$, and $y_d \in \mathbb{R}^q$ are the state-variable vector, the input vector, the proportional output vector, and the derivative output vector, respectively, and A, $E \in \mathbb{R}^{n \times n}$, $B \in \mathbb{R}^{n \times r}$, $C_p \in \mathbb{R}^{m \times n}$, and $C_d \in \mathbb{R}^{q \times n}$ are known matrices satisfying Assumption 5.3.

When an *output plus partial-state derivative feedback controller*

$$u = K_p y_p + K_d y_d = K_p C_p x + K_d C_d \dot{x} \tag{5.25}$$

is applied to system (5.24), the closed-loop system is obtained as

$$(E - BK_dC_d)\dot{x} = (A + BK_pC_p)x. \tag{5.26}$$

In this section, we investigate the problem of designing a feedback controller in the form of (5.25) such that the closed-loop system (5.26) is regular. Therefore, the following assumption is needed.

Assumption 5.4. System (5.24) is regularizable via the output plus partial-state derivative feedback controller (5.25).

Based on the above preparation, the regularization problem to be solved in the section can be stated as follows.

Problem 5.4. Given the descriptor linear system (5.24) satisfying Assumptions 5.3 and 5.4, seek a pair of feedback gain matrices $K_d \in \mathbb{R}^{r \times q}$ and $K_p \in \mathbb{R}^{r \times m}$ such that the matrix pair $(E - BK_dC_d, A + BK_pC_p)$ is regular.

To solve this problem, let us first present some preliminary results.

5.3.2 Preliminaries

The following definition and lemmas related to Zariski open sets are used in our treatment. They are taken from Wang and Soh (1999) and Potter et al. (1979).

Definition 5.4. A subset of $\mathbb{R}^{m \times n}$ (respectively, $\mathbb{C}^{m \times n}$) is a Zariski open set of $\mathbb{R}^{m \times n}$ (respectively, $\mathbb{C}^{m \times n}$) if it is nonempty and its complement is formed by the solutions to finitely many polynomial equations in $\mathbb{R}^{m \times n}$ (respectively, $\mathbb{C}^{m \times n}$).

Lemma 5.1. *Zariski open sets admit the following properties:*

1. *Zariski open sets in $\mathbb{R}^{m \times n}$ (respectively, $\mathbb{C}^{m \times n}$) are open and dense in $\mathbb{R}^{m \times n}$ (respectively, $\mathbb{C}^{m \times n}$).*
2. *Each Zariski open set in $\mathbb{C}^{m \times n}$ contains a largest subset, which is a Zariski open set in $\mathbb{R}^{m \times n}$.*
3. *The intersection of two Zariski open sets in $\mathbb{R}^{m \times n}$ (respectively, $\mathbb{C}^{m \times n}$) is also a Zariski open set in $\mathbb{R}^{m \times n}$ (respectively, $\mathbb{C}^{m \times n}$).*
4. *Any union of Zariski open sets in $\mathbb{R}^{m \times n}$ (respectively, $\mathbb{C}^{m \times n}$) is also a Zariski open set in $\mathbb{R}^{m \times n}$ (respectively, $\mathbb{C}^{m \times n}$).*

The importance of Zariski open sets lies in the following fact.

Proposition 5.1. *Let $\mathscr{F}$ be some field, and P be some property defined on $\mathscr{F}$. Define*

$$\Gamma = \{p| \ p \in \mathscr{F} \text{ and possesses property } P\}.$$

If Γ is a Zariski open set, then "almost all" $p \in \mathscr{F}$ possesses the property P.

The following lemma is due to Wang and Soh (1999), which performs an important function in this section.

Lemma 5.2. *Let $A \in \mathbb{R}^{m \times n}$, $B \in \mathbb{R}^{m \times h}$, and $C \in \mathbb{R}^{l \times n}$ be some fixed real matrices. Further, denote*

$$r(A, B, C) = \max_{K \in \mathbb{R}^{h \times l}} \text{rank}\,(A + BKC).$$

Then

$$r(A, B, C) - \min \left\{ \text{rank}\,[\, A \;\; B \,], \; \text{rank} \begin{bmatrix} A \\ C \end{bmatrix} \right\},$$

and the set

$$S = \left\{ K \in \mathbb{R}^{h \times l} \,\middle|\, \text{rank}\,(A + BKC) = r(A, B, C) \right\}$$

forms a Zariski open set, or equivalently, $\text{rank}\,(A + BKC)$ *reaches its maximum for "almost all" $K \in \mathbb{R}^{h \times l}$.*

5.3.3 The Conclusion

Let

$$\mathscr{S} = \left\{ (K_\text{d}, K_\text{p}) \,\middle|\, (K_\text{d}, K_\text{p}) \in \mathbb{R}^{r \times q} \times \mathbb{R}^{r \times m} \text{ and } (5.26) \text{ is regular} \right\}, \qquad (5.27)$$

then $\mathscr{S}$ is the set of all matrix pairs $(K_\text{d}, K_\text{p}) \in \mathbb{R}^{r \times q} \times \mathbb{R}^{r \times m}$ such that the closed-loop system (5.26) is regular. If this set is a Zariski open set, then for "almost all" matrix pairs (K_d, K_p) in $\mathbb{R}^{r \times q} \times \mathbb{R}^{r \times m}$ the closed-loop system (5.26) is regular. In this case, a "trial and test" procedure would be sufficient to find a particular matrix pair (K_d, K_p) such that the controller (5.25) regularizes the open-loop system (5.24). The following theorem tells us that the set $\mathscr{S}$ is indeed a Zariski open set.

Theorem 5.8. *Suppose system (5.24) satisfies Assumptions 5.3 and 5.4. Then the set $\mathscr{S}$ given by (5.27) is a Zariski open set.*

Proof. Since system (5.24) is regularizable via output plus partial-state derivative feedback, then there exist $K_\text{d} \in \mathbb{R}^{r \times q}$ and $K_\text{p} \in \mathbb{R}^{r \times m}$ such that $(E - BK_\text{d}C_\text{d}, A + BK_\text{p}C_\text{p})$ is regular. This is equivalent to the existence of $(\alpha, \beta) \in \mathbb{C}^2$ such that

$$\det \left(\alpha(E - BK_\text{d}C_\text{d}) - \beta(A + BK_\text{p}C_\text{p}) \right) \neq 0, \qquad (5.28)$$

or, equivalently,

$$\det\left((\alpha E - \beta A) - B\begin{bmatrix} \alpha K_d & \beta K_p \end{bmatrix}\begin{bmatrix} C_d \\ C_p \end{bmatrix}\right) \neq 0. \tag{5.29}$$

Further, according to Theorem 5.7, this is equivalent to (5.18). Let

$$\Lambda = \left\{(\alpha, \beta) \,|\, (\alpha, \beta) \in \mathbb{C}^2 \text{ satisfies } (5.18)\right\},$$

and denote, for any $(\alpha, \beta) \in \Lambda$,

$$\mathscr{S}_{(\alpha,\beta)} = \left\{(K_d, K_p) \,|\, (K_d, K_p) \text{ satisfies } (5.29)\right\}.$$

Then

$$\mathscr{S} = \bigcup_{(\alpha,\beta) \in \Lambda} S_{(\alpha,\beta)}.$$

For arbitrarily fixed $(\alpha, \beta) \in \Lambda$, it follows from (5.18) that

$$\min\left\{\operatorname{rank}\begin{bmatrix} \alpha E - \beta A & B \end{bmatrix}, \operatorname{rank}\begin{bmatrix} \alpha E - \beta A \\ C_d \\ C_p \end{bmatrix}\right\} = n. \tag{5.30}$$

Using (5.30) and applying Lemma 5.2 to the matrices

$$A_0 = \alpha E - \beta A, \quad B_0 = B, \quad C_0 = \begin{bmatrix} C_d \\ C_p \end{bmatrix},$$

yields

$$r\left(\alpha E - \beta A, B, \begin{bmatrix} C_d \\ C_p \end{bmatrix}\right) = n,$$

and implies the fact that the set

$$\left\{(K_d, K_p) \,|\, (K_d, K_p) \text{ satisfies } (5.29)\right\}$$

is a Zariski open set, or equivalently, the set $S_{(\alpha,\beta)}$ is a Zariski open set. This, together with the fourth condition of Lemma 5.1, implies that the set S is also a Zariski open set. □

The above theorem reveals, in view of Proposition 5.1, a very important fact, that is, "almost all" pairs of matrices $K_d \in \mathbb{R}^{r \times q}$ and $K_p \in \mathbb{R}^{r \times m}$ can regularize the open-loop system (5.24) if system (5.24) is regularizable via output plus partial-state derivative feedback. Therefore, a regularizing feedback controller for system (5.24) can, when it exists, be easily sought by a trial-and-test procedure. From this sense, the Problem 5.4 is solved. Such a fact tells us that regularization, as a requirement for descriptor linear systems design, is very easy to achieve, and often can be ignored if the system under consideration is regularizable.

Example 5.5. Consider the three-link planar manipulator system (1.29) in Sect. 1.2.4. The system is in the form of (5.24) with

$$E = \begin{bmatrix} I & 0 & 0 \\ 0 & M_0 & 0 \\ 0 & 0 & 0 \end{bmatrix}, \quad A = \begin{bmatrix} 0 & I & 0 \\ -K_0 & -D_0 & F_0^{\mathrm{T}} \\ F_0 & 0 & 0 \end{bmatrix}, \quad B = \begin{bmatrix} 0 \\ S_0 \\ 0 \end{bmatrix},$$

and

$$M_0 = \begin{bmatrix} 18.7532 & -7.94493 & 7.94494 \\ -7.94493 & 31.8182 & -26.8182 \\ 7.94494 & -26.8182 & 26.8182 \end{bmatrix},$$

$$D_0 = \begin{bmatrix} -1.52143 & -1.55168 & 1.55168 \\ 3.22064 & 3.28467 & -3.28467 \\ -3.22064 & -3.28467 & 3.28467 \end{bmatrix},$$

$$K_0 = \begin{bmatrix} 67.4894 & 69.2393 & -69.2392 \\ 69.8124 & 1.68624 & -1.68617 \\ -69.8123 & -1.68617 & -68.2707 \end{bmatrix},$$

$$S_0 = \begin{bmatrix} -0.216598 & -0.338060 & 0.554659 \\ 0.458506 & -0.845153 & 0.386648 \\ -0.458506 & 0.845153 & 0.613353 \end{bmatrix},$$

$$F_0 = \begin{bmatrix} 1 & 0 & 0 \\ 0 & 0 & 1 \end{bmatrix}, \quad C_{\mathrm{p}} = \begin{bmatrix} 0 & 1 & 0 & 0 & 0 & 0 & 0 & 0 \\ 0 & 0 & 0 & 0 & 0 & 0 & 1 & 0 \\ 0 & 0 & 0 & 0 & 0 & 0 & 0 & 1 \end{bmatrix}.$$

Thus, we have

$$E = \begin{bmatrix} 1 & 0 & 0 & 0 & 0 & 0 & 0 & 0 \\ 0 & 1 & 0 & 0 & 0 & 0 & 0 & 0 \\ 0 & 0 & 1 & 0 & 0 & 0 & 0 & 0 \\ 0 & 0 & 0 & 18.7532 & -7.9449 & 7.9449 & 0 & 0 \\ 0 & 0 & 0 & -7.9449 & 31.8182 & -26.8182 & 0 & 0 \\ 0 & 0 & 0 & 7.9449 & -26.8112 & 26.8112 & 0 & 0 \\ 0 & 0 & 0 & 0 & 0 & 0 & 0 & 0 \\ 0 & 0 & 0 & 0 & 0 & 0 & 0 & 0 \end{bmatrix},$$

$$
A = \begin{bmatrix}
0 & 0 & 0 & 1 & 0 & 0 & 0 & 0 \\
0 & 0 & 0 & 0 & 1 & 0 & 0 & 0 \\
0 & 0 & 0 & 0 & 0 & 1 & 0 & 0 \\
-67.4894 & -69.2393 & 69.2393 & 1.5214 & 1.5517 & -1.5517 & 1 & 0 \\
-69.8124 & -1.6862 & 1.6862 & -3.2206 & -3.2847 & 3.2847 & 0 & 0 \\
69.8123 & 1.6862 & 68.2707 & 3.2206 & 3.2847 & -3.2847 & 0 & 1 \\
1 & 0 & 0 & 0 & 0 & 0 & 0 & 0 \\
0 & 0 & 1 & 0 & 0 & 0 & 0 & 0
\end{bmatrix},
$$

$$
B = \begin{bmatrix}
0 & 0 & 0 \\
0 & 0 & 0 \\
0 & 0 & 0 \\
-0.2166 & -0.0338 & 0.5547 \\
0.4585 & -0.8452 & 0.3866 \\
-0.4585 & 0.8452 & 0.6134 \\
0 & 0 & 0 \\
0 & 0 & 0
\end{bmatrix}.
$$

For this system, we have $n = 8$, $r = 3$ and $m = 3$. Noting that the matrix A is nonsingular in this example, (5.22) holds. Therefore, it follows from Corollary 5.7 that this system is regularizable via output plus partial-state derivative feedback for arbitrarily given C_d. Further, according to Theorem 5.8, "almost all" pairs of matrices $K_d \in \mathbb{R}^{r \times q}$ and $K_p \in \mathbb{R}^{r \times m}$ can regularize this system. In fact, when taking

$$
C_d = \begin{bmatrix}
1 & 1 & 0 & 0 & 0 & 0 & 0 & 0 \\
0 & 0 & -1 & 1 & 0 & 0 & 0 & 0
\end{bmatrix},
$$

using the Matlab command fix(100*rand(m,n)), we chose 10,000 pairs of matrices $K_d \in \mathbb{R}^{3 \times 2}$ and $K_p \in \mathbb{R}^{3 \times 3}$, with entries being integers randomly chosen from $[0\ 100]$, and found that every matrix pair regularizes the system.

Example 5.6. Consider the system treated in Example 5.4, which has the following parameters

$$
E = \begin{bmatrix}
1 & 0 & 0 & 0 \\
0 & 1 & 1 & 0 \\
0 & 0 & 0 & 0 \\
0 & 0 & 0 & 1
\end{bmatrix}, \quad
A = \begin{bmatrix}
1 & -1 & 0 & 0 \\
0 & 1 & -1 & 1 \\
1 & -2 & 0 & -1 \\
0 & 0 & 0 & 0
\end{bmatrix},
$$

$$
B = \begin{bmatrix}
0 & 0 \\
0 & 1 \\
1 & 0 \\
0 & 0
\end{bmatrix}, \quad
C_p = \begin{bmatrix}
0 & 1 & 0 & 0 \\
1 & 0 & 0 & -1 \\
0 & 0 & 1 & 1
\end{bmatrix}, \quad
C_d = \begin{bmatrix}
1 & 0 & 0 & 0 \\
0 & 0 & 0 & -1
\end{bmatrix}.
$$

As shown in Example 5.4 that this system is regularizable via output plus partial-state derivative feedback. According to Theorem 5.8, "almost all" pairs of matrices $K_d \in \mathbb{R}^{r \times q}$ and $K_p \in \mathbb{R}^{r \times m}$ can regularize the system. Using the Matlab command `fix(100*rand(m,n))` again, we chose randomly 10,000 pairs of matrices $K_d \in \mathbb{R}^{2 \times 2}$ and $K_p \in \mathbb{R}^{2 \times 3}$, with entries being integers randomly chosen from $[0 \quad 100]$, and found that every matrix pair regularizes the system.

Remark 5.1. Theorem 5.8 is about regularization of descriptor linear systems via output plus partial-state derivative feedback controllers. Since this theorem does not require any other assumptions except some rank requirements on some of the coefficient matrices, it naturally holds for the special cases of $C_p = I$ and/or $C_d = I$, $C_p = 0$ or $C_d = 0$ as well as $C_p = C_d$. Therefore, similar conclusions hold for all the other types of static feedback controllers that have been referred to in this chapter.

5.4 Proof of Theorem 5.7

To prove Theorem 5.7, we need some preliminary results.

5.4.1 Preliminary Results

The following lemma is an obvious corollary of Lemma 5.2.

Lemma 5.3. *Suppose $\Phi \in \mathbb{R}^{n \times n}$, $\Psi \in \mathbb{R}^{n \times r}$, and $\Omega \in \mathbb{R}^{q \times n}$ are given matrices. Then there exists $X \in \mathbb{R}^{r \times q}$ such that* rank$(\Phi - \Psi X \Omega) = n$ *if and only if*

$$\text{rank} \begin{bmatrix} \Phi & \Psi \end{bmatrix} = \text{rank} \begin{bmatrix} \Phi \\ \Omega \end{bmatrix} = n.$$

The following result states that the rank of a matrix is robust in the sense that it remains unchanged when small parameter perturbation is added.

Lemma 5.4. *Suppose $M \in \mathbb{R}^{n \times m}$ with* rank$M = n$. *Then, for any given matrix $N \in \mathbb{R}^{n \times m}$, there exists a nonzero $k_N \in \mathbb{C}$ with scalar $|k_N|$ sufficiently small such that*

$$\text{rank}(M - k_N N) = n.$$

Proof. It follows from rank$M = n$ that there exists a nonsingular matrix $P \in \mathbb{R}^{m \times m}$ such that

$$M = \begin{bmatrix} I_n & 0 \end{bmatrix} P. \tag{5.31}$$

For any given matrix $N \in \mathbb{R}^{n \times m}$, let

$$[N_1 \ N_2] = NP^{-1}$$

with $N_1 \in \mathbb{R}^{n \times n}$. Then

$$N = [N_1 \ N_2] P. \tag{5.32}$$

Thus, it follows from (5.31) to (5.32) that for any $k \in \mathbb{C}$, there holds

$$\begin{aligned} M - kN &= [I \ 0] P - k [N_1 \ N_2] P \\ &= [I - kN_1 \quad -kN_2] P. \end{aligned} \tag{5.33}$$

Further, notice that there exists nonzero $k_N \in \mathbb{C}$ with $|k_N|$ sufficiently small such that $I - k_N N_1$ is nonsingular. We have from (5.33) that there exists $k_N \in \mathbb{C}$ with $|k_N|$ sufficiently small such that

$$\operatorname{rank}(M - k_N N) = \operatorname{rank}[I - k_N N_1 \quad -k_N N_2] = n.$$

With this, we complete the proof. $\qquad \square$

Lemma 5.5. *Suppose $M, N \in \mathbb{R}^{n \times m}$ and $m \geq n$. Then there exists some $\alpha \in \mathbb{C}$ such that*

$$\operatorname{rank}(\alpha M - N) = n,$$

if and only if for arbitrarily given distinct scalars $\gamma_i \in \mathbb{C}$, $i = 1, \ldots, n + 1$, there holds

$$\max \{\operatorname{rank}(\gamma_i M - N), \quad i = 1, \ldots, n + 1\} = n.$$

Proof. The sufficiency is obvious. To show the necessity, we need only to prove that

$$\operatorname{rank}(\gamma M - N) < n, \quad \forall \gamma \in \mathbb{C} \tag{5.34}$$

if there exist some distinct $\gamma_i \in \mathbb{C}$, $i = 1, \ldots, n + 1$, satisfying

$$\operatorname{rank}(\gamma_i M - N) < n, \quad i = 1, \ldots, n + 1. \tag{5.35}$$

We proceed by reduction to absurdity. Suppose that there exist some distinct $\gamma_i \in \mathbb{C}$, $i = 1, \ldots, n + 1$, satisfying (5.35), but (5.34) does not hold. Then, there exist some $\gamma_0 \in \mathbb{C}$ such that

$$\operatorname{rank}(\gamma_0 M - N) = n.$$

Therefore, there exists a permutation matrix $P \in \mathbb{R}^{m \times m}$ such that

$$M = [M_1 \ M_2] P, \quad N = [N_1 \ N_2] P \tag{5.36}$$

and

$$\text{rank}\,(\gamma_0 M_1 - N_1) = n,$$

where $M_1, N_1 \in \mathbb{R}^{n \times n}$. This implies that

$$\det\,(\gamma_0 M_1 - N_1) \neq 0. \tag{5.37}$$

On the other hand, combining (5.35) and (5.36) yields

$$\text{rank}\,(\gamma_i M_1 - N_1) < n, \quad i = 1, \ldots, n + 1,$$

which is clearly equivalent to

$$\det\,(\gamma_i M_1 - N_1) = 0, \quad i = 1, \ldots, n + 1. \tag{5.38}$$

Since $\deg \det\,(s M_1 - N_1) \leq n$, it follows from (5.38) that

$$\det\,(s M_1 - N_1) \equiv 0.$$

This contradicts with (5.37). Therefore, (5.34) holds when the conditions in (5.35) are met.   □

5.4.2   Proof of Theorem 5.7

Equivalence between the First Two Conclusions

It follows from Definition 5.3 that system (5.11) is regularizable via output plus partial-state derivative feedback if and only if there exist $\alpha,\ \beta \in \mathbb{C}$, $K_d \in \mathbb{R}^{r \times q}$ and $K_p \in \mathbb{R}^{r \times m}$ such that (5.28), or equivalently, (5.29) holds.

In view of (5.29) and applying Lemma 5.3 to the matrices

$$\Phi = (\alpha E - \beta A), \quad \Psi = B, \quad \Omega = \begin{bmatrix} C_d \\ C_p \end{bmatrix},$$

it is easily seen that (5.29) holds for some $K_d \in \mathbb{R}^{r \times q}$ and $K_p \in \mathbb{R}^{r \times m}$ if and only if (5.18) is met. Thus, the first two conclusions are equivalent.

Equivalence between the Second and the Third Conclusions

Let the third conclusion of the theorem holds, that is, the condition (5.19) holds. Then it is obvious that (5.18) holds for $\alpha = \gamma$ and $\beta = 1$. Therefore, the third conclusion implies the second one.

Now we prove that the second conclusion implies the third. To do this, we assume that (5.18) is valid for some $(\alpha, \beta) \in \mathbb{C}^2$, and show that (5.19) holds. The proof is divided into three cases.

Case 1. If $\beta \neq 0$ and $\alpha \neq 0$, by taking $\gamma = \frac{\alpha}{\beta}$, and through dividing both sides of (5.18) by β, condition (5.18) turns into

$$\operatorname{rank}\begin{bmatrix} \gamma E - A & B \end{bmatrix} = \operatorname{rank}\begin{bmatrix} \gamma E - A \\ C_d \\ C_p \end{bmatrix} = n. \tag{5.39}$$

Condition (5.39) is obviously equivalent to (5.19) in view of $\gamma \neq 0$.

Case 2. When $\beta = 0$, (5.18) turns into

$$\operatorname{rank}\begin{bmatrix} \alpha E & B \end{bmatrix} = \operatorname{rank}\begin{bmatrix} \alpha E \\ C \end{bmatrix} = n \tag{5.40}$$

for some nonzero $\alpha \in \mathbb{C}$. Due to (5.40), by applying Lemma 5.4 to matrices $M = \begin{bmatrix} \alpha E & B \end{bmatrix}$ and $N = \begin{bmatrix} A & 0 \end{bmatrix}$, we have

$$\operatorname{rank}\begin{bmatrix} \alpha E - \beta_1 A & B \end{bmatrix} = n \tag{5.41}$$

for some nonzero $\alpha, \beta_1 \in \mathbb{C}$ with $|\beta_1|$ sufficiently small.

Similarly, by applying Lemma 5.4 to the matrices

$$M = \begin{bmatrix} \alpha E \\ C \end{bmatrix}^{\mathrm{T}} \quad \text{and} \quad N = \begin{bmatrix} A \\ 0 \end{bmatrix}^{\mathrm{T}},$$

we have

$$\operatorname{rank}\begin{bmatrix} \alpha E - \beta_2 A \\ C \end{bmatrix} = n \tag{5.42}$$

for some nonzero $\alpha, \beta_2 \in \mathbb{C}$ with $|\beta_2|$ sufficiently small. Combining (5.41) and (5.42)

$$\operatorname{rank}\begin{bmatrix} \alpha E - \beta_0 A & B \end{bmatrix} = \operatorname{rank}\begin{bmatrix} \alpha E - \beta_0 A \\ C \end{bmatrix} = n, \tag{5.43}$$

where

$$\beta_0 = \min\{\beta_1, \ \beta_2\}.$$

Finally, letting

$$\gamma = \frac{\alpha}{\beta_0},$$

and dividing both sides of (5.43) by β_0, condition (5.43) can be converted into (5.39), which is equivalent to condition (5.19) since $\gamma \neq 0$.

Case 3. When $\alpha = 0$, the proof is similar to Case 2.

Combining the above three cases completes the proof of the equivalence between the second and the third conclusions.

Equivalence between the Second and the Fourth Conclusions

Similar to the proof of the equivalence between the second and the third conclusions, the proof of the equivalence between the second and the fourth can be easily carried out.

Equivalence between the Third and the Fifth Conclusions

It is clearly seen that the third conclusion implies the fifth one. Therefore, it suffices to show that the fifth implies the third.

Due to Assumption 5.3, we can assume, without loss of generality, that

$$
B = \begin{bmatrix} B_1 \\ 0 \end{bmatrix}, \quad A = \begin{bmatrix} A_1 \\ A_2 \end{bmatrix}, \quad E = \begin{bmatrix} E_1 \\ E_2 \end{bmatrix}, \tag{5.44}
$$

where $B_1 \in \mathbb{R}^{r \times r}$ is nonsingular, and $A_1, E_1 \in \mathbb{R}^{r \times n}$.

First, let us show that there exists a nonzero $\gamma \in \mathbb{C}$ such that rank $[\gamma E - A \ B] = n$ if (5.20) holds.

Let (5.20) hold. Substituting (5.44) into (5.20) produces

$$
\max \left\{ \operatorname{rank} \begin{bmatrix} \gamma_i^c E_1 - A_1 & B_1 \\ \gamma_i^c E_2 - A_2 & 0 \end{bmatrix}, \ i = 1, \ldots, n - r + 1 \right\} = n.
$$

This implies that

$$
\max \left\{ \operatorname{rank} \left(\gamma_i^c E_2 - A_2 \right), \ i = 1, \ldots, n - r + 1 \right\} = n - r. \tag{5.45}
$$

Based on (5.45) and Lemma 5.5, we can find a $\gamma^c \in \mathbb{C}$ such that

$$
\operatorname{rank} \left(\gamma^c E_2 - A_2 \right) = n - r.
$$

This, together with (5.44), gives

$$
\operatorname{rank} \left[\gamma^c E - A \ B \right] = \operatorname{rank} \begin{bmatrix} \gamma^c E_1 - A_1 \ B_1 \\ \gamma^c E_2 - A_2 \ 0 \end{bmatrix}
$$

$$
= \operatorname{rank} \left(\gamma^c E_2 - A_2 \right) + \operatorname{rank} B_1
$$

$$
= n. \tag{5.46}
$$

Second, let us show that there exists a nonzero $\gamma \in \mathbb{C}$ such that

$$\text{rank} \begin{bmatrix} \gamma E - A \\ C \end{bmatrix} = n$$

if (5.21) holds.

It is easy to see that there exists a nonsingular matrix P such that

$$C = P \begin{bmatrix} C_1 \\ 0 \end{bmatrix}, \tag{5.47}$$

with $C_1 \in \mathbb{R}^{l \times n}$, $l = \text{rank} C_1$. Using (5.47), we have

$$\text{rank} \begin{bmatrix} \gamma_i^o E - A \\ C \end{bmatrix} = \text{rank} \begin{bmatrix} \gamma_i^o E - A \\ P \begin{bmatrix} C_1 \\ 0 \end{bmatrix} \end{bmatrix}$$

$$= \text{rank} \begin{bmatrix} \gamma_i^o E - A \\ C_1 \end{bmatrix},$$

$$i = 1, \ldots, n - l + 1.$$

Since (5.21) holds, it follows from the above relation and (5.21) that

$$\max \left\{ \text{rank} \begin{bmatrix} \gamma_i^o E - A \end{bmatrix}, \ i = 1, \ldots, n - l + 1 \right\} = n - l.$$

Using Lemma 5.5 again, we can find a $\gamma^o \in \mathbb{C}$ such that

$$\text{rank} \begin{bmatrix} \gamma^o E - A \end{bmatrix} = n - l.$$

Therefore,

$$\text{rank} \begin{bmatrix} \gamma^o E - A \\ C \end{bmatrix} = \text{rank} \begin{bmatrix} \gamma^o E - A \\ P \begin{bmatrix} C_1 \\ 0 \end{bmatrix} \end{bmatrix}$$

$$= \text{rank} \begin{bmatrix} \gamma^o E - A \\ C_1 \end{bmatrix}$$

$$= n. \tag{5.48}$$

Finally, note that once the relation (5.46) (or (5.48)) holds for one point, there are only a finite number of points, that do not satisfy this relation. Therefore, a common $\gamma \in \mathbb{C}$ can be found to satisfy (5.19). With this, we complete the proof.   □

5.5 Notes and References

Regularity is an important property for descriptor linear systems. It guarantees the existence and the uniqueness of the solution to a descriptor linear system (see Chap. 3 and also Yip and Sincovec 1981). Quite a few reported results for descriptor linear systems have assumed open-loop regularity, see, for example, Verghese (1981), Lewis (1986), and Dai (1989b), and the bibliographies therein. However, this assumption is unnecessarily strong since it limits the analysis of a number of practical physical systems (Lewis 1986; Ozcaldiran and Lewis 1990). Due to this practical reason, the problem of regularizing descriptor linear systems using various feedbacks has attracted much attention.

In this chapter, we studied regularization of descriptor linear systems by feedback. The problem can be generally stated as follows:

Given a typical descriptor linear system, find a feedback controller of certain type for the system such that the closed-loop system is regular.

A Brief Overview

Ozcaldiran and Lewis (1990) and Duan (1998b) studied regularization of descriptor linear systems via state feedback controllers and presented various necessary and sufficient conditions for the problem, while Owens and Askarpour (2001) and Duan and Patton (1997) concentrated on regularization of descriptor linear systems using proportional plus state derivative feedback controllers. In Fletcher (1988) and Duan et al. (1999b), regularization of descriptor linear systems via proportional output feedback was considered, together with eigenstructure assignment, and again necessary and sufficient conditions were given. Chu et al. (1999) considered regularization of a descriptor linear system using an output proportional plus derivative feedback and established necessary and sufficient conditions.

Besides giving necessary and sufficient conditions for the regularization problem, many researchers have also developed numerical algorithms for finding the regularizing feedback controllers, that guarantee certain desired characteristics of the closed-loop systems. One type of such problems is to design a feedback controller for a descriptor linear system such that the closed-loop system is regular and has index at most one (see Bunse-Gerstner et al. 1992, 1994, 1999; Chu et al. 1998, 1999; Chu and Ho 1999; Chu and Cai 2000). For this type of problems, Bunse-Gerstner et al. (1992) considered the state feedback case, while Bunse-Gerstner et al. (1994, 1999), and Chu et al. (1998) concentrated on the output feedback case. In Chu and Ho (1999), Bunse-Gerstner et al. (1999) and Chu and Cai (2000), the more general output proportional plus derivative feedback controllers were adopted. Different from Bunse-Gerstner et al. (1999) and Chu and Cai (2000), Chu et al. (1999) considered the problem of designing a feedback controller for a descriptor linear system such that the closed-loop system is regular, has index at most one, and moreover, possesses a desired dynamical order. Besides the above, the problems of designing a feedback controller for a descriptor linear system such that the closed-loop system is regular and impulse-free (Lovass-Nagy et al. 1994, 1996; Wang and Soh 1999) or regular and strongly controllable and strongly observable (Chu et al. 1998; Bunse-Gerstner et al. 1994) have also been investigated. Wang and Soh (1999) were focused on the existence of a decentralized output feedback control law or a P-D decentralized output feedback control law for a descriptor linear system such that the closed-loop system is regular and impulse-free, and at the same time, possesses a maximal dynamical order. In addition, the regularization problem of linear descriptor systems with variable coefficients has also been studied (see Byers et al. 1997; Kunkel et al. 2001).

Although regularization of descriptor linear systems has been researched by many authors as stated above, the problem of seeking an output plus partial-state derivative feedback for a descriptor linear system such that the closed-loop system is regular has not been investigated so far. Moreover, most of the reported results are based on certain decomposed forms of descriptor linear systems, and as a consequence, the derived conditions are not expressed by directly the original system coefficient matrices.

Further Comments

In this chapter, we have considered regularization for the cases of proportional feedback, derivative feedback, and P-D feedback, and have established necessary and sufficient conditions for the corresponding problems. The content is mainly based on the author's work in Duan and Zhang (2002c, 2003). The necessary and sufficient conditions established for the proposed regularization problems are neat and simple, and are dependent only on the open-loop system coefficient matrices. Regarding solutions of the regularizing output plus partial-state derivative feedback, it is shown based on the concept of Zariski open set that for a descriptor linear system which is regularizable via output plus partial-state derivative feedback, "almost all" output plus partial-state derivative feedback controllers can regularize the system. This indicates the important fact that such a regularizing controller can be easily obtained by a "trial-and-test" procedure, and in many design applications, the issue of closed-loop regularity can often be ignored.

"Regularizability" has been given different meanings in the history. The "regularizability" defined in this chapter has the same sense as those in Ozcaldiran and Lewis (1990), Owens and Askarpour (2001), Duan and Patton (1997), Fletcher (1988), and Fletcher et al. (1986), but is different from those defined in Bunse-Gerstner et al. (1999) and Mukundan and Dayawansa (1983). In Bunse-Gerstner et al. (1999), the special case of $C_d = C_p = C$ is treated, and the regularity of the system is defined in a different way: the matrix pair $\left(E - BK_dC, A + BK_pC \right)$ is required to be regular and has index at most one for some matrices $K_d \in \mathbb{R}^{r \times q}$ and $K_p \in \mathbb{R}^{r \times m}$. While in Mukundan and Dayawansa (1983), the special case of $C_d = C_p = I$ is treated, and the system is said to be regularizable if $E - BK_d$ is nonsingular for some $K_d \in \mathbb{R}^{r \times q}$. This is actually in our framework the problem of normalizability of descriptor systems using state derivative feedback (see Sect. 6.5 in Chap. 6).

Chapter 6
Dynamical Order Assignment and Normalization

It has been demonstrated by classical control that derivative feedback is very essential for improving the stability and the performance of a control system. For descriptor systems, derivative feedback is even more important since it can alter many properties of a descriptor system, which a pure proportional state feedback can not. One of such properties is the dynamical order, and this leads to the problem of dynamical order assignment, which is fundamental in descriptor systems theory.

In this chapter, we investigate the problem of dynamical order assignment in descriptor linear systems via full and partial state derivative feedback. In Sect. 6.1, two basic results about assignable dynamical orders in descriptor linear systems via full state derivative feedback and partial state derivative feedback are stated. The problem of dynamical order assignment via full state feedback is investigated in Sect. 6.2, where the set of all the state feedback gains which assign a certain dynamical order to the closed-loop system is characterized. Based on this result, the problem of dynamical order assignment via full state derivative feedback with minimum Frobenius norm is further studied in Sect. 6.3, and the set of optimal solutions is established. Section 6.4 deals with dynamical order assignment in descriptor linear systems with partial state derivative feedback. Based on the main results presented in Sects. 6.2 and 6.3, the problem is converted into dynamical order assignment via full state feedback. Normalization of descriptor linear systems via derivative feedback is a special problem of dynamical order assignment, and is investigated in Sect. 6.5. Some notes and comments follow in Sect. 6.6.

The main materials of this chapter are taken from the authors' work (Duan and Zhang (2002a,b)). It is worth being pointed out that this whole chapter, except Sect. 6.5, deals with nonsquare descriptor linear systems.

6.1 Assignable Dynamical Orders

In this section, we state two basic results about dynamical order assignment in linear nonsquare descriptor systems via state derivative feedback.

G.-R. Duan, *Analysis and Design of Descriptor Linear Systems*, Advances
in Mechanics and Mathematics 23, DOI 10.1007/978-1-4419-6397-0_6,
© Springer Science+Business Media, LLC 2010

6.1.1 Full-State Derivative Feedback

A nonsquare descriptor system of the linear, time-invariant type is described by

$$E\dot{x}(t) = Ax(t) + Bu(t), \tag{6.1}$$

where $x(t) \in \mathbb{R}^m$ is the descriptor vector, $u(t) \in \mathbb{R}^r$ is the control vector, and A, $E \in \mathbb{R}^{n \times m}$ and $B \in \mathbb{R}^{n \times r}$ are known real coefficient matrices.

Recall that rank E is the dynamical order of the system (6.1). Clearly, a proportional (descriptor-variable) feedback controller has no influence on the dynamical order since the matrix E is the same in both the open-loop and the closed-loop systems. However, when a derivative feedback controller of the form

$$u(t) = -K\dot{x}(t) \tag{6.2}$$

is applied to (6.1), a closed-loop system is obtained as

$$(E + BK)\dot{x}(t) = Ax(t). \tag{6.3}$$

The dynamical order of the above closed-loop system (6.3) is given by rank($E + BK$), which is dependent on the feedback gain matrix K. The problem of dynamical order assignment in the descriptor linear system (6.1) via the full-state derivative feedback controller (6.2) is to seek the feedback gain K such that the matrix $E + BK$ possesses a desired rank. For convenience, we give the following definition.

Definition 6.1. An integer p is called a dynamical order assignable to the system (6.1) or the matrix pair (E, B) via the full-state derivative feedback (6.2) if there exists a matrix $K \in \mathbb{R}^{r \times m}$ such that

$$\text{rank}(E + BK) = p.$$

In this case, it is also said that the integer p is an allowable dynamical order of the closed-loop system (6.3) or the matrix pair (E, B).

Regarding the assignable dynamical orders for the open-loop system (6.1), we have the following theorem.

Theorem 6.1. *Let $E \in \mathbb{R}^{n \times m}$ and $B \in \mathbb{R}^{n \times r}$ be fixed real matrices, and define*

$$\begin{cases} n_2 = \min \{\text{rank} \begin{bmatrix} E & B \end{bmatrix}, m\} \\ n_1 = \text{rank} \begin{bmatrix} E & B \end{bmatrix} - \text{rank} B \end{cases} . \tag{6.4}$$

Then, p is an allowable dynamical order assignable to system (6.1) by the full-state derivative feedback (6.2) if and only if

$$n_1 \leq p \leq n_2. \tag{6.5}$$

Proof. Denote

$$s = \operatorname{rank} B, \tag{6.6}$$

then there exist two nonsingular matrices $P \in \mathbb{R}^{n \times n}$ and $Q \in \mathbb{R}^{r \times r}$ satisfying

$$B = P \begin{bmatrix} I_s & 0 \\ 0 & 0 \end{bmatrix} Q. \tag{6.7}$$

Furthermore, let

$$P^{-1} E = \begin{bmatrix} E_1 \\ E_2 \end{bmatrix}, \quad E_1 \in \mathbb{R}^{s \times m}. \tag{6.8}$$

Combining (6.7) and (6.8) yields

$$
\begin{aligned}
\operatorname{rank} \begin{bmatrix} E & B \end{bmatrix} &= \operatorname{rank} \begin{bmatrix} P \begin{bmatrix} E_1 \\ E_2 \end{bmatrix} & P \begin{bmatrix} I_s & 0 \\ 0 & 0 \end{bmatrix} Q \end{bmatrix} \\
&= \operatorname{rank} \begin{bmatrix} E_1 & I_s & 0 \\ E_2 & 0 & 0 \end{bmatrix} \\
&= s + \operatorname{rank} E_2. \tag{6.9}
\end{aligned}
$$

This, together with (6.6) and (6.4), implies

$$n_1 = \operatorname{rank} E_2 \tag{6.10}$$

and

$$n_2 = \min\{n_1 + s, m\}. \tag{6.11}$$

It follows from (6.10) that there exist another pair of nonsingular matrices $P_1 \in \mathbb{R}^{(n-s) \times (n-s)}$ and $Q_1 \in \mathbb{R}^{m \times m}$ satisfying

$$E_2 = P_1 \begin{bmatrix} I_{n_1} & 0 \\ 0 & 0 \end{bmatrix} Q_1. \tag{6.12}$$

Using the matrix $Q_1 \in \mathbb{R}^{m \times m}$, we can define

$$\begin{bmatrix} E_{11} & E_{12} \end{bmatrix} = E_1 Q_1^{-1}, \quad E_{11} \in \mathbb{R}^{s \times n_1}. \tag{6.13}$$

With the above preparation, we can now show the necessity and the sufficiency of the conclusion.

Necessity: For any fixed $K \in \mathbb{R}^{r \times m}$, let

$$QK = \begin{bmatrix} K_1 \\ K_2 \end{bmatrix}, \quad K_1 \in \mathbb{R}^{s \times m}. \tag{6.14}$$

Combining (6.14), (6.7), and (6.8) clearly gives

$$\text{rank}\,(E + BK) = \text{rank}\left(P \begin{bmatrix} E_1 \\ E_2 \end{bmatrix} + P \begin{bmatrix} I_s & 0 \\ 0 & 0 \end{bmatrix} \begin{bmatrix} K_1 \\ K_2 \end{bmatrix} \right)$$

$$= \text{rank} \begin{bmatrix} E_1 + K_1 \\ E_2 \end{bmatrix}. \tag{6.15}$$

Furthermore, letting

$$\begin{bmatrix} K_{11} & K_{12} \end{bmatrix} = K_1 Q_1^{-1}, \quad K_{11} \in \mathbb{R}^{s \times n_1}, \tag{6.16}$$

and substituting (6.13), (6.12), and (6.16) into (6.15) yields

$$\text{rank}\,(E + BK) = \text{rank} \begin{bmatrix} \begin{bmatrix} E_{11} & E_{12} \end{bmatrix} Q_1 + \begin{bmatrix} K_{11} & K_{12} \end{bmatrix} Q_1 \\ P_1 \begin{bmatrix} I_{n_1} & 0 \\ 0 & 0 \end{bmatrix} Q_1 \end{bmatrix}$$

$$= \text{rank} \begin{bmatrix} E_{11} + K_{11} & E_{12} + K_{12} \\ I_{n_1} & 0 \\ 0 & 0 \end{bmatrix}$$

$$= n_1 + \text{rank}\,(E_{12} + K_{12}). \tag{6.17}$$

Since

$$\text{rank}\,(E_{12} + K_{12}) \le \min\{s, \, m - n_1\},$$

it follows from (6.17) that

$$n_1 \le \text{rank}\,(E + BK) \le n_1 + \min\{s, \, m - n_1\},$$

or, equivalently,

$$n_1 \le \text{rank}\,(E + BK) \le \min\{n_1 + s, \, m\}. \tag{6.18}$$

Combining (6.18) and (6.11) gives (6.5). This completes the proof of necessity.
Sufficiency: For any p satisfying (6.5), it is easy to see from (6.10) and (6.11) that

$$n_1 \le p \le \min\{n_1 + s, \, m\},$$

or, equivalently,

$$0 \le p - n_1 \le \min\{s, \, m - n_1\}.$$

Choosing

$$K_\text{p} = Q^{-1} \begin{bmatrix} \begin{bmatrix} 0_{s \times n_1} & -E_{12} + K_0 \end{bmatrix} Q_1 \\ 0_{(r-s) \times m} \end{bmatrix} \tag{6.19}$$

with

$$K_0 = \begin{bmatrix} I_{p-n_1} & 0 \\ 0 & 0_{(s-p+n_1)\times(m-p)} \end{bmatrix}, \tag{6.20}$$

then, by using the relations (6.7), (6.8), (6.12), (6.13), and (6.19), we can obtain

$$\begin{aligned}
E + BK_p &= P \begin{bmatrix} E_1 \\ E_2 \end{bmatrix} + P \begin{bmatrix} I_s & 0 \\ 0 & 0 \end{bmatrix} QK_p \\
&= P \begin{bmatrix} E_1 + \begin{bmatrix} 0_{s\times n_1} & -E_{12} + K_0 \end{bmatrix} Q_1 \\ E_2 \end{bmatrix} \\
&= P \begin{bmatrix} \begin{bmatrix} E_{11} & E_{12} \end{bmatrix} Q_1 + \begin{bmatrix} 0_{s\times n_1} & -E_{12} + K_0 \end{bmatrix} Q_1 \\ P_1 \begin{bmatrix} I_{n_1} & 0 \\ 0 & 0 \end{bmatrix} Q_1 \end{bmatrix} \\
&= P \begin{bmatrix} I_s & 0 \\ 0 & P_1 \end{bmatrix} \begin{bmatrix} E_{11} & K_0 \\ I_{n_1} & 0 \\ 0 & 0 \end{bmatrix} Q_1.
\end{aligned}$$

Combining the above with (6.20) clearly gives

$$\mathrm{rank}\,(E + BK_p) = n_1 + \mathrm{rank}K_0 = p,$$

i.e., p is an allowable dynamical order for system (6.3).   □

The above Theorem 6.1 is a generalization of the corresponding result reported in Chu et al. (1998), Chu et al. (1999), and Moor and Golub (1991), which is for square descriptor linear systems only.

The following lemma about a necessary and sufficient condition for $n_1 = n_0$ (respectively, $n_2 = n_0$), which corresponds to the case that a state feedback controller cannot further reduce (respectively, further increase) the dynamical order of the open-loop system (6.1).

Lemma 6.1. *Let $E \in \mathbb{R}^{n\times m}$ and $B \in \mathbb{R}^{n\times r}$ satisfy $0 < \mathrm{rank}E = n_0 \le \min\{m,n\}$ and $\mathrm{rank}B = r$, integers n_1 and n_2 be defined as in (6.4), and $U_1 \in \mathbb{R}^{n\times n}$ and $V_1 \in \mathbb{R}^{m\times m}$ be two orthogonal matrices and $D_1 \in \mathbb{R}^{n_0\times n_0}$ a positive definite diagonal matrix satisfying the following singular value decomposition*

$$E = U_1 \begin{bmatrix} D_1 & 0 \\ 0 & 0 \end{bmatrix} V_1^{\mathrm{T}}. \tag{6.21}$$

Furthermore, define matrices $B_1 \in \mathbb{R}^{n_0\times r}$ and $B_2 \in \mathbb{R}^{(n-n_0)\times r}$ by

$$\begin{bmatrix} B_1 \\ B_2 \end{bmatrix} = U_1^{\mathrm{T}} B. \tag{6.22}$$

Then

$$n_1 = n_0 \iff \mathrm{rank}\, B_2 = r \tag{6.23}$$

and

$$n_2 = n_0 \iff \mathrm{rank}\, B_2 = n_1 + r - n_2. \tag{6.24}$$

Proof. It follows from (6.21) and (6.22) that

$$
\begin{aligned}
\mathrm{rank}\begin{bmatrix} E & B \end{bmatrix} &= \mathrm{rank}\left[\, U_1 \begin{bmatrix} D_1 & 0 \\ 0 & 0 \end{bmatrix} V_1^{\mathrm{T}} \quad U_1 \begin{bmatrix} B_1 \\ B_2 \end{bmatrix} \right] \\
&= \mathrm{rank}\left(U_1 \begin{bmatrix} D_1 & 0 & B_1 \\ 0 & 0 & B_2 \end{bmatrix} \begin{bmatrix} V_1^{\mathrm{T}} & 0 \\ 0 & I \end{bmatrix} \right) \\
&= \mathrm{rank}\begin{bmatrix} D_1 & 0 & B_1 \\ 0 & 0 & B_2 \end{bmatrix} \\
&= n_0 + \mathrm{rank}\, B_2.
\end{aligned}
$$

This, together with (6.4), implies (6.23) and (6.24).   □

6.1.2 Partial-State Derivative Feedback

Consider the following nonsquare descriptor linear system:

$$\begin{cases} E\dot{x}(t) = Ax(t) + Bu(t) \\ z(t) = C\dot{x}(t) \end{cases}, \tag{6.25}$$

where $x(t) \in \mathbb{R}^m$ is the descriptor state vector, $u(t) \in \mathbb{R}^r$ is the control vector, $z(t) \in \mathbb{R}^q$ is the measured state derivative vector. $A,\ E \in \mathbb{R}^{n \times m}$, $B \in \mathbb{R}^{n \times r}$, and $C \in \mathbb{R}^{q \times m}$ are known real coefficient matrices.

When a partial-state derivative feedback controller of the form

$$u(t) = -Kz(t) = -KC\dot{x}(t), \quad K \in \mathbb{R}^{r \times q} \tag{6.26}$$

is applied to system (6.25), a closed-loop system is obtained as

$$(E + BKC)\dot{x}(t) = Ax(t). \tag{6.27}$$

The dynamical order of the above closed-loop system (6.27) is given by $\mathrm{rank}(E + BKC)$, which is dependent on the feedback gain matrix K. Parallel to Definition 6.1, we have the following definition.

Definition 6.2. An integer p is called a dynamical order assignable to the system (6.25) or the matrix triple (E, B, C) via the partial-state derivative feedback controller (6.26) if there exists a matrix K such that

$$\text{rank}(E + BKC) = p. \tag{6.28}$$

For simplicity, it is also said that p is an allowable dynamical order for the closed-loop system (6.27) or the matrix triple (E, B, C).

It is clear from the above definition that n_0 is an allowable dynamical order for the closed-loop system (6.27). Furthermore, regarding the allowable dynamical order for the matrix triple (E, B, C), the following result holds.

Theorem 6.2. *Let $E \in \mathbb{R}^{n \times m}$, $B \in \mathbb{R}^{n \times r}$ and $C \in \mathbb{R}^{q \times m}$, and define*

$$\begin{cases} \chi_1 = \text{rank} \begin{bmatrix} E & B \end{bmatrix} + \text{rank} \begin{bmatrix} E \\ C \end{bmatrix} - \text{rank} \begin{bmatrix} E & B \\ C & 0 \end{bmatrix} \\ \chi_2 = \min \left\{ \text{rank} \begin{bmatrix} E & B \end{bmatrix}, \text{rank} \begin{bmatrix} E \\ C \end{bmatrix} \right\} \end{cases}. \tag{6.29}$$

Then an integer p is a dynamical order assignable to the matrix triple (E, B, C) via the partial-state derivative feedback controller (6.26) if and only if

$$\chi_1 \leq p \leq \chi_2. \tag{6.30}$$

Proof. According to the proof of Theorem 6.1, we can assume that the relations (6.6)–(6.13) hold.

Letting

$$QK = \begin{bmatrix} K_1 \\ K_2 \end{bmatrix}, \quad K_1 \in \mathbb{R}^{s \times q}, \quad \forall K \in \mathbb{R}^{r \times q}, \tag{6.31}$$

it then follows from (6.7), (6.8), and (6.31) that

$$\text{rank}(E + BKC) = \text{rank} \left(P \begin{bmatrix} E_1 \\ E_2 \end{bmatrix} + P \begin{bmatrix} I_s & 0 \\ 0 & 0 \end{bmatrix} \begin{bmatrix} K_1 \\ K_2 \end{bmatrix} C \right)$$

$$= \text{rank} \begin{bmatrix} E_1 + K_1 C \\ E_2 \end{bmatrix}, \quad \forall K \in \mathbb{R}^{r \times q}. \tag{6.32}$$

Furthermore, letting

$$\begin{bmatrix} C_1 & C_2 \end{bmatrix} = CQ_1^{-1}, \quad C_1 \in \mathbb{R}^{q \times n_1}, \tag{6.33}$$

and substituting (6.13), (6.12), and (6.33) into (6.32) yields

$$\operatorname{rank}(E + BKC) = \operatorname{rank} \begin{bmatrix} \begin{bmatrix} E_{11} & E_{12} \end{bmatrix} Q_1 + K_1 \begin{bmatrix} C_1 & C_2 \end{bmatrix} Q_1 \\ P_1 \begin{bmatrix} I_{n_1} & 0 \\ 0 & 0 \end{bmatrix} Q_1 \end{bmatrix}$$

$$= \operatorname{rank} \begin{bmatrix} E_{11} + K_1 C_1 & E_{12} + K_1 C_2 \\ I_{n_1} & 0 \\ 0 & 0 \end{bmatrix}$$

$$= n_1 + \operatorname{rank}(E_{12} + K_1 C_2)$$

$$= n_1 + \operatorname{rank}(E_{12}^{\mathrm{T}} + C_2^{\mathrm{T}} K_1^{\mathrm{T}}), \quad \forall K \in \mathbb{R}^{r \times q}.$$

This, together with the arbitrariness of K and K_1, implies that an integer p is an allowable dynamical order for the matrix triple (E, B, C) if and only if $p - n_1$ is an allowable dynamical order for the matrix pair $(E_{12}^{\mathrm{T}}, C_2^{\mathrm{T}})$ by full-state derivative feedback. By applying Theorem 6.1, this is further equivalent to

$$\operatorname{rank} \begin{bmatrix} E_{12}^{\mathrm{T}} & C_2^{\mathrm{T}} \end{bmatrix} - \operatorname{rank} C_2^{\mathrm{T}} \leq p - n_1 \leq \min \left\{ \operatorname{rank} \begin{bmatrix} E_{12}^{\mathrm{T}} & C_2^{\mathrm{T}} \end{bmatrix}, s \right\},$$

i.e.,

$$n_1 + \operatorname{rank} \begin{bmatrix} E_{12} \\ C_2 \end{bmatrix} - \operatorname{rank} C_2 \leq p \leq n_1 + \min \left\{ \operatorname{rank} \begin{bmatrix} E_{12} \\ C_2 \end{bmatrix}, s \right\}.$$

Therefore, to complete the proof, it suffices to show in the following

$$\chi_1 = n_1 + \operatorname{rank} \begin{bmatrix} E_{12} \\ C_2 \end{bmatrix} - \operatorname{rank} C_2 \tag{6.34}$$

and

$$\chi_2 = n_1 + \min \left\{ \operatorname{rank} \begin{bmatrix} E_{12} \\ C_2 \end{bmatrix}, s \right\}. \tag{6.35}$$

First, it follows from (6.8), (6.13), (6.12), and (6.33) that

$$\operatorname{rank} \begin{bmatrix} E \\ C \end{bmatrix} = \operatorname{rank} \begin{bmatrix} P \begin{bmatrix} \begin{bmatrix} E_{11} & E_{12} \end{bmatrix} Q_1 \\ P_1 \begin{bmatrix} I_{n_1} & 0 \\ 0 & 0 \end{bmatrix} Q_1 \end{bmatrix} \\ \begin{bmatrix} C_1 & C_2 \end{bmatrix} Q_1 \end{bmatrix}$$

$$= \operatorname{rank} \begin{bmatrix} E_{11} & E_{12} \\ I_{n_1} & 0 \\ 0 & 0 \\ C_1 & C_2 \end{bmatrix}$$

$$= n_1 + \operatorname{rank} \begin{bmatrix} E_{12} \\ C_2 \end{bmatrix}. \tag{6.36}$$

Substituting (6.36) and (6.9)–(6.10) into the second expression in (6.29) yields (6.35).

Second, it follows from (6.8) and (6.7) that

$$\text{rank}\begin{bmatrix} E & B \\ C & 0 \end{bmatrix} = \text{rank}\begin{bmatrix} P\begin{bmatrix} E_1 \\ E_2 \end{bmatrix} & P\begin{bmatrix} I_s & 0 \\ 0 & 0 \end{bmatrix}Q \\ C & 0 \end{bmatrix}$$

$$= \text{rank}\begin{bmatrix} E_1 & I_s & 0 \\ E_2 & 0 & 0 \\ C & 0 & 0 \end{bmatrix}$$

$$= s + \text{rank}\begin{bmatrix} E_2 \\ C \end{bmatrix}.$$

Furthermore, using (6.12) and (6.33) gives

$$\text{rank}\begin{bmatrix} E & B \\ C & 0 \end{bmatrix} = s + \text{rank}\begin{bmatrix} P_1\begin{bmatrix} I_{n_1} & 0 \\ 0 & 0 \end{bmatrix}Q_1 \\ [\, C_1 & C_2 \,]Q_1 \end{bmatrix}$$

$$= s + \text{rank}\begin{bmatrix} I_{n_1} & 0 \\ 0 & 0 \\ C_1 & C_2 \end{bmatrix}$$

$$= s + n_1 + \text{rank}C_2. \tag{6.37}$$

Substituting (6.9)–(6.10), (6.36) and (6.37) into the first expression of (6.29) yields the relation (6.34). This proof is then completed. □

The above Theorem 6.2 is a generalization of the corresponding result reported in Chu et al. (1998), Chu et al. (1999), and Moor and Golub (1991), which is for square descriptor linear systems only. Unlike the full-state feedback case, it is interesting to point out that this result for partial-state feedback remains exactly the same form in both the square and the nonsquare cases.

6.2 Dynamical Order Assignment via Full-State Derivative Feedback

In this section, the dynamical order assignment problem for linear nonsquare descriptor systems via full-state derivative feedback is studied. A new, direct, and simple parametric approach is presented, which assigns via full-state derivative feedback any desired allowable dynamical order to the system. A simple, general direct and complete parametric expression for all the feedback gains which assign the desired dynamical order to the closed-loop system is established. The proposed approach is convenient to use and possesses good numerical reliability since it mainly involves only a series of singular value decompositions and the inverses of some positive definite diagonal matrices only.

6.2.1 Problem Formulation

Consider again the nonsquare descriptor linear system

$$E\dot{x}(t) = Ax(t) + Bu(t), \tag{6.38}$$

where $x(t) \in \mathbb{R}^m$ is the descriptor vector, $u(t) \in \mathbb{R}^r$ is the control vector, and $A, E \in \mathbb{R}^{n \times m}$ and $B \in \mathbb{R}^{n \times r}$ are known real coefficient matrices. Without loss of generality, we propose the following assumption:

Assumption 6.1. $\mathrm{rank}\, E = n_0 \le \min\{m, n\}$ and $\mathrm{rank}\, B = r$.

When a full-state derivative feedback controller

$$u(t) = -K\dot{x}(t) \tag{6.39}$$

is applied to system (6.38), the closed-loop system is obtained as

$$(E + BK)\dot{x}(t) = Ax(t). \tag{6.40}$$

The dynamical order of the above closed-loop system (6.40) is given by $\mathrm{rank}(E + BK)$, which is dependent on the feedback gain matrix K. It follows from Theorem 6.1 that

$$\mathrm{rank}(E + BK) = p \tag{6.41}$$

holds for some $K \in \mathbb{R}^{r \times m}$ if and only if

$$n_1 \le p \le n_2, \tag{6.42}$$

where n_1 and n_2 are given by (6.4). In view of this fact, the problem of dynamical order assignment in system (6.38) via the state derivative feedback controller (6.39) can be precisely stated as follows.

Problem 6.1. [*Dynamical order assignment*] Let matrices $E \in \mathbb{R}^{n \times m}$ and $B \in \mathbb{R}^{n \times r}$ satisfy Assumption 6.1, and integers n_1 and n_2 be defined as in (6.4). For an arbitrary integer p satisfying (6.42), find a parameterization for all the matrices $K \in \mathbb{R}^{r \times m}$ satisfying (6.41). In other words, characterize the set

$$\mathscr{S}_p = \{K \mid K \in \mathbb{R}^{r \times m}, \ \mathrm{rank}(E + BK) = p\}.$$

6.2.2 Preliminary Results

Our development in this section is based on the system equivalent form for derivative feedback presented in Sect. 2.2.

Recall that Algorithm 2.2 produces a nonsingular matrix $Q_3 \in \mathbb{R}^{n \times n}$ and two orthogonal matrices $P_3 \in \mathbb{R}^{m \times m}$ and $U_3 \in \mathbb{R}^{r \times r}$ based on the pair of matrices $E \in \mathbb{R}^{n \times m}$ and $B \in \mathbb{R}^{n \times r}$, and it follows from Theorem 2.6 that under the transformation (P_3, Q_3) the system (E, A, B) is transformed into a third restricted equivalent form for derivative feedback. For convenience, we here give a slightly altered version of Theorem 2.6.

Theorem 6.3. *Let $E \in \mathbb{R}^{n \times m}$, $B \in \mathbb{R}^{n \times r}$ with $\operatorname{rank} B = r$, and let the nonsingular matrix $Q_3 \in \mathbb{R}^{n \times n}$ and the two orthogonal matrices $P_3 \in \mathbb{R}^{m \times m}$ and $U_3 \in \mathbb{R}^{r \times r}$ be obtained by Algorithm 2.2. Put*

$$N = Q_3^{-1}, \quad V = P_3^{\mathrm{T}}, \quad T = U_3^{\mathrm{T}},$$

then

$$B = N \begin{bmatrix} 0_{n_1 \times r} \\ I_r \\ 0 \end{bmatrix} T, \tag{6.43}$$

$$E = N \begin{bmatrix} I_{n_1} & 0 \\ 0 & M \\ 0 & 0 \end{bmatrix} V, \tag{6.44}$$

where the matrix $M \in \mathbb{R}^{r \times (m - n_1)}$ possesses the following form:

$$M = \begin{cases} 0 & \text{if } n_0 = n_1 \\ \operatorname{diag}(D, 0_{(r + n_1 - n_0) \times (m - n_0)}) & \text{if } n_0 > n_1 \end{cases}, \tag{6.45}$$

$$D = \operatorname{diag}(d_1, \ldots, d_{n_0 - n_1}),$$

$$0 < d_1 \leq \cdots \leq d_{n_0 - n_1}. \tag{6.46}$$

Remark 6.1. It is clear that the matrix M in the above theorem satisfies

$$\operatorname{rank} M = n_0 - n_1.$$

Remark 6.2. It will be seen that the nonsingular matrix N in the above theorem does not appear in the general expression for the solution to Problem 6.1. Therefore, it does not need to be computed if we are only interested in the solution to Problem 6.1.

6.2.3 Solution to the Problem

Now let us consider the solution to Problem 6.1. Based on Theorem 6.3, the solution to Problem 6.1 can be stated as follows.

Theorem 6.4. *Given matrices $E \in \mathbb{R}^{n \times m}$, $B \in \mathbb{R}^{n \times r}$ satisfying Assumption 6.1, and the matrices $V \in \mathbb{R}^{m \times m}$, $T \in \mathbb{R}^{r \times r}$, and $M \in \mathbb{R}^{r \times (m-n_1)}$ as in Theorem 6.3, for an arbitrary integer p satisfying (6.42), a general form for all the gain matrices $K \in \mathbb{R}^{r \times m}$ satisfying (6.41) (i.e., $K \in S_p$) is given as follows*

$$K = T^{\mathrm{T}} \begin{bmatrix} K_1 & K_2 - M \end{bmatrix} V, \tag{6.47}$$

where $K_1 \in \mathbb{R}^{r \times n_1}$ is an arbitrary parameter matrix, and $K_2 \in \mathbb{R}^{r \times (m-n_1)}$ is a parameter matrix satisfying the following constraint

$$\operatorname{rank} K_2 = p - n_1. \tag{6.48}$$

Proof. Under given conditions, it follows from Theorem 6.3 that

$$
\begin{aligned}
E + BK &= N \begin{bmatrix} I_{n_1} & 0 \\ 0 & M \\ 0 & 0 \end{bmatrix} V + N \begin{bmatrix} 0 \\ I_r \\ 0_{(n-n_2) \times r} \end{bmatrix} TK \\
&= N \left(\begin{bmatrix} I_{n_1} & 0 \\ 0 & M \\ 0 & 0 \end{bmatrix} + \begin{bmatrix} 0 \\ I_r \\ 0_{(n-n_2) \times r} \end{bmatrix} TKV^{\mathrm{T}} \right) V \\
&= N \left(\begin{bmatrix} I_{n_1} & 0 \\ 0 & M \\ 0 & 0 \end{bmatrix} + \begin{bmatrix} 0 \\ TKV^{\mathrm{T}} \\ 0_{(n-n_2) \times r} \end{bmatrix} \right) V.
\end{aligned} \tag{6.49}
$$

Let

$$TKV^{\mathrm{T}} = \begin{bmatrix} K_1 & K_2 - M \end{bmatrix}, \tag{6.50}$$

where $K_1 \in \mathbb{R}^{r \times n_1}$ and $K_2 \in \mathbb{R}^{r \times (m-n_1)}$. Then (6.49) becomes

$$E + BK = N \begin{bmatrix} I_{n_1} & 0 \\ K_1 & K_2 \\ 0 & 0 \end{bmatrix} V.$$

This, together with (6.41), implies (6.48). While (6.50) clearly gives (6.47).   □

When $p = n_1$, constraint (6.48) becomes $K_2 = 0$. Thus, the following corollary immediately holds.

Corollary 6.1. *When $p = n_1$, a general form for all the matrices $K \in \mathbb{R}^{r \times m}$ satisfying (6.41) (i.e., $K \in \mathscr{S}_p$) is given as follows*

$$K = T^{\mathrm{T}} \begin{bmatrix} K_1 & -M \end{bmatrix} V, \tag{6.51}$$

where $K_1 \in \mathbb{R}^{r \times n_1}$ is an arbitrarily chosen parameter matrix.

Obviously, for any given integer p satisfying $n_1 \leq p \leq n_2$, all the matrices K given by (6.47) comprise the set $\mathscr{S}_p$. Specially, all those given by (6.51) comprise the set $\mathscr{S}_{n_1}$.

Remark 6.3. For any p satisfying (6.42), it is very easy to determine using the proposed direct parametric approach (Theorem 6.4) a general form for all the gain matrices $K \in \mathbb{R}^{r \times m}$ satisfying (6.41). Furthermore, since all the main operations are realized via only some singular value decompositions and inverses of positive definite diagonal matrices only, the presented approach is also very numerically reliable.

Remark 6.4. We have parameterized in Theorem 6.4 all the feedback gain matrices K assigning the desired dynamical order to the closed-loop system (6.40). The parameter matrices $K_1 \in \mathbb{R}^{r \times n_1}$ and $K_2 \in \mathbb{R}^{r \times (m-n_1)}$ represent the degrees of freedom in the general parameterization. These parameter matrices can be further properly chosen to meet some other design specification requirements beyond dynamical order assignment.

To finish this section, let us finally illustrate the result using an example.

Example 6.1. Consider a system in the form of (6.38) with the following parameters (see Fahmy and Tantawy 1990; Owens and Askarpour 2001):

$$E = \begin{bmatrix} 2 & 1 & 1 & -2 \\ 3 & 6 & -3 & -12 \\ -1 & -1 & 0 & 2 \\ 2 & 3 & -1 & -6 \end{bmatrix},$$

$$A = \begin{bmatrix} 3 & 3 & 9 & 3 \\ 0 & -6 & 9 & 18 \\ 0 & 4 & -9 & -15 \\ 0 & -4 & 9 & 15 \end{bmatrix}, \quad B = \begin{bmatrix} 2 & 1 \\ -3 & 3 \\ 1 & -2 \\ -1 & 2 \end{bmatrix}.$$

For this system, $n = m = 4$, $r = 2$ and $n_0 = 2$, and thus

$$n_2 = \mathrm{rank}\,[E \ B] = 3,$$
$$n_1 = n_2 - \mathrm{rank}\,B = 1.$$

Therefore, the allowable range for the dynamical order p is $1 \leq p \leq 3$.

By applying Algorithm 2.2, we obtain the matrices M, $T = U_3^T$ and $V = P_3^T$ as follows:

$$T = \begin{bmatrix} 0.7071 & -0.7071 \\ 0.7071 & 0.7071 \end{bmatrix}, \quad M = \begin{bmatrix} 0 & 0 & 0 \\ 1 & 0 & 0 \end{bmatrix}$$

and

$$V = \begin{bmatrix} -0.2132 & -0.4264 & 0.2132 & 0.8528 \\ 0.7071 & -0.0000 & 0.7071 & -0.0000 \\ 0.6684 & -0.2505 & -0.6684 & 0.2090 \\ -0.0880 & -0.8692 & 0.0880 & -0.4786 \end{bmatrix},$$

Therefore, it can be obtained following Theorem 6.4 that, for $1 \leq p \leq 3$,

$$\mathscr{S}_p = \left\{ T^T \begin{bmatrix} k_1 & k_3 & k_5 & k_7 \\ k_2 & k_4 - 1 & k_6 & k_8 \end{bmatrix} V \right|$$

$$\left. \operatorname{rank} \begin{bmatrix} k_3 & k_5 & k_7 \\ k_4 & k_6 & k_8 \end{bmatrix} = p - 1, k_i \in \mathbb{R}, 1 \leq i \leq 8 \right\}.$$

Specially, we have

$$\mathscr{S}_1 = \left\{ T^T \begin{bmatrix} k_1 & 0 & 0 & 0 \\ k_2 & -1 & 0 & 0 \end{bmatrix} V \right| k_i \in \mathbb{R}, 1 \leq i \leq 2 \right\}.$$

6.3 Dynamical Order Assignment via State Derivative Feedback with Minimum Norm

In this section, dynamical order assignment for nonsquare linear descriptor systems via state derivative feedback with minimum Frobenius norm is completely solved based on the main results proposed in Sect. 6.2. A simple, general direct, and complete parametric expression for all the feedback gains which assign the desired dynamical order to the closed-loop system and possess the minimum Frobenius norm is established. The proposed approach is convenient to use and possesses good numerical reliability since it mainly involves only some orthogonal transformations and singular value decompositions.

6.3.1 Problem Formulation

Again consider the nonsquare descriptor linear system (6.38) satisfying Assumption 6.1. When a state derivative feedback controller in the form of (6.39) is applied to

(6.38), the closed-loop system is obtained as (6.40). The dynamical order of (6.40) is given by $\text{rank}(E + BK)$, which is dependent on the feedback gain matrix K.

For an allowable dynamical order p of the closed-loop system (6.40) (i.e., p is an integer satisfying (6.5)), in Sect. 6.2, we have characterized the set

$$\mathscr{S}_p = \{K \mid K \in \mathbb{R}^{r \times m}, \ \text{rank}(E + BK) = p\}, \tag{6.52}$$

which is composed of all the state derivative feedback gains such that the closed-loop system (6.40) possesses the dynamical order p. Furthermore, define a number

$$\gamma_p = \min_{K \in S_p} \|K\|_{\text{F}} \tag{6.53}$$

and a set

$$\widetilde{\mathscr{F}_p} = \big\{W \mid W \in S_p \text{ and } \|W\|_{\text{F}} \le \|K\|_{\text{F}}, \quad \forall K \in S_p\big\}, \tag{6.54}$$

where $\|\cdot\|_{\text{F}}$ represents the Frobenius norm. Then it is easy to observe the following:

1. The number γ_p may not exist since $\mathscr{S}_p$ may contain an element which has an arbitrarily small Frobenius norm.
2. $\widetilde{\mathscr{F}_p} = \emptyset$ (the empty set) if and only if γ_p does not exist.
3. When γ_p exists, the set $\widetilde{\mathscr{F}_p}$ can be more clearly expressed as

$$\widetilde{\mathscr{F}_p} = \big\{W \mid W \in \mathscr{S}_p \text{ and } \|W\|_{\text{F}} = \gamma_p\big\}. \tag{6.55}$$

It follows from (6.55) that the set $\widetilde{\mathscr{F}_p}$ is formed by all the state derivative feedback gains which assign a fixed dynamical order p to the closed-loop system (6.40) and meanwhile, possess the minimum Frobenius norm. The purpose of this section is to give a complete parametric characterization of this set $\widetilde{\mathscr{F}_p}$. The problem can be precisely stated as follows.

Problem 6.2. [Minimum gain problem] Let the matrices $E \in \mathbb{R}^{n \times m}$ and $B \in \mathbb{R}^{n \times r}$ satisfy Assumption 6.1, p be an arbitrary integer satisfying (6.42), and γ_p, $\mathscr{S}_p$ and $\widetilde{\mathscr{F}_p}$ be defined as in (6.52)–(6.54).

1. Determine a necessary and sufficient condition for $\widetilde{S_p} \ne \emptyset$, or equivalently, for the existence of the number γ_p.
2. When $\widetilde{\mathscr{F}_p} \ne \emptyset$, or γ_p exists, find the minimum value γ_p and give a general complete parameterization for all the matrices $W \in \widetilde{\mathscr{F}_p}$.

6.3.2 A Preliminary Result

Consider the following minimization problem

$$\min_{X \in \mathbb{R}^{r \times (m-n_1)}_{p-n_1}} \|X - M\|_{\text{F}}, \tag{6.56}$$

where p, $n_1 < p < n_0$, is a prespecified integer and M is the matrix defined by (6.45)–(6.46). Please note that the cases of $p = n_1$ and $p = n_0$ are not included because in the first case the above minimization problem (6.56) obviously has a unique solution $X = 0$, and in the latter case the above minimization problem (6.56) clearly has a unique solution $X = M$.

Applying the Theorem B.3 in Appendix B, we have the following result about solution to the minimization (6.56).

Theorem 6.5. *Suppose that the matrix M is defined as in (6.45)–(6.46). Furthermore, let p be a prespecified integer satisfying $n_1 < p < n_0$, and k_{n_0-p} and $l_{k_{n_0-p}}$ be integers defined as in Lemma B.3. Then for arbitrary matrices $X \in \mathbb{R}_{p-n_1}^{r \times (m-n_1)}$, there holds*

$$\|X - M\|_{\mathrm{F}} \geq (d_1^2 + d_2^2 + \cdots + d_{n_0-p}^2)^{1/2}. \tag{6.57}$$

Furthermore, the equality in (6.57) holds if and only if

$$X = \begin{bmatrix} D + P\tilde{X}D & 0 \\ 0_{(n_2-n_0)\times(n_0-n_1)} & 0 \end{bmatrix}, \tag{6.58}$$

$$\tilde{X} = \begin{bmatrix} 0 & 0_{(p-n_1)\times(n_0-n_1-l_{k_{n_0-p}})} \\ \widetilde{\Gamma} & 0 \end{bmatrix},$$

$$\widetilde{\Gamma} = \mathrm{diag}\left(Y_1, Y_2, \ldots, Y_{k_{n_0-p}}\right), \tag{6.59}$$

where $P \in \mathbb{R}^{(n_0-n_1)\times(n_0-n_1)}$ is an arbitrary orthogonal matrix, $Y_i \in \mathbb{R}^{m_i \times m_i}$, $i = 1, \ldots, k_{n_0-p}-1$, and $Y_{k_{n_0-p}} \in \mathbb{R}^{(n_0-p-l_{k_{n_0-p}}-1)\times m_{k_{n_0-p}}}$ are full-row rank upper triangular matrices satisfying

$$Y_i Y_i^{\mathrm{T}} = I, \quad i = 1, 2, \ldots, k_{n_0-p}.$$

In particular, if (6.46) turns into

$$0 < d_1 < d_2 < \cdots < d_{n_0-n_1}, \tag{6.60}$$

then the matrix $\widetilde{\Gamma}$ in (6.59) turns into a diagonal involutory matrix (i.e., with all the diagonal elements being 1 or −1).

Remark 6.5. The term of full-row rank upper triangular matrix appeared in the above theorem refers to any matrix $A \in \mathbb{R}^{m \times n}$ possessing the form of

$$A = \begin{bmatrix} A_1 & A_2 \end{bmatrix},$$

where $A_1 \in \mathbb{R}^{m \times m}$ is nonsingular upper triangular.

6.3.3   Solution to the Problem

Now let us consider the solution to the Problem 6.2. Based on Theorems 6.4 and 6.5 and Corollary 6.1, the following theorem can be derived, which provides the complete solution to Problem 6.2.

Theorem 6.6. *Let the matrices $E \in \mathbb{R}^{n \times m}$ and $B \in \mathbb{R}^{n \times r}$ satisfy Assumption 6.1, p be an allowable dynamical order of the closed-loop system (6.40), and the matrices $M \in \mathbb{R}^{r \times (m-n_1)}$, $V = P_3^{\mathrm{T}} \in \mathbb{R}^{m \times m}$, and $T = U_3^{\mathrm{T}} \in \mathbb{R}^{r \times r}$ be obtained by Algorithm 2.2. Furthermore, let γ_p, $\mathscr{S}_p$ and $\widetilde{\mathscr{S}_p}$ be defined as in (6.52)–(6.54).*

1. *When $p > n_0$, for an arbitrarily small positive scalar ϵ there exists $K \in S_p$ such that $\| K \|_{\mathrm{F}} = \varepsilon$, i.e., γ_p does not exist and $\widetilde{\mathscr{S}_p} = \emptyset$.*
2. *When $p = n_0$,*

$$\gamma_{n_0} = 0 \tag{6.61}$$

   and

$$\widetilde{\mathscr{S}_{n_0}} = \{0\}.$$

3. *When $p = n_1$,*

$$\gamma_{n_1} = (d_1^2 + d_2^2 + \cdots + d_{n_0-n_1}^2)^{1/2} \tag{6.62}$$

   and

$$\widetilde{\mathscr{S}_{n_1}} = \{T^{\mathrm{T}} \begin{bmatrix} 0 & -M \end{bmatrix} V \}. \tag{6.63}$$

4. *When $n_1 < p < n_0$,*

$$\gamma_p = (d_1^2 + d_2^2 + \cdots + d_{n_0-p}^2)^{1/2},$$

   and a general form for the matrix $K \in \widetilde{\mathscr{S}_p}$ is given by

$$K = T^{\mathrm{T}} \begin{bmatrix} 0_{r \times n_1} & X \end{bmatrix} V,$$

   where $X \in \mathbb{R}^{r \times (m-n_1)}$ is defined as in Theorem 6.5.

Proof. Proof of conclusion *1*. Let $p > n_0$, then $p - n_0 > 0$. Thus, we can define

$$x = \frac{\varepsilon}{\sqrt{p - n_0}}, \quad \varepsilon > 0.$$

Specially choosing in (6.47) $K_1 = 0_{r \times n_1}$ and

$$K_2 = M + \begin{bmatrix} 0 & \begin{bmatrix} xI_{p-n_0} & 0 \\ 0 & 0 \end{bmatrix} \\ 0_{(n_0-n_1) \times (n_0-n_1)} & 0 \end{bmatrix},$$

gives the following special element of $\mathscr{S}_p$:

$$K(x) = T^{\mathrm{T}} \begin{bmatrix} 0 & \begin{bmatrix} xI_{p-n_0} & 0 \\ 0 & 0 \end{bmatrix} \\ 0_{(n_0-n_1)\times n_0} & 0 \end{bmatrix} V.$$

Furthermore, in view of the fact that, for any orthogonal matrices Y and Z, there holds

$$\|XZ\|_{\mathrm{F}} = \|YX\|_{\mathrm{F}} = \|X\|_{\mathrm{F}},$$

we can derive

$$\|K(x)\|_{\mathrm{F}} = \left\| \begin{bmatrix} 0 & \begin{bmatrix} xI_{p-n_0} & 0 \\ 0 & 0 \end{bmatrix} \\ 0_{(n_0-n_1)\times n_0} & 0 \end{bmatrix} \right\|_{\mathrm{F}}$$

$$= \|xI_{p-n_0}\|_{\mathrm{F}}$$

$$= \varepsilon.$$

This proves the first conclusion.

Proof of conclusion 2. The relation (6.61) follows by choosing $K_1 = 0$ and $K_2 = M$ in (6.47). Due to (6.55), we have for any $K \in \widetilde{\mathscr{S}_{n_0}}$,

$$K \in \mathscr{S}_{n_0} \quad \text{and} \quad \|K\|_{\mathrm{F}} = 0.$$

This implies $K = 0$ and the second conclusion is thus proven.

Proof of conclusion 3. It follows from $p = n_1$ and Corollary 6.1 that for any $K \in \mathscr{S}_{n_1}$,

$$\|K\|_{\mathrm{F}}^2 = \|K_1\|_{\mathrm{F}}^2 + \|M\|_{\mathrm{F}}^2$$
$$\geq \|M\|_{\mathrm{F}}^2,$$

and the inequality turns into an equality if and only if

$$K_1 = 0. \tag{6.64}$$

Therefore, γ_{n_1} defined by (6.53) exists and

$$\gamma_{n_1} = \|M\|_{\mathrm{F}}. \tag{6.65}$$

Combining (6.65) and (6.45) yields (6.62), and further substituting (6.64) into (6.47) gives

$$K = T^{\mathrm{T}} \begin{bmatrix} 0 & -M \end{bmatrix} V.$$

Therefore, (6.63) holds and the third conclusion is then proven.

Proof of conclusion 4. First, it follows from Theorem 6.4 and $n_1 < p < n_0$ that

$$\|K\|_F^2 = \|K_1\|_F^2 + \|K_2 - M\|_F^2$$
$$\geq \|K_2 - M\|_F^2,$$

and the inequality turns into an equality if and only if (6.64) holds.

Second, applying Theorem 6.5 yields

$$\|K_2 - M\|_F \geq (d_1^2 + d_2^2 + \cdots + d_{n_0-p}^2)^{1/2},$$

and the equality holds if and only if K_2 possesses the form of the X defined in (6.58).

By combining the above two aspects, the proof of the fourth conclusion is completed.   □

Remark 6.6. The above Theorem 6.6 gives a complete solution to the minimum gain problem, Problem 6.2. It can be easily observed that the proposed complete solution to the minimum gain problem can be easily obtained by only carrying out some simple singular value decompositions, and thus is both numerically very simple and reliable.

Remark 6.7. Owens and Askarpour (2001) demonstrated through an example that their approach can determine, by properly selecting the degree of freedom, state derivative feedback gains with smaller Frobenius norms. Compared with the approach proposed in Owens and Askarpour (2001), our approach presented in the above Theorem 6.6 is not only much simpler, but also provides the complete answer to this minimum gain problem in the sense of giving both the minimum Frobenius norm and the complete parameterization of all the optimal solutions.

To finish this section, let us demonstrate the above result using the system considered in Example 6.1.

Example 6.2. Consider the system treated in Example 6.1, which has been studied by Fahmy and Tantawy (1990) and Owens and Askarpour (2001). It has been shown in Example 6.1 that for this system, $n = m = 4$, $r = 2$, $n_0 = 2$, and $n_1 = 1$ and $n_2 = 3$, and thus the allowable range for p is $1 \leq p \leq 3$.

By applying Algorithm 2.2, we have obtained in Example 6.1 the matrices M, $T = U_3^T$ and $V = P_3^T$ as follows:

$$T = \begin{bmatrix} 0.7071 & -0.7071 \\ 0.7071 & 0.7071 \end{bmatrix}, \quad M = \begin{bmatrix} 0 & 0 & 0 \\ 1 & 0 & 0 \end{bmatrix}$$

and

$$V = \begin{bmatrix} -0.2132 & -0.4264 & 0.2132 & 0.8528 \\ 0.7071 & -0.0000 & 0.7071 & -0.0000 \\ 0.6684 & -0.2505 & -0.6684 & 0.2090 \\ -0.0880 & -0.8692 & 0.0880 & -0.4786 \end{bmatrix}.$$

Using the second and the third conclusions of Theorem 6.6 gives

$$\gamma_1 = 1 \text{ and } \widetilde{\mathscr{F}}_1 = \left\{ T^{\mathrm{T}} \begin{bmatrix} 0 & 0 & 0 & 0 \\ 0 & -1 & 0 & 0 \end{bmatrix} V \right\},$$

$$\gamma_2 = 0 \text{ and } \widetilde{\mathscr{F}}_2 = \{0\}.$$

Since $p = 3 > 2 = n_0$, it follows from the first conclusion of Theorem 6.6 that γ_3 does not exist, and thus $\widetilde{\mathscr{F}}_3 = \emptyset$.

For this example system, Fahmy and Tantawy (1990) have also considered assignment of the dynamical orders $p = 1, 2$, and 3 using state derivative feedback, and give the following gains:

$$K = \begin{cases} \begin{bmatrix} -3 & -5 & 2 & 10 \\ -4 & -7 & 3 & 14 \end{bmatrix}, & p = 1 \\ \begin{bmatrix} 2 & 3 & 1 & -4 \\ -4 & -8 & 5 & 17 \end{bmatrix}, & p = 2. \\ \begin{bmatrix} 3 & 5 & -4 & -15 \\ 1 & 1 & -1 & -5 \end{bmatrix}, & p = 3 \end{cases}$$

Owens and Askarpour (2001) considered for this example system assignment of the dynamical orders $p = 1, 3$, and got the following gains:

$$K = \begin{cases} \begin{bmatrix} 0 & 1 & -1 & -2 \\ 0 & 1 & -1 & -2 \end{bmatrix}, & p = 1 \\ \begin{bmatrix} 0 & 0.3333 & -1 & -1 \\ 0 & 0 & 0 & 0 \end{bmatrix}, & p = 3 \end{cases}.$$

The Frobenius norms of our solutions as well as those given by Fahmy and Tantawy (1990) and Owens and Askarpour (2001) are shown in Table 6.1 together

Table 6.1 Frobenius norms of the state derivative feedback gains

Item	$p = 1$	$p = 2$	$p = 3$
Fahmy and Tantawy (1990)	20.1983	0.4280	17.3945
Owens and Askarpour (2001)	3.4641	–	1.4530
Minimum values	$\gamma_1 = 1$	$\gamma_2 = 0$	γ_3, arbitrarily small

with their actual minimum values derived above. It is clearly seen from this table that the solutions in Fahmy and Tantawy (1990) and Owens and Askarpour (2001) are far from optimal.

6.4 Dynamical Order Assignment via Partial-State Derivative Feedback

In this section, dynamical order assignment for nonsquare descriptor linear systems via partial-state derivative feedback is studied. Based on Theorem 6.4 proposed in Sect. 6.2 for dynamical order assignment in nonsquare descriptor linear systems via full-state derivative feedback, it is proven that the problem of dynamical order assignment for a nonsquare descriptor linear system through a partial-state derivative feedback controller can be converted into a problem of dynamical order assignment for some descriptor linear system through a full-state derivative feedback controller. The conversion process involves only several singular value decompositions and inverses of some diagonal positive definite matrices, and is thus numerically very simple and reliable.

6.4.1 Problem Formulation

Consider the following nonsquare descriptor linear system:

$$\begin{cases} E\dot{x}(t) = Ax(t) + Bu(t) \\ z(t) = C\dot{x}(t) \end{cases}, \qquad (6.66)$$

where $x(t) \in \mathbb{R}^m$ is the descriptor state vector, $u(t) \in \mathbb{R}^r$ is the control vector, $z(t) \in \mathbb{R}^q$ is the measured state derivative vector. $A, E \in \mathbb{R}^{n \times m}$, $B \in \mathbb{R}^{n \times r}$, and $C \in \mathbb{R}^{q \times m}$ are known real coefficient matrices, which satisfy the following assumption:

Assumption 6.2. $\text{rank}E = n_0 \leq \min\{m, n\}$, $\text{rank}B = r$ and $\text{rank}C = q$.

When a partial-state derivative feedback controller of the form

$$u = -Kz(t) = -KC\dot{x}(t), \quad K \in \mathbb{R}^{r \times q} \qquad (6.67)$$

is applied to system (6.66), the closed-loop system is obtained as

$$(E + BKC)\dot{x}(t) = Ax(t). \qquad (6.68)$$

The dynamical order of the above closed-loop system (6.68) is given by $\text{rank}(E + BKC)$, which is dependent on the feedback gain matrix K. It follows from

Theorem 6.2 that an integer p is a dynamical order assignable to the matrix triple (E, B, C) by partial-state derivative feedback if and only if

$$\chi_1 \leq p \leq \chi_2, \tag{6.69}$$

where χ_1 and χ_2 are two integers given by (6.29).

For convenience, we introduce the following definition.

Definition 6.3. Let m, n, r, q, and p be a group of integers, and $X \in \mathbb{R}^{n \times m}$, $Y \in \mathbb{R}^{n \times r}$, $Z \in \mathbb{R}^{q \times m}$. Then, the following set

$$\Gamma_{(X,Y,Z,p)} = \{ K \mid K \in \mathbb{R}^{r \times q}, \ \mathrm{rank}(X + YKZ) = p \}$$

is called the rank-p set associated with the matrix triple (X, Y, Z).

Based on the above concept, the problem of assigning an allowable dynamical order to the descriptor linear system (6.66) via the partial-state derivative feedback controller (6.67) can be precisely stated as follows.

Problem 6.3. [Dynamical order assignment] Given the system (6.66) satisfying Assumption 6.2, and an integer p satisfying (6.69), find a parameterization for all the matrices $K \in \mathbb{R}^{r \times q}$ satisfying (6.28). In other words, characterize the rank-p set $\Gamma_{(E,B,C,p)}$ associated with the matrix triple (E, B, C).

Obviously, the set $\Gamma_{(E,B,C,p)}$ is not empty if and only if p is a dynamical order assignable to the matrix triple (E, B, C). In other words, the relation (6.69) holds.

6.4.2   Preliminary Results

Using the concept of rank-p set, Theorem 6.4 can be reinterpreted as follows.

Lemma 6.2. Let $E \in \mathbb{R}^{n \times m}$, $B \in \mathbb{R}^{n \times r}$ with $\mathrm{rank} B = r$, and p be a dynamical order assignable to the matrix triple (E, B, I). Furthermore, let the matrices M, $V = P_3^T$ and $T = U_3^T$ be obtained using Algorithm 2.2, and n_1 be defined as in (6.4). Then $W \in \Gamma_{(E,B,I,p)}$ if and only if it possesses the following form:

$$W = T^T \begin{bmatrix} W_1 & W_2 - M \end{bmatrix} V,$$

where $W_1 \in \mathbb{R}^{r \times n_1}$ is an arbitrary parameter matrix, and $W_2 \in \mathbb{R}^{r \times (m - n_1)}$ is a parameter matrix satisfying the following constraint:

$$\mathrm{rank} W_2 = p - n_1.$$

The main result in this section is a conversion, which converts a dynamical order assignment problem via partial-state derivative feedback into one via full-state derivative feedback. To introduce this conversion, we need to find, based on the system (6.66), four matrices $E_0 \in \mathbb{R}^{(m-n_1) \times r}$, $B_0 \in \mathbb{R}^{(m-n_1) \times n_3}$, $C_0 \in \mathbb{R}^{q \times (m-n_1)}$, and $P \in \mathbb{R}^{q \times q}$ according to the following algorithm.

Algorithm 6.1. Conversion for dynamical order assignment. [Given the system (6.66) satisfying Assumption 6.2 and $n_1 < m$, find four matrices E_0, B_0, C_0, and P]

Step 1   Obtain the four matrices $V = P_3^T \in \mathbb{R}^{m \times m}$, $T = U_3^T \in \mathbb{R}^{r \times r}$, $N \in \mathbb{R}^{n \times n}$, and $M \in \mathbb{R}^{r \times (m-n_1)}$ according to Algorithm 2.2.

Step 2   Calculate matrix $E_0 \in \mathbb{R}^{(m-n_1) \times r}$ according to

$$E_0 = M^T T. \tag{6.70}$$

Step 3   Calculate matrices $C_1 \in \mathbb{R}^{q \times n_1}$ and $C_0 \in \mathbb{R}^{q \times (m-n_1)}$ according to

$$\begin{bmatrix} C_1 & C_0 \end{bmatrix} = C V^T. \tag{6.71}$$

Step 4   Let

$$n_3 = \operatorname{rank} C_0, \tag{6.72}$$

and find a pair of orthogonal matrices $P \in \mathbb{R}^{q \times q}$ and $Q \in \mathbb{R}^{(m-n_1) \times (m-n_1)}$ and a positive definite diagonal matrix $\Sigma \in \mathbb{R}^{n_3 \times n_3}$ satisfying the following singular value decomposition

$$C_0 = P \begin{bmatrix} \Sigma & 0 \\ 0 & 0 \end{bmatrix} Q^T. \tag{6.73}$$

Step 5   Calculate the full-column rank matrix $B_0 \in \mathbb{R}^{(m-n_1) \times n_3}$ according to

$$B_0 = Q \begin{bmatrix} \Sigma \\ 0 \end{bmatrix}. \tag{6.74}$$

Based on the matrices M, T, and C_0 obtained by the above algorithm, the following lemma is immediately obtained.

Lemma 6.3. *Let the system (6.66) satisfy Assumption 6.2, and the integer n_1 be defined as in (6.4). Furthermore, let the matrices E_0, B_0, and P be obtained using Algorithm 6.1, and assume that $n_1 < m$. Then for arbitrary $K \in \mathbb{R}^{r \times q}$, there holds*

$$\operatorname{rank}(E + BKC) = n_1 + \operatorname{rank}(E_0 + B_0 K_0),$$

where $K_0 \in \mathbb{R}^{n_3 \times r}$ is determined by

$$K_0^T = KP \begin{bmatrix} I_r \\ 0 \end{bmatrix}. \tag{6.75}$$

Proof. It follows from Theorem 6.3 and (6.71) that, for arbitrary $K \in \mathbb{R}^{r \times q}$,

$$
E + BKC = N \begin{bmatrix} I_{n_1} & 0 \\ 0 & M \\ 0 & 0 \end{bmatrix} V + N \begin{bmatrix} 0_{n_1 \times r} \\ I_r \\ 0 \end{bmatrix} TK \begin{bmatrix} C_1 & C_0 \end{bmatrix} V
$$

$$
= N \left(\begin{bmatrix} I_{n_1} & 0 \\ 0 & M \\ 0 & 0 \end{bmatrix} + \begin{bmatrix} 0_{n_1 \times r} \\ I_r \\ 0 \end{bmatrix} TK \begin{bmatrix} C_1 & C_0 \end{bmatrix} \right) V
$$

$$
= N \begin{bmatrix} I_{n_1} & 0 \\ TKC_1 & M + TKC_0 \\ 0 & 0 \end{bmatrix} V. \tag{6.76}
$$

Since the matrix N is nonsingular, and the matrix V is orthogonal, (6.76) immediately implies

$$
\operatorname{rank}(E + BKC) = n_1 + \operatorname{rank}(M + TKC_0), \quad \forall K \in \mathbb{R}^{r \times q}.
$$

Therefore, in the following it suffices only to show

$$
\operatorname{rank}(M + TKC_0) = \operatorname{rank}(E_0 + B_0 K_0), \quad \forall K \in \mathbb{R}^{r \times q}.
$$

Using (6.70), (6.73), (6.74), and (6.75), we have

$$
\operatorname{rank}(M + TKC_0) = \operatorname{rank}(T^{\mathrm{T}} M + T^{\mathrm{T}} TKC_0)
$$

$$
= \operatorname{rank}(E_0^{\mathrm{T}} + KC_0)
$$

$$
= \operatorname{rank}\left(E_0^{\mathrm{T}} + KP \begin{bmatrix} \Sigma & 0 \\ 0 & 0 \end{bmatrix} Q^{\mathrm{T}} \right)
$$

$$
= \operatorname{rank}\left(E_0^{\mathrm{T}} + KP \begin{bmatrix} B_0^{\mathrm{T}} \\ 0 \end{bmatrix} \right)
$$

$$
= \operatorname{rank}(E_0^{\mathrm{T}} + K_0^{\mathrm{T}} B_0^{\mathrm{T}})
$$

$$
= \operatorname{rank}(E_0 + B_0 K_0).
$$

The proof is then completed.   □

6.4.3 Solution to the Problem

This subsection considers solution to Problem 6.3 proposed in Sect. 6.4.1. Based on the results in Sect. 6.2, the following main result of this section can be obtained, which converts a dynamical order assignment problem via partial-state derivative feedback into one via full-state derivative feedback.

Theorem 6.7. *Let the system (6.66) satisfy Assumption 6.2, and the integer n_1 be defined as in (6.4). Furthermore, let the matrices E_0, B_0, and P be obtained using Algorithm 6.1.*

1. When $n_1 = m$, there holds

$$n_0 = \text{rank} E = m \leq n - r \tag{6.77}$$

and, in this case, the system (6.68) has a unique allowable dynamical order m and $\Gamma_{(E,B,C,m)} = \mathbb{R}^{r \times q}$.
2. When $n_1 < m$, $K \in \Gamma_{(E,B,C,p)}$ if and only if it possesses the following form:

$$K = \begin{bmatrix} K_0^{\text{T}} & Y \end{bmatrix} P^{\text{T}}, \quad K_0 \in \Gamma_{(E_0,B_0,I,p-n_1)}, \tag{6.78}$$

where $Y \in \mathbb{R}^{r \times (q-n_3)}$, with $n_3 = \text{rank} B_0$, is an arbitrary parameter matrix.

Proof. Suppose $n_1 = m$. Then (6.44) turns into

$$E = N \begin{bmatrix} I_m \\ 0_{r \times m} \\ 0 \end{bmatrix} V. \tag{6.79}$$

This indicates that

$$n \geq m + r$$

and

$$n_0 = \text{rank} E = m.$$

Combining the above two relations yields (6.77). Furthermore, using (6.79) and (6.43) gives

$$E + BKC = N \begin{bmatrix} I_m \\ 0_{r \times m} \\ 0 \end{bmatrix} V + N \begin{bmatrix} 0_{m \times r} \\ I_r \\ 0 \end{bmatrix} TKC_1 V$$

$$= N \begin{bmatrix} I_m \\ TKC_1 \\ 0 \end{bmatrix} V, \quad \forall K \in \mathbb{R}^{r \times q},$$

and hence

$$\text{rank}\,(E + BKC) = m, \quad \forall K \in \mathbb{R}^{r \times q}.$$

This implies the first conclusion in this theorem.

When $n_1 < m$, the second conclusion directly follows from Lemma 6.3. $\qquad\square$

Since

$$\text{rank}\,E_0 = \text{rank}\,M = n_0 - n_1,$$

$\text{rank}\,E_0 = p - n_1$ clearly implies $p = n_0$. Thus, in the special case of $B_0 = 0$, it is clear to see that

$$\Gamma_{(E_0, B_0, I,\, p-n_1)} = \begin{cases} \mathbb{R}^{n_3 \times r}, & \text{when } p = n_0 \\ \varnothing, & \text{when } p \neq n_0 \end{cases}.$$

This, together with the second conclusion of Theorem 6.7, gives

$$\Gamma_{(E, B, C,\, p)} = \begin{cases} \mathbb{R}^{r \times q}, & \text{when } p = n_0 \\ \varnothing, & \text{when } p \neq n_0 \end{cases}.$$

With this observation, we clearly have the following corollary of Theorem 6.7.

Corollary 6.2. *Let the system (6.66) satisfy Assumption 6.2, and the integer n_1 be defined as in (6.4) satisfying $n_1 < m$. Furthermore, let the matrices E_0, B_0, and P be obtained using Algorithm 6.1. If $B_0 = 0$, then the system (6.68) has a unique allowable dynamical order n_0 and $\Gamma_{(E, B, C,\, n_0)} = \mathbb{R}^{r \times q}$.*

In the special case of $n_3 = q$, the parameter matrix Y in (6.78) vanishes. For this case, we have the following corollary.

Corollary 6.3. *Let the system (6.66) satisfy Assumption 6.2 and p be a dynamical order assignable to the matrix triple (E, B, C). Furthermore, let the matrices E_0, B_0, and P be obtained using Algorithm 6.1, and assume*

$$n_3 = \text{rank}\,B_0 = q. \tag{6.80}$$

Then $K \in \Gamma_{(E, B, C,\, p)}$ if and only if

$$K^{\mathrm{T}} \in \Gamma_{(E_0, B_0 P^T, I,\, p-n_1)}. \tag{6.81}$$

Proof. Due to (6.80), (6.78) turns into

$$K = K_0^{\mathrm{T}} P^{\mathrm{T}}, \quad K_0 \in \Gamma_{(E_0, B_0, I,\, p-n_1)}. \tag{6.82}$$

Obviously, (6.82) is equivalent to (6.81). Therefore, (6.78) is equivalent to (6.81) when (6.80) holds, i.e., $K \in \Gamma_{(E, B, C,\, p)}$ if and only if the relation (6.81) holds. $\qquad\square$

Remark 6.8. The above theorem and corollaries tell us that to characterize the rank-p set $\Gamma_{(E,B,C,\,p)}$, it suffices only to characterize the set $\Gamma_{(E_0,B_0,I,\,p-n_1)}$ when $n_1 < m$ and $B_0 \neq 0$, which is related to the dynamical order assignment in the descriptor linear system

$$E_0 \dot{y} = y + B_0 v$$

via the full-state derivative feedback controller

$$v = -K_0 \dot{y}.$$

Applying Algorithm 2.2 and the Lemma 6.2 in Sect. 6.4.2 to the matrix triple (E_0, B_0, I), the set $\Gamma_{(E_0,B_0,I,\,p-n_1)}$ can be completely characterized. Therefore, the rank-p set $\Gamma_{(E,B,C,\,p)}$ can be further completely characterized through the above Theorem 6.7.

Remark 6.9. Among all the elements in the rank-p set $\Gamma_{(E,B,C,\,p)}$, the one with a minimum Frobenius norm is of special interest. It is clear from Theorem 6.7 and Corollary 6.2 that this minimum Frobenius norm problem does not have a solution when $n_1 = m$ or $B_0 = 0$. Now we give a method to solve this minimum Frobenius norm problem when $n_1 < m$ and $B_0 \neq 0$. Since the matrix P is orthogonal, we have from (6.78) that

$$\begin{aligned}
\|K\|_F^2 &= \left\| [K_0^T \ \ Y] \right\|_F^2 \\
&= \|K_0\|_F^2 + \|Y\|_F^2 \\
&\geq \|K_0\|_F^2,
\end{aligned}$$

and, in view of the arbitrariness of the parameter matrix Y, we further have

$$\|K\|_F = \|K_0\|_F \iff Y = 0.$$

The above fact states that the minimization problem

$$\begin{cases} \min \ \|K\|_F \\ \text{s.t.} \ \ K \in \Gamma_{(E,B,C,p)} \end{cases} \tag{6.83}$$

is equivalent to

$$\begin{cases} \min \ \|K_0\|_F \\ \text{s.t.} \ \ K_0 \in \Gamma_{(E_0,B_0,I,p-n_1)} \end{cases} \tag{6.84}$$

in the sense that once a solution $\widetilde{K}_0$ to the minimization (6.84) is obtained, a solution to the minimization (6.83) is given by

$$K = \left[\ \widetilde{K}_0^T \ \ 0 \ \right] P^T.$$

With regard to the solution to the minimization (6.84), the complete solution presented in Sect. 6.3 can be readily used.

6.4.4 The Example

Consider a system in the form of (6.66) with the following parameters:

$$
E = \begin{bmatrix} 2 & 1 & 1 & -2 \\ 3 & 6 & -3 & -12 \\ -1 & -1 & 0 & 2 \\ 2 & 3 & -1 & -6 \end{bmatrix}, \quad
B = \begin{bmatrix} 2 & 1 \\ -3 & 3 \\ 1 & -2 \\ -1 & 2 \end{bmatrix},
$$

$$
A = \begin{bmatrix} 3 & 3 & 9 & 3 \\ 0 & -6 & 9 & 18 \\ 0 & 4 & -9 & -15 \\ 0 & -4 & 9 & 15 \end{bmatrix}, \quad
C = \begin{bmatrix} 1 & 0 & 0 & 0 \\ 0 & 0 & 1 & 1 \end{bmatrix},
$$

where the matrices E, A, and B are taken from Examples 6.1 and 6.2 (see also, Duan and Zhang 2002a; Fahmy and Tantawy 1990; Owens and Askarpour 2001). For this system, $n = m = 4$, $r = 2$, $q = 2$ and $n_0 = \operatorname{rank} E = 2$, and thus, following from (6.4) and (6.29), we have

$$
n_1 = \operatorname{rank}[E \ B] - \operatorname{rank} B = 1
$$

and

$$
\begin{cases}
\chi_1 = \operatorname{rank}\begin{bmatrix} E & B \end{bmatrix} + \operatorname{rank}\begin{bmatrix} E \\ C \end{bmatrix} - \operatorname{rank}\begin{bmatrix} E & B \\ C & 0 \end{bmatrix} = 2 \\[2mm]
\chi_2 = \min\left\{ \operatorname{rank}\begin{bmatrix} E & B \end{bmatrix}, \ \operatorname{rank}\begin{bmatrix} E \\ C \end{bmatrix} \right\} = 3
\end{cases}
$$

It therefore follows from (6.69) that the allowable dynamical orders for the system are $p = 2$ and $p = 3$.

Following Algorithm 2.2, we obtain the matrices M, $V = P_3^{\mathrm{T}}$, and $T = U_3^{\mathrm{T}}$ in the first step of Algorithm 6.1 as follows:

$$
T = \begin{bmatrix} 0.7071 & -0.7071 \\ 0.7071 & 0.7071 \end{bmatrix}, \quad
M = \begin{bmatrix} 0 & 0 & 0 \\ 1 & 0 & 0 \end{bmatrix},
$$

$$
V = \begin{bmatrix}
-0.2132 & -0.4264 & 0.2132 & 0.8528 \\
0.7071 & -0.0000 & 0.7071 & -0.0000 \\
0.6684 & -0.2505 & -0.6684 & 0.2090 \\
-0.0880 & -0.8692 & 0.0880 & -0.4786
\end{bmatrix}.
$$

According to (6.72), we clearly have $n_3 = 2$. Carrying on with the other steps of Algorithm 6.1, we finally obtain the matrices P, B_0, and E_0 as

$$E_0 = \begin{bmatrix} 0.7071 & 0.7071 \\ 0 & 0 \\ 0 & 0 \end{bmatrix},$$

$$P = \begin{bmatrix} -0.7733 & -0.6340 \\ -0.6340 & 0.7733 \end{bmatrix},$$

and

$$B_0 = \begin{bmatrix} -0.9951 & 0.0985 \\ -0.2256 & -0.7791 \\ 0.3157 & -0.2462 \end{bmatrix},$$

and $n_3 = \text{rank} B_0 = 2$. It thus follows from Theorem 6.7 that we need only to characterize $K_0 \in \Gamma_{(E_0, B_0, I_2, p-1)}$.

Noting Remark 6.8 and applying Algorithm 2.2 to the matrix triple (E_0, B_0, I), we obtain for this matrix triple the matrices $T = T_0$, $V = V_0$ and $M = M_0$ as follows:

$$T_0 = \begin{bmatrix} 1 & 0 \\ 0 & 1 \end{bmatrix}, \quad M_0 = \begin{bmatrix} 0 \\ 0 \end{bmatrix},$$

$$V_0 = \begin{bmatrix} 0.7071 & 0.7071 \\ -0.7071 & 0.7071 \end{bmatrix}.$$

Furthermore, using Lemma 6.2, or Theorem 6.4, yields that $K_0 \in \Gamma_{(E_0, B_0, I_2, p-1)}$ if and only if

$$K_0 = \begin{bmatrix} k_1 & k_3 \\ k_2 & k_4 \end{bmatrix} V_0,$$

where $k_i \in \mathbb{R}$, $1 \le i \le 4$, with k_1 and k_2 arbitrary and k_3 and k_4 satisfying

$$\text{rank} \begin{bmatrix} k_3 \\ k_4 \end{bmatrix} = p - 2.$$

This implies that

$$\Gamma_{(E_0, B_0, I_2, 1)} = \left\{ \begin{bmatrix} k_1 & 0 \\ k_2 & 0 \end{bmatrix} V_0 \,\middle|\, k_i \in \mathbb{R}, 1 \le i \le 2 \right\} \tag{6.85}$$

and

$$\Gamma_{(E_0, B_0, I_2, 2)} = \left\{ \begin{bmatrix} k_1 & k_3 \\ k_2 & k_4 \end{bmatrix} V_0 \,\middle|\, k_3 \ne 0 \text{ or } k_4 \ne 0 \right\}. \tag{6.86}$$

Noting $n_3 = \text{rank} B_0 = q = 2$, the parameter matrix Y in (6.78) vanishes. Thus it follows from Theorem 6.7 that for any $2 \le p \le 3$, $K \in \Gamma_{(E, B, C, p)}$ if and only if $P^T K^T \in \Gamma_{(E_0, B_0, I_2, p-1)}$. Using (6.85) and (6.86) yields that

$$\Gamma_{(E, B, C, 2)} = \left\{ V_0^T \begin{bmatrix} k_1 & k_2 \\ 0 & 0 \end{bmatrix} P^T \,\middle|\, k_i \in \mathbb{R}, 1 \le i \le 2 \right\}$$

and

$$\Gamma_{(E,B,C,3)} = \left\{ V_0^T \begin{bmatrix} k_1 & k_2 \\ k_3 & k_4 \end{bmatrix} P^T \;\middle|\; k_3 \neq 0 \text{ or } k_4 \neq 0 \right\}.$$

Finally, in view of Remark 6.9 and (6.85), we clearly have

$$\min_{K \in \Gamma_{(E,B,C,2)}} \|K\|_F = 0,$$

and the optimal solution to the above minimization is uniquely given by $K = 0$. As for $\Gamma_{(E,B,C,3)}$, it follows from Remark 6.9 and (6.86) that for an arbitrary small positive scalar ε there always exists some $\widetilde{K} \in \Gamma_{(E,B,C,3)}$ such that $\|\widetilde{K}\|_F = \varepsilon$. Furthermore, all such gain matrices are given by

$$\widetilde{K} = V_0^T \begin{bmatrix} k_1 & k_2 \\ k_3 & k_4 \end{bmatrix} P^T$$

with $k_i \in \mathbb{R}$, $1 \leq i \leq 4$, satisfying

$$k_1^2 + k_2^2 + k_3^2 + k_4^2 = \varepsilon^2$$

and

$$k_3^2 + k_4^2 \neq 0.$$

6.5 Normalization of Descriptor Linear Systems

In this section, we study a special dynamical order assignment problem for square descriptor linear systems, which aims to assign a dynamical order that is equal to the system dimension n. Such a problem is called normalization. As a typical procedure, let us first consider the problem of normalizability, which looks into the conditions for normalization.

6.5.1 Normalizability

Again, let us consider the descriptor linear system in the following form:

$$\begin{cases} E\dot{x}(t) = Ax(t) + Bu(t) \\ z(t) = C\dot{x}(t) \end{cases}, \tag{6.87}$$

where $x(t) \in \mathbb{R}^n$ is the descriptor state vector, $u(t) \in \mathbb{R}^r$ is the control vector, $z(t) \in \mathbb{R}^q$ is the measured state derivative vector. A, $E \in \mathbb{R}^{n \times n}$, $B \in \mathbb{R}^{n \times r}$, and $C \in \mathbb{R}^{q \times n}$ are real coefficient matrices. Please note that, unlike the preceding three sections, the system is now square. Without loss of generality, the system coefficient matrices are required to satisfy the following assumption:

Assumption 6.3. $\mathrm{rank}E = n_0 < n$, $\mathrm{rank}B = r$ and $\mathrm{rank}C = q$.

Definition 6.4. System (6.87) is called normalizable by partial-state derivative feedback if a partial-state derivative feedback controller

$$u = -Kz(t) = -KC\dot{x}(t), \quad K \in \mathbb{R}^{r \times q}$$

can be found such that the resulted closed-loop system

$$(E + BKC)\dot{x}(t) = Ax(t)$$

is normal, i.e.,

$$\det(E + BKC) \neq 0.$$

In the literature, normalizability of system (6.87), or the matrix pair (E, B) refers to a special case of the above definition. To make things clearer, we give this definition below.

Definition 6.5. System (6.87), or the matrix pair (E, B), is called normalizable if a full-state derivative feedback controller

$$u = -K\dot{x}, \quad K \in \mathbb{R}^{r \times n}$$

can be found such that the resulted closed-loop system

$$(E + BK)\dot{x}(t) = Ax(t)$$

is normal, i.e.,

$$\det(E + BK) \neq 0.$$

A great advantage of a normal linear system is that it has n finite poles. As a consequence, it has no infinite poles and hence no impulsive behavior. Under the assumption of normalizability, some appropriate static derivative feedback controller can be chosen for the system such that the closed-loop system is normal, and hence enabling the application of certain analysis and design results in normal linear systems theory.

Regarding criterion for normalizability, the following result obviously holds following Theorem 6.1.

Theorem 6.8. *System (6.87) is normalizable if and only if*

$$\mathrm{rank}\begin{bmatrix} E & B \end{bmatrix} = n. \tag{6.88}$$

Definition 6.6. System (6.87), or the matrix pair (E, C), is called dual normalizable if its dual system, or the matrix pair (E^T, C^T), is normalizable.

Combining the above definition with Theorem 6.8, immediately gives the following theorem.

Theorem 6.9. *System (6.87) is dual normalizable if and only if*

$$\text{rank}\begin{bmatrix} E \\ C \end{bmatrix} = n. \qquad (6.89)$$

It follows from Theorems 4.10 and 4.11 that conditions (6.88) and (6.89) are necessary conditions for the C-controllability and C-observability of system (6.87), respectively. Now we see that they determine the normalizability of the matrix pair (E, B) and the dual normalizability of the matrix pair (E, C). It further follows from Theorems 4.14 and 4.15 that conditions (6.88) and (6.89) imply the I-controllability and I-observability of system (6.87), respectively. These observations clearly give the following corollary.

Corollary 6.4. *The regular system (E, A, B, C) is*

1. *C-controllable if and only if it is R-controllable and (E, B) is normalizable, and C-observable if and only if it is R-observable and (E, C) is dual normalizable;*
2. *I-controllable if (E, B) is normalizable, and I-observable if (E, C) is dual normalizable.*

In view of Theorem 6.2, the following result about normalization via partial-state derivative feedback can be given.

Theorem 6.10. *System (6.87) is normalizable by partial-state derivative feedback if and only if*

$$\text{rank}\begin{bmatrix} E & B \end{bmatrix} = \text{rank}\begin{bmatrix} E \\ C \end{bmatrix} = n, \qquad (6.90)$$

that is, system (6.87) is normalizable by partial-state derivative feedback if and only if (E, B) is normalizable and (E, C) is dual normalizable.

It follows from the above Theorem 6.10 and Theorems 4.10 and 4.11 that normalizability and dual normalizability determine the controllability and observability, respectively, of the fast subsystem. The most important fact about normalization is that a normalizable descriptor system can be changed into a normal one via suitably selecting a derivative feedback controller.

Example 6.3. Consider the three-link planar manipulator system in Example 5.5 again. Since

$$\text{rank}\begin{bmatrix} E & B \end{bmatrix} = \text{rank}\begin{bmatrix} I & 0 & 0 & 0 \\ 0 & M_0 & 0 & S_0 \\ 0 & 0 & 0 & 0 \end{bmatrix} = 6 < n = 8,$$

this system is not normalizable according to Theorem 6.8.

6.5.2 Normalizing Controllers

Simply letting $p = n$ in Theorem 6.4, we obtain the following result about solution of the normalizing full-state derivative feedback controller.

Theorem 6.11. *Given the normalizable matrix pair (E, B) with $E \in \mathbb{R}^{n \times n}$, $B \in \mathbb{R}^{n \times r}$ satisfying Assumption 6.1, and the matrices $V \in \mathbb{R}^{n \times n}$, $T \in \mathbb{R}^{r \times r}$ and $M \in \mathbb{R}^{r \times (n-n_1)}$ as in Theorem 6.3, a general form for all the normalizing full-state derivative feedback controllers is given as follows:*

$$K = T^{\mathrm{T}} \begin{bmatrix} K_1 & K_2 - M \end{bmatrix} V,$$

where $K_1 \in \mathbb{R}^{r \times n_1}$ is an arbitrary parameter matrix, and $K_2 \in \mathbb{R}^{r \times (n-n_1)}$ is a parameter matrix satisfying the constraint

$$\mathrm{rank} K_2 = n - n_1.$$

Similarly, letting $p = n$ in Theorem 6.7, we obtain the following result for solving the normalizing partial-state derivative feedback controllers.

Theorem 6.12. *Let the system (6.87) satisfy Assumption 6.3, and be normalizable by partial-state derivative feedback. Let integer n_1 be defined as in (6.4). Furthermore, let the matrices E_0, B_0, and P be obtained using Algorithm 6.1. Then, $K \in \Gamma_{(E,B,C,n)}$ if and only if it possesses the following form:*

$$K = \begin{bmatrix} K_0^{\mathrm{T}} & Y \end{bmatrix} P^{\mathrm{T}}, \quad K_0 \in \Gamma_{(E_0,B_0,I,n-n_1)},$$

where $Y \in \mathbb{R}^{r \times (q-n_3)}$, with $n_3 = \mathrm{rank} B_0$, is an arbitrary parameter matrix.

Clearly, this theorem converts a normalization problem via partial-state derivative feedback into a dynamical order assignment problem via full-state derivative feedback.

The above Theorems 6.11 and 6.12 provide all the normalizing controllers for the full- and partial-state derivative feedback cases, respectively. In the case that only a single solution is of interest, the following theorem can be used.

Theorem 6.13. *Let the system (6.87) be normalizable and satisfy Assumption 6.3, and let*

$$QEP = \mathrm{diag}(\Sigma, 0) \tag{6.91}$$

be a singular value decomposition of the matrix E, where Q and P are two orthogonal matrices, and Σ is a positive definite diagonal matrix. Then the following hold:

1. *Define*

$$\begin{bmatrix} B_1 \\ B_2 \end{bmatrix} = QB,$$

then

$$u = \begin{bmatrix} 0 & B_2^T \end{bmatrix} P^{-1} \dot{x} \tag{6.92}$$

is a normalizing full-state derivative controller for system (6.87).

2. *Furthermore, when condition (6.90) is met, define*

$$[C_1 \ C_2] = CP,$$

then

$$u = B_2^T C_2^T \dot{z} \tag{6.93}$$

is a normalizing partial-state derivative controller for system (6.87).

Proof. It follows from Theorems 6.8 and 6.9 that, under the assumption of normalizability and dual normalizability, matrices B_2 and C_2 are of full-row and column rank, respectively. Therefore, both $B_2 B_2^T$ and $C_2^T C_2$ are nonsingular.

Since

$$\det \left(E + B \begin{bmatrix} 0 & B_2^T \end{bmatrix} P^{-1} \right) = \det \left(Q^{-1} \right) \det(P^{-1}) \det \left(QEP + QB \begin{bmatrix} 0 & B_2^T \end{bmatrix} \right)$$

$$= \det \left(Q^{-1} \right) \det(P^{-1}) \det \begin{bmatrix} \Sigma & B_1 B_2^T \\ 0 & B_2 B_2^T \end{bmatrix}$$

$$= \det \left(Q^{-1} \right) \det(P^{-1}) \det \Sigma \det \left(B_2 B_2^T \right)$$

$$\neq 0,$$

(6.92) is indeed a normalizing full-state derivative controller for system (6.87).

Furthermore, noting that

$$\det \left(B_2 B_2^T \right) \neq 0, \quad \det \left(C_2^T C_2 \right) \neq 0,$$

we have

$$\det \left(B_2 B_2^T C_2^T C_2 \right) \neq 0,$$

and

$$\Theta = \left(B_1 B_2^T C_2^T C_2 \right) \left(B_2 B_2^T C_2^T C_2 \right)^{-1} B_2 B_2^T C_2^T C_1$$

$$= \left(B_1 B_2^T C_2^T C_2 \right) \left(C_2^T C_2 \right)^{-1} \left(B_2 B_2^T \right)^{-1} B_2 B_2^T C_2^T C_1$$

$$= B_1 B_2^T C_2^T C_1.$$

Thus, by the Theorem A.2 in Appendix A, we have

$$\det \begin{bmatrix} \Sigma + B_1 B_2^T C_2^T C_1 & B_1 B_2^T C_2^T C_2 \\ B_2 B_2^T C_2^T C_1 & B_2 B_2^T C_2^T C_2 \end{bmatrix}$$

$$= \det \left(B_2 B_2^T C_2^T C_2 \right) \det \left(\Sigma + B_1 B_2^T C_2^T C_1 - \Theta \right)$$

$$= \det \left(B_2 B_2^T C_2^T C_2 \right) \det \left(\Sigma + B_1 B_2^T C_2^T C_1 - B_1 B_2^T C_2^T C_1 \right)$$

$$= \det \left(B_2 B_2^T C_2^T C_2 \right) \det \Sigma$$

$$\neq 0.$$

Using the above relation, we can derive

$$\det\left(E + BB_2^{\mathrm{T}}C_2^{\mathrm{T}}C\right)$$
$$= \det\left(Q^{-1}\right)\det\left(P^{-1}\right)\det\left(QEP + QBB_2^{\mathrm{T}}C_2^{\mathrm{T}}CP\right)$$
$$= \det\left(Q^{-1}\right)\det\left(P^{-1}\right)\det\begin{bmatrix} \Sigma + B_1 B_2^{\mathrm{T}}C_2^{\mathrm{T}}C_1 & B_1 B_2^{\mathrm{T}}C_2^{\mathrm{T}}C_2 \\ B_2 B_2^{\mathrm{T}}C_2^{\mathrm{T}}C_1 & B_2 B_2^{\mathrm{T}}C_2^{\mathrm{T}}C_2 \end{bmatrix}$$
$$\neq 0.$$

Therefore, (6.93) is indeed a normalizing partial-state derivative controller for system (6.87). □

Example 6.4. Consider the system (4.80) in Example 4.7, where

$$E = \begin{bmatrix} 1 & 0 & 0 \\ 0 & 1 & 0 \\ 0 & 0 & 0 \end{bmatrix}, \quad B = \begin{bmatrix} 0 & 0 \\ 1 & 0 \\ 0 & 1 \end{bmatrix}.$$

Since

$$\mathrm{rank}\begin{bmatrix} E & B \end{bmatrix} = 3,$$

the system is normalizable. It is easy to see that a solution to the singular value decomposition (6.91) is given by $P = Q = I_3$, and

$$\Sigma = \mathrm{diag}(1, 1).$$

Therefore, by (6.92), we obtain a normalizing full-state derivative feedback controller gain for the system as

$$K = \begin{bmatrix} 0 & B_2^{\mathrm{T}} \end{bmatrix} P^{-1}$$
$$= \begin{bmatrix} 0 & 0 & 0 \\ 0 & 0 & 1 \end{bmatrix}.$$

In fact, with this gain we have

$$\det\left(E + BK\right) = \det\begin{bmatrix} 1 & 0 & 0 \\ 0 & 1 & 0 \\ 0 & 0 & 1 \end{bmatrix} = 1 \neq 0.$$

Thus, the derived K is indeed a normalizing controller.

To finish this section, we finally present a result which reveals a property of the set of normalizing state derivative feedback controllers.

Theorem 6.14. *Let system (6.87) be normalizable by partial-state derivative feedback. Then, the set of normalizing derivative feedback controllers $\Gamma_{(E,B,C,n)}$ for system (6.87) is a Zariski open set.*

Proof. This is a special case of Lemma 5.2.   □

The above theorem indicates the important fact that when the system (6.87) is normalizable by partial-state derivative feedback, almost all gains in $\mathbb{R}^{r \times q}$ normalize system (6.87).

6.6 Notes and References

This chapter studies the problem of dynamical order assignment in descriptor linear systems, which can be stated as follows:

Given a descriptor linear system in the typical form and an appropriate integer p, find a derivative feedback for the system such that the closed-loop system has dynamical order p.

The above problem has been studied by several authors (see, e.g., Wang and Soh 1999; Fahmy and Tantawy 1990; Owens and Askarpour 2001; Duan and Zhang 2002a,b). Fahmy and Tantawy (1990), Owens and Askarpour (2001) and Duan and Zhang (2002b) all studied dynamical order assignment in square descriptor linear systems via full-state derivative feedback, while Wang and Soh (1999) considered maximal dynamical order assignment in square descriptor linear systems via partial-state derivative feedback. Different from Wang and Soh (1999), the other researchers, Fahmy and Tantawy (1990), Owens and Askarpour (2001), and Duan and Zhang (2002a,b) all considered assignment of an arbitrary allowable dynamical order to the closed-loop system. The main results in this chapter are taken from Duan and Zhang (2002a,b).

Fahmy and Tantawy (1990) and Owens and Askarpour (2001) both considered dynamical order assignment in square linear descriptor systems via state derivative feedback. They first converted the problem into an eigenstructure assignment problem by embedding the descriptor-space problem in an equivalent state-space that can be conceived in simple terms, and then determined the feedback K by solving a matrix equation of the following form:

$$KV = F, \qquad\qquad (6.94)$$

where the matrices V and F are constructed by designers and contain some design parameters. The difference between the two approaches lies in that the matrix V is of full-column rank in Fahmy and Tantawy's approach while is nonsingular in Owens and Askarpour's approach. With their approaches, the design parameters in V and F generally cannot appear distinctly in the expression for the feedback gain K which is solved from (6.94).

Using eigenstructure assignment is a natural way for solving problems of dynamical order assignment. The basic idea underlying this eigenstructure assignment approach is to assign the eigenstructure of the matrix $E + BK$ by choosing the matrix K (see, Duan et al. 1992, 1999b; Duan 1995, 1998b; Duan and Patton 1997). The rank of this matrix $E + BK$ can be easily determined by assigning certain numbers of zero eigenvalues to the matrix. However, this often gives more complicated solutions. As with most eigenstructure assignment approaches, including those in Fahmy and Tantawy (1990), Owens and Askarpour (2001), Duan et al. (1992, 1999b), Duan (1998b), and Duan and Patton (1997), matrix inverses are inevitably involved, which may give problems with large computational load or poor numerical stability. Furthermore, the eigenstructure assignment approach is generally not applicable to nonsquare descriptor linear systems. Comparatively, the approach presented in this chapter is much simpler and complete in the sense that it provides all the solutions to the problem.

Fahmy and Tantawy (1990) and Owens and Askarpour (2001) used the same example to demonstrate their approaches, and Owens and Askarpour (2001) claimed that the state derivative feedback gain obtained for this example by their approach has a smaller Frobenius norm than that obtained in Fahmy and Tantawy (1990). However, both papers did not give a complete solution to the minimum gain problem since using their approaches,

- It is generally impossible to conclude whether the Frobenius norm of the state derivative feedback gain obtained has reached the minimum value.
- It is hard to obtain a complete parameterization for all the state derivative feedback gains which have the minimum Frobenius norm.

In this chapter, we have completely solved the problem of dynamical order assignment in descriptor linear systems via derivative feedback in the sense of providing all the solutions to the problem. Furthermore, we have also completely solved the problem of dynamical order assignment in descriptor linear systems via derivative feedback with minimum Frobenius norms by presenting both the minimum Frobenius norm and the whole set of optimal solutions.

Normalization is a special dynamical order assignment problem: the dynamical order to be assigned equal to the system dimension. As a consequence, we have also given in this chapter a complete solution for normalization using derivative feedback.

In the literature, some researchers, such as Mukundan and Dayawansa (1983), Wang and Wang (1986), and Zhou et al. (1987), have used the term normalizability to describe regularity. In this book, we convent that regularity of a typical descriptor linear system, or the matrix pencil (E, A), means the existence of $s \in \mathbb{C}$ such that $\det(sE - A) \neq 0$, and normalizability of a typical descriptor linear system, or the matrix pair (E, B), means the existence of a gain matrix K such that $\det(E + BK) \neq 0$.

Chapter 7
Impulse Elimination

It is seen from Sect. 3.4 that the response of a descriptor linear system may contain impulse terms. These terms may cause saturation of control and may even destroy the system, and hence are not expected to exist. In this chapter, we study the problem of eliminating the impulsive behavior of a descriptor linear system via certain feedback control. As we will soon find out, the problem is closely related with normalization of descriptor linear systems.

Section 7.1 presents several conditions for a descriptor linear system to be impulse-free. Based on these conditions, the problem of impulse elimination in descriptor linear systems is studied in Sects. 7.2–7.3 and 7.5. Section 7.2 is focused on the case of state feedback, while Sects. 7.3 and 7.5 are about impulse elimination using output feedback and P-D feedback, respectively. Before treating the problem of impulse elimination using P-D feedback, in Sect. 7.4 the concepts of I-controllablizability and I-observablizability are introduced.

7.1 The Impulse-Free Property

It has been known that for descriptor linear systems, there may be impulse terms in their responses. In a practical system, the impulse term and its derivatives are usually not expected in the state response since strong impulse behavior may saturate the state response or even destroy the system. Before studying the problem of impulse elimination in descriptor linear systems, in this section we first gain a better understanding about the impulse behavior of the regular descriptor linear system

$$E\dot{x} = Ax + Bu, \tag{7.1}$$

where E, $A \in \mathbb{R}^{n \times n}$ and $B \in \mathbb{R}^{n \times r}$. Our purpose is to answer the following question: *Under what conditions the state response of the descriptor linear system (7.1) does not contain impulse terms?*

In view of Definition 3.3, the above question can be interpreted as:
Under what conditions is a descriptor linear system impulse-free?

G.-R. Duan, *Analysis and Design of Descriptor Linear Systems*, Advances
in Mechanics and Mathematics 23, DOI 10.1007/978-1-4419-6397-0_7,
© Springer Science+Business Media, LLC 2010

7.1.1 Basic Criteria

For the descriptor system (7.1), there exist nonsingular matrices Q and P such that

$$QEP = \text{diag}\left(I_{n'_1}, N\right), \quad QAP = \text{diag}\left(A_1, I_{n'_2}\right), \quad QB = \begin{bmatrix} B_1 \\ B_2 \end{bmatrix}, \quad (7.2)$$

where $n'_1 + n'_2 = n$, $B_1 \in \mathbb{R}^{n'_1 \times r}$, $B_2 \in \mathbb{R}^{n'_2 \times r}$, N is nilpotent. Thus, the system is r.s.e. to the following standard decomposition form:

$$\begin{cases} \dot{x}_1 = A_1 x_1 + B_1 u \\ N\dot{x}_2 = x_2 + B_2 u \end{cases}.$$

The following theorem summarizes some basic conditions for a regular descriptor linear system to be impulse-free. This is to say that part of the state variables of the impulse-free descriptor linear systems can always be solved out, and the system can then be converted into a normal linear system with a suitable dimension.

Theorem 7.1. *The regular descriptor linear system (7.1) is impulse-free if and only if one of the following conditions holds:*

1. *The nilpotent matrix N in its fast subsystem is a zero matrix.*
2. *$\deg \det (sE - A) = \text{rank} E$.*
3. *The system has $\text{rank} E$ finite poles.*

Proof. The first conclusion is given in Corollary 3.3. We need only to show the last two conclusions.

In view of (7.2), we have

$$Q(sE - A) P = \text{diag}(sI - A_1, sN - I).$$

It is obvious that $N = 0$ if and only if

$$\deg \det (sE - A) = \deg \det (sI - A_1) = \text{rank} E.$$

This shows the equivalence between the first and the second statements. Finally, noting that the second and the third statements are clearly equivalent, we complete the proof.   □

It follows from the above theorem that the standard decomposition form of an impulse-free descriptor linear system becomes a dynamics decomposition form of the system.

The following theorem, which is due to Dai (1989a), gives a simple rank criterion for a descriptor linear system to be impulse-free.

Theorem 7.2. *The regular system (7.1) is impulse-free if and only if the following relation holds:*

$$\text{rank}\begin{bmatrix} E & 0 \\ A & E \end{bmatrix} = n + \text{rank}E. \tag{7.3}$$

Proof. For the regular descriptor system (7.1), there exist nonsingular matrices Q and P such that (7.2) holds. Thus, we have

$$\text{rank}E = \text{rank}QEP = \text{rank}\begin{bmatrix} I_{n_1'} & 0 \\ 0 & N \end{bmatrix} = n_1' + \text{rank}N,$$

and

$$\text{rank}\begin{bmatrix} E & 0 \\ A & E \end{bmatrix} = \text{rank}\begin{bmatrix} QEP & 0 \\ QAP & QEP \end{bmatrix}$$

$$= \text{rank}\begin{bmatrix} I_{n_1'} & 0 & 0 & 0 \\ 0 & N & 0 & 0 \\ A_1 & 0 & I_{n_1'} & 0 \\ 0 & I_{n_2'} & 0 & N \end{bmatrix}.$$

Further, note that

$$\text{rank}\begin{bmatrix} N & 0 \\ I_{n_2'} & N \end{bmatrix} = \text{rank}\left(\begin{bmatrix} I_{n_2'} & -N \\ 0 & I_{n_2'} \end{bmatrix} \begin{bmatrix} N & 0 \\ I_{n_2'} & N \end{bmatrix} \begin{bmatrix} I_{n_2'} & -N \\ 0 & I_{n_2'} \end{bmatrix} \right)$$

$$= \text{rank}\begin{bmatrix} 0 & -N^2 \\ I_{n_2'} & 0 \end{bmatrix}$$

$$= n_2' + \text{rank}N^2,$$

we have,

$$\text{rank}\begin{bmatrix} E & 0 \\ A & E \end{bmatrix} = n + n_1' + \text{rank}N^2.$$

Thus, (7.3) is equivalent to

$$n + n_1' + \text{rank}N^2 = n + n_1' + \text{rank}N,$$

that is,

$$\text{rank}N^2 = \text{rank}N.$$

Note that N is a nilpotent matrix. The above equation holds if and only if $N = 0$. Therefore, the conclusion holds according to Theorem 7.1.   □

Example 7.1. Consider a descriptor system in the form of (7.1) with the following parameters

$$E = \begin{bmatrix} 0 & 1 & 0 & 0 \\ 0 & 0 & 0 & 1 \\ 0 & 0 & 0 & 0 \\ 0 & 0 & 0 & 0 \end{bmatrix}, \quad A = \begin{bmatrix} 1 & 0 & 0 & 0 \\ 0 & 1 & 2 & 0 \\ 0 & 0 & 1 & 1 \\ 0 & 1 & 1 & 0 \end{bmatrix}, \quad B = \begin{bmatrix} 0 & 0 \\ 1 & 0 \\ 0 & 1 \\ 0 & 0 \end{bmatrix}.$$

Thus, we have $n_0 = 2$. In the following, we check the impulse behavior of the system using Theorems 7.1 and 7.2.

It can be easily checked that with the following transformation matrices

$$Q = \begin{bmatrix} 0 & -1 & 1 & 1 \\ -1 & 1 & 0 & -2 \\ 0 & 0 & 1 & -1 \\ 0 & 0 & -1 & 2 \end{bmatrix}, \quad P = \begin{bmatrix} 1 & -1 & 1 & 0 \\ -1 & 0 & -1 & 0 \\ 1 & 0 & 2 & 1 \\ -1 & 0 & 0 & 0 \end{bmatrix}, \quad (7.4)$$

the system is converted into its standard decomposition form with

$$A_1 = -1, \quad N = \begin{bmatrix} 0 & 1 & 0 \\ 0 & 0 & 0 \\ 0 & 0 & 0 \end{bmatrix}.$$

Since the nilpotent matrix N is not zero, according to Theorem 7.1 the system is not impulse-free.

It is easily verified that

$$\mathrm{rank} \begin{bmatrix} E & 0 \\ A & E \end{bmatrix} = 5 \neq 4 + 2,$$

thus this system is not impulse-free according to Theorem 7.2.

7.1.2 Criteria Based on Equivalent Forms

First, we give the criterion based on the dynamics decomposition form. For system (7.1), there exist a pair of orthogonal matrices T_1 and T_2 such that

$$T_1 E T_2 = \mathrm{diag}\,(\Sigma,\ 0), \quad \Sigma = \mathrm{diag}\,(\sigma_1,\ \sigma_2,\ \ldots,\ \sigma_{n_0}) > 0. \quad (7.5)$$

Denoting

$$T_1 A T_2 = \begin{bmatrix} A_{11} & A_{12} \\ A_{21} & A_{22} \end{bmatrix}, \quad T_1 B = \begin{bmatrix} B_1 \\ B_2 \end{bmatrix}, \quad (7.6)$$

where the partitions are in consistent dimensions, we have the following result.

Theorem 7.3. *Let the descriptor linear system (7.1) be regular, and the associated relations (7.5) and (7.6) hold. Then system (7.1) is impulse-free if and only if*

$$\det A_{22} \neq 0. \tag{7.7}$$

Proof. It follows from Theorem 7.2 that the system (7.1) is impulse-free if and only if (7.3) holds.

Using (7.5) and (7.6), we have

$$\text{rank}\begin{bmatrix} E & 0 \\ A & E \end{bmatrix} = \text{rank}\begin{bmatrix} T_1 E T_2 & 0 \\ T_1 A T_2 & T_1 E T_2 \end{bmatrix}$$

$$= \text{rank}\begin{bmatrix} \Sigma & 0 & 0 & 0 \\ 0 & 0 & 0 & 0 \\ A_{11} & A_{12} & \Sigma & 0 \\ A_{21} & A_{22} & 0 & 0 \end{bmatrix}$$

$$= \text{rank}\begin{bmatrix} \Sigma & 0 & 0 & 0 \\ 0 & 0 & \Sigma & 0 \\ 0 & A_{22} & 0 & 0 \end{bmatrix}$$

$$= 2n_0 + \text{rank}A_{22}.$$

Thus, the relation (7.3) is converted into

$$2n_0 + \text{rank}A_{22} = n_0 + n,$$

which is equivalent to (7.7).   □

The next theorem gives a criterion for a descriptor system to be impulse-free based on the inverse form.

Theorem 7.4. *Let (4.99) be an inverse form of system (7.1). Then system (7.1) is impulse-free if and only if the following relation holds:*

$$\text{rank}\hat{E}^2 = \text{rank}\hat{E}.$$

Proof. It follows from Theorem 7.2 that the system (7.1) is impulse-free if and only if (7.3) holds.

Using (4.100), we have

$$\text{rank}\begin{bmatrix} E & 0 \\ A & E \end{bmatrix} = \text{rank}\begin{bmatrix} \hat{E} & 0 \\ \gamma\hat{E} - I_n & \hat{E} \end{bmatrix}$$

$$= \text{rank}\begin{bmatrix} \hat{E} & 0 \\ -I_n & \hat{E} \end{bmatrix}$$

$$= \text{rank}\begin{bmatrix} 0 & \hat{E}^2 \\ -I_n & \hat{E} \end{bmatrix}$$

$$= n + \text{rank}\hat{E}^2.$$

Thus, the relation (7.3) is converted into

$$n + \operatorname{rank}\hat{E}^2 = n + \operatorname{rank}\hat{E},$$

that is $\operatorname{rank}\hat{E}^2 = \operatorname{rank}\hat{E}$.   □

The following theorem, which is due to Duan and Wu (2005d), gives an alternative criterion for a regular descriptor linear system to be impulse-free. This result is based on the third canonical equivalent form for derivative feedback.

Theorem 7.5. *Consider the descriptor linear system (7.1) with* $\operatorname{rank}E = n_0$ *and* $\operatorname{rank}B = r$. *Let the nonsingular matrix* $Q \in \mathbb{R}^{n\times n}$ *and the two orthogonal matrices* $P \in \mathbb{R}^{n\times n}$ *and* $U_3 \in \mathbb{R}^{r\times r}$ *be obtained by Algorithm 4.1. Further, let*

$$\Theta_L = \begin{bmatrix} 0_{(n_2-n_0)\times n_0} & I_{n_2-n_0} & 0_{(n_2-n_0)\times(n-n_2)} \end{bmatrix} Q, \tag{7.8}$$

$$\Theta_R = P \begin{bmatrix} 0_{(n_2-n_0)\times(n-n_2+n_0)} & I_{n_2-n_0} \end{bmatrix}^{\mathrm{T}}. \tag{7.9}$$

Then the system (7.1) is impulse-free if and only if the following relation holds:

$$\det(\Theta_L A \Theta_R) \neq 0. \tag{7.10}$$

Proof. It follows from Theorem 7.2 that the system (7.1) is impulse-free if and only if (7.3) holds.

Obviously, an impulse-free descriptor linear system is I-controllable. Thus, the relations (4.115)–(4.117) hold. In this case, we have

$$\operatorname{rank} \begin{bmatrix} E & 0 \\ A & E \end{bmatrix} = \operatorname{rank} \begin{bmatrix} QEP & 0 \\ QAP & QEP \end{bmatrix}$$

$$= \operatorname{rank} \begin{bmatrix} \Pi_1 & 0 & 0 & 0 & 0 \\ 0 & 0 & 0 & 0 & 0 \\ A_{11} & A_{12} & A_{13} & \Pi_1 & 0 \\ A_{21} & A_{22} & A_{23} & 0 & 0 \\ A_{31} & \Delta & 0 & 0 & 0 \end{bmatrix}$$

$$= \operatorname{rank} \begin{bmatrix} \Pi_1 & 0 & 0 & 0 \\ 0 & 0 & 0 & \Pi_1 \\ 0 & 0 & A_{23} & 0 \\ 0 & \Delta & 0 & 0 \end{bmatrix}$$

$$= 2n_0 + n - n_2 + \operatorname{rank}A_{23}.$$

Thus, the relation (7.3) is converted into

$$2n_0 + n - n_2 + \operatorname{rank}A_{23} = n + n_0,$$

which implies

$$\det A_{23} \neq 0.$$

In view of the partition of (4.116), we have the following relation

$$\Theta_L A \Theta_R = A_{23}.$$

Combining this with (4.116) gives the conclusion.   □

Example 7.2. Consider the system treated in Example 7.1. By some computations, we can obtain

$$T_1 = I_4,$$

$$T_2 = \begin{bmatrix} 0 & 0 & 0 & 1 \\ 1 & 0 & 0 & 0 \\ 0 & 0 & 1 & 0 \\ 0 & 1 & 0 & 0 \end{bmatrix},$$

and

$$A_{22} = \begin{bmatrix} 1 & 0 \\ 1 & 0 \end{bmatrix}.$$

It is obvious that $\det A_{22} = 0$, so the system is not impulse-free according to Theorem 7.3.

When we choose $\gamma = 2$, the matrix $(2E - A)$ is invertible, and in this case, we have

$$\hat{E} = (2E - A)^{-1}E$$

$$= \begin{bmatrix} 0 & -1 & 0 & \frac{2}{3} \\ 0 & 0 & 0 & \frac{1}{3} \\ 0 & 0 & 0 & -\frac{1}{3} \\ 0 & 0 & 0 & \frac{1}{3} \end{bmatrix}.$$

It is obvious that $\operatorname{rank}\hat{E} = 2$ and $\operatorname{rank}\hat{E}^2 = 1$. So this system is not impulse-free according to Theorem 7.4.

Following the steps in Algorithm 4.1, we can obtain

$$Q = I_4, \quad U_3 = I_2,$$

$$P = \begin{bmatrix} 0 & 0 & 0 & 1 \\ 1 & 0 & 0 & 0 \\ 0 & 0 & 1 & 0 \\ 0 & 1 & 0 & 0 \end{bmatrix}.$$

In addition,

$$\Theta_L = \begin{bmatrix} 0 & 0 & 1 & 0 \end{bmatrix}, \quad \Theta_R = \begin{bmatrix} 1 & 0 & 0 & 0 \end{bmatrix}^{\mathrm{T}}.$$

Then we have

$$\det(\Theta_L A \Theta_R) = \det[0] = 0.$$

The system is therefore not impulse-free according to Theorem 7.5.

7.2 Impulse Elimination by State Feedback

In this section, we consider impulse elimination in descriptor linear systems via state feedback. The problem can be stated as follows.

Problem 7.1. Given the regular descriptor linear system

$$E\dot{x} = Ax + Bu, \tag{7.11}$$

where E, $A \in \mathbb{R}^{n \times n}$, $B \in \mathbb{R}^{n \times r}$, with $\mathrm{rank}\, E = n_0$, find a proportional state feedback controller

$$u = Kx \tag{7.12}$$

such that the closed-loop system

$$E\dot{x} = (A + BK)x \tag{7.13}$$

is impulse-free.

Remark 7.1. In view of Theorems 7.1 and 7.2, the requirement in the above problem can be replaced by any of the following ones:

- the closed-loop system (7.13) has $\mathrm{rank}\, E$ finite poles;
- $\deg \det (sE - (A + BK)) = \mathrm{rank}\, E$;
- $\mathrm{rank} \begin{bmatrix} E & 0 \\ A + BK & E \end{bmatrix} = n + \mathrm{rank}\, E$.

7.2.1 Solution Based on Dynamics Decomposition Forms

The following theorem gives a solution to Problem 7.1 based on the dynamics decomposition forms of descriptor linear systems.

Theorem 7.6. *Let the descriptor linear system (7.11) be regular, and the associated relations (7.5) and (7.6) hold.*

1. *There exists a state feedback controller (7.12) such that the closed-loop system (7.13) is impulse-free if and only if system (7.11) is I-controllable.*

2. *When system (7.11) is I-controllable, there holds*

$$\text{rank} \begin{bmatrix} A_{22} & B_2 \end{bmatrix} = n - n_0. \tag{7.14}$$

Furthermore, a general expression for all the state feedback controllers in the form of (7.12), which makes the closed-loop system (7.13) impulse-free, is given by

$$K = \begin{bmatrix} K_1 & K_2 \end{bmatrix} T_2^T, \tag{7.15}$$

with $K_1 \in \mathbb{R}^{r \times n_0}$ and

$$K_2 \in \mathcal{N}_{(A_{22}, B_2, I_{n-n_0})} = \{ K \,|\, \det (A_{22} + B_2 K) \neq 0 \}. \tag{7.16}$$

Proof. Noting that T_2 is orthogonal, we have from (7.15) that

$$K T_2 = \begin{bmatrix} K_1 & K_2 \end{bmatrix}.$$

Due to (7.5) and (7.6), and the above relation, it follows from Theorem 7.3 that the system is impulse-free if and only if

$$\begin{aligned}
&\exists K_2, \text{ s.t. } \det (A_{22} + B_2 K_2) \neq 0 \\
&\Leftrightarrow \text{ the normalizability of } (A_{22}, B_2) \\
&\Leftrightarrow \text{rank} \begin{bmatrix} A_{22} & B_2 \end{bmatrix} = n - \text{rank} E \\
&\Leftrightarrow \text{ system (7.1) is I-controllable.}
\end{aligned} \tag{7.17}$$

With this, we complete the proof of the first conclusion.

When system (7.11) is I-controllable, it follows from (7.17) that (7.14) holds. It also follows from the above proof of the first conclusion that every matrix K making the closed-loop system (7.13) impulse-free is given by (7.15) with K_1 being an arbitrary real matrix, while K_2 being a real matrix satisfying

$$\det (A_{22} + B_2 K_2) \neq 0. \tag{7.18}$$

The second conclusion of the theorem therefore becomes obvious. □

This theorem further reveals the relationship between the impulse controllability and the impulse terms of the system. By this theorem, we have clearly seen that impulse controllability guarantees the ability to eliminate the impulse terms in the state response of the closed-loop system resulted in by state feedback control.

Using the notations in Sect. 6.4, (7.16) is equivalent to

$$K_2 \in \Gamma_{(A_{22}, B_2, I_{n-n_0})},$$

that is, K_2 is a state derivative feedback gain matrix, which normalizes the following descriptor linear system

$$A_{22} \dot{x} = x + B_2 u.$$

Further, in view of Theorem 6.14, we have the following corollary.

Corollary 7.1. *Assume that the regular descriptor linear system (7.11) is I-controllable. Then the set of all state feedback controllers, which make the closed-loop system (7.13) impulse-free is a Zariski open set.*

The above corollary tells us that almost all state feedback controllers can make the closed-loop system (7.13) impulse-free when the open-loop system is I-controllable. This is a very good conclusion since it allows us to solve the problem of impulse elimination by a "trial-and-test" procedure. Now we give an algorithm for impulse elimination via state feedback.

Algorithm 7.1. Impulse elimination based on dynamics decomposition.

Step 1 Perform the singular value decomposition (7.5) and find the matrices A_{22} and B_2 according to (7.6).

Step 2 Find a matrix K_2 satisfying (7.18), that is $K_2 \in \mathcal{N}_{(A_{22}, B_2, I_{n-n_0})}$. Under the assumption of impulse controllability, $\text{rank}\begin{bmatrix} A_{22} & B_2 \end{bmatrix} = n - \text{rank}E$, such a K_2 is solvable.

Step 3 Arbitrarily choose a $K_1 \in \mathbb{R}^{r \times n_0}$ and form the matrix K according to (7.15).

Example 7.3 (Chen and Chang 1993; Duan and Patton 1997; Duan 1998a). Consider a system with the following parameters

$$
E = \begin{bmatrix} 1 & 0 & 0 & 0 & 0 & 0 \\ 0 & 1 & 0 & 0 & 0 & 0 \\ 0 & 0 & 1 & 0 & 0 & 0 \\ 0 & 0 & 0 & 0 & 1 & 0 \\ 0 & 0 & 0 & 0 & 0 & 0 \\ 1 & 0 & 0 & 0 & 0 & 0 \end{bmatrix}, A = \begin{bmatrix} 0 & 0 & 1 & 0 & 0 & 0 \\ 1 & 0 & 0 & 0 & 0 & 0 \\ 0 & 1 & 0 & 1 & 0 & 0 \\ 0 & 0 & 0 & 1 & 0 & 0 \\ 0 & 0 & 0 & 0 & 1 & 0 \\ 1 & 0 & 0 & 0 & 0 & 1 \end{bmatrix}, B = \begin{bmatrix} 1 & 0 \\ 0 & 0 \\ 0 & 0 \\ 0 & 0 \\ 1 & 1 \\ 0 & 0 \end{bmatrix}.
$$

$$(7.19)$$

For this system, $n = 6$, $\text{rank}E = 4$, and

$$
\text{rank}\begin{bmatrix} E & 0 \\ A & E \end{bmatrix} = 9 \neq n + \text{rank}E.
$$

Thus, impulse terms are certain to appear in the state response.

However, it can be verified that the system is I-controllable. Therefore, almost any state feedback controller can eliminate the impulse in the system. For example, choose

$$
K = \begin{bmatrix} 0 & 0 & 0 & -1 & 0 & 0 \\ 0 & 0 & 0 & 0 & 0 & 1 \end{bmatrix},
$$

which satisfies

$$\deg \det (sE - (A + BK)) = 4 = \operatorname{rank} E.$$

For this K, the closed-loop system is

$$
\begin{bmatrix}
1 & 0 & 0 & 0 & 0 & 0 \\
0 & 1 & 0 & 0 & 0 & 0 \\
0 & 0 & 1 & 0 & 0 & 0 \\
0 & 0 & 0 & 0 & 1 & 0 \\
0 & 0 & 0 & 0 & 0 & 0 \\
1 & 0 & 0 & 0 & 0 & 0
\end{bmatrix}
\dot{x} =
\begin{bmatrix}
0 & 0 & 1 & -1 & 0 & 0 \\
1 & 0 & 0 & 0 & 0 & 0 \\
0 & 1 & 0 & 1 & 0 & 0 \\
0 & 0 & 0 & 1 & 0 & 0 \\
0 & 0 & 0 & -1 & 1 & 1 \\
1 & 0 & 0 & 0 & 0 & 1
\end{bmatrix}
x. \qquad (7.20)
$$

It can be easily verified that the closed-loop system (7.20) has the characteristic polynomial

$$\det (sE - (A + BK)) = 2s^4 - 3s^3 + s^2 - s + 1,$$

which implies

$$\sigma(E, A + BK) = \left\{ 1, 1, \frac{-1 + \sqrt{7}i}{4}, \frac{-1 - \sqrt{7}i}{4} \right\}.$$

Therefore, system (7.20) has $\operatorname{rank} E$ finite poles, hence is impulse-free.

Example 7.4. Consider the system in Example 1.4 again, whose parameters are given as

$$
E =
\begin{bmatrix}
1 & 0 & 0 & 0 & 0 \\
0 & 1 & 0 & 0 & 0 \\
0 & 0 & 1 & 0 & 0 \\
0 & 0 & 0 & 1 & 0 \\
0 & 0 & 0 & 0 & 0
\end{bmatrix},
A =
\begin{bmatrix}
0 & 0 & 1 & 0 & 0 \\
0 & 0 & 0 & 1 & 0 \\
-2 & 0 & -1 & -1 & 1 \\
0 & -1 & -1 & -1 & 1 \\
1 & 1 & 0 & 0 & 0
\end{bmatrix},
B =
\begin{bmatrix}
0 \\
0 \\
-1 \\
1 \\
0
\end{bmatrix}. \qquad (7.21)
$$

For this system, $n = 5$, $\operatorname{rank} E = 4$, and

$$\operatorname{rank} \begin{bmatrix} E & 0 & 0 \\ A & E & B \end{bmatrix} = 8 < 9 = n + \operatorname{rank} E.$$

Thus, system (7.21) is not I-controllable. Therefore, it follows from the above theorem that there does not exist a state feedback $u = Kx$ for this system such that the corresponding closed-loop system is impulse-free.

7.2.2 Solution Based on Standard Decomposition

In the preceding subsection, we have considered solution to the impulse elimination problem based on the dynamics decomposition form. In this subsection, we investigate the solution to the same problem based on the standard decomposition.

Recall that there exist two nonsingular matrices Q and P, such that the regular system (7.11) is r.s.e. to the following standard decomposition form:

$$\begin{cases} \dot{x}_1 = A_1 x_1 + B_1 u \\ N\dot{x}_2 = x_2 + B_2 u \end{cases},$$ (7.22)

where $x_1 \in \mathbb{R}^{n'_1}$, $x_2 \in \mathbb{R}^{n'_2}$, $n'_1 + n'_2 = n$, and

$$x = P \begin{bmatrix} x_1 \\ x_2 \end{bmatrix}.$$

The relations among the coefficients are given by (7.2).

Under decomposition (7.22), the state feedback $u = Kx + v$ becomes

$$u = \bar{K}_1 x_1 + \bar{K}_2 x_2 + v,$$ (7.23)

where v is an external signal, and $\bar{K}_1 \in \mathbb{R}^{r \times n'_1}$, $\bar{K}_2 \in \mathbb{R}^{r \times n'_2}$ are determined by

$$KP = \begin{bmatrix} \bar{K}_1 & \bar{K}_2 \end{bmatrix}.$$

For convenience, let us introduce the following definition.

Definition 7.1. Consider the descriptor linear system (7.22) in standard decomposition form.

1. The general state feedback control (7.23), which includes both the slow substate and the fast substate, is called a combined slow-fast substate feedback for the system, or simply, a state feedback hereafter.
2. When $\bar{K}_2 = 0$, the state feedback (7.23) becomes

$$u = \bar{K}_1 x_1 + v,$$ (7.24)

which involves only the substate x_1 and is called a slow substate feedback, a slow state feedback hereafter.
3. When $\bar{K}_1 = 0$, the state feedback control (7.23) becomes

$$u = \bar{K}_2 x_2 + v,$$ (7.25)

which involves only the fast substate x_2 and is called a fast substate feedback control, a fast state feedback hereafter.

While the standard decomposition characterizes the system's inner structure, the slow and fast feedbacks show the effect of a partial-state feedback on the whole system structure.

Theorem 7.7. *Consider the regular linear system (7.11).*

1. There exists a state feedback control (7.12) such that the closed-loop system (7.13) is impulse-free if and only if there exists a fast feedback (7.25) for its fast subsystem such that the resulted closed-loop system is impulse-free, that is, there exists a $\bar{K}_2$ matrix such that

$$\deg \det \left(sN - \left(I + B_2\bar{K}_2\right)\right) = \mathrm{rank}N. \tag{7.26}$$

2. When this condition is satisfied, all the solutions to the problem are characterized by

$$K = \begin{bmatrix} \bar{K}_1 & \bar{K}_2 \end{bmatrix} P^{-1}, \tag{7.27}$$

where P is the transformation matrix satisfying (7.2), $\bar{K}_1 \in \mathbb{R}^{r \times n_1'}$ is an arbitrary real parameter matrix, while $\bar{K}_2 \in \mathbb{R}^{r \times n_2'}$ is an arbitrary matrix satisfying (7.26).

Proof. By Theorem 7.6, the I-controllability of (7.11) is the necessary and sufficient condition for the existence of a state feedback in the form of (7.12) to eliminate impulse in the closed-loop system (7.13). While the I-controllability of (7.11) is equivalent to the impulse controllability of the fast subsystem. Thus, it follows from Theorem 7.1 that system (7.11) is I-controllable if and only if there exists a matrix $\bar{K}_2$ satisfying (7.26). The first conclusion is proven.

Now let $\bar{K}_2$ be an arbitrary matrix satisfying (7.26), and K be given by (7.27), then

$$\begin{aligned}
\deg \det &(sE - (A + BK)) \\
&= \deg \det (sQEP - Q(A + BK)P) \\
&= \deg \det \begin{bmatrix} sI_{n_1'} - A_1 - B_1\bar{K}_1 & -B_1\bar{K}_2 \\ -B_2\bar{K}_1 & sN - I - B_2\bar{K}_2 \end{bmatrix} \\
&= n_1' + \deg \det \left(sN - \left(I + B_2\bar{K}_2\right)\right) \\
&= n_1' + \mathrm{rank}N \\
&= \mathrm{rank}E.
\end{aligned}$$

This indicates that any feedback gain K given by (7.27) makes the closed-loop system impulse-free. □

Inspired by this proof, we can give the following algorithm for solution to Problem 7.1.

Algorithm 7.2. Impulse elimination based on standard decomposition.

Step 1 Find a pair of transformation matrices P and Q, and transform the system (7.11) into a standard decomposition form (7.22).

Step 2 Solve the state feedback impulse elimination problem for the fast subsystem, that is, find a matrix $\bar{K}_2$ satisfying (7.26).

Step 3 Find the matrix K according to (7.27), with $\bar{K}_1$ being an arbitrary real parameter matrix.

The above Step 2 can be carried out by applying Algorithm 7.1 to the fast subsystem

$$N\dot{x}_2 = x_2 + B_2 u.$$

Due to the special structure of the nilpotent matrix N, the first step in Algorithm 7.1 can be easily carried out. It is suggested that readers deduce a general formula for the feedback controller eliminating the impulse in a fast subsystem.

Example 7.5. Consider the following system

$$\begin{bmatrix} 0 & 0 & 0 & 1 \\ 0 & 0 & 1 & 0 \\ 0 & 0 & 0 & 0 \\ -1 & 0 & 0 & 1 \end{bmatrix} \dot{x} = \begin{bmatrix} 0 & 1 & 0 & 0 \\ 1 & 0 & 0 & 0 \\ -1 & 0 & 0 & 1 \\ 0 & 1 & 1 & 1 \end{bmatrix} x + \begin{bmatrix} 0 \\ 0 \\ 1 \\ 0 \end{bmatrix} u. \qquad (7.28)$$

If we choose the following transformation matrices

$$Q = \begin{bmatrix} 1 & 0 & 1 & -1 \\ 0 & 1 & 0 & 0 \\ 0 & 0 & -1 & 1 \\ 0 & 0 & 1 & 0 \end{bmatrix}, \quad P = \begin{bmatrix} 1 & 0 & 0 & 0 \\ -1 & -1 & 1 & 0 \\ 0 & 1 & 0 & 0 \\ 1 & 0 & 0 & 1 \end{bmatrix}$$

and the state transformation

$$P^{-1}x = \begin{bmatrix} x_1 \\ x_2 \end{bmatrix}, x_1 \in \mathbb{R}^2, x_2 \in \mathbb{R}^2,$$

the standard decompositions form for the system (7.28) can then be obtained as

$$\begin{cases} \dot{x}_1 = \begin{bmatrix} -1 & -1 \\ 1 & 0 \end{bmatrix} x_1 + \begin{bmatrix} 1 \\ 0 \end{bmatrix} u \\ \begin{bmatrix} 0 & 1 \\ 0 & 0 \end{bmatrix} \dot{x}_2 = x_2 + \begin{bmatrix} -1 \\ 1 \end{bmatrix} u \end{cases}$$

Then the matrix

$$\bar{K}_2 = \begin{bmatrix} 1 & 0 \end{bmatrix}$$

satisfies

$$\deg \det \left(sN - \left(I + B_2 \bar{K}_2\right)\right) = 1 = \text{rank} N.$$

Thus, the gain matrix K can be obtained as

$$K = \begin{bmatrix} 0 & 0 & \bar{K}_2 \end{bmatrix} P^{-1} = \begin{bmatrix} 1 & 1 & 1 & 0 \end{bmatrix}.$$

7.2.3 Solution Based on Canonical Equivalent Form for Derivative Feedback

In this subsection, we investigate the solution to impulse elimination by state feedback based on the third canonical equivalent form for derivative feedback. The main results are taken from Duan and Wu (2005d).

Applying Theorem 7.5 we can prove the following theorem.

Theorem 7.8. *Consider the descriptor linear system (7.11) with* $\text{rank} E = n_0$ *and* $\text{rank} B = r$. *Let the nonsingular matrix* $Q \in \mathbb{R}^{n \times n}$ *and the two orthogonal matrices* $P \in \mathbb{R}^{n \times n}$ *and* $U_3 \in \mathbb{R}^{r \times r}$ *be obtained by Algorithm 4.1, and* $\tilde{E}$, $\tilde{A}$ *and* $\tilde{B}$ *be given by (4.115)–(4.117), and integers* n_2 *and* n_1 *be defined in (6.4). When the system (7.11) is I-controllable, all the state feedback controllers (7.12) making the closed-loop system (7.13) impulse-free are given by*

$$K = U_3 \begin{bmatrix} K_{11} & K_{12} \\ K_{21} & K_{22} \end{bmatrix} P^{\text{T}}, \qquad (7.29)$$

where $K_{11} \in \mathbb{R}^{(n_0 - n_1) \times (n + n_0 - n_2)}$, $K_{12} \in \mathbb{R}^{(n_0 - n_1) \times (n_2 - n_0)}$ *and* $K_{21} \in \mathbb{R}^{(n_2 - n_0) \times (n + n_0 - n_2)}$ *are some arbitrary parameter matrices, and* $K_{22} \in \mathbb{R}^{(n_2 - n_0) \times (n_2 - n_0)}$ *is the parameter matrix satisfying the following constraint:*

$$\det (K_{22} + A_{23}) \neq 0. \qquad (7.30)$$

Proof. Let

$$U_3^{\text{T}} K P = \begin{bmatrix} K_{11} & K_{12} \\ K_{21} & K_{22} \end{bmatrix}, \quad K_{22} \in \mathbb{R}^{(n_2 - n_0) \times (n_2 - n_0)}, \qquad (7.31)$$

then we have

$$Q(A + BK)P = \begin{bmatrix} A_{11} & A_{12} & A_{13} \\ A_{21} & A_{22} & A_{23} \\ A_{31} & \Delta & 0 \end{bmatrix} + \begin{bmatrix} \Pi_2 & 0 & \\ 0 & I_{n_2 - n_0} & \\ 0 & & 0 \end{bmatrix} \begin{bmatrix} K_{11} & K_{12} \\ K_{21} & K_{22} \end{bmatrix}$$

$$= \begin{bmatrix} A_{11} & A_{12} & A_{13} \\ A_{21} & A_{22} & A_{23} \\ A_{31} & \Delta & 0 \end{bmatrix} + \begin{bmatrix} 0 & 0 \\ K_{11} & K_{12} \\ K_{21} & K_{22} \\ 0 & 0 \end{bmatrix}.$$

Let the matrix Θ_L and Θ_R be defined by (7.8) and (7.9), then

$$\Theta_L(A + BK)\Theta_R = K_{22} + A_{23}.$$

Thus, by Theorem 7.5 the closed-loop system is impulse-free if and only if (7.30) is met. Due to the relation (7.31), all the gain matrix K making the closed-loop system (7.13) impulse-free can be parameterized by (7.29).   □

Example 7.6. Consider the system mentioned in Example 7.1. For this example system, $n_0 = 2$, $n_2 = 3$. From Example 7.1 it is known that this system is not impulse-free. Our aim is to design a state feedback controller in the form of (7.12) such that the resulted closed-loop system is impulse-free.

It has been known from Example 7.2 that

$$Q = I_4, \quad P = \begin{bmatrix} 0 & 0 & 0 & 1 \\ 1 & 0 & 0 & 0 \\ 0 & 0 & 1 & 0 \\ 0 & 1 & 0 & 0 \end{bmatrix}, \quad U_3 = I_2.$$

So

$$Q_3 A P_3 = \begin{bmatrix} 0 & 0 & 0 & 1 \\ 1 & 0 & 2 & 0 \\ 0 & 1 & 1 & 0 \\ 1 & 0 & 1 & 0 \end{bmatrix},$$

$$A_{23} = 0.$$

According to Theorem 7.8, all the controllers which make the closed-loop system impulse-free are given by

$$K = U_3 \begin{bmatrix} K_{11} & K_{12} \\ K_{21} & K_{22} \end{bmatrix} P^{\mathrm{T}}.$$

with K_{22} satisfying

$$\det(0 + K_{22}) \neq 0. \tag{7.32}$$

Let

$$\begin{bmatrix} K_{11} & K_{12} \\ K_{21} & K_{22} \end{bmatrix} = \begin{bmatrix} k_{11} & k_{12} & k_{13} & k_{10} \\ k_{21} & k_{22} & k_{23} & k_{20} \end{bmatrix},$$

then (7.32) is equivalent to $k_{20} \neq 0$. So all the controllers (7.12) making the resulted closed-loop system impulse-free are parameterized as

$$K = \begin{bmatrix} k_{10} & k_{11} & k_{13} & k_{12} \\ k_{20} & k_{21} & k_{23} & k_{22} \end{bmatrix}, \; k_{20} \neq 0,$$

where $k_{ij}, i = 1, 2, j = 1, 2, 3$, and k_{10} are arbitrary real scalars except k_{20}, which is restricted to be different from zero.

7.3 Impulse Elimination by Output Feedback

In this section, we study the problem of impulse elimination in descriptor linear systems via static output feedback.

7.3.1 Problem Formulation

Consider the following descriptor linear system:

$$\begin{cases} E\dot{x} = Ax + Bu \\ y = Cx \end{cases}, \tag{7.33}$$

where $x(t) \in \mathbb{R}^n$ is the descriptor state vector, $u(t) \in \mathbb{R}^r$ is the control vector, $y(t) \in \mathbb{R}^m$ is the measured output vector. $A, E \in \mathbb{R}^{n \times n}$, $B \in \mathbb{R}^{n \times r}$ and $C \in \mathbb{R}^{m \times n}$ are known real coefficient matrices.

A static output feedback controller for system (7.33) is of the form

$$u = Ky + v, \quad K \in \mathbb{R}^{r \times m}, \tag{7.34}$$

where $v(t)$ is the external or reference signal. The static output feedback is easy to realize but it contains less information about the state vector of the system and hence has more limitations than state feedback. In certain cases, it may not be able to meet the design requirements.

When the static output feedback law (7.34) is applied to system (7.33), the closed-loop system is obtained as

$$\begin{cases} E\dot{x} = (A + BKC)x + Bv \\ y = Cx \end{cases}. \tag{7.35}$$

Problem 7.2. Given the regular descriptor linear system (7.33), find for the system a static output feedback controller in the form of (7.34) such that the closed-loop system (7.35) is impulse-free.

7.3.2 The Solution

By performing the singular value decomposition of the matrix E, we obtain

$$Q_1 E P_1 = \text{diag} (\Sigma, \, 0), \quad \Sigma = \text{diag} \left(\sigma_1, \, \sigma_2, \, \ldots, \, \sigma_{n_0} \right) > 0, \tag{7.36}$$

where Q_1 and P_1 are two orthogonal matrices. Denoting

$$Q_1 B = \begin{bmatrix} B_1 \\ B_2 \end{bmatrix}, \quad C P_1 = [C_1 \; C_2], \tag{7.37}$$

and

$$Q_1 A P_1 = \begin{bmatrix} A_{11} & A_{12} \\ A_{21} & A_{22} \end{bmatrix}. \tag{7.38}$$

Then the main result in this section can be given as follows.

Theorem 7.9. *Let the descriptor linear system (7.33) be regular, and the associated relations (7.36)–(7.38) hold.*

1. *There exists an output feedback (7.34) for this system such that the closed-loop system (7.35) is impulse-free, if and only if the system is both I-controllable and I-observable.*
2. *When system (7.33) is I-controllable and I-observable, there holds*

$$\text{rank} \begin{bmatrix} A_{22} & B_2 \end{bmatrix} = \text{rank} \begin{bmatrix} A_{22} \\ C_2 \end{bmatrix} = n - n_0. \tag{7.39}$$

Furthermore, a general form for all the output feedback controllers in the form of (7.34), which make the closed-loop system (7.35) impulse-free, is given by

$$K \in \mathcal{N}_{(A_{22}, B_2, C_2)} = \{ K \mid \det (A_{22} + B_2 K C_2) \neq 0 \}.$$

Proof. Proof of conclusion 1. First let us show the necessity. Suppose that there exists an output feedback (7.34) for this system such that the closed-loop system (7.35) is impulse-free, then

$$\deg \det (sE - (A + BKC)) = \text{rank} E,$$

hence

$$\deg \det \left(sE^{\text{T}} - \left(A^{\text{T}} + C^{\text{T}} K^{\text{T}} B^{\text{T}} \right) \right) = \text{rank} E^{\text{T}}.$$

From these two equations, we know that the system (E, A, B, C) and its dual system $(E^{\text{T}}, A^{\text{T}}, C^{\text{T}}, B^{\text{T}})$ are both I-controllable. Therefore, the system (7.33) is both I-controllable and I-observable.

Now we prove the sufficiency. By assumption, system (7.33) is I-controllable and I-observable. Thus, by Theorems 4.18 and 4.19, the relation (7.39) holds. With these relations, Lemma 5.3 indicates that a matrix K may be chosen such that

$$\det (A_{22} + B_2 K C_2) \neq 0. \tag{7.40}$$

This implies that the closed-loop system is impulse-free according to Theorem 7.3.
Proof of conclusion 2. This clearly follows from the sufficiency part of the above proof of the first conclusion.   □

It is seen from the above problem that, like the case of state feedback, impulse elimination using output feedback is again converted into a normalization problem. The solution to this problem involves only two steps: first solve the matrices A_{22}, B_2 and C_2, and then solve a gain matrix from the set

$$\mathscr{N}_{(A_{22}, B_2, C_2)} = \{K \mid \det (A_{22} + B_2 K C_2) \neq 0\}.$$

Recall that $\mathscr{N}_{(A_{22}, B_2, C_2)}$ is a Zariski open set, such a gain matrix can often be easily sought by a "trial-and-test" procedure.

Example 7.7. Consider the system in Example 3.5 again, where $n = 4$, rank$E = 3$ and

$$E = \begin{bmatrix} 1 & 0 & 0 & 0 \\ 0 & 1 & 0 & 0 \\ 0 & 0 & 1 & 0 \\ 0 & 0 & 0 & 0 \end{bmatrix}, \quad A = \begin{bmatrix} 0 & 0 & 1 & 0 \\ 1 & 0 & 0 & 0 \\ 0 & 1 & 0 & 1 \\ 0 & 0 & 1 & 0 \end{bmatrix},$$

$$B^{\mathrm{T}} = \begin{bmatrix} 1 & 1 & 0 & 0 \\ 0 & -1 & 1 & 0 \\ 0 & 2 & 0 & 1 \end{bmatrix}, \quad C = \begin{bmatrix} 0 & 1 & 0 & 0 \\ 0 & 0 & 0 & 1 \end{bmatrix}.$$

Since

$$\mathrm{rank} \begin{bmatrix} E & 0 & 0 \\ A & E & B \end{bmatrix} = 7 = n + \mathrm{rank}E,$$

$$\mathrm{rank} \begin{bmatrix} E & A \\ 0 & E \\ 0 & C \end{bmatrix} = 7 = n + \mathrm{rank}E,$$

the system is both I-controllable and I-observable. Therefore, an output feedback $u = Ky + v$ exists such that its closed-loop system is impulse-free. Next, let us find such a matrix K.

Here, we have

$$A_{22} = 0, \quad B_2 = \begin{bmatrix} 0 & 0 & 1 \end{bmatrix}, \quad C_2 = \begin{bmatrix} 0 & 1 \end{bmatrix}^{\mathrm{T}}.$$

Choose

$$K^{\mathrm{T}} = \begin{bmatrix} 1 & 1 & 1 \\ 1 & 1 & 1 \end{bmatrix}.$$

It clearly satisfies

$$\det (A_{22} + B_2 K C_2) = 1 \neq 0.$$

Corresponding to (7.35), the closed-loop system is

$$
\left\{
\begin{array}{l}
\begin{bmatrix} 1 & 0 & 0 & 0 \\ 0 & 1 & 0 & 0 \\ 0 & 0 & 1 & 0 \\ 0 & 0 & 0 & 0 \end{bmatrix} \dot{x} =
\begin{bmatrix} 0 & 1 & 1 & 1 \\ 1 & 2 & 0 & 2 \\ 0 & 2 & 0 & 2 \\ 0 & 1 & 1 & 1 \end{bmatrix} x +
\begin{bmatrix} 1 & 0 & 0 \\ 1 & -1 & 2 \\ 0 & 1 & 0 \\ 0 & 0 & 1 \end{bmatrix} v \\[20pt]
y = \begin{bmatrix} 0 & 1 & 0 & 0 \\ 0 & 0 & 0 & 1 \end{bmatrix} x
\end{array}
\right. .
\qquad (7.41)
$$

It is easy to verify that

$$\deg \det (sE - (A + BKC)) = \deg \left(-s^3 - 2s^2\right) = 3 = \operatorname{rank} E.$$

Thus, no impulse terms exist in the state response of the closed-loop system (7.41).

7.4 I-Controllablizability and I-Observablizability

Consider the following descriptor linear system:

$$
\left\{
\begin{array}{l}
E\dot{x} = Ax + Bu \\
y_p = C_p x \\
y_d = C_d \dot{x}
\end{array}
\right. ,
\qquad (7.42)
$$

where $x(t) \in \mathbb{R}^n$ is the descriptor state vector, $u(t) \in \mathbb{R}^r$ is the control vector, $y_p(t) \in \mathbb{R}^{m_p}$ and $y_d(t) \in \mathbb{R}^{m_d}$ are the measured proportional output and derivative output, respectively. A, $E \in \mathbb{R}^{n \times n}$, $B \in \mathbb{R}^{n \times r}$, $C_p \in \mathbb{R}^{m_p \times n}$, and $C_d \in \mathbb{R}^{m_d \times n}$ are known real matrices.

Applying the state-feedback control law

$$u = K_p x + v$$

with v being the external input to the system, (7.42) results in the following closed-loop system

$$E\dot{x} = (A + BK_p)x + Bv. \qquad (7.43)$$

The following proposition shows that the open-loop system (7.42) and the closed-loop system (7.43) possess the same I-controllability.

Proposition 7.1. *A pure proportional state feedback does not change the I-controllability of a descriptor linear system.*

Proof. Since

$$
\mathrm{rank} \begin{bmatrix} E & 0 & 0 \\ A + BK_p & E & B \end{bmatrix} = \mathrm{rank} \left(\begin{bmatrix} E & 0 & 0 \\ A & E & B \end{bmatrix} \begin{bmatrix} I & 0 & 0 \\ 0 & I & 0 \\ K_p & 0 & I \end{bmatrix} \right)
$$

$$
= \mathrm{rank} \begin{bmatrix} E & 0 & 0 \\ A & E & B \end{bmatrix},
$$

the conclusion clearly follows from Theorem 4.14.   □

It is seen from Proposition 7.1 that a pure proportional state feedback does not change the I-controllability of a descriptor linear system. Now applying the state derivative feedback law

$$
u = -K_d \dot{x} + v \tag{7.44}
$$

with v being the external input, to the system (7.42) we obtain

$$
(E + BK_d)\dot{x} = Ax + Bv. \tag{7.45}
$$

It is well known that a state derivative feedback does. Due to this fact, let us introduce the following definition.

Definition 7.2. The regular descriptor linear system (7.42), or the matrix triple (E, A, B), is said to be I-controllablizable if there exists a matrix $K_d \in \mathbb{R}^{r \times n}$ such that the matrix triple $(E + BK_d, A, B)$, or the system (7.45), is I-controllable. In this case, (7.44) is called an I-controllablizing controller for the system.

Dually, we introduce the concept of I-observablizability.

Definition 7.3. The regular descriptor linear system (7.42), or the matrix (E, A, C_d), is said to be I-observablizable if there exists a matrix $L_d \in \mathbb{R}^{n \times m_p}$ such that the matrix triple $(E + L_d C_d, A, C_d)$, is I-observablizable.

7.4.1 Basic Criterion

In this subsection, we will give a matrix-rank criterion for I-controllablizability of the system (7.11) in terms of the original system coefficient matrices. Before obtaining the result, we give a proposition.

Proposition 7.2. *Applying the state derivative feedback law (7.44) to the system (7.42) gives the following closed-loop system:*

$$
(E + BK_d)\dot{x} = Ax + Bv. \tag{7.46}
$$

Then the following are equivalent:

1. *The system (7.46) is I-controllable.*
2. *The following relation holds:*

$$\text{rank} \begin{bmatrix} E + BK_d & 0 & 0 \\ A & E & B \end{bmatrix} = n + \text{rank}\,(E + BK_d). \tag{7.47}$$

Proof. It follows from Theorem 4.14 that the system (7.46) is I-controllable if and only if

$$\text{rank} \begin{bmatrix} E + BK_d & 0 & 0 \\ A & E + BK_d & B \end{bmatrix} = n + \text{rank}\,(E + BK_d). \tag{7.48}$$

Since

$$\text{rank} \begin{bmatrix} E + BK_d & 0 & 0 \\ A & E + BK_d & B \end{bmatrix} = \text{rank}\left(\begin{bmatrix} E + BK_d & 0 & 0 \\ A & E & B \end{bmatrix} \begin{bmatrix} I & 0 & 0 \\ 0 & I & 0 \\ 0 & K_d & I \end{bmatrix} \right)$$

$$= \text{rank} \begin{bmatrix} E + BK_d & 0 & 0 \\ A & E & B \end{bmatrix},$$

condition (7.48) is clearly equivalent to (7.47). Therefore, the system (7.46) is I-controllable if and only if (7.47) is met. □

Based on the above proposition, we can give the following conclusion.

Theorem 7.10. *The system (7.42) is I-controllablizable if and only if the following relation holds:*

$$\text{rank} \begin{bmatrix} E & 0 & 0 & B \\ A & E & B & 0 \end{bmatrix} = n + \text{rank} \begin{bmatrix} E & B \end{bmatrix}. \tag{7.49}$$

Proof. According to the proof of Theorem 6.1, we can assume that the relations (6.6)–(6.13) hold, with n_1 and n_2 be defined by (6.4). Letting

$$QK_d = \begin{bmatrix} K_1 \\ K_2 \end{bmatrix}, \quad K_1 \in \mathbb{R}^{s \times n},$$

and

$$P^{-1}A = \begin{bmatrix} A_1 \\ A_2 \end{bmatrix}, \quad A_1 \in \mathbb{R}^{s \times n},$$

we have

$$\text{rank} \begin{bmatrix} E + BK_d & 0 & 0 \\ A & E & B \end{bmatrix} = \text{rank} \begin{bmatrix} P^{-1}(E + BK_d) & 0 & 0 \\ P^{-1}A & P^{-1}E & P^{-1}BQ^{-1} \end{bmatrix}$$

$$= \operatorname{rank} \begin{bmatrix} E_1 + K_1 & 0 & 0 & 0 \\ E_2 & 0 & 0 & 0 \\ A_1 & E_1 & I_s & 0 \\ A_2 & E_2 & 0 & 0 \end{bmatrix}$$

$$= s + \operatorname{rank} \begin{bmatrix} E_1 + K_1 & 0 \\ E_2 & 0 \\ A_2 & E_2 \end{bmatrix}.$$

Further, letting

$$\begin{bmatrix} K_{11} & K_{12} \end{bmatrix} = K_1 Q_1^{-1}, \quad K_{11} \in \mathbb{R}^{s \times n_1},$$

and

$$A_2 = P_1 \begin{bmatrix} A_2^{11} & A_2^{12} \\ A_2^{21} & A_2^{22} \end{bmatrix} Q_1, \quad A_2^{22} \in \mathbb{R}^{(n-n_2) \times (n-n_1)},$$

we have

$$\operatorname{rank} \begin{bmatrix} E + BK_d & 0 & 0 \\ A & E & B \end{bmatrix}$$

$$= s + \operatorname{rank} \begin{bmatrix} \begin{bmatrix} E_{11} & E_{12} \end{bmatrix} Q_1 + \begin{bmatrix} K_{11} & K_{12} \end{bmatrix} Q_1 & 0 \\ P_1 \begin{bmatrix} I_{n_1} & 0 \\ 0 & 0 \end{bmatrix} Q_1 & 0 \\ P_1 \begin{bmatrix} A_2^{11} & A_2^{12} \\ A_2^{21} & A_2^{22} \end{bmatrix} Q_1 & P_1 \begin{bmatrix} I_{n_1} & 0 \\ 0 & 0 \end{bmatrix} \end{bmatrix}$$

$$= s + \operatorname{rank} \begin{bmatrix} 0 & E_{12} + K_{12} & 0 \\ I_{n_1} & 0 & 0 \\ 0 & 0 & I_{n_1} \\ 0 & A_2^{22} & 0 \end{bmatrix}$$

$$= s + 2n_1 + \operatorname{rank} \begin{bmatrix} E_{12} + K_{12} \\ A_2^{22} \end{bmatrix}.$$

Similar to the above derivation, we have

$$\operatorname{rank}(E + BK_d) = n_1 + \operatorname{rank}(E_{12} + K_{12}).$$

Therefore, according to Proposition 7.2 we know that the regular descriptor linear system (7.11) is I-controllablizable if and only if there exists a matrix K_{12} satisfying the following formulas:

$$s + 2n_1 + \operatorname{rank} \begin{bmatrix} E_{12} + K_{12} \\ A_2^{22} \end{bmatrix} = n + n_1 + \operatorname{rank}(E_{12} + K_{12}).$$

This gives

$$\text{rank}\begin{bmatrix} E_{12} + K_{12} \\ A_2^{22} \end{bmatrix} = n - n_2 + \text{rank}(E_{12} + K_{12}). \qquad (7.50)$$

Thus, it is clear that there exists a matrix K_{12} satisfying (7.50) if and only if A_2^{22} is of full-row rank, that is

$$\text{rank}A_2^{22} = n - n_2. \qquad (7.51)$$

Next, we show that (7.49) is equivalent to (7.51). We consider

$$\text{rank}\begin{bmatrix} E & 0 & 0 & B \\ A & E & B & 0 \end{bmatrix}$$

$$= \text{rank}\begin{bmatrix} P\begin{bmatrix} E_1 \\ E_2 \end{bmatrix} & 0 & 0 & P\begin{bmatrix} I_s & 0 \\ 0 & 0 \end{bmatrix}Q \\ P\begin{bmatrix} A_1 \\ A_2 \end{bmatrix} & P\begin{bmatrix} E_1 \\ E_2 \end{bmatrix} & P\begin{bmatrix} I_s & 0 \\ 0 & 0 \end{bmatrix}Q & 0 \end{bmatrix}$$

$$= \text{rank}\begin{bmatrix} E_1 & 0 & 0 & I_s \\ E_2 & 0 & 0 & 0 \\ A_1 & E_1 & I_s & 0 \\ A_2 & E_2 & 0 & 0 \end{bmatrix}$$

$$= \text{rank}\begin{bmatrix} 0 & 0 & 0 & I_s \\ E_2 & 0 & 0 & 0 \\ 0 & 0 & I_s & 0 \\ A_2 & E_2 & 0 & 0 \end{bmatrix}$$

$$= 2s + \text{rank}\begin{bmatrix} E_2 & 0 \\ A_2 & E_2 \end{bmatrix}$$

$$= 2s + \text{rank}\begin{bmatrix} P_1\begin{bmatrix} I_{n_1} & 0 \\ 0 & 0 \end{bmatrix}Q_1 & 0 \\ P_1\begin{bmatrix} A_2^{11} & A_2^{12} \\ A_2^{21} & A_2^{22} \end{bmatrix}Q_1 & P_1\begin{bmatrix} I_{n_1} & 0 \\ 0 & 0 \end{bmatrix}Q_1 \end{bmatrix}$$

$$= 2s + 2n_1 + \text{rank}A_2^{22}$$

$$= 2n_2 + \text{rank}A_2^{22}.$$

Thus, (7.49) turns to

$$2n_2 + \text{rank}A_2^{22} = n + n_2.$$

This is clearly equivalent to (7.51).   □

Example 7.8. Consider again the system in Example 7.1. It is easily verified that

$$\text{rank}\begin{bmatrix} E & 0 & 0 & B \\ A & E & B & 0 \end{bmatrix} = 7,$$

$$\text{rank}\begin{bmatrix} E & B \end{bmatrix} = 3.$$

This implies that the system is I-controllablizable according to the theorem proposed in this section.

7.4.2 Criteria Based on Equivalent Forms

In the subsection above, we have given a basic criterion for I-controllablizability. In this section, we investigate criteria based on canonical equivalent forms of a descriptor linear system. First, we give the criterion based on the dynamics decomposition form.

Theorem 7.11. *Let the descriptor linear system (7.42) be regular, and the associated relations (7.5) and (7.6) hold. Then system (7.42) is I-controllablizable if and only if*

$$\text{rank}\begin{bmatrix} 0 & 0 & B_2 \\ A_{22} & B_2 & A_{21}B_1 \end{bmatrix} = n - n_0 + \text{rank}\,B_2. \tag{7.52}$$

Proof. It follows from Theorem 7.10 that the system (7.1) is I-controllablizable if and only if (7.49) holds. Using (7.5) and (7.6), we have

$$\text{rank}\begin{bmatrix} E & 0 & 0 & B \\ A & E & B & 0 \end{bmatrix} = \text{rank}\begin{bmatrix} I_{n_0} & 0 & 0 & 0 & 0 & B_1 \\ 0 & 0 & 0 & 0 & 0 & B_2 \\ A_{11} & A_{12} & I_{n_0} & 0 & B_1 & 0 \\ A_{21} & A_{22} & 0 & 0 & B_2 & 0 \end{bmatrix}$$

$$= \text{rank}\begin{bmatrix} I_{n_0} & 0 & 0 & 0 & 0 & B_1 \\ 0 & 0 & 0 & 0 & 0 & B_2 \\ 0 & 0 & I_{n_0} & 0 & 0 & 0 \\ A_{21} & A_{22} & 0 & 0 & B_2 & 0 \end{bmatrix}$$

$$= n_0 + \text{rank}\begin{bmatrix} I_{n_0} & 0 & 0 & B_1 \\ 0 & 0 & 0 & B_2 \\ A_{21} & A_{22} & B_2 & 0 \end{bmatrix}$$

$$= n_0 + \text{rank}\begin{bmatrix} I_{n_0} & 0 & 0 & B_1 \\ 0 & 0 & 0 & B_2 \\ 0 & A_{22} & B_2 & A_{21}B_1 \end{bmatrix}$$

$$= 2n_0 + \text{rank}\begin{bmatrix} 0 & 0 & B_2 \\ A_{22} & B_2 & A_{21}B_1 \end{bmatrix},$$

and

$$\text{rank}\begin{bmatrix} E & B \end{bmatrix} = \text{rank}\begin{bmatrix} I_{n_0} & 0 & B_1 \\ 0 & 0 & B_2 \end{bmatrix} = n_0 + \text{rank}\, B_2.$$

So the relation (7.49) is equivalent to (7.52).   □

The following theorem provides the criterion based on the standard decomposition form.

Theorem 7.12. *Let the descriptor system (7.42) be regular, and Q and P be two nonsingular matrices satisfying the relation (7.2). Then the descriptor system (7.42) is I-controllablizable if and only if*

$$\text{rank}\, B_2 = \text{rank}\begin{bmatrix} N & B_2 \end{bmatrix}. \tag{7.53}$$

Proof. Due to the relation (7.2), we have

$$\text{rank}\begin{bmatrix} E & 0 & 0 & B \\ A & E & B & 0 \end{bmatrix} = \text{rank}\begin{bmatrix} I_{n_1'} & 0 & 0 & 0 & 0 & B_1 \\ 0 & N & 0 & 0 & 0 & B_2 \\ A_1 & 0 & I_{n_1'} & 0 & B_1 & 0 \\ 0 & I_{n-n_1'} & 0 & N & B_2 & 0 \end{bmatrix}$$

$$= \text{rank}\begin{bmatrix} I_{n_1'} & 0 & 0 & 0 & 0 & 0 \\ 0 & 0 & 0 & 0 & 0 & B_2 \\ 0 & 0 & I_{n_1'} & 0 & 0 & 0 \\ 0 & I_{n-n_1'} & 0 & 0 & 0 & 0 \end{bmatrix}$$

$$= n + n_1' + \text{rank}\, B_2,$$

and

$$\text{rank}\begin{bmatrix} E & B \end{bmatrix} = \text{rank}\begin{bmatrix} I_{n_1'} & 0 & B_1 \\ 0 & N & B_2 \end{bmatrix} = n_1' + \text{rank}\begin{bmatrix} N & B_2 \end{bmatrix}.$$

Therefore, the system is I-controllablizable if and only if

$$n + n_1' + \text{rank}\, B_2 = n + n_1' + \text{rank}\begin{bmatrix} N & B_2 \end{bmatrix},$$

which is clearly equivalent to (7.53).   □

The following theorem provides the criterion based on the inverse form.

Theorem 7.13. *Let (4.99) be an inverse form of system (7.42). Then system (7.42) is I-controllablizable if and only if the following relation holds:*

$$\text{rank}\begin{bmatrix} \hat{E}^2 & \hat{E}\hat{B} & \hat{B} \end{bmatrix} = \text{rank}\begin{bmatrix} \hat{E} & \hat{B} \end{bmatrix}. \tag{7.54}$$

Proof. Using (4.100), we have

$$\text{rank}\begin{bmatrix} E & 0 & 0 & B \\ A & E & B & 0 \end{bmatrix} = \text{rank}\begin{bmatrix} \hat{E} & 0 & 0 & \hat{B} \\ \gamma\hat{E} - I_n & \hat{E} & \hat{B} & 0 \end{bmatrix}$$

$$= \text{rank}\begin{bmatrix} \hat{E} & 0 & 0 & \hat{B} \\ -I_n & \hat{E} & \hat{B} & 0 \end{bmatrix}$$

$$= \text{rank}\begin{bmatrix} 0 & \hat{E}[\hat{E} \ \hat{B}] & \hat{B} \\ -I_n & 0 & 0 \end{bmatrix}$$

$$= n + \text{rank}\begin{bmatrix} \hat{E}^2 & \hat{E}\hat{B} & \hat{B} \end{bmatrix},$$

and

$$\text{rank}\begin{bmatrix} E & B \end{bmatrix} = \text{rank}\begin{bmatrix} \hat{E} & \hat{B} \end{bmatrix}.$$

Applying Theorem 7.10 gives the conclusion. □

This subsection ends with the criterion based on canonical equivalent form for derivative feedback.

It follows from Theorem 2.6 that, for any descriptor linear system (7.42), or (E, A, B), with $\text{rank} E = n_0$, there exist a nonsingular matrix $Q_3 \in \mathbb{R}^{n \times n}$ and two orthogonal matrices $P_3 \in \mathbb{R}^{n \times n}$ and $U_3 \in \mathbb{R}^{r \times r}$ such that

$$(E, A, B) \overset{(P_3, Q_3)}{\Longleftrightarrow} (\tilde{E}, \tilde{A}, \tilde{B}),$$

where $\tilde{E}, \tilde{A}$ and $\tilde{B}$ are given by these formulas (4.107)–(4.109), with M given in (2.32). Using this equivalent form, we can obtain the following I-controllablizability criterion.

Theorem 7.14. *Consider the descriptor linear system (7.42) with $\text{rank} E = n_0$ and $\text{rank} B = r$. Let the nonsingular matrix $Q_3 \in \mathbb{R}^{n \times n}$ and the two orthogonal matrices $P_3 \in \mathbb{R}^{n \times n}$ and $U_3 \in \mathbb{R}^{r \times r}$ be obtained by Algorithm 2.2, and $\tilde{E}, \tilde{A}$ and $\tilde{B}$ be given by (4.107)–(4.109), with M given in (2.32) and integers n_2 and $l (= n_1)$ defined in (4.110). Then the system (7.42) is I-controllablizable if and only if A_{32} is of full-row rank, that is,*

$$\text{rank} A_{32} = n - n_2. \tag{7.55}$$

Proof. Note that

$$\text{rank}\begin{bmatrix} E & 0 & 0 & B \\ A & E & B & 0 \end{bmatrix} = \text{rank}\begin{bmatrix} \tilde{E} & 0 & 0 & \tilde{B}U_3 \\ \tilde{A} & \tilde{E} & \tilde{B}U_3 & 0 \end{bmatrix}$$

$$= \text{rank}\begin{bmatrix} I_{n_1} & 0_{n_1 \times (n-n_1)} & 0 & 0 & 0 & 0 \\ 0_{r \times n_1} & M & 0 & 0 & 0 & I_r \\ 0_{(n-n_2) \times n_1} & 0 & 0 & 0 & 0 & 0 \\ A_{11} & A_{12} & I_{n_1} & 0_{n_1 \times (n-n_1)} & 0 & 0 \\ A_{21} & A_{22} & 0 & M & I_r & 0 \\ A_{31} & A_{32} & 0 & 0_{(n-n_2) \times (n-n_1)} & 0 & 0 \end{bmatrix}$$

$$= \text{rank} \begin{bmatrix} I_{n_1} & 0_{n_1 \times (n-n_1)} & 0 & 0 & 0 & 0 \\ 0_{r \times n_1} & 0 & 0 & 0 & 0 & I_r \\ 0 & 0 & I_{n_1} & 0_{n_1 \times (n-n_1)} & 0 & 0 \\ 0 & 0 & 0 & 0 & I_r & 0 \\ A_{31} & A_{32} & 0 & 0_{(n-n_2) \times (n-n_1)} & 0 & 0 \end{bmatrix}$$

$$= n_1 + 2r + \text{rank} \begin{bmatrix} I_{n_1} & 0_{n_1 \times (n-n_1)} \\ A_{31} & A_{32} \end{bmatrix}$$

$$= 2n_1 + 2r + \text{rank} A_{32}$$

$$= 2n_2 + \text{rank} A_{32}.$$

Thus, (7.49) turns to

$$2n_2 + \text{rank} A_{32} = n + n_2.$$

This is clearly equivalent to (7.55).   □

Example 7.9. Consider again the system in Example 7.1. For this system, we have $n_0 = 2$ and

$$T_1 = I_4, \quad T_2 = \begin{bmatrix} 0 & 0 & 0 & 1 \\ 1 & 0 & 0 & 0 \\ 0 & 0 & 1 & 0 \\ 0 & 1 & 0 & 0 \end{bmatrix},$$

$$A_{21} = \begin{bmatrix} 0 & 1 \\ 1 & 0 \end{bmatrix}, \quad A_{22} = \begin{bmatrix} 1 & 0 \\ 1 & 0 \end{bmatrix}, \quad B_1 = \begin{bmatrix} 0 & 0 \\ 1 & 0 \end{bmatrix}, \quad B_2 = \begin{bmatrix} 0 & 1 \\ 0 & 0 \end{bmatrix}.$$

Thus,

$$\text{rank} \begin{bmatrix} 0 & 0 & B_2 \\ A_{22} & B_2 & A_{21} B_1 \end{bmatrix} = 3 = n - n_0 + \text{rank} B_2.$$

The system is thus I-controllablizable according to Theorem 7.11.

It follows from Example 7.1 that under the transformation of (7.4), the system has a standard decomposition form (7.2) with

$$N = \begin{bmatrix} 0 & 1 & 0 \\ 0 & 0 & 0 \\ 0 & 0 & 0 \end{bmatrix}, \quad B_2 = \begin{bmatrix} 1 & 0 \\ 0 & 1 \\ 0 & -1 \end{bmatrix}.$$

Thus, we have

$$\text{rank} B_2 = 3 = \text{rank} \begin{bmatrix} N & B_2 \end{bmatrix}.$$

So the system is I-controllablizable according to Theorem 7.12. When we choose $\gamma = 1$, the matrix $(E - A)$ is invertible, and in this case,

$$\hat{E} = (E - A)^{-1} E = \begin{bmatrix} 0 & -1 & 0 & \frac{1}{2} \\ 0 & 0 & 0 & \frac{1}{2} \\ 0 & 0 & 0 & -\frac{1}{2} \\ 0 & 0 & 0 & \frac{1}{2} \end{bmatrix},$$

$$\hat{B} = (E - A)^{-1} E = \begin{bmatrix} \frac{1}{2} & \frac{1}{2} \\ \frac{1}{2} & \frac{1}{2} \\ -\frac{1}{2} & -\frac{1}{2} \\ \frac{1}{2} & -\frac{1}{2} \end{bmatrix},$$

$$\text{rank} \begin{bmatrix} \hat{E}^2 & \hat{E}\hat{B} & \hat{B} \end{bmatrix} = 3 = \text{rank} \begin{bmatrix} \hat{E} & \hat{B} \end{bmatrix}.$$

Thus, the system is I-controllablizable according to Theorem 7.13.

Under the following transformation

$$Q_3 = I_4, \ P_3 = \begin{bmatrix} 0 & 0 & 1 & 0 \\ 1 & 0 & 0 & 0 \\ 0 & 0 & 0 & 1 \\ 0 & 1 & 0 & 0 \end{bmatrix}, \ U_3 = I_2,$$

we have

$$n_2 = \text{rank} \begin{bmatrix} E & B \end{bmatrix} = 3,$$

and

$$A_{32} = \begin{bmatrix} 0 & 0 & 1 \end{bmatrix}.$$

So

$$\text{rank} A_{32} = 1 = n - n_2.$$

Thus, the system is I-controllablizable according to Theorem 7.14.

Regarding the I-observablizability of descriptor linear systems, criteria corresponding to the above ones for I-controllablizability of descriptor linear systems can be easily obtained using the well-known dual principle.

7.5 Impulsive Elimination by P-D Feedback

In this section, we consider the problem of impulse elimination via state P-D feedback for the following descriptor linear system:

$$E\dot{x} = Ax + Bu, \quad (7.56)$$

where $E, A \in \mathbb{R}^{n \times n}$, $B \in \mathbb{R}^{n \times r}$, and $\text{rank} E = n_0$. The problem is described as follows.

Problem 7.3. Given the regular descriptor linear system (7.56), find a general state P-D feedback controller

$$u = K_p x - K_d \dot{x} + v, \quad K_p, \ K_d \in \mathbb{R}^{r \times n}, \tag{7.57}$$

such that the closed-loop system

$$(E + BK_d)\dot{x} = (A + BK_p)x + Bv \tag{7.58}$$

is impulse-free.

For the solution of the above problem, we have the following proposition.

Proposition 7.3. *Problem 7.3 has a solution if and only if the system (7.56) is I-controllablizable.*

Proof. When system (7.56) is I-controllablizable, then there exists a gain matrix $K_d \in \mathbb{R}^{r \times n}$ making the matrix triple $(E + BK_d, \ A, \ B)$ I-controllable. In this case, there exists a gain matrix K_p such that the system (7.58) is impulse-free. Conversely, if the system is not I-controllablizable, then there does not exist a matrix K_d making $(E + BK_d, \ A, \ B)$ I-controllable, then according to Theorem 7.6, for arbitrary $K_d \in \mathbb{R}^{r \times n}$, there does not exist a gain matrix K_p making the system (7.58) impulse-free. Therefore, Problem 7.3 has a solution if and only if there exists a matrix K_d making the system (7.46) I-controllable. □

In the next two subsections, we give two methods for solving Problem 7.3.

7.5.1 Method I

From Proposition 7.3, it follows that for solving Problem 7.3, we can first give a gain matrix K_d making the matrix triple $(E + BK_d, \ A, \ B)$ I-controllable, then give the gain matrix K_p making the closed-loop system (7.58) impulse-free by using the previously proposed method to solve the impulse elimination problem by state feedback. The following theorem gives a parameterization for all the I-controllablizing controller gain K_d.

Theorem 7.15. *Consider the descriptor linear system (7.56) with* $\mathrm{rank}E = n_0$ *and* $\mathrm{rank}B = r$. *Let the nonsingular matrix* $Q_3 \in \mathbb{R}^{n \times n}$ *and the two orthogonal matrices* $P_3 \in \mathbb{R}^{n \times n}$ *and* $U_3 \in \mathbb{R}^{r \times r}$ *satisfying (4.107)–(4.109), with* M *given in (2.32) and integers* n_2 *and* n_1 *defined in (6.4). When the system (7.56) is I-controllablizable, all the I-controllablizing controller gains* K_d *are given by*

$$K_d = U_3 \left[\ K_1 \ \left[\ K_2 P \quad K_2 \ \right] Q_r^{\mathrm{T}} - M \ \right] P_3^{\mathrm{T}}, \tag{7.59}$$

where $K_1 \in \mathbb{R}^{r \times n_1}$, $K_2 \in \mathbb{R}^{r \times r}$, and $P \in \mathbb{R}^{r \times (n-n_2)}$ are three arbitrary parameter matrices, and $Q_r^T \in \mathbb{R}^{(n-n_1) \times (n-n_1)}$ are an orthogonal matrix satisfying the following singular value decomposition:

$$Q_1 A_{32} Q_r = \left[\Delta \ 0_{(n-n_2) \times r} \right], \tag{7.60}$$

where Δ, $Q_l \in \mathbb{R}^{(n-n_2) \times (n-n_2)}$, and Δ is a diagonal positive definite matrix, Q_l is also orthogonal. Furthermore,

$$p = n_1 + \mathrm{rank} K_2 \tag{7.61}$$

gives the dynamical order of the closed-loop system (7.58).

Proof. Let p is an allowable dynamical order of (7.46). According to Theorem 6.4 that all the derivative feedback gain K_d making

$$\mathrm{rank}\,(E + BK_d) = p$$

can be parameterized as

$$K_d = U_3 \left[K_1 \ \bar{K}_2 - M \right] P_3^T,$$

with the matrix $\bar{K}_2 \in \mathbb{R}^{r \times (n-n_1)}$ satisfying

$$\mathrm{rank}\,\bar{K}_2 = p - n_1. \tag{7.62}$$

It follows from the proof of Theorem 6.4 that

$$E + BK_d = Q_3^{-1} \begin{bmatrix} I_{n_1} & 0_{n_1 \times (n-n_1)} \\ K_1 & \bar{K}_2 \\ 0_{(n-n_2) \times n_1} & 0_{(n-n_2) \times (n-n_1)} \end{bmatrix} P_3^T. \tag{7.63}$$

Using (4.107)–(4.109) and (7.63), we have

$$\mathrm{rank} \begin{bmatrix} E + BK_d & 0 & 0 \\ A & E & B \end{bmatrix}$$

$$= \mathrm{rank} \begin{bmatrix} \tilde{E} + \tilde{B} K_d & 0 & 0 \\ \tilde{A} & \tilde{E} & \tilde{B} \end{bmatrix}$$

$$= \mathrm{rank} \begin{bmatrix} I_{n_1} & 0_{n_1 \times (n-n_1)} & 0 & 0 & 0 \\ K_1 & \bar{K}_2 & 0 & 0 & 0 \\ 0_{(n-n_2) \times n_1} & 0_{(n-n_2) \times (n-n_1)} & 0 & 0 & 0 \\ A_{11} & A_{12} & I_{n_1} & 0_{n_1 \times (n-n_1)} & 0_{n_1 \times r} \\ A_{21} & A_{22} & 0_{r \times n_1} & M & I_r \\ A_{31} & A_{32} & 0 & 0_{(n-n_2) \times (n-n_1)} & 0 \end{bmatrix}$$

$$= \text{rank} \begin{bmatrix} I_{n_1} & O_{n_1 \times (n-n_1)} & 0 & 0 & 0 \\ K_1 & \bar{K}_2 & 0 & 0 & 0 \\ 0 & 0 & I_{n_1} & O_{n_1 \times (n-n_1)} & O_{n_1 \times r} \\ 0 & 0 & O_{r \times n_1} & 0 & I_r \\ A_{31} & A_{32} & 0 & O_{(n-n_2) \times (n-n_1)} & 0 \end{bmatrix}$$

$$= \text{rank} \begin{bmatrix} I_{n_1} & O_{n_1 \times (n-n_1)} & 0 & 0 \\ O_{r \times n_1} & \bar{K}_2 & 0 & 0 \\ 0 & 0 & I_{n_1} & O_{n_1 \times r} \\ 0 & 0 & O_{r \times n_1} & I_r \\ O_{(n-n_2) \times n_1} & A_{32} & 0 & 0 \end{bmatrix}$$

$$= 2n_1 + r + \text{rank} \begin{bmatrix} \bar{K}_2 \\ A_{32} \end{bmatrix}$$

$$= n_1 + n_2 + \text{rank} \begin{bmatrix} \bar{K}_2 \\ A_{32} \end{bmatrix}.$$

On the other hand,

$$\text{rank}(E + BK_d) = \text{rank} \begin{bmatrix} I_{n_1} & O_{n_1 \times (n-n_1)} \\ K_1 & \bar{K}_2 \\ O_{(n-n_2) \times n_1} & O_{(n-n_2) \times (n-n_1)} \end{bmatrix}$$

$$= \text{rank} \begin{bmatrix} I_{n_1} & O_{n_1 \times (n-n_1)} \\ O_{r \times n_1} & \bar{K}_2 \\ O_{(n-n_2) \times n_1} & O_{(n-n_2) \times (n-n_1)} \end{bmatrix}$$

$$= n_1 + \text{rank} \bar{K}_2.$$

Thus, $(E + BK_d, A)$ is I-controllable if and only if

$$n_1 + n_2 + \text{rank} \begin{bmatrix} \bar{K}_2 \\ A_{32} \end{bmatrix} = n + n_1 + \text{rank} \bar{K}_2,$$

that is

$$\text{rank} \begin{bmatrix} \bar{K}_2 \\ A_{32} \end{bmatrix} = n - n_2 + \text{rank} \bar{K}_2. \tag{7.64}$$

Let

$$\bar{K}_2 Q_r = \begin{bmatrix} K_{21} & K_2 \end{bmatrix}, \quad K_2 \in \mathbb{R}^{r \times r}, \tag{7.65}$$

then the relation (7.64) is converted into

$$\text{rank} \begin{bmatrix} K_{21} & K_2 \\ \Delta & O_{(n-n_2) \times r} \end{bmatrix} = n - n_2 + \text{rank} \begin{bmatrix} K_{21} & K_2 \end{bmatrix},$$

which is equivalent to

$$\mathrm{rank}\, K_2 = \mathrm{rank}\begin{bmatrix} K_{21} & K_2 \end{bmatrix}. \tag{7.66}$$

This relation holds if and only if the following relation holds

$$K_{21} = K_2 P \tag{7.67}$$

with $P \in \mathbb{R}^{r\times(n-n_2)}$ be an arbitrary matrix. By combining this with (7.65) and (7.67), it follows that all the I-controllablizing controllers are given by (7.59), with $K_1 \in \mathbb{R}^{r\times n_1}$, $K_2 \in \mathbb{R}^{r\times r}$, and $P \in \mathbb{R}^{r\times(n-n_2)}$ being some arbitrary parameter matrices. Furthermore, combining (7.65) and (7.66) gives

$$\mathrm{rank}\,\bar{K}_2 = \mathrm{rank}\, K_2.$$

Combining this with (7.62) gives the relation (7.61). □

Example 7.10. Consider a descriptor linear system in the form of (7.56) with the following parameters:

$$E = \begin{bmatrix} 0 & 1 & 0 & 0 \\ 0 & 0 & 0 & 1 \\ 0 & 0 & 0 & 0 \\ 0 & 0 & 0 & 0 \end{bmatrix}, \quad A = \begin{bmatrix} 1 & 0 & 0 & 0 \\ 0 & 1 & 2 & 0 \\ 0 & 0 & 1 & 1 \\ 0 & 1 & 0 & 1 \end{bmatrix}, \quad B = \begin{bmatrix} 0 & 0 \\ 1 & 0 \\ 0 & 1 \\ 0 & 0 \end{bmatrix}.$$

For this example system, we will only give all the I-controllablizing gains K_d. It is easily known that $n_0 = 2$, $n_2 = \mathrm{rank}\begin{bmatrix} E & B \end{bmatrix} = 3$, and $n_1 = 1$. So the allowable dynamical order p is 1, 2 or 3. Correspondingly, $\mathrm{rank}\, K_2$ takes the value of 0, 1, or 2. By some computations, we have

$$Q_3 = I_4, \quad P_3 = \begin{bmatrix} 0 & 0 & 1 & 0 \\ 1 & 0 & 0 & 0 \\ 0 & 0 & 0 & 1 \\ 0 & 1 & 0 & 0 \end{bmatrix}, \quad U_3 = I_2,$$

and

$$M = \begin{bmatrix} 1 & 0 & 0 \\ 0 & 0 & 0 \end{bmatrix}, \quad A_{32} = \begin{bmatrix} 1 & 0 & 0 \end{bmatrix}.$$

Since $A_{32} = \begin{bmatrix} 1 & 0 & 0 \end{bmatrix}$ is of full-row rank, the system is I-controllablizable. Obviously, we have $Q_r = I_3$. In the following, we denote

$$K_1 = \begin{bmatrix} k_{11} \\ k_{21} \end{bmatrix}, \quad K_2 = \begin{bmatrix} t_{11} & t_{12} \\ t_{21} & t_{22} \end{bmatrix}, \quad P = \begin{bmatrix} p_1 \\ p_2 \end{bmatrix}.$$

According to Theorem 7.15, all the I-controllablizing controller gains are given by

$$K_d = U_3 \Big[K_1 \big[K_2 P \ K_2 \big] - M \Big] P_3^T$$

$$= \begin{bmatrix} t_{11} & k_{11} & t_{12} & p_1 t_{11} + p_2 t_{12} - 1 \\ t_{21} & k_{21} & t_{22} & p_1 t_{21} + p_2 t_{22} \end{bmatrix}. \tag{7.68}$$

When $\operatorname{rank} K_2 = 0$, that is, $t_{ij} = 0$, i, $j = 1, 2$, the closed-loop dynamical order p is assigned to 1. In this case, the I-controllablizing controller gains are given by

$$K_d = \begin{bmatrix} 0 & k_{11} & 0 & -1 \\ 0 & k_{21} & 0 & 0 \end{bmatrix}.$$

When $t_{11}t_{22} = t_{12}t_{21}$ and there exists a $t_{ij} \neq 0$, i, $j = 1, 2$, the resulted closed-loop dynamical order p is 2. When $t_{11}t_{22} \neq t_{12}t_{21}$, the dynamical order p is 3.

7.5.2   Method II

In this subsection, the proposed method to solve Problem 7.3 can simultaneously offer the parameterizations of the proportional gain K_p and the derivative gain K_d.

Theorem 7.16. *Consider the descriptor linear system (7.56) with* $\operatorname{rank} E = n_0$ *and* $\operatorname{rank} B = r$. *Let the nonsingular matrix* $Q_3 \in \mathbb{R}^{n \times n}$ *and the two orthogonal matrices* $P_3 \in \mathbb{R}^{n \times n}$ *and* $U_3 \in \mathbb{R}^{r \times r}$ *be the matrices satisfying (4.107)–(4.109). In addition, let* $Q_r \in \mathbb{R}^{(n-n_1) \times (n-n_1)}$ *be the orthogonal matrix satisfying the singular value decomposition (7.60). Let*

$$Q = \begin{bmatrix} I_{n_1} & 0 \\ 0 & Q_r \end{bmatrix}. \tag{7.69}$$

Then all the P-D controllers (7.57) making the closed-loop system (7.58) impulse-free are given by

$$K_p = U_3 \big[K_{p1} \ K_{p2} \big] Q^T P_3^T, \tag{7.70}$$

$$K_d = U_3 \Big[K_{d1} \ \big[K_{d2} P \ K_{d2} \big] Q_r^T - M \Big] P_3^T, \tag{7.71}$$

where $K_{p1} \in \mathbb{R}^{r \times (n-r)}$ *and* $P \in \mathbb{R}^{r \times (n-n_2)}$ *are two arbitrary parameter matrices, and* $K_{p2} \in \mathbb{R}^{r \times r}$, $K_{d1} \in \mathbb{R}^{r \times n_1}$, *and* $K_{d2} \in \mathbb{R}^{r \times r}$ *are the parameter matrices satisfying the following constraint*

$$\operatorname{rank} \begin{bmatrix} K_{p2} + (A_{22} - K_{d1} A_{12}) Q_r \Theta & K_{d2} \\ K_{d2} & 0 \end{bmatrix} = r + \operatorname{rank} K_{d2}, \tag{7.72}$$

with

$$\Theta = \begin{bmatrix} 0_{r\times(n-n_2)} & I_r \end{bmatrix}^{\mathrm{T}}.$$

Furthermore,

$$p = n_1 + \mathrm{rank}K_{d2} \tag{7.73}$$

gives the dynamical order of the closed-loop system (7.58).

Proof. It follows from Proposition 7.3 that the derivative gain K_d which makes that the closed-loop system (7.58) impulse-free must be a I-controllablizing controller gain. According to Theorem 7.15, such a gain is given by (7.59) and the closed-loop dynamical order is given by (7.73). In addition, it follows from Theorem 7.2 that the closed-loop system (7.58) is impulse-free if and only if the following relation holds:

$$\mathrm{rank}\begin{bmatrix} E + BK_d & 0 \\ A + BK_p & E + BK_d \end{bmatrix} = n + \mathrm{rank}(E + BK_d). \tag{7.74}$$

Let

$$U_3^T K_p P_3 = \begin{bmatrix} \bar{K}_{p1} & \bar{K}_{p2} \end{bmatrix}, \tag{7.75}$$

where

$$\bar{K}_{p1} \in \mathbb{R}^{r\times n_1}, \qquad \bar{K}_{p2} \in \mathbb{R}^{r\times(n-n_1)},$$

then it follows from (4.107) to (4.109) and (7.59) that

$$\mathrm{rank}\begin{bmatrix} E + BK_d & 0 \\ A + BK_p & E + BK_d \end{bmatrix}$$

$$= \mathrm{rank}\begin{bmatrix} Q_3(E + BK_d)P_3 & 0 \\ Q_3(A + BK_p)P_3 & Q_3(E + BK_d)P_3 \end{bmatrix}$$

$$= \mathrm{rank}\begin{bmatrix} I_{n_1} & 0_{n_1\times(n-n_1)} & 0 & 0 \\ K_{d1} & \begin{bmatrix} K_{d2}P & K_{d2} \end{bmatrix}Q_{\mathrm{r}}^T & 0 & 0 \\ 0_{(n-n_2)\times n_1} & 0_{(n-n_2)\times(n-n_1)} & 0 & 0 \\ A_{11} & A_{12} & I_{n_1} & 0_{n_1\times(n-n_1)} \\ \bar{K}_{p1} + A_{21} & \bar{K}_{p2} + A_{22} & K_{d1} & \begin{bmatrix} K_{d2}P & K_{d2} \end{bmatrix}Q_{\mathrm{r}}^T \\ A_{31} & A_{32} & 0 & 0_{(n-n_2)\times(n-n_1)} \end{bmatrix}$$

$$= \mathrm{rank}\begin{bmatrix} I_{n_1} & 0_{n_1\times(n-n_1)} & 0 & 0 \\ 0 & \begin{bmatrix} K_{d2}P & K_{d2} \end{bmatrix}Q_{\mathrm{r}}^T & 0 & 0 \\ 0 & 0_{(n-n_2)\times(n-n_1)} & 0 & 0 \\ 0 & A_{12} & I_{n_1} & 0_{n_1\times(n-n_1)} \\ 0 & \bar{K}_{p2} + A_{22} & K_{d1} & \begin{bmatrix} K_{d2}P & K_{d2} \end{bmatrix}Q_{\mathrm{r}}^T \\ 0 & A_{32} & 0 & 0_{(n-n_2)\times(n-n_1)} \end{bmatrix}$$

$$= n_1 + \mathrm{rank} \begin{bmatrix} A_{12} & I_{n_1} & 0_{n_1 \times (n-n_1)} \\ \bar{K}_{p2} + A_{22} & K_{d1} & \begin{bmatrix} K_{d2}P & K_{d2} \end{bmatrix} Q_{\mathrm{r}}^T \\ \begin{bmatrix} K_{d2}P & K_{d2} \end{bmatrix} Q_{\mathrm{r}}^T & 0 & 0 \\ A_{32} & 0_{(n-n_2) \times n_1} & 0_{(n-n_2) \times (n-n_1)} \end{bmatrix}$$

$$= n_1 + \mathrm{rank} \begin{bmatrix} 0 & I_{n_1} & 0_{n_1 \times (n-n_1)} \\ \bar{K}_{p2} + A_{22} - K_{d1}A_{12} & K_{d1} & \begin{bmatrix} K_{d2}P & K_{d2} \end{bmatrix} Q_{\mathrm{r}}^T \\ \begin{bmatrix} K_{d2}P & K_{d2} \end{bmatrix} Q_{\mathrm{r}}^T & 0 & 0 \\ A_{32} & 0_{(n-n_2) \times n_1} & 0_{(n-n_2) \times (n-n_1)} \end{bmatrix}$$

$$= 2n_1 + \mathrm{rank} \begin{bmatrix} (\bar{K}_{p2} + A_{22} - K_{d1}A_{12}) Q_{\mathrm{r}} & \begin{bmatrix} K_{d2}P & K_{d2} \end{bmatrix} \\ \begin{bmatrix} K_{d2}P & K_{d2} \end{bmatrix} & 0 \\ Q_1 A_{32} Q_{\mathrm{r}} & 0_{(n-n_2) \times (n-n_1)} \end{bmatrix}.$$

It follows from Theorem 7.15 that $\mathrm{rank}(E + BK_d) = p$, with p given by (7.73). In addition, partition the matrix Q_{r} into

$$Q_{\mathrm{r}} = \begin{bmatrix} Q_{\mathrm{r}1} & Q_{\mathrm{r}2} \end{bmatrix}, \quad Q_{r2} \in \mathbb{R}^{(n-n_1) \times r},$$

and let

$$\bar{K}_{p2} Q_{\mathrm{r}} = \begin{bmatrix} K_{p21} & K_{p2} \end{bmatrix}, \quad K_{p2} \in \mathbb{R}^{r \times r}, \tag{7.76}$$
$$\Psi_1 = K_{p21} + (A_{22} - K_{d1}A_{12}) Q_{\mathrm{r}1},$$
$$\Psi_2 = K_{p2} + (A_{22} - K_{d1}A_{12}) Q_{\mathrm{r}2},$$

then (7.74) is equivalent to

$$2n_1 + \mathrm{rank} \begin{bmatrix} \Psi_1 & \Psi_2 & K_{d2}P & K_{d2} \\ K_{d2}P & K_{d2} & 0 & 0 \\ \Delta & 0_{(n-n_2) \times r} & 0_{(n-n_2) \times (n-n_2)} & 0_{(n-n_2) \times r} \end{bmatrix} = n + n_1 + \mathrm{rank} K_{d2},$$

that is,

$$2n_1 + n - n_2 + \mathrm{rank} \begin{bmatrix} K_{p2} + (A_{22} - K_{d1}A_{12}) Q_{r2} & K_{d2} \\ K_{d2} & 0 \end{bmatrix} = n + n_1 + \mathrm{rank} K_{d2}. \tag{7.77}$$

This is exactly the constraint (7.72) in view of

$$Q_{\mathrm{r}2} = Q_{\mathrm{r}} \begin{bmatrix} 0_{r \times (n-n_2)} \\ I_r \end{bmatrix}.$$

Combining (7.75) and (7.76) gives

$$K_p = U_3 \begin{bmatrix} \bar{K}_{p1} & \begin{bmatrix} K_{p21} & K_{p2} \end{bmatrix} Q_{\mathrm{r}}^{\mathrm{T}} \end{bmatrix} P_3^{\mathrm{T}}$$
$$= U_3 \begin{bmatrix} \bar{K}_{p1} & K_{p21} & K_{p2} \end{bmatrix} Q^{\mathrm{T}} P_3^{\mathrm{T}}, \tag{7.78}$$

where Q is definition by (7.69). Let

$$K_{p1} = \begin{bmatrix} \bar{K}_{p1} & K_{p21} \end{bmatrix} \in \mathbb{R}^{r \times (n-r)},$$

the expression (7.70) is immediately obtained from (7.78). The proof is completed.

□

Example 7.11. Consider a descriptor linear system in the form of (7.56) with the following parameters:

$$E = \begin{bmatrix} 0 & 1 & 0 \\ 0 & 0 & 1 \\ 0 & 0 & 0 \end{bmatrix}, \quad A = \begin{bmatrix} 1 & 0 & 0 \\ 0 & 1 & 0 \\ 0 & 0 & 1 \end{bmatrix}, \quad B = \begin{bmatrix} 0 \\ 1 \\ 0 \end{bmatrix}.$$

Thus, we have $n_0 = 2$, $n_2 = \text{rank} \begin{bmatrix} E & B \end{bmatrix} = 2$. $n_1 = 1$. So the allowable dynamical order p is either 1 or 2, correspondingly rankK_{d2} takes the value of 0 or 1. By some computations, we have

$$Q_3 = I, \quad P_3 = \begin{bmatrix} 0 & 0 & 1 \\ 1 & 0 & 0 \\ 0 & 1 & 0 \end{bmatrix}, U_3 = 1,$$

and

$$A_{12} = \begin{bmatrix} 0 & 1 \end{bmatrix}, \quad A_{22} = \begin{bmatrix} 0 & 0 \end{bmatrix}, \quad A_{32} = \begin{bmatrix} 1 & 0 \end{bmatrix}, \quad M = \begin{bmatrix} 1 & 0 \end{bmatrix}.$$

Obviously, we have

$$Q_r = I_2, \quad Q = I_4.$$

Since A_{32} is of full-row rank, the system is I-controllablizable. In the following, we denote

$$K_{d1} = k_{d1}, \quad K_{d2} = k_{d2},$$
$$K_{p1} = \begin{bmatrix} k_{p11} & k_{p12} \end{bmatrix}, \quad K_{p2} = k_{p2}.$$

According to Theorem 7.16, all the P-D controllers such that the resulted closed-loop system impulse-free are given by :

$$\begin{aligned} K_p &= U_3 \begin{bmatrix} k_{p11} & k_{p12} & k_{p2} \end{bmatrix} P_3^{\mathrm{T}} \\ &= \begin{bmatrix} k_{p2} & k_{p11} & k_{p12} \end{bmatrix}, \end{aligned} \quad (7.79)$$

$$\begin{aligned} K_d &= U_3 \begin{bmatrix} k_{d1} & \begin{bmatrix} P k_{d2} & k_{d2} \end{bmatrix} - M \end{bmatrix} P_3^{\mathrm{T}} \\ &= \begin{bmatrix} k_{d2} & k_{d1} & P k_{d2} - 1 \end{bmatrix}, \end{aligned} \quad (7.80)$$

where k_{p11}, k_{p12} and P are three arbitrary real scalars, and k_{d1}, k_{d2}, and k_{p2} are the parameters satisfying the following constraint:

$$\text{rank} \begin{bmatrix} k_{p2} - k_{d1} & k_{d2} \\ k_{d2} & 0 \end{bmatrix} = 1 + \text{rank} k_{d2}. \tag{7.81}$$

When $k_{d2} = 0$, the dynamical order of the closed-loop system is chosen to be $p = 1$. In this case, the constraint (7.81) is $k_{p2} \neq k_{d1}$, and the P-D controllers are given by

$$\begin{cases} K_p = \begin{bmatrix} k_{p2} & k_{p11} & k_{p12} \end{bmatrix} \\ K_d = \begin{bmatrix} 0 & k_{d1} & -1 \end{bmatrix} \end{cases}, \; k_{p2} \neq k_{d1}.$$

When $k_{d2} \neq 0$, the dynamical order of the closed-loop system is chosen to be $p = 2$. In this case, the corresponding P-D controller is given by

$$\begin{cases} K_p = \begin{bmatrix} k_{p2} & k_{p11} & k_{p12} \end{bmatrix} \\ K_d = \begin{bmatrix} k_{d2} & K_{d1} & pk_{d2} - 1 \end{bmatrix} \end{cases}, \; k_{d2} \neq 0.$$

7.6 Notes and References

This chapter is focused on the following impulse elimination problem:

Given a descriptor linear system in the typical form, find a certain type of feedback controller for the system such that the closed-loop system is impulse-free.

To solve the above problem, conditions for a descriptor linear system to be impulse-free is first proposed. It is seen in Sect. 7.1 that a descriptor linear system in the typical form is impulse-free if and only if one of the following statements is true:

- the nilpotent matrix N in its fast subsystem is a zero matrix;
- $\deg \det (sE - A) = \text{rank} E$;
- the system has $\text{rank} E$ finite poles;
- $\text{rank} \begin{bmatrix} E & 0 \\ A & E \end{bmatrix} = n + \text{rank} E$;
- $\det (\Theta_L A \Theta_R) \neq 0$, where

$$\Theta_L = \begin{bmatrix} 0_{(n_2-n_0) \times n_0} & I_{n_2-n_0} & 0_{(n_2-n_0) \times (n-n_2)} \end{bmatrix} Q,$$

$$\Theta_R = P \begin{bmatrix} 0_{(n_2-n_0) \times (n-n_2+n_0)} & I_{n_2-n_0} \end{bmatrix}^{\mathrm{T}},$$

and Q and P are obtained via Algorithm 4.1.

Based on these conditions, the problems of impulse elimination by state feedback, by output feedback, and by P-D feedback are studied. Conditions for existence of solutions to these problems are presented, which are summarized in Table 7.1. Algorithms for deriving the corresponding controllers are also given.

Among the several impulse elimination problems, the one by state feedback is very fundamental. As is revealed, the condition for the problem is that the system

Table 7.1 Conditions for impulse elimination

Controllers	Conditions for solvability
State feedback	I-controllability
Output feedback	I-controllability and I-observability
State P-D feedback	I-controllablizability

is I-controllable. Regarding solutions of the state feedback controllers, three methods are presented, the first one is based on the dynamics decomposition form, with which the set of all such controllers are established. The second method is based on the standard decomposition, while the third one is based on the system equivalent form for derivative feedback. It is shown that the solution is completely determined by such controllers that makes the fast-subsystem impulse-free.

For the case of output feedback, it is first shown that the problem has a solution if and only if the system is both I-controllable and I-observable, and under such a condition, all the output feedback controllers which eliminate the impulse in the closed-loop system are parameterized.

For the case of P-D feedback, we have introduced the definition of I-controllablizability and converted the problem of impulse elimination into an I-controllablization problem. Furthermore, based on the canonical equivalent form for derivative feedback and a necessary and sufficient condition for I-controllability of descriptor linear systems, two sufficient and necessary conditions for I-controllablization are proposed, and a general parameterization of all the I-controllablizing controllers are presented. The solution also has the advantage of being able to give a desired closed-loop dynamical order.

For main references related to impulses elimination in descriptor linear systems, readers may refer to Cobb (1984), Dai (1989a), Yip and Sincovec (1981), Hou (2004), Duan and Wu (2005b,c,d), and Liu and Duan (2006). Very recently, a relevant problem, which is called impulsive-mode controllablizability in descriptor linear systems, has been proposed and considered by Wu and Duan (2007b).

Chapter 8
Pole Assignment and Stabilization

It is wellknown from normal linear systems theory that pole assignment and stabilization are two very basic closely related design problems in linear systems theory. In this chapter, we investigate these two problems for descriptor linear systems.

Pole assignment problems are studied in Sect. 8.1 and 8.2. Section 8.1 considers pole assignment in descriptor linear systems via state feedback, and gives conditions for the problem in terms of R- and S-controllabilities. Section 8.2 treats pole assignment in descriptor linear systems via state proportional plus derivative feedback, and gives conditions for the problem in terms of C-controllability. To investigate the problem of stabilization, two important concepts, namely, stabilizability and detectability, are first introduced in Sect. 8.3, necessary and sufficient conditions are also presented. Based on the concept of stabilizability, the problem of designing the stabilizing controllers for a stabilizable descriptor linear system via state feedback is examined in Sect. 8.4, several methods are presented.

8.1 Pole Assignment by State Feedback

Distribution of poles, especially the finite poles, has a profound effect on the dynamic and static properties of a descriptor linear system. This consequently makes the problem of pole assignment very important.

8.1.1 Problems Formulation

Consider the square descriptor linear system

$$E\dot{x} = Ax + Bu, \tag{8.1}$$

where $x \in \mathbb{R}^n$, $u \in \mathbb{R}^r$ are the state vector and control input vector, respectively, and $\mathrm{rank}\,E = n_0$. With the following state feedback control

$$u = Kx + v, \tag{8.2}$$

G.-R. Duan, *Analysis and Design of Descriptor Linear Systems*, Advances in Mechanics and Mathematics 23, DOI 10.1007/978-1-4419-6397-0_8, © Springer Science+Business Media, LLC 2010

the closed-loop system is obtained as

$$E\dot{x} = (A + BK)x + Bv, \tag{8.3}$$

where $v \in \mathbb{R}^r$ is an external signal.

When the system (8.1) is regular, there always exist two nonsingular matrices Q and P to transform it into the following standard decomposition form:

$$\begin{cases} \dot{x}_1 = A_1 x_1 + B_1 u \\ N\dot{x}_2 = x_2 + B_2 u \end{cases}, \tag{8.4}$$

where N is nilpotent, $x_1 \in \mathbb{R}^{n_1}$, $x_2 \in \mathbb{R}^{n_2}$, with

$$n_1 = \deg \det(sE - A), \ n_2 = n - n_1.$$

The relation between the state vectors is given by

$$P^{-1}x = \begin{bmatrix} x_1 \\ x_2 \end{bmatrix},$$

while the relations among the coefficient matrices are given as follows:

$$\begin{cases} QEP = \operatorname{diag}(I, \ N) \\ QAP = \operatorname{diag}(A_1, \ I) \\ QB = \begin{bmatrix} B_1 \\ B_2 \end{bmatrix} \end{cases}, \tag{8.5}$$

where $n_1 + n_2 = n$ and $N \in \mathbb{R}^{n_2 \times n_2}$ is nilpotent.

The Pole Assignment Problem

Remember that the set of open-loop finite poles of the system (8.1) is given by

$$\sigma(E, A) = \sigma(A_1) = \{\lambda_1, \lambda_2, \ldots, \lambda_{n_1}\}. \tag{8.6}$$

Our first problem is to find out whether these open-loop finite poles can be arbitrarily assigned by state feedback.

Problem 8.1. [Pole assignment] Given the regular system (8.1), and a set of complex numbers

$$\Gamma_1 = \{s_1, s_2, \ldots, s_{n_1}\}, \tag{8.7}$$

which is symmetric about the real axis, find a state feedback controller in the form of (8.2) such that Γ_1 is the set of finite poles of the closed-loop system (8.3), that is

$$\sigma(E, A + BK) = \Gamma_1 = \{s_1, s_2, \ldots, s_{n_1}\}. \tag{8.8}$$

Pole Assignment with Impulse Elimination

It is well known that

$$n_1 = \deg \det(sE - A) \le \operatorname{rank} E = n_0.$$

When (8.8) holds, we also have

$$n_1 = \deg \det(sE - (A + BK)) \le \operatorname{rank} E = n_0.$$

This indicates that with a solution to the above Problem 8.1, the closed-loop system may still contain infinite poles and hence possesses impulsive behavior. Our second problem is to assign arbitrarily $\operatorname{rank} E$ finite poles to the closed-loop system to make sure that the closed-loop system has no infinite poles and hence no impulse terms in its state response.

Problem 8.2. [Pole assignment with impulse elimination] Given the regular system (8.1), and a set of complex numbers

$$\Gamma_0 = \{s_1, s_2, \ldots, s_{n_0}\}, \tag{8.9}$$

which is symmetric about the real axis, find a state feedback controller in the form of (8.2) such that Γ_0 is the set of finite poles of the closed-loop system (8.3), that is

$$\sigma(E, A + BK) = \Gamma_0 = \{s_1, s_2, \ldots, s_{n_0}\}. \tag{8.10}$$

8.1.2 Pole Assignment under R-Controllability

This subsection provides the solution to Problem 8.1, the pole assignment problem. Before presenting the solution, let us restate a known result.

Lemma 8.1. *The regular descriptor linear system (8.1) with its standard decomposition form given by (8.4) is R-controllable if and only if the matrix pair (A_1, B_1) is controllable.*

It therefore follows from normal linear systems theory and the above lemma that for an arbitrary set Γ_1 given as in (8.7), there exists some K_1 such that

$$\sigma(A_1 + B_1 K_1) = \Gamma_1 \tag{8.11}$$

holds when system (8.1) is R-controllable.

Theorem 8.1. *Let (8.4) be the standard decomposition form of the regular descriptor linear system (8.1) under the transformation (P, Q). Then the following hold:*

1. Problem 8.1 has solutions if and only if the system (8.1) is R-controllable.

2. *When the system (8.1) is R-controllable, every member of the following set*

$$\mathscr{K}_1 = \{K| \; K = [K_1 \; 0] \, P^{-1}, \; \sigma \, (A_1 + B_1 K_1) = \Gamma_1\}$$

is a solution to Problem 8.1.

Proof. Now we proof the first conclusion.

Let the system (8.1) be R-controllable. Then it follows from Lemma 8.1 that (A_1, B_1) is controllable. Therefore, for arbitrary set Γ_1 given by (8.7), there exists some K_1 such that the relation (8.11) holds.

Let the feedback control be

$$u = K_1 x_1 + v = [K_1 \; 0] \, P^{-1} x + v,$$

the closed-loop system can be obtained as

$$\begin{cases} \dot{x}_1 = (A_1 + B_1 K_1) \, x_1 + B_1 v \\ N \dot{x}_2 = x_2 + B_2 K_1 x_1 + B_2 v \end{cases}. \tag{8.12}$$

Letting

$$K = [K_1 \; 0] \, P^{-1},$$

and paying attention to the nilpotent property of N, we can obtain the set of finite closed-loop poles of system (8.12) (and thus system (8.3)) as

$$\sigma \, (E, A + BK) = \sigma \, (QEP, Q \, (A + BK) \, P)$$

$$= \sigma \left(\begin{bmatrix} I_{n_1} & 0 \\ 0 & N \end{bmatrix}, \begin{bmatrix} A_1 + B_1 K_1 & 0 \\ B_2 K_1 & I \end{bmatrix} \right)$$

$$= \sigma \, (A_1 + B_1 K_1)$$

$$= \Gamma_1.$$

Thus, R-controllability is sufficient for the existence of solutions of Problem 8.1.

Conversely, if system (8.1) is not R-controllable, there exists a finite pole $\lambda \in \sigma(E, A)$ such that

$$\text{rank} \, [\lambda E - A \; B] < n.$$

Note that for arbitrary K,

$$\text{rank} \, [\lambda E - A - BK \; B]$$

$$= \text{rank} \left([\lambda E - A \; B] \begin{bmatrix} I & 0 \\ -K & I \end{bmatrix} \right)$$

$$= \text{rank} \, [\lambda E - A \; B].$$

Thus, for arbitrary K, there holds

$$\text{rank}\,[\lambda E - A - BK \quad B] < n,$$

or equivalently

$$\lambda \in \sigma(E, A + BK).$$

This is to say that the system always has a finite pole λ that cannot be replaced by any state feedback. Therefore, Problem 8.1 has a solution only if the system (8.1) is R-controllable. By now the first conclusion is proven.

The second conclusion clearly follows from the above proof of the first conclusion. □

By the above theorem, pole assignment in a regular and R-controllable descriptor linear system by state feedback can be converted into pole assignment in a standard normal linear systems by state feedback. However, it should be also aware that the main disadvantage of such feedbacks is that, there may exist impulse terms in the state response of the closed-loop system (8.3) since $n_1 \le \text{rank}\,E$ and N may not be zero.

Example 8.1. Consider the system (7.19) in Example 7.3. With the transformation matrix (P, Q),

$$Q = \begin{bmatrix} 1 & 1 & 1 & -1 & -1 & 0 \\ 2 & -1 & -1 & 1 & -2 & 0 \\ 0 & \sqrt{3} & -\sqrt{3} & \sqrt{3} & 0 & 0 \\ 0 & 0 & 0 & 1 & 0 & 0 \\ 0 & 0 & 0 & 0 & 1 & 0 \\ -1 & 0 & 0 & 0 & 0 & 1 \end{bmatrix},$$

$$P = \begin{bmatrix} \frac{1}{3} & \frac{1}{3} & 0 & 0 & 0 & 0 \\ \frac{1}{3} & -\frac{1}{6} & \frac{\sqrt{3}}{6} & 0 & 0 & 0 \\ \frac{1}{3} & -\frac{1}{6} & -\frac{\sqrt{3}}{6} & 0 & 1 & 0 \\ 0 & 0 & 0 & 1 & 0 & 0 \\ 0 & 0 & 0 & 0 & 1 & 0 \\ 0 & -\frac{1}{2} & -\frac{\sqrt{3}}{6} & 0 & 1 & 1 \end{bmatrix},$$

we can obtain for system (7.19) the standard decomposition with the following parameters

$$N = \begin{bmatrix} 0 & 1 & 0 \\ 0 & 0 & 0 \\ 0 & 0 & 0 \end{bmatrix}, \quad A_1 = \begin{bmatrix} 1 & 0 & 0 \\ 0 & -\frac{1}{2} & -\frac{\sqrt{3}}{2} \\ 0 & \frac{\sqrt{3}}{2} & -\frac{1}{2} \end{bmatrix},$$

$$B_1 = \begin{bmatrix} 0 & -1 \\ 0 & -2 \\ 0 & 0 \end{bmatrix}, \quad B_2 = \begin{bmatrix} 0 & 0 \\ 1 & 1 \\ -1 & 0 \end{bmatrix}.$$

This system can be easily shown to be R-controllable. Assume that $\Gamma_1 = \{-1, -2, -3\}$, then there exists K_1 such that

$$\sigma(A_1 + B_1 K_1) = \{-1, -2, -3\}.$$

Such a gain matrix can be easily found as

$$K_1 = \begin{bmatrix} 0 & 0 & 0 \\ 8 & -1 & \frac{2\sqrt{3}}{3} \end{bmatrix}.$$

In this case, we have for system (7.19) the following pole assignment controller

$$K = \begin{bmatrix} K_1 & 0 \end{bmatrix} P^{-1} = \begin{bmatrix} 0 & 0 & 0 & 0 & 0 & 0 \\ 6 & 11 & 7 & 0 & -7 & 0 \end{bmatrix}.$$

Since

$$\begin{aligned} &\deg\det(sE - (A + BK)) \\ &= \deg(-s^3 - 6s^2 - 11s - 6) \\ &= 3 < 4 = \text{rank}(E), \end{aligned}$$

we know that impulse terms may still appear in the state response of the closed-loop system.

8.1.3 Pole Assignment under S-Controllability

In this subsection, let us look at the solution to Problem 8.2, the problem with impulse elimination. This problem aims to impose a state feedback, which has control over both the finite poles and the infinite poles. Since rank E finite poles are assigned to the closed-loop system, impulse terms are eliminated in the state response of the closed-loop system.

Lemma 8.2. *System (8.1) is R-controllable if and only if (8.3) is R-controllable for any feedback gain matrix* $K \in \mathbb{R}^{r \times n}$.

Proof. Noting

$$[sE - (A + BK) \quad B] = [sE - A \quad B]\begin{bmatrix} I & 0 \\ -K & I \end{bmatrix}, \quad \forall s \in \mathbb{C}, \ K \in \mathbb{R}^{r \times n},$$

we have

$$\text{rank}[sE - (A + BK) \quad B] = \text{rank}[sE - A \quad B], \quad \forall s \in \mathbb{C}, \ K \in \mathbb{R}^{r \times n}.$$

Therefore, the conclusion holds.   □

Regarding the solution to Problem 8.2, we have the following result.

Theorem 8.2. *Problem 8.2 has a solution if and only if the system (8.1) is S-controllable.*

The necessity of the above theorem is obvious. Suppose Problem 8.2 has a solution K, then the closed-loop system has $n_0 = \text{rank} E$ finite eigenvalues, that is

$$\deg \det(sE - (A + BK)) = \text{rank} E = n_0.$$

Therefore, the open-loop system (8.1) must be I-controllable. Furthermore, noting that Problem 8.2 has a solution K means that the eigenvalues of the closed-loop system are arbitrarily assignable, we know that the closed-loop system is R-controllable. In view of the above Lemma 8.2, so is the open-loop system (8.1).

The following algorithm finds a solution to Problem 8.2 under the S-controllability of system (8.1). The procedure can also be viewed as a constructive proof of Theorem 8.2. The basic idea is to first eliminate the possible impulse in the open-loop system, and then to replace $n_0 = \text{rank} E$ finite eigenvalues with desired ones. Please note that the first three steps are actually the realization of the impulse elimination Algorithm 7.2.

Algorithm 8.1. Pole assignment via state feedback under S-controllability.

Step 1. Finding a dynamics decomposition   Find a pair of orthogonal matrices Q_1 and P_1 satisfying the following singular value decomposition

$$Q_1 E P_1 = \text{diag}\,(\Sigma,\ 0)\,, \quad \Sigma = \text{diag}\,(\sigma_1, \sigma_2, \dots, \sigma_{n_0})\,,$$

and carry out the matrix products and their partitions

$$Q_1 A P_1 = \begin{bmatrix} A_{11} & A_{12} \\ A_{21} & A_{22} \end{bmatrix}, \quad Q_1 B = \begin{bmatrix} B_1 \\ B_2 \end{bmatrix}, \quad \bar{x} = P_1^{-1} x.$$

Then, system (8.1) is r.s.e. to the following dynamics decomposition form

$$\begin{bmatrix} \Sigma & 0 \\ 0 & 0 \end{bmatrix} \dot{\bar{x}} = \begin{bmatrix} A_{11} & A_{12} \\ A_{21} & A_{22} \end{bmatrix} \bar{x} + \begin{bmatrix} B_1 \\ B_2 \end{bmatrix} u. \qquad (8.13)$$

Step 2. Normalization of (A_{22}, B_2) system   Accounting for the S-controllability assumption, system (8.1) is I-controllable. Thus, according to Theorem 4.18,

$$\text{rank}\,[A_{22}\ \ B_2] = n - n_0,$$

which indicates that (A_{22}, B_2) is normalizable. Thus, a matrix $K_2 \in \mathbb{R}^{r \times (n-n_0)}$ may be found such that

$$\det\,(A_{22} + B_2 K_2) \neq 0. \qquad (8.14)$$

Step 3. Realizing impulse elimination   Setting

$$\bar{K} = [\ 0 \quad K_2 \ ] P_1^{-1}$$

and

$$u = \bar{K}x + u_1 = [0 \ K_2] \bar{x} + u_1, \tag{8.15}$$

in which $u_1(t)$ is a new input for the system, the closed-loop system formed by (8.15) and (8.13) is

$$\begin{bmatrix} \Sigma & 0 \\ 0 & 0 \end{bmatrix} \dot{\bar{x}} = \begin{bmatrix} A_{11} & A_{12} + B_1 K_2 \\ A_{21} & A_{22} + B_2 K_2 \end{bmatrix} \bar{x} + \begin{bmatrix} B_1 \\ B_2 \end{bmatrix} u_1. \tag{8.16}$$

It follows from Theorem 7.6 that the system (8.16) is impulse-free.

Step 4. Finding the standard decomposition form   From condition (8.14), we know $(A_{22} + B_2 K_2)^{-1}$ exists, define

$$Q_2 = \begin{bmatrix} I_{n_0} & -(A_{12} + B_1 K_2)(A_{22} + B_2 K_2)^{-1} \\ 0 & I_{n-n_0} \end{bmatrix},$$

$$P_2 = \begin{bmatrix} \Sigma^{-1} & 0 \\ -(A_{22} + B_2 K_2)^{-1} A_{21} \Sigma^{-1} & (A_{22} + B_2 K_2)^{-1} \end{bmatrix}.$$

Direct computation shows that

$$\begin{cases} Q_2 \text{diag}(\Sigma, \ 0) P_2 = \text{diag}(I_{n_0}, \ 0) \\ Q_2 \begin{bmatrix} A_{11} & A_{12} + B_1 K_2 \\ A_{21} & A_{22} + B_2 K_2 \end{bmatrix} P_2 = \text{diag}(\tilde{A}_1, \ I_{n-n_0}), \\ Q_2 \begin{bmatrix} B_1 \\ B_2 \end{bmatrix} = \begin{bmatrix} \tilde{B}_1 \\ \tilde{B}_2 \end{bmatrix} \end{cases} \tag{8.17}$$

where

$$\begin{cases} \tilde{A}_1 = A_{11} \Sigma^{-1} - (A_{12} + B_1 K_2)(A_{22} + B_2 K_2)^{-1} A_{21} \Sigma^{-1} \\ \tilde{B}_1 = B_1 - (A_{12} + B_1 K_2)(A_{22} + B_2 K_2)^{-1} B_2 \\ \tilde{B}_2 = B_2 \end{cases} \ .$$

Thus, under the state transformation

$$\tilde{x} = P_2^{-1} \bar{x} = P_2^{-1} P_1^{-1} x,$$

system (8.16) is transformed into the following equivalent form

$$\text{diag}\left(I_{n_0}, 0\right) \dot{\tilde{x}} = \text{diag}(\tilde{A}_1, I_{n-n_0})\tilde{x} + \begin{bmatrix} \tilde{B}_1 \\ \tilde{B}_2 \end{bmatrix} u_1, \qquad (8.18)$$

which is still R-controllable according to Lemma 8.2.

Step 5. *Assignment of pole set Γ_0* Since system (8.18) is R-controllable, the matrix pair $(\tilde{A}_1, \tilde{B}_1)$ is controllable. Therefore, for any symmetric set Γ_0 with rank$E$ elements, a matrix $\tilde{K}_1$ can be chosen such that

$$\sigma\left(\tilde{A}_1 + \tilde{B}_1 \tilde{K}_1\right) = \Gamma_0.$$

Step 6. *Finding the gain matrix* Define

$$u_1 = [\tilde{K}_1 \ 0]\tilde{x} + v.$$

The overall feedback control is

$$u = \bar{K}x + u_1 = Kx + v,$$

in which

$$K = \bar{K} + \left[\tilde{K}_1 \ 0\right] P_2^{-1} P_1^{-1}.$$

For such a feedback gain matrix K, it is easy to verify that the closed-loop system (8.3) has the set of finite poles as

$$\sigma\left(E, A + BK\right) = \sigma(\tilde{A}_1 + \tilde{B}_1 \tilde{K}_1) = \Gamma_0.$$

Example 8.2. Consider again the system (4.80) in Example 4.7. This system is in the dynamical decomposition from, which means that $P_1 = Q_1 = I_3$ and

$$A_{11} = \begin{bmatrix} 0 & 1 \\ 0 & 0 \end{bmatrix}, \quad A_{12} = \begin{bmatrix} 0 \\ 1 \end{bmatrix}, \quad B_1 = \begin{bmatrix} 0 & 0 \\ 1 & 0 \end{bmatrix},$$

$$A_{22} = -1, \quad A_{21} = \begin{bmatrix} 0 & 0 \end{bmatrix}, \quad B_2 = \begin{bmatrix} 0 & 1 \end{bmatrix}.$$

It is easily verified that system is S-controllable. Let $K_2 = \begin{bmatrix} 1 & 0 \end{bmatrix}^T$. Then

$$\det\left(A_{22} + B_2 K_2\right) = -1 \neq 0,$$

and

$$\tilde{A}_1 = A_{11} - (A_{12} + B_1 K_2)(A_{22} + B_2 K_2)^{-1} A_{21} = \begin{bmatrix} 0 & 1 \\ 0 & 0 \end{bmatrix},$$

$$\tilde{B}_1 = B_1 - (A_{12} + B_1 K_2)(A_{22} + B_2 K_2)^{-1} B_2 = \begin{bmatrix} 0 & 0 \\ 1 & 2 \end{bmatrix},$$

$$P_2 = \begin{bmatrix} I_2 & 0 \\ -(A_{22} + B_2 K_2)^{-1} A_{21} & (A_{22} + B_2 K_2)^{-1} \end{bmatrix} = \begin{bmatrix} 1 & 0 & 0 \\ 0 & 1 & 0 \\ 0 & 0 & -1 \end{bmatrix}.$$

Since

$$\text{rank}[\tilde{B}_1 \ \tilde{A}_1 \tilde{B}_1] = \text{rank} \begin{bmatrix} 0 & 0 & 1 & 2 \\ 1 & 2 & 0 & 0 \end{bmatrix} = 2,$$

the matrix pair $(\tilde{A}_1, \tilde{B}_1)$ is indeed controllable. The closed-loop system is assumed to have $\text{rank}E = 2$ duplicate stable finite poles -1 or

$$\Gamma_0 = \{-1, -1\}.$$

By setting

$$\tilde{K}_1 = \begin{bmatrix} -1 & -2 \\ 0 & 0 \end{bmatrix},$$

we have

$$\det\left(sI - (\tilde{A}_1 + \tilde{B}_1 \tilde{K}_1)\right) = (s+1)^2.$$

Thus,

$$\sigma(\tilde{A}_1 + \tilde{B}_1 \tilde{K}_1) = \Gamma_0 = \{-1, -1\}.$$

In this case, the feedback gain matrix K for system (4.80) is

$$K = \begin{bmatrix} 0 & K_2 \end{bmatrix} P_1^{-1} + [\tilde{K}_1 \ 0] P_2^{-1} P_1^{-1} = \begin{bmatrix} -1 & -2 & 1 \\ 0 & 0 & 0 \end{bmatrix},$$

and the corresponding closed-loop system (8.3) for this system becomes

$$\begin{bmatrix} 1 & 0 & 0 \\ 0 & 1 & 0 \\ 0 & 0 & 0 \end{bmatrix} \dot{x} = \begin{bmatrix} 0 & 1 & 0 \\ -1 & -2 & 2 \\ 0 & 0 & -1 \end{bmatrix} x + \begin{bmatrix} 0 & 0 \\ 1 & 0 \\ 0 & 1 \end{bmatrix} v.$$

It is easy to verify that the closed-loop pole set is the expected $\Gamma_0 = \{-1, -1\}$ and no impulse portions exist in the closed-loop system response.

8.2 Pole Assignment by P-D Feedback

In this section, we further investigate pole assignment in the descriptor linear system (8.1) via P-D feedback.

8.2.1 Problem Formulation

Consider the following state proportional plus derivative feedback control

$$u(t) = K_{\mathrm{p}} x(t) - K_{\mathrm{d}} \dot{x}(t) + v(t), \qquad (8.19)$$

where $K_{\mathrm{p}}, K_{\mathrm{d}} \in \mathbb{R}^{r \times n}$ are the feedback gains, and $v \in \mathbb{R}^{r}$ is an external signal. When this controller is applied to system (8.1), the closed-loop system is obtained as

$$(E + BK_{\mathrm{d}}) \dot{x} = (A + BK_{\mathrm{p}}) x + Bv. \qquad (8.20)$$

Since derivative signal is used, we can consider the assignment of n closed-loop eigenvalues. Therefore, the pole assignment problem studied in this section can be precisely stated as follows.

Problem 8.3. Given the regular system (8.1), and a set of complex numbers

$$\Gamma_{\mathrm{c}} = \{s_1, s_2, \ldots, s_n\}, \qquad (8.21)$$

which is symmetric about the real axis, find a state P-D feedback controller in the form of (8.19) such that Γ_{c} is the set of finite poles of the closed-loop system (8.20), that is

$$\sigma\left(E + BK_{\mathrm{d}}, A + BK_{\mathrm{p}}\right) = \Gamma_{\mathrm{c}} = \{s_1, s_2, \ldots, s_n\}. \qquad (8.22)$$

8.2.2 The Solution

To present the solution to Problem 8.3, we need the following lemma.

Lemma 8.3. *The regular descriptor linear system (8.1) is R-controllable if and only if the system $(E + BK_{\mathrm{d}}, A, B)$ is R-controllable for any $K_{\mathrm{d}} \in \mathbb{R}^{r \times n}$.*

Proof. By the definition of R-controllability and Theorem 4.12, system (8.1) is R-controllable if and only if

$$\mathrm{rank}\,[sE - A \ \ B] = n, \quad \forall s \in \mathbb{C}, s \text{ finite}.$$

On the other hand,

$$\text{rank} \begin{bmatrix} s\,(E + BK_d) - A & B \end{bmatrix}$$

$$= \text{rank}\, [sE - A \;\; B] \begin{bmatrix} I & 0 \\ K_d s & I \end{bmatrix}$$

$$= \text{rank}\, [sE - A \;\; B], \;\; \forall s \in \mathbb{C}, s \text{ finite.}$$

Thus, the conclusion holds.   □

Remark 8.1. It is shown in Chap. 7 that state derivative feedback may change the impulse controllability of a system. Lemma 8.3, together with Lemma 8.2, clarifies that neither proportional nor state derivative feedback control changes the R-controllability of a system.

Based on the above lemma, the main result in this section can be given below.

Theorem 8.3. *Problem 8.3 has a solution for an arbitrarily given set Γ_c if and only if the system (8.1) is C-controllable.*

Proof. Sufficiency: Under the assumption of C-controllability, system (8.1) is R-controllable and normalizable. Due to the normalizability of the system, there exists a matrix K_d such that

$$\det (E + BK_d) \neq 0. \tag{8.23}$$

Therefore, the system $(E + BK_d, A, B)$ is a normal one. Furthermore, noting that system $(E + BK_d, A, B)$ is also R-controllable by Lemma 8.3, we know that the matrix pair $(\check{A}, \check{B})$ with

$$\check{A} = (E + BK_d)^{-1} A, \;\; \check{B} = (E + BK_d)^{-1} B \tag{8.24}$$

is controllable. Consequently, for any Γ_c in the form of (8.21), a matrix K_p can be found such that

$$\sigma\left(\check{A} + \check{B} K_p\right) = \sigma\left((E + BK_d)^{-1} A + (E + BK_d)^{-1} BK_p\right)$$

$$= \sigma\left((E + BK_d)^{-1} (A + BK_p)\right)$$

$$= \Gamma_c.$$

Hence, the closed-loop system (8.20) has the finite pole set:

$$\sigma\left(E + BK_d, A + BK_p\right)$$

$$= \sigma\left((E + BK_d)^{-1} (A + BK_p)\right)$$

$$= \Gamma_c.$$

Necessity: The existence of n finite poles in (8.20) shows that (8.23) holds. Thus, system (8.1) is normalizable. Further, since Problem 8.3 has a solution for an arbitrarily given set Γ_c, the R-controllability of system (8.1) clearly follows.   □

It follows from the above theorem that under the C-controllability assumption, feedback control (8.19) may not only drive all the infinite poles to finite positions, but also make the closed-loop system normal. This feature could not be achieved via pure proportional feedback (8.2).

The sufficiency part of the above proof of the theorem clearly suggests the following algorithm for solving Problem 8.3.

Algorithm 8.2. Pole assignment via state P-D feedback.

Step 1 Find a derivative feedback gain matrix K_d satisfying (8.23) using certain normalization method.
Step 2 Compute matrices $\check{A}$ and $\check{B}$ according to (8.24).
Step 3 Find a proportional feedback gain matrix K_p satisfying $\sigma(\check{A} + \check{B}K_p) = \Gamma_c$ using certain pole assignment technique for normal linear systems.

Example 8.3. Consider again the system (4.80). It has been shown in Example 4.7 that the system is C-controllable. Let $\Gamma_c = \{-1, -1, -1\}$ be the expected closed-loop pole set, which will be assigned by a proportional plus derivative feedback in the form of (8.19).

Let

$$K_d = \begin{bmatrix} 0 & 0 & 0 \\ 0 & 0 & 1 \end{bmatrix}.$$

Then it is obvious that

$$\det(E + BK_d) \neq 0.$$

Thus, the feedback

$$u = -K_d \dot{x} + u_1$$

results in the following normal closed-loop system

$$\dot{x} = \begin{bmatrix} 0 & 1 & 0 \\ 0 & 0 & 1 \\ 0 & 0 & -1 \end{bmatrix} x + \begin{bmatrix} 0 & 0 \\ 1 & 0 \\ 0 & 1 \end{bmatrix} u_1. \qquad (8.25)$$

Choose for the above system (8.25) a feedback controller in the form of

$$u_1 = K_p x + v, \quad K_p = \begin{bmatrix} k_{11} & k_{12} & k_{13} \\ k_{21} & k_{22} & k_{23} \end{bmatrix},$$

then the closed-loop system characteristic polynomial is

$$\begin{aligned} \Delta_c(s) = \; &s^3 - (k_{23} - 1 + k_{12})\, s^2 \\ &+ (k_{23}k_{12} - k_{12} - k_{22}k_{13} - k_{22} - k_{11})\, s \\ &+ (k_{11}k_{23} - k_{11} - k_{21}k_{13} - k_{21}). \end{aligned}$$

Assume that the expected closed-loop pole set is $\{-1,-1,-1\}$, then it can be obtained that

$$k_{11} = -1, \; k_{12} = -2, \; k_{13} = 0, \; k_{21} = 0, \; k_{22} = 0, \; k_{23} = 0.$$

Therefore, the controller is

$$u = \begin{bmatrix} 0 & 0 & 0 \\ 0 & 0 & -1 \end{bmatrix} \dot{x} + \begin{bmatrix} -1 & -2 & 0 \\ 0 & 0 & 0 \end{bmatrix} x + v.$$

The corresponding closed-loop system is

$$\begin{bmatrix} 1 & 0 & 0 \\ 0 & 1 & 0 \\ 0 & 0 & 1 \end{bmatrix} \dot{x} = \begin{bmatrix} 0 & 1 & 0 \\ -1 & -2 & 1 \\ 0 & 0 & -1 \end{bmatrix} x + \begin{bmatrix} 0 & 0 \\ 1 & 0 \\ 0 & 1 \end{bmatrix} v,$$

which can be verified to have the finite pole set $\{-1,-1,-1\}$.

8.3 Stabilizability and Detectability

Stabilizability and detectability are a pair of concepts in linear systems, which are weaker than controllability and observability, respectively. They perform a fundamental role in the stabilization and observer design problems, respectively.

Again, let us consider the descriptor linear system

$$\begin{cases} E\dot{x} = Ax + Bu \\ y = Cx \end{cases},$$
(8.26)

where $x \in \mathbb{R}^n$, $u \in \mathbb{R}^r$, $y \in \mathbb{R}^m$ are the state vector, the control input vector, and the measured output vector, respectively; E, $A \in \mathbb{R}^{n \times n}$, $B \in \mathbb{R}^{n \times r}$, and $C \in \mathbb{R}^{m \times n}$ are constant matrices.

Under the state feedback controller

$$u = Kx + v, \quad K \in \mathbb{R}^{r \times n},$$
(8.27)

the closed-loop system is

$$\begin{cases} E\dot{x} = (A + BK)x + Bv \\ y = Cx \end{cases}.$$
(8.28)

The problem of stabilization is to seek a controller (8.27) such that the closed-loop system (8.28) is stable. In this section, we focus on the existence condition of controller (8.27), which makes the closed-loop system (8.28) stable. In the next section, we will present several ways for designing such controllers.

8.3.1 Stabilizability

Stabilizability of the descriptor linear system (8.26) is defined as follows.

Definition 8.1. System (8.26) is called stabilizable if there exists a state feedback (8.27) such that the resulted closed-loop system (8.28) is stable. In this case, the feedback controller (8.27) is called a stabilizing state feedback controller for system (8.26).

According to the above definition, it is obvious that system (8.26) is stabilizable if it is stable.

The purpose of this subsection is to determine the conditions for stabilizability of the given regular descriptor linear system (8.26). Using Definition 8.1 and Theorem 3.14, we can prove the following theorem.

Theorem 8.4. *The regular system (8.26) is stabilizable if and only if*

$$\text{rank}\,[sE - A \ \ B] = n, \quad \forall s \in \bar{\mathbb{C}}^+, s \text{ finite}, \tag{8.29}$$

where $\bar{\mathbb{C}}^+ = \{s \mid s \in \mathbb{C}, \ \text{Re}\,(s) \geq 0\}$ *is the closed right half complex plane.*

Proof. Necessity: According to Definition 8.1 and Theorem 3.14, system (8.26) is stabilizable if and only if there exists a $K \in \mathbb{R}^{r \times n}$ such that

$$\sigma\,(E, A + BK) \subset \mathbb{C}^-,$$

or, equivalently,

$$\text{rank}\,[sE - (A + BK)] = n, \quad \forall s \in \bar{\mathbb{C}}^+, \ s \text{ finite}. \tag{8.30}$$

On the other side,

$$\text{rank}\,[sE - (A + BK)] = \text{rank}\left([sE - A \ \ B] \begin{bmatrix} I \\ -K \end{bmatrix} \right)$$

$$\leq \min\left\{ \text{rank}\,[sE - A \ \ B],\ \text{rank}\begin{bmatrix} I \\ -K \end{bmatrix} \right\}$$

$$= \text{rank}\,[sE - A \ \ B].$$

This, together with (8.30), clearly implies (8.29).

Sufficiency: Assume that (8.29) holds. Then it follows from Lemma 5.3 that there exists a matrix K such that (8.30) holds. The proof is done.   □

Combining the above Theorem 8.4 with Definition 2.3 immediately gives the following corollary.

Corollary 8.1. *The regular descriptor system (8.26) is stabilizable if and only if all its input transmission zeros are stable.*

Recall that system (8.26) is R-controllable if and only if

$$\text{rank}\,[sE - A \quad B] = n, \quad \forall s \in \mathbb{C}, \ s \text{ finite,}$$

and is regularizable via the state feedback if there exists some nonzero $\gamma \in \mathbb{C}$ such that

$$\text{rank}\left[\gamma E - A \quad B\right] = n. \tag{8.31}$$

Comparing the above condition with (8.29), immediately gives the following corollary.

Corollary 8.2. *System (8.26) is stabilizable if it is R-controllable, and is regularizable if it is stabilizable.*

It should be noted that the inverse of the above corollary is not true. The following lemma tells us that state feedback does not change the stabilizability of a descriptor linear system.

Lemma 8.4. *System (8.26) is stabilizable if and only if (8.28) is stabilizable for any feedback gain matrix $K \in \mathbb{R}^{r \times n}$.*

Proof. Noting

$$[sE - (A + BK) \quad B] = [sE - A \quad B]\begin{bmatrix} I & 0 \\ -K & I \end{bmatrix}, \ \forall s \in \bar{\mathbb{C}}^+, \ K \in \mathbb{R}^{r \times n},$$

we have

$$\text{rank}\,[sE - (A + BK) \quad B] = \text{rank}\,[sE - A \quad B], \ \forall s \in \bar{\mathbb{C}}^+, \ K \in \mathbb{R}^{r \times n}.$$

Therefore, the conclusion holds.   □

Example 8.4. Consider again the system (4.80). It has been verified that the system is R-controllable in Example 4.7. In fact, for this system we need only to set

$$K_1 = \begin{bmatrix} 0 & -1 & 0 \\ -1 & -1 & 0 \end{bmatrix}.$$

The feedback control is then

$$u_e(t) = K_1 x + v$$

$$= \begin{bmatrix} -x_2 \\ -x_1 - x_2 \end{bmatrix} + v,$$

where $v(t)$ is an external input. Direct computation shows

$$\sigma\,(E, A + BK_1) = \{-1, -1\} \subset \mathbb{C}^-.$$

Thus, the closed-loop system is stable.

8.3.2 Detectability

Stabilizability characterizes the system's stability for control. It has a dual concept, which is called detectability.

Definition 8.2. System (8.26) is detectable if its dual system (E^T, A^T, C^T, B^T) is stabilizable.

A direct result of Definition 8.2 and Theorem 8.4 is the following.

Theorem 8.5. *System (8.26) is detectable if and only if*

$$\text{rank}\begin{bmatrix} sE - A \\ C \end{bmatrix} = n, \quad \forall s \in \bar{\mathbb{C}}^+, \; s \text{ finite}.$$

Parallel to Corollaries 8.1 and 8.2, we have

Corollary 8.3. *The regular descriptor linear system (8.26) is detectable if and only if all its output transmission zeros are stable.*

Corollary 8.4. *System (8.26) is detectable if it is R-observable, and is dual regularizable if it is detectable.*

This corollary shows that detectability can be deduced from R-observability, but the inverse is not true. It follows from this corollary that detectability is also determined by the slow subsystem. This indicates the fact that stabilizability and detectability only characterize the properties of the slow subsystem, although they are defined for the whole system.

Parallel to Lemma 8.4, we have the following lemma.

Lemma 8.5. *System (8.1) is detectable if and only if the output feedback system* $(E, A + BKC, B, C)$ *is detectable for any feedback gain matrix* $K \in \mathbb{R}^{r \times m}$.

Proof. Noting

$$\begin{bmatrix} sE - (A + BKC) \\ C \end{bmatrix} = \begin{bmatrix} I & -BK \\ 0 & I \end{bmatrix}\begin{bmatrix} sE - A \\ C \end{bmatrix}, \quad \forall s \in \bar{\mathbb{C}}^+, \; K \in \mathbb{R}^{r \times m},$$

we have

$$\text{rank}\begin{bmatrix} sE - (A + BKC) \\ C \end{bmatrix} = \text{rank}\begin{bmatrix} sE - A \\ C \end{bmatrix}, \quad \forall s \in \bar{\mathbb{C}}^+, \; K \in \mathbb{R}^{r \times m}.$$

Therefore, the conclusion holds. □

Example 8.5. Consider the system

$$\begin{cases} \begin{bmatrix} 1 & 0 & -1 \\ 0 & 1 & 0 \\ -1 & 0 & 1 \end{bmatrix} \dot{x} = \begin{bmatrix} 2 & 0 & -2 \\ 0 & 3 & 0 \\ -2 & 0 & 3 \end{bmatrix} x + \begin{bmatrix} 1 \\ 1 \\ -1 \end{bmatrix} u \\ y = \begin{bmatrix} 0 & 0 & 1 \end{bmatrix} x \end{cases}$$

Since

$$\mathrm{rank} \begin{bmatrix} sE - A \\ C \end{bmatrix} = \mathrm{rank} \begin{bmatrix} s-2 & 0 & -s+2 \\ 0 & s-3 & 0 \\ -s+2 & 0 & s-3 \\ 0 & 0 & 1 \end{bmatrix}$$

$$= 2 < n, \text{ when } s = 2,$$

the system is not detectable by Theorem 8.5.

Example 8.6. Consider the system

$$\begin{cases} \begin{bmatrix} 1 & 0 & 0 & 1 \\ 0 & 1 & 1 & 0 \\ 0 & 0 & 1 & 0 \\ 0 & -1 & -1 & 0 \end{bmatrix} \dot{x} = \begin{bmatrix} -1 & 1 & 0 & 0 \\ 0 & -2 & -2 & 0 \\ 0 & 0 & -1 & 0 \\ 0 & 3 & 2 & 1 \end{bmatrix} x + \begin{bmatrix} 1 \\ 4 \\ 3 \\ -4 \end{bmatrix} u \\ y = \begin{bmatrix} 0 & 1 & 0 & 1 \end{bmatrix} x \end{cases}$$

Since

$$\mathrm{rank} \begin{bmatrix} sE - A \\ C \end{bmatrix}$$

$$= \mathrm{rank} \begin{bmatrix} s+1 & -1 & 0 & s \\ 0 & s+2 & s+2 & 0 \\ 0 & 0 & s+1 & 0 \\ 0 & -s-3 & -s-2 & -1 \\ 0 & 1 & 0 & 1 \end{bmatrix}$$

$$= \mathrm{rank} \begin{bmatrix} 1 & 0 & 0 & 1 & 0 \\ 0 & 1 & 0 & 0 & 0 \\ 0 & 0 & 1 & 0 & 0 \\ 0 & -1 & 0 & 1 & 0 \\ 0 & 0 & 0 & 0 & 1 \end{bmatrix} \begin{bmatrix} s+1 & 0 & 0 & 0 \\ 0 & s+2 & 0 & 0 \\ 0 & 0 & s+1 & 0 \\ 0 & 0 & 0 & -1 \\ 0 & 0 & 0 & 1 \end{bmatrix} \begin{bmatrix} 1 & 0 & 0 & 1 \\ 0 & 1 & 1 & 0 \\ 0 & 0 & 1 & 0 \\ 0 & 1 & 0 & 1 \end{bmatrix}$$

$$= \text{rank} \begin{bmatrix} s+1 & 0 & 0 & 0 \\ 0 & s+2 & 0 & 0 \\ 0 & 0 & s+1 & 0 \\ 0 & 0 & 0 & -1 \\ 0 & 0 & 0 & 1 \end{bmatrix}$$

$$= 4 = n, \ \forall s \in \bar{\mathbb{C}}^+, s \text{ finite},$$

the system is thus detectable by Theorem 8.5. However, since

$$\text{rank} \begin{bmatrix} sE - A \\ C \end{bmatrix}_{s=-2} = \text{rank} \begin{bmatrix} -1 & 0 & 0 & 0 \\ 0 & 0 & 0 & 0 \\ 0 & 0 & -1 & 0 \\ 0 & 0 & 0 & -1 \\ 0 & 0 & 0 & 1 \end{bmatrix}$$

$$= 3 < n,$$

the system is actually not R-observable.

So far, we have introduced various concepts in descriptor systems theory, including the various controllabilities and observabilities, normalizability and dual normalizability, stabilizability, detectability and regularizability. Among these concepts, there exist both similarities and differences. Their relationships are complicated, but they characterize descriptor linear systems from different points of view. As for their intrinsic properties, C-controllability and C-observability characterize the properties of the whole system, while R-controllability, R-observability, stabilizability, and detectability characterize the properties of the slow subsystem only; and impulse controllability, impulse observability, normalizability, and dual normalizability characterize those of the fast subsystem. These concepts cannot be substituted for each other.

To sum up, the relationships among these concepts may be described by the following diagrams, where A⇒B represents that B can be deduced from A, and A ⟺ B indicates the equivalence between A and B.

C-controllability ⟺ R-controllability + Normalizability
⇓   ⇓
S-controllability ⟺ R-controllability + I-controllability
⇓   ⇓
Stabilizability   I-controllablizability
⇓
Regularizability

Diagram (a): Relations among controllabilities, normalizability, and stabilizability

C-observability $\Longleftrightarrow$ R-observability $+$ Dual normalizability

$\Downarrow$ $\Downarrow$

S-observability $\Longleftrightarrow$ R-observability $+$ I-observability

$\Downarrow$ $\Downarrow$

Detectability   I-observablizability

$\Downarrow$

Dual regularizability

Diagram (b): Relations among observabilities, dual normalizability, and detectability

8.4 Stabilizing Controller Design

In the preceding section, we have established necessary and sufficient conditions for stabilizability and detectability of descriptor linear systems. In this section, we further look at the problem of stabilizing controller design for descriptor linear systems.

Problem 8.4. Given the regular and stabilizable descriptor system (8.26), find a state feedback controller in the form of (8.27) such that the closed-loop system (8.28) is stable.

8.4.1 Design Based on Standard Decomposition

When the system (8.26) is regular, it is well known that there exist nonsingular matrices Q and P, such that system (8.26) is r.s.e. to the following standard decomposition form

$$\begin{cases} \dot{x}_1 = A_1 x_1 + B_1 u \\ N \dot{x}_2 = x_2 + B_2 u \quad , \\ y = C_1 x_1 + C_2 x_2 \end{cases} \tag{8.32}$$

which is linked with the original system by the following relations:

$$QEP = \text{diag}\,(I,\ N)\,, \quad QAP = \text{diag}\,(A_1,\ I)\,,$$

$$QB = \begin{bmatrix} B_1 \\ B_2 \end{bmatrix}, \quad CP = [C_1\ C_2]\,,$$

$$x = P \begin{bmatrix} x_1 \\ x_2 \end{bmatrix}, \quad x_1 \in \mathbb{R}^{n_1}, \quad x_2 \in \mathbb{R}^{n_2}, \quad n_1 + n_2 = n.$$

Theorem 8.6. *Consider the regular descriptor linear system (8.26) with its standard decomposition form (8.32).*

1. *System (8.26) is stabilizable if and only if the matrix pair (A_1, B_1) is stabilizable.*
2. *When system (8.26) is stabilizable, any*

$$K \in \{K \mid K = [K_1 \ 0] \, P^{-1}, \ \sigma (A_1 + B_1 K_1) \subset \mathbb{C}^-\} \qquad (8.33)$$

is a stabilizing state feedback gain.

Proof. By Theorem 8.4, system (8.26) is stabilizable if and only if (8.29) holds. Paying attention to the nilpotent property of N, we have

$$
\begin{aligned}
&\text{rank} \, [sE - A \ \ B] \\
&= \text{rank} \, [sQEP - QAP \ \ QB] \\
&= \text{rank} \begin{bmatrix} sI - A_1 & 0 & B_1 \\ 0 & sN - I & B_2 \end{bmatrix} \\
&= n_2 + \text{rank} \, [sI - A_1 \ \ B_1].
\end{aligned}
$$

Thus, (8.29) holds if and only if

$$\text{rank} \, [sI - A_1 \ \ B_1] = n - n_2 = n_1, \ \forall s \in \bar{\mathbb{C}}^+.$$

With this, we complete the proof of the first conclusion.

Under the given condition, a matrix $K_1 \in \mathbb{R}^{r \times n_1}$ may be found such that

$$\sigma (A_1 + B_1 K_1) \subset \mathbb{C}^-.$$

With K being given by (8.33), direct computation results in

$$
\begin{aligned}
\sigma (E, A + BK) &= \sigma (QEP, Q (A + BK) P) \\
&= \sigma \left(\begin{bmatrix} I & 0 \\ 0 & N \end{bmatrix}, \begin{bmatrix} A_1 + B_1 K_1 & 0 \\ B_2 K_1 & I \end{bmatrix} \right) \\
&= \sigma (A_1 + B_1 K_1) \\
&\subset \mathbb{C}^-.
\end{aligned}
$$

The system is thus stabilizable. The second conclusion is also proven. $\square$

The above theorem shows that the stabilization property is determined uniquely by its slow subsystem. Furthermore, it suggests the following algorithm for finding a stabilizing state feedback controller for the system (8.26).

Algorithm 8.3. Stabilization based on standard decomposition form.

Step 1 Convert the system (8.26) into the standard decomposition canonical form (8.32).

Step 2 Find a stabilizing controller gain K_1 for the controllable system (A_1, B_1) by using some pole assignment method for normal linear systems.

Step 3 Compute the stabilizing controller gain for system (8.26) by

$$K = [K_1 \ 0] P^{-1}.$$

Example 8.7. Consider the following system

$$\begin{bmatrix} 0 & 0 & 0 & 1 \\ 0 & 0 & 1 & 0 \\ 0 & 0 & 0 & 0 \\ -1 & 0 & 0 & 1 \end{bmatrix} \dot{x} = \begin{bmatrix} 0 & 1 & 0 & 0 \\ 1 & 0 & 0 & 0 \\ -1 & 0 & 0 & 1 \\ 0 & 1 & 1 & 1 \end{bmatrix} x + \begin{bmatrix} 0 \\ 0 \\ 1 \\ 0 \end{bmatrix} u.$$

If we choose the following transformation matrices

$$Q = \begin{bmatrix} 1 & 0 & 1 & -1 \\ 0 & 1 & 0 & 0 \\ 0 & 0 & -1 & 1 \\ 0 & 0 & 1 & 0 \end{bmatrix}, \quad P = \begin{bmatrix} 1 & 0 & 0 & 0 \\ -1 & -1 & 1 & 0 \\ 0 & 1 & 0 & 0 \\ 1 & 0 & 0 & 1 \end{bmatrix}$$

and the state transformation

$$P^{-1}x = \begin{bmatrix} x_1 \\ x_2 \end{bmatrix}, x_1 \in \mathbb{R}^2, x_2 \in \mathbb{R}^2,$$

the standard decomposition form for this system can be obtained as

$$\begin{cases} \dot{x}_1 = \begin{bmatrix} -1 & -1 \\ 1 & 0 \end{bmatrix} x_1 + \begin{bmatrix} 1 \\ 0 \end{bmatrix} u \\ \begin{bmatrix} 0 & 1 \\ 0 & 0 \end{bmatrix} \dot{x}_2 = x_2 + \begin{bmatrix} -1 \\ 1 \end{bmatrix} u \end{cases},$$

and

$$y = \begin{bmatrix} 0 & 1 \end{bmatrix} x_1 + \begin{bmatrix} 0 & 0 \end{bmatrix} x_2.$$

Then the matrix

$$K_1 = \begin{bmatrix} -1 & -1 \end{bmatrix}$$

satisfies

$$\sigma (A_1 + B_1 K_1) = \{-1 + i, -1 - i\} \subset \mathbb{C}^-.$$

Thus, the matrix K can be obtained as

$$K = \begin{bmatrix} K_1 & 0 & 0 \end{bmatrix} P^{-1} = \begin{bmatrix} -1 & 0 & -1 & 0 \end{bmatrix}.$$

8.4.2 Design Based on Controllability Canonical Forms

As has been pointed out in Sect. 4.7, for regular system (8.26) there exist nonsingular matrices Q_1 and P_1 such that (8.26) is r.s.e. to the following controllability canonical form:

$$\begin{cases} \begin{bmatrix} E_{11} & E_{12} \\ 0 & E_{22} \end{bmatrix} \begin{bmatrix} \dot{x}_1 \\ \dot{x}_2 \end{bmatrix} = \begin{bmatrix} A_{11} & A_{12} \\ 0 & A_{22} \end{bmatrix} \begin{bmatrix} x_1 \\ x_2 \end{bmatrix} + \begin{bmatrix} B_1 \\ 0 \end{bmatrix} u \\ y = [C_1 \ C_2] \begin{bmatrix} x_1 \\ x_2 \end{bmatrix} \end{cases} \tag{8.34}$$

where

$$Q_1 E P_1 = \begin{bmatrix} E_{11} & E_{12} \\ 0 & E_{22} \end{bmatrix}, \ Q_1 A P_1 = \begin{bmatrix} A_{11} & A_{12} \\ 0 & A_{22} \end{bmatrix},$$

$$Q_1 B = \begin{bmatrix} B_1 \\ 0 \end{bmatrix}, \ C P_1 = [C_1 \ C_2], \ x = P_1 \begin{bmatrix} x_1 \\ x_2 \end{bmatrix},$$

and (E_{11}, A_{11}, B_1) is C-controllable.

Recall again Theorem 8.4 we know that the system (8.26) is stabilizable if and only if (8.29) holds. Noting that

$$\begin{aligned} & [sE - A \ B] \\ & = [sQ_1 E P_1 - Q_1 A P_1 \ Q_1 B] \\ & = \begin{bmatrix} sE_{11} - A_{11} & sE_{12} - A_{12} & B_1 \\ 0 & sE_{22} - A_{22} & 0 \end{bmatrix}, \end{aligned}$$

we can immediately obtain the following theorem.

Theorem 8.7. *Consider the regular system (8.26) with its controllability canonical form given in (8.34).*

1. *System (8.26) is stabilizable if and only if $\sigma(E_{22}, A_{22}) \subset \mathbb{C}^-$.*
2. *When system (8.26) is stabilizable, all the stabilizing state feedback gains for the system are given by*

$$K = [K_1 \ K_2] P_1^{-1} \tag{8.35}$$

with K_2 being an arbitrary parameter matrix of proper dimension, and K_1 being an arbitrary stabilizing controller gain for the C-controllable subsystem (E_{11}, A_{11}, B_1).

The above theorem suggests the following algorithm for finding a stabilizing state feedback controller for the system (8.26).

Algorithm 8.4. Stabilization based on the controllability form.

Step 1 Convert the system (8.26) into the controllability canonical form (8.34).
If the matrix pair (E_{22}, A_{22}) has unstable finite eigenvalues, the system is
not stabilizable, terminate algorithm. If the matrix pair (E_{22}, A_{22}) is stable,
carry on with the next step.

Step 2 Find a stabilizing controller gain K_1 for the C-controllable system
(E_{11}, A_{11}, B_1) by using some pole assignment method.

Step 3 Compute the stabilizing controller gain for system (8.26) by (8.35) with K_2
being an arbitrary parameter matrix of proper dimension.

Example 8.8. Consider the system

$$
\left\{
\begin{aligned}
\begin{bmatrix}
0 & 0 & 0 & 1 & 2 & 1 \\
0 & 0 & 1 & 1 & 0 & 1 \\
0 & 0 & 0 & 0 & 0 & 0 \\
-1 & -1 & 0 & 1 & 1 & 0 \\
0 & 0 & 0 & 0 & 1 & 1 \\
0 & 0 & 0 & 0 & 0 & 1
\end{bmatrix} \dot{x} &=
\begin{bmatrix}
0 & 1 & 1 & 0 & 2 & 2 \\
1 & 1 & 0 & 0 & 0 & 0 \\
-1 & -1 & 0 & 1 & 1 & 1 \\
0 & 1 & 2 & 2 & 1 & 0 \\
0 & 0 & 0 & 0 & -1 & -1 \\
0 & 0 & 0 & 0 & 0 & -1
\end{bmatrix} x +
\begin{bmatrix}
0 \\
0 \\
1 \\
0 \\
0 \\
0
\end{bmatrix} u . \\
y &= \begin{bmatrix} 0 & 0 & 1 & 1 & 0 & 0 \end{bmatrix} x
\end{aligned}
\right.
$$

In the following, we will solve a stabilizing controller for the system using the above
algorithm. To complete Step 1, we choose the following transformation matrices

$$
Q = I \text{ and } P =
\begin{bmatrix}
1 & -1 & 1 & -1 & 1 & -1 \\
0 & 1 & -1 & 1 & -1 & 1 \\
0 & 0 & 1 & -1 & 1 & -1 \\
0 & 0 & 0 & 1 & -1 & 1 \\
0 & 0 & 0 & 0 & 1 & -1 \\
0 & 0 & 0 & 0 & 0 & 1
\end{bmatrix},
$$

and the controllability canonical form for this system can be obtained as

$$
\left\{
\begin{aligned}
\begin{bmatrix} E_{11} & E_{12} \\ 0 & E_{22} \end{bmatrix} \dot{\tilde{x}} &=
\begin{bmatrix} A_{11} & A_{12} \\ 0 & A_{22} \end{bmatrix} \tilde{x} +
\begin{bmatrix} B_1 \\ 0 \end{bmatrix} u \\
y &= \begin{bmatrix} C_1 & C_2 \end{bmatrix} \tilde{x}
\end{aligned}
\right.
$$

with

$$
E_{11} =
\begin{bmatrix}
0 & 0 & 0 & 1 \\
0 & 0 & 1 & 0 \\
0 & 0 & 0 & 0 \\
-1 & 0 & 0 & 1
\end{bmatrix},
E_{12} =
\begin{bmatrix}
1 & 0 \\
0 & 1 \\
0 & 0 \\
0 & 0
\end{bmatrix},
E_{22} =
\begin{bmatrix}
1 & 0 \\
0 & 1
\end{bmatrix},
$$

$$A_{11} = \begin{bmatrix} 0 & 1 & 0 & 0 \\ 1 & 0 & 0 & 0 \\ -1 & 0 & 0 & 1 \\ 0 & 1 & 1 & 1 \end{bmatrix}, A_{12} = \begin{bmatrix} 2 & 0 \\ 0 & 0 \\ 0 & 1 \\ 0 & 0 \end{bmatrix}, A_{22} = \begin{bmatrix} -1 & 0 \\ 0 & -1 \end{bmatrix},$$

$$B_1 = \begin{bmatrix} 0 \\ 0 \\ 1 \\ 0 \end{bmatrix}, C_1 = \begin{bmatrix} 0 & 0 & 1 & 0 \end{bmatrix}, C_2 = \begin{bmatrix} 0 & 0 \end{bmatrix}.$$

Since $\sigma(E_{22}, A_{22}) = \{-1, -1\}$, the system is stabilizable.

To complete Step 2, for the C-controllable system (E_{11}, A_{11}, B_1), we choose the following stabilizing controller gain

$$K_1 = \begin{bmatrix} -1 & 0 & -1 & 0 \end{bmatrix}.$$

Finally, to complete Step 3, we set $K_2 = 0$, thus the stabilizing controller gain K can be obtained as

$$K = \begin{bmatrix} K_1 & 0 & 0 \end{bmatrix} P^{-1} = \begin{bmatrix} -1 & -1 & -1 & -1 & 0 & 0 \end{bmatrix},$$

and the set of finite closed-loop eigenvalues is

$$\sigma(E, A + BK) = \{-1, -1, -1 + i, -1 - i\}.$$

Since there are four finite eigenvalues assigned to the closed-loop system, while rank$E = 5$, possible impulse terms may exist in the closed-loop system response. However, if we choose in Step 2

$$K_1 = \begin{bmatrix} -1 & \frac{1}{2} & -1 & 0 \end{bmatrix},$$

in Step 3 the corresponding stabilizing controller gain K can be obtained as

$$K = \begin{bmatrix} K_1 & 0 & 0 \end{bmatrix} P^{-1} = \begin{bmatrix} -1 & -\frac{1}{2} & -\frac{1}{2} & -1 & 0 & 0 \end{bmatrix}.$$

With this stabilizing controller gain K, the set of finite closed-loop eigenvalues is

$$\sigma(E, A + BK) = \left\{ -1, -1, -1, -\frac{1}{2} - \frac{\sqrt{15}}{2}i, -\frac{1}{2} + \frac{\sqrt{15}}{2}i \right\},$$

which has rank$E = 5$ elements, the corresponding closed-loop system is therefore impulse-free.

8.4.3 Design Based on Lyapunov Theory

For system (8.26), we introduce the generalized Riccati matrix equation

$$E^T VA + A^T VE + E^T WE - E^T VBR^{-1}B^T VE = 0, \tag{8.36}$$

where $W = W^T > 0$, $R = R^T > 0$. Regarding the solution to this matrix equation, we have the following lemma.

Lemma 8.6. *If the system (8.26) is regular, impulse-free, and R-controllable, then for each $W > 0$, there exists $V > 0$ satisfying the Riccati equation (8.36).*

Proof. Since the system (8.26) is regular, according to standard decomposition for descriptor systems, there exist two nonsingular matrices Q and P satisfying (8.5). Since the system (8.26) is impulsefree, we actually have $N = 0$. Further, denote

$$\tilde{V} = Q^{-T}VQ^{-1} = \begin{bmatrix} V_{11} & V_{12} \\ V_{12}^T & V_{22} \end{bmatrix} \tag{8.37}$$

and

$$\tilde{W} = Q^{-T}WQ^{-1} = \begin{bmatrix} W_{11} & W_{12} \\ W_{12}^T & W_{22} \end{bmatrix}, \tag{8.38}$$

pre-multiplying by P^T and post-multiplying by P both sides of the Riccati equation (8.36) yields

$$\tilde{E}^T \tilde{V} \tilde{A} + \tilde{A}^T \tilde{V} \tilde{E} + \tilde{E}^T \tilde{W} \tilde{E} - \tilde{E}^T \tilde{V} \tilde{B} R^{-1} \tilde{B}^T \tilde{V} \tilde{E} = 0. \tag{8.39}$$

In view of (8.5), (8.37), and (8.38), we have

$$\tilde{E}^T \tilde{V} \tilde{A} = \begin{bmatrix} I_{n_1} & 0 \\ 0 & 0 \end{bmatrix} \begin{bmatrix} V_{11} & V_{12} \\ V_{12}^T & V_{22} \end{bmatrix} \begin{bmatrix} A_1 & 0 \\ 0 & I_{n_2} \end{bmatrix}$$

$$= \begin{bmatrix} V_{11}A_1 & V_{12} \\ 0 & 0 \end{bmatrix},$$

$$\tilde{E}^T \tilde{W} \tilde{E} = \begin{bmatrix} I_{n_1} & 0 \\ 0 & 0 \end{bmatrix} \begin{bmatrix} W_{11} & W_{12} \\ W_{12}^T & W_{22} \end{bmatrix} \begin{bmatrix} I_{n_1} & 0 \\ 0 & 0 \end{bmatrix}$$

$$= \begin{bmatrix} W_{11} & 0 \\ 0 & 0 \end{bmatrix},$$

and

$$\tilde{E}^T \tilde{V} \tilde{B} R^{-1} \tilde{B}^T \tilde{V} \tilde{E}$$

$$= \begin{bmatrix} I_{n_1} & 0 \\ 0 & 0 \end{bmatrix} \begin{bmatrix} V_{11} & V_{12} \\ V_{12}^T & V_{22} \end{bmatrix} \begin{bmatrix} B_1 \\ B_2 \end{bmatrix} R^{-1}$$

$$\times [\, B_1^{\mathsf T} \quad B_2^{\mathsf T} \,] \begin{bmatrix} V_{11} & V_{12} \\ V_{12}^{\mathsf T} & V_{22} \end{bmatrix} \begin{bmatrix} I_{n_1} & 0 \\ 0 & 0 \end{bmatrix}$$

$$= \begin{bmatrix} (V_{11}B_1 + V_{12}B_2)R^{-1}(B_1^{\mathsf T}V_{11} + B_2^{\mathsf T}V_{12}) & 0 \\ 0 & 0 \end{bmatrix}.$$

Therefore, (8.39) is equivalent to

$$V_{11}A_1 + A_1^{\mathsf T}V_{11} + W_{11} - (V_{11}B_1 + V_{12}B_2)R^{-1}(B_1^{\mathsf T}V_{11} + B_2^{\mathsf T}V_{12}) = 0 \quad (8.40)$$

and

$$V_{12} = 0. \tag{8.41}$$

Since the system (8.26) is R-controllable, the matrix pair (A_1, B_1) is controllable. Therefore, it follows from the Riccati equation theory that (8.40) has a unique solution $V_{11} > 0$ for any $W_{11} > 0$. Choose arbitrarily a $V_{22} > 0$, and in view of (8.41), we know that

$$\tilde{V} = \begin{bmatrix} V_{11} & 0 \\ 0 & V_{22} \end{bmatrix} > 0$$

is a solution to the converted Riccati equation (8.39). Therefore,

$$V = Q^{\mathsf T} \begin{bmatrix} V_{11} & 0 \\ 0 & V_{22} \end{bmatrix} Q > 0$$

is a solution to the original Riccati equation (8.36).   □

Let V be a positive definite solution to the Riccati equation (8.36), and define the state feedback controller

$$u = -\frac{1}{2}R^{-1}B^{\mathsf T}VEx + v, \tag{8.42}$$

where v is a new r-dimensional input. Then, with the help of the above lemma we can show that the above controller is a stabilizing controller for system (8.26).

Theorem 8.8. *Let system (8.26) be regular, impulsefree, and R-controllable, and $V > 0$ be a solution to the Riccati equation (8.36) for given $W > 0$ and $R > 0$. Then (8.42) is a state feedback stabilizing controller for system (8.26), and the closed-loop descriptor system*

$$E\dot{x} = \left(A - \frac{1}{2}BR^{-1}B^{\mathsf T}VE\right)x + Bv$$

is also impulse-free.

Proof. Since the system is regular, impulsefree, and R-controllable, according to Lemma 8.6 the Riccati equation (8.36) has a solution $V > 0$ for given $W > 0$ and $R > 0$. Denote

$$A_c = A - \frac{1}{2} BR^{-1} B^T VE.$$

Then by using the Riccati equation (8.36), we have

$$E^T V A_c + A_c^T VE$$

$$= E^T V \left(A - \frac{1}{2} BR^{-1} B^T VE \right) + \left(A - \frac{1}{2} BR^{-1} B^T VE \right)^T VE$$

$$= E^T VA + A^T VE - E^T VBR^{-1} B^T VE$$

$$= -E^T WE.$$

It thus follows from Theorem 3.16 that the system (E, A_c) is stable and impulse-free. $\qquad\square$

Example 8.9. Consider again the system in Example 3.4:

$$\left\{ \begin{array}{l} \begin{bmatrix} 1 & 0 & 0 & 0 \\ 0 & 0 & 1 & 0 \\ 0 & 0 & 0 & 0 \\ 0 & 0 & 0 & 0 \end{bmatrix} \dot{x} = \begin{bmatrix} 0 & 1 & 0 & 0 \\ 1 & 0 & 0 & 0 \\ -1 & 0 & 0 & 1 \\ 0 & 1 & 1 & 1 \end{bmatrix} x + \begin{bmatrix} 0 \\ 0 \\ 0 \\ -1 \end{bmatrix} u. \\ y = \begin{bmatrix} 0 & 0 & 1 & 0 \end{bmatrix} x \end{array} \right.$$

It is seen in Example 3.4 that with the following transformation matrices

$$Q = \begin{bmatrix} 1 & 0 & 1 & -1 \\ 0 & 1 & 0 & 0 \\ 0 & 0 & -1 & 1 \\ 0 & 0 & 1 & 0 \end{bmatrix} \quad \text{and} \quad P = \begin{bmatrix} 1 & 0 & 0 & 0 \\ -1 & -1 & 1 & 0 \\ 0 & 1 & 0 & 0 \\ 1 & 0 & 0 & 1 \end{bmatrix},$$

and the state transformation

$$P^{-1} x = \begin{bmatrix} x_1 \\ x_2 \end{bmatrix}, x_1 \in \mathbb{R}^2, x_2 \in \mathbb{R}^2,$$

the standard decomposition form for this system can then be obtained as

$$\left\{ \begin{array}{l} \dot{x}_1 = \begin{bmatrix} -1 & -1 \\ 1 & 0 \end{bmatrix} x_1 + \begin{bmatrix} 1 \\ 0 \end{bmatrix} u \\ 0 = x_2 + \begin{bmatrix} -1 \\ 0 \end{bmatrix} u \\ y = \begin{bmatrix} 0 & 1 \end{bmatrix} x_1 + \begin{bmatrix} 0 & 0 \end{bmatrix} x_2 \end{array} \right.$$

So

$$A_1 = \begin{bmatrix} -1 & -1 \\ 1 & 0 \end{bmatrix}, B_1 = \begin{bmatrix} 1 \\ 0 \end{bmatrix}.$$

If we choose

$$W_{11} = \begin{bmatrix} 1 & -1 \\ -1 & 0 \end{bmatrix} \text{ and } R = -1,$$

the Riccati equation

$$V_{11}A_1 + A_1^T V_{11} + W_{11} - V_{11}B_1 R^{-1} B_1^T V_{11} = 0$$

has a unique solution

$$V_{11} = \begin{bmatrix} 1 & 0 \\ 0 & 2 \end{bmatrix}.$$

The matrix $V_{22} = I$ clearly satisfies $V_{22} > 0$, therefore

$$V = Q^T \begin{bmatrix} V_{11} & 0 \\ 0 & V_{22} \end{bmatrix} Q = \begin{bmatrix} 1 & 0 & 1 & -1 \\ 0 & 2 & 0 & 0 \\ 1 & 0 & 3 & -2 \\ -1 & 0 & -2 & 2 \end{bmatrix} > 0$$

is a solution to the original Riccati equation (8.36). Thus, the stabilizing feedback gain is

$$K = -\frac{1}{2}R^{-1}B^T V E = \begin{bmatrix} \frac{1}{2} & 0 & 0 & 0 \end{bmatrix},$$

and the closed-loop matrix A_c can be obtained as

$$A_c = A - \frac{1}{2}BR^{-1}B^T V E = \begin{bmatrix} 0 & 1 & 0 & 0 \\ 1 & 0 & 0 & 0 \\ -1 & 0 & 0 & 1 \\ -\frac{1}{2} & 1 & 1 & 1 \end{bmatrix}.$$

8.5 Notes and References

This chapter investigated the problems of pole assignment in descriptor linear systems and stabilization of descriptor linear systems by state feedback.

Pole Assignment

The problem of finite pole assignment in a linear system in the typical form by a certain type of feedback controllers is explored. Regarding conditions and the capacity of the feedback controllers, we have the following conclusions:

- When the system is R-controllable, there exists a state feedback controller such that the set of open-loop finite poles can be replaced in the closed-loop by $\deg \det (sE - A)$ arbitrarily prescribed self-conjugate complex numbers. Solutions of such a controller can be obtained based on the standard decomposition, and the problem is finally converted into a pole assignment problem for the normal system, or the slow subsystem (A_1, B_1).
- When the system is S-controllable, there exists a state feedback such that the closed-loop system has $\mathrm{rank}\, E$ arbitrarily prescribed finite eigenvalues. Solutions of such a controller can be obtained following two steps. The first step is to realize impulse elimination in the system using a state feedback, consequently, we obtain a system with $\mathrm{rank}\, E$ finite poles. The second step is to replace these $\mathrm{rank}\, E$ finite poles with desired ones using the method for pole assignment under R-controllability.
- When the system is C-controllable, there exists a state P-D feedback such that the closed-loop system has n arbitrarily prescribed finite eigenvalues. Solution of such controllers can obviously be derived following two steps. The first step is to normalize the system using a state derivative feedback controller, the second step is to realize pole assignment in the normalized system.

There are many papers on the pole placement problem for descriptor systems. For references on this problem, one may refer to AI-Nasr et al. (1983b), Armentano (1984), Bajic (1986), Bender (1985), Campbell (1982), Christodoulou and Paraskevopoulos (1984), Cobb (1981), Cullen (1986b), Kautsky and Nichols (1986), Lovass-Nagy et al. (1986a), Mukundan and Dayawansa (1983), Pandolfi (1980), Wang and Dai (1987b), Zhou et al. (1987), and Wang (2000). Papers on pole structure placement include Van Dooren (1981), Fletcher et al. (1986), Lewis and Ozcaldiran (1985), and Ozcaldiran and Lewis (1987). Particularly, Wang et al. (2004) considered the problem of infinite eigenvalue assignment in descriptor systems via state variable feedback, which finds a feedback gain such that the close-loop descriptor system possesses only infinite eigenvalues.

Pole assignment in descriptor linear systems via output feedback is not considered in this book. Readers who are interested in this topic may refer to AI-Nasr et al. (1983a), Duan (1995, 1999), and Duan and Lam (2000).

Stabilization

For simplicity, in this chapter only stabilization by state feedback is studied. The problem is to find a state feedback controller for a given descriptor linear system in the typical form such that the closed-loop system is stable. When such a problem has a solution, the system is called stabilizable. It is shown in this chapter that a direct necessary and sufficient condition for stabilizability is

$$\mathrm{rank}\, [sE - A \quad B] = n, \quad \forall s \in \bar{\mathbb{C}}^+, s \text{ finite.}$$

Such a condition indicates that stabilization is a property only concerned with the slow subsystem since this condition can be easily shown to be equivalent to

$$\text{rank}\,[sI - A_1 \quad B_1] = n_1, \quad \forall s \in \bar{\mathbb{C}}^+, s \text{ finite.}$$

Regarding solutions of the stabilizing state feedback controllers, three methods are presented. The first two methods are based on the standard decomposition form and the controllability form of the system, respectively, while the third one is based on Lyapunov stability theory.

Detectability is also introduced in this chapter, which is the dual concept of stabilizability. We will see in Chap. 11 that detectability ensures the existence of a full-order state observer for the system.

For simplicity, in this chapter we have only considered the very basic stabilization problems, while in the literature there are many complicated ones. The problem of robust stability and stabilization of uncertain continuous descriptor systems has been dealt with by Chaabane et al. (2006). The stabilization problem of the class of continuous-time singular linear systems with time-delay in the state vector using a state feedback controller was considered in Boukas (2007), while the problem of robust stabilization for uncertain continuous descriptor system with both state and control delays has been investigated in Piao et al. (2006). Furthermore, global asymptotic stabilization for a class of delayed bilinear descriptor systems has been studied by Lu and Ho (2006b), and a BMI approach for admissibly stabilization in a linear descriptor linear system has been proposed by Huang and Huang (2005).

Robust Pole Assignment

It is clearly seen from the pole assignment results provided in this chapter that certain degrees of freedom exist in pole assignment for multivariable linear systems. More specifically, the state feedback gain, which assigns a set of closed-loop eigenvalues, is generally nonunique. The degrees of freedom in a pole assignment design can be further utilized to achieve additional system performances. Such an idea gives rise to many design problems, and one of them is the so-called robust pole assignment problem, which aims at selecting the design freedom such that the closed-loop eigenvalues are as insensitive as possible to perturbations in the components of the closed-loop system coefficient matrices. As is well known, the stability and the transient response of a linear system are mainly determined by the eigenvalues, or poles, of the system. For a linear system with certain eigenvalues very sensitive to perturbations in certain elements of the system coefficient matrices, a very small perturbation in those elements of the system coefficient matrices may significantly affect the dynamical property of the system, or even make the system unstable. Therefore, the stability and the dynamical property of a linear system with smaller eigenvalue sensitivities are more robust in the sense that they are less affected by perturbations in the system coefficient matrices. Due to such an importance, the problem of robust pole assignment has been intensively studied in the last three decades. Most of the results are obtained for the case of normal linear systems (for e.g. Kautsky et al. 1985; Sun 1987; Kautsky and Nichols 1990; Owens and O'Reilly 1989; Duan et al. 1992; Duan 1993a; Lam et al. 1997; Liu and Patton 1998; Lam and Tam 2000). For the case of descriptor systems, there are also some reported results (Kautsky et al. 1989; Kautsky and Nichols 1986; Syrmos and Lewis 1992; Duan and Patton 1999; Duan and Lam 2000, 2004; Duan et al. 2001b, 2002b; Duan and Wu 2005e).

Chapter 9
Eigenstructure Assignment

This chapter studies the problem of eigenstructure assignment in descriptor linear systems by state feedback. The complete parametric approach proposed in Duan (1998a) for eigenstructure assignment in S-controllable descriptor linear systems via static state feedback is introduced. General complete parametric expressions in direct closed forms for the right closed-loop eigenvectors associated with the finite closed-loop eigenvalues, and two simple complete parametric solutions for the feedback gain matrix K are established. The approach possesses the following features:

- It arbitrarily assigns rank$E$ finite closed-loop eigenvalues with arbitrary given algebraic and geometric multiplicities, and hence realizes elimination of all possible initial time impulsive responses.
- It is very simple and convenient, and possesses good numerical reliability.
- It does not require the open-loop system to be regular, but guarantees the closed-loop regularity.
- It does not impose any conditions on the closed-loop finite eigenvalues, such as the distinctness of the finite closed-loop eigenvalues, and the distinctness of the finite closed-loop eigenvalues from the open-loop ones.
- It allows the closed-loop eigenvalues to be zeros, and thus suits both continuous-time and discrete-time descriptor linear systems.
- It provides all the degrees of design freedom, which are represented by a group of parameter vectors and a parameter matrix.
- It allows the closed-loop eigenvalues to be also set undetermined and utilized as a part of degrees of freedom.
- When the closed-loop eigenvalues are previously prescribed, the approach adopts mainly a series of singular value decompositions and is therefore numerically very simple and reliable.

Based on this approach, general parametric expression for the closed-loop left eigenvector matrix associated with the finite closed-loop eigenvalues are also established, and it is shown that the closed-loop system can be transformed equivalently into a canonical form, which is clearly composed of a dynamical part and a

G.-R. Duan, *Analysis and Design of Descriptor Linear Systems*, Advances
in Mechanics and Mathematics 23, DOI 10.1007/978-1-4419-6397-0_9,
© Springer Science+Business Media, LLC 2010

non-dynamical part. With the help of this canonical form, general analytical parametric expression of the closed-loop system is presented in terms of the closed-loop eigenvector matrices and the closed-loop Jordan matrix. The "Method of Companion" is introduced from the computational point of view to cope with inverses of all the matrices involved.

In Sect. 9.1, the problem of eigenstructure assignment in descriptor linear systems via state feedback control is proposed, and is linked to a type of generalized Sylvester matrix equations. As a consequence, the problem is divided into two separate subproblems, respectively, involving general parametric solutions of the closed-loop eigenvectors and the state feedback gain. The general complete parametric solution to the problem of eigenstructure assignment is provided in Sect. 9.2. In Sect. 9.3, general parametric solutions for the closed-loop left eigenvector matrix associated with the finite closed-loop eigenvalues are further established. Section 9.4 presents the canonical form and the general parametric response solution for the closed-loop system. An illustrative example is given in Sect. 9.5. Some notes and comments follow in Sect. 9.6.

Please note that the general, complete parametric solution introduced in Appendix C to the type of generalized Sylvester matrix equations involved in the eigenstructure assignment performs a fundamental role in the development adopted in Sect. 9.2.

9.1 The Problem

Consider the following time-invariant linear descriptor system

$$E\dot{x} = Ax + Bu, \tag{9.1}$$

where $x \in \mathbb{R}^n$ and $u \in \mathbb{R}^r$ are, respectively, the descriptor-variable vector and the input vector; $A, E \in \mathbb{R}^{n \times n}$, and $B \in \mathbb{R}^{n \times r}$ are known matrices with $\mathrm{rank}\, E = n_0 \leq n$, $\mathrm{rank}\, B = r$, and satisfy the following controllability assumption.

Assumption 9.1. System (9.1) is S-controllable.

It follows from the definition of S-controllability that the system (9.1) is both R-controllable and I-controllable. Due to the R-controllability of the system, we have

$$\mathrm{rank}[sE - A \quad B] = n \quad \text{for all} \quad s \in \mathbb{C}.$$

Because of the I-controllability of the system, arbitrary $\mathrm{rank}\, E$ finite eigenvalues can be assigned to the closed-loop system by state feedback, in other words, there exists a state feedback for the system such that the closed-loop system is impulse-free.

9.1.1 The Problem

When the following state feedback control law

$$u = Kx + u_e, \ K \in \mathbb{R}^{r \times n} \tag{9.2}$$

is applied to system (9.1), the closed-loop system is obtained in the following form:

$$E\dot{x} = A_c x + B u_e, \quad A_c = A + BK, \tag{9.3}$$

where u_e is the external signal.

Let

$$\Gamma = \left\{ s_i \mid s_i \in \mathbb{C}, \ i = 1, 2, \ldots, n_0', \ 1 \le n_0' \le n_0 \right\}, \tag{9.4}$$

which is symmetric about the real axis, be the set of relative finite eigenvalues of the matrix pair (E, A_c), and denote the algebraic and geometric multiplicities of s_i by m_i and q_i, respectively. Then in the Jordan form J determined by the relative finite eigenvalues of the matrix pair (E, A_c), there are q_i Jordan blocks associated with s_i. Denote the orders of the q_i number Jordan blocks associated with s_i by p_{ij}, $j = 1, 2, \ldots, q_i$, so that the following relation holds:

$$p_{i1} + p_{i2} + \cdots + p_{iq_i} = m_i. \tag{9.5}$$

In order to eliminate a possible impulsive response, it is necessary to require

$$m_1 + m_2 + \cdots + m_{n_0'} = n_0. \tag{9.6}$$

Note that this is feasible because the open-loop system (9.1) is I-controllable. Furthermore, denote the relative eigenvector chains associated with eigenvalue s_i by $v_{ij}^k \in \mathbb{C}^n$, $k = 1, 2, \ldots, p_{ij}$, $j = 1, 2, \ldots, q_i$. We then have the following equations by definition:

$$(A + BK - s_i E) v_{ij}^k = E v_{ij}^{k-1}, \quad v_{ij}^0 = 0, \tag{9.7}$$

$$k = 1, 2, \ldots, p_{ij}, \ j = 1, 2, \ldots, q_i, \ i = 1, 2, \ldots, n_0'.$$

Let the infinite eigenvalue of the closed-loop system be denoted by s_∞. Then, due to (9.6), s_∞ is a multiple eigenvalue with both geometric and algebraic multiplicities being equal to $n - n_0$. The $n - n_0$ number eigenvectors associated with s_∞, denoted by $v_{\infty j}$, $j = 1, 2, \ldots, n - n_0$, are of the first grade and is a group of linearly independent vectors simply defined by:

$$E v_{\infty j} = 0, \ j = 1, 2, \ldots, n - n_0. \tag{9.8}$$

With the above notations, the problem of eigenstructure assignment via the state feedback controller (9.2) for the linear descriptor system (9.1) can be stated as follows.

Problem 9.1. Given

- the system (9.1) satisfying Assumption 9.1;
- the closed-loop finite eigenvalue set Γ defined as in (9.4); and
- integers p_{ij}, m_i, q_i, $j = 1, 2, \ldots, q_i$, $i = 1, 2, \ldots, n_0'$, satisfying (9.5) and (9.6),

determine the general parametric forms for a real matrix $K \in \mathbb{R}^{r \times n}$, and a group of vectors $v_{ij}^k \in \mathbb{C}^n$, $k = 1, 2, \ldots, p_{ij}$, $j = 1, 2, \ldots, q_i$, $i = 1, 2, \ldots, n_0'$, such that the following three requirements are simultaneously met:

1. All the equations in (9.7) hold.
2. The vectors v_{ij}^k, $k = 1, 2, \ldots, p_{ij}$, $j = 1, 2, \ldots, q_i$, $i = 1, 2, \ldots, n_0'$, and $v_{\infty j}$, $j = 1, 2, \ldots, n - n_0$, are linearly independent.
3. The closed-loop system is regular, that is, $\det(sE - A - BK)$ is not identically zero.

As a consequence of the first two requirements in the above Problem 9.1, the resulting closed-loop system (9.3) inherently possesses

$$d = \deg \det(sE - A_c) = \operatorname{rank} E = n_0 \qquad (9.9)$$

finite assigned eigenvalues and $(n - n_0)$ infinite eigenvalues, or a multiple eigenvalue at infinity with both geometric and algebraic multiplicities being equal to $n - n_0$, associated with a nondynamical response. It is clear that these are, respectively, the maximum number of finite eigenvalues and the minimum number of infinite eigenvalues, which can be assigned to the closed-loop system. Due to (9.9), elimination of possible initial time impulsive response is achieved. The third requirement in Problem 9.1 on the closed-loop regularity is necessary since it ensures uniqueness of solutions of descriptor linear systems.

9.1.2 Interpretations of Requirements

In this subsection, we further give some interpretations of the three requirements in Problem 9.1.

The First Requirement

First, let us consider the equations in (9.7). It will be seen that they can be converted into the type of Sylvester matrix equations studied in Appendix C.

It follows from the description in Sect. 9.1.1 that the Jordan matrix J, determined by the relative finite closed-loop eigenvalues s_i, $i = 1, 2, \ldots, n_0'$, is in the following form

$$
\begin{cases}
J = \mathrm{diag}\left(J_1, J_2, \ldots, J_{n_0'}\right) \\[4pt]
J_i = \mathrm{diag}\left(J_{i1}, J_{i2}, \ldots, J_{iq_i}\right) \\[4pt]
J_{ij} = \begin{bmatrix} s_i & 1 & & \\ & s_i & \ddots & \\ & & \ddots & 1 \\ & & & s_i \end{bmatrix}_{(p_{ij} \times p_{ij})}
\end{cases}
$$

For convenience, we make the following three conventions used in this chapter.

Cv1. Integers k, j, i given by $k=1, 2, \ldots, p_{ij}$, $j=1, 2, \ldots, q_i$, $i = 1, 2, \ldots, n_0'$, are simply denoted by $(k, j, i) \in \pi$.

Cv2. Any group of vectors x_{ij}^k, $(k, j, i) \in \pi$, is simply denoted by $\{x_{ij}^k\}$.

Cv3. Any group of vectors $\{x_{ij}^k\}$ uniquely corresponds to a matrix $X = [x_{ij}^k]$ in the following manner:

$$
\begin{cases}
X = [X_1 \quad X_2 \quad \cdots \quad X_{n_0'}] \\
X_i = [X_{i1} \quad X_{i2} \quad \cdots \quad X_{iq_i}] \\
X_{ij} = [x_{ij}^1 \quad x_{ij}^2 \quad \cdots \quad x_{ij}^{p_{ij}}]
\end{cases}
$$

Based on the above conventions and the format of the above Jordan matrix J, the following simple fact, concerning the first requirement in Problem 9.1 and the real property of the feedback gain K, can be easily verified.

Lemma 9.1. *Let Assumption 9.1 be valid. Then the following hold:*

1. *All the equations in (9.7) can be written equivalently in the following compact matrix form:*

$$
AV + BKV = EVJ, \tag{9.10}
$$

which can be converted into the following generalized Sylvester matrix equation

$$
AV + BW = EVJ \tag{9.11}
$$

by letting

$$
W = KV. \tag{9.12}
$$

2. *The matrix K satisfying (9.12) is real if and only if*

$$
v_{ij}^k = \bar{v}_{lj}^k, \quad w_{ij}^k = \bar{w}_{lj}^k \quad \text{when} \quad s_i = \bar{s}_l. \tag{9.13}
$$

In view of the above lemma, the first requirement in Problem 9.1 is equivalent to finding three matrices W, V, and K, of proper dimensions, satisfying the generalized Sylvester matrix equation (9.11), (9.12), and condition (9.13).

The Second Requirement

Let

$$V_\infty = \begin{bmatrix} v_{\infty 1} & v_{\infty 2} & \cdots & v_{\infty,n-n_0} \end{bmatrix}, \tag{9.14}$$

then V_∞ is actually the part of right closed-loop eigenvector matrix associated with the infinite eigenvalue s_∞, which satisfies

$$E V_\infty = 0, \quad \mathrm{rank} V_\infty = n - n_0. \tag{9.15}$$

Obviously, the second requirement is equivalent to

$$\det \begin{bmatrix} V & V_\infty \end{bmatrix} \neq 0. \tag{9.16}$$

Regarding the second requirement in Problem 9.1, we have the following lemma.

Lemma 9.2. *Let E, A, B, J be given matrices, and V and W be two matrices satisfying the generalized Sylvester matrix equation (9.11). Then the second requirement in Problem 9.1 is met if and only if*

$$\mathrm{rank}(EV) = n_0. \tag{9.17}$$

Proof. It suffices to show that conditions (9.17) and (9.16) are equivalent.
    (9.17)$\Longrightarrow$(9.16): Noting that

$$\mathrm{rank}(EV) \leq \max\{\mathrm{rank} E, \mathrm{rank} V\},$$

we have

$$\mathrm{rank}(EV) = \mathrm{rank} V = n_0.$$

This implies that V spans Image E, the image of the matrix E. Furthermore, noting that V_∞ spans $\ker E$, and

$$\ker E \cap \mathrm{Image}\ E = \{0\},$$

we clearly have (9.16).
    (9.16)$\Longrightarrow$(9.17): When (9.16) holds, we have $\mathrm{rank} V = n_0$. Since

$$E \begin{bmatrix} V & V_\infty \end{bmatrix} = \begin{bmatrix} EV & 0 \end{bmatrix},$$

there holds, in view of (9.16),

$$\mathrm{rank}(EV) = \mathrm{rank} E = n_0.$$

The proof is then completed. $\square$

The Third Requirement

Define the part of left closed-loop eigenvector matrix $T_\infty \in \mathbb{R}^{n \times (n-n_0)}$ associated with the infinite eigenvalue s_∞ as follows:

$$T_\infty^T E = 0, \quad \text{rank} T_\infty = n - n_0, \tag{9.18}$$

then the following lemma about a sufficient and necessary condition for closed-loop regularity can be proven.

Lemma 9.3. *Let the matrices E, A, B, J be as prescribed previously, and matrices V_∞ and T_∞ be defined by (9.15) and (9.18), respectively. Furthermore, let V and W be two matrices satisfying the generalized Sylvester matrix equation (9.11) and condition (9.17). Then the matrix K satisfying (9.12) makes the matrix pencil $[sE - (A + BK)]$ regular if and only if*

$$\det(T_\infty^T (A + BK) V_\infty) \neq 0. \tag{9.19}$$

Proof. In view of (9.18), we know that there exists a matrix $T \in \mathbb{C}^{n \times n_0}$ that makes the matrix

$$\tilde{T} = [T \ T_\infty] \tag{9.20}$$

nonsingular. For this matrix T we have, using (9.18),

$$\tilde{T}^T E V = \begin{bmatrix} T^T E V \\ T_\infty^T E V \end{bmatrix} = \begin{bmatrix} T^T E V \\ 0 \end{bmatrix}.$$

Since $\tilde{T}$ is nonsingular, and the matrix EV is of full-column rank according to given conditions, we have from the above relation that the matrix $T^T E V$ is also nonsingular.

Again using the full-column rank property of matrix EV, and Lemma 9.2, we know that the compound matrix

$$\tilde{V} = [V \ V_\infty] \tag{9.21}$$

is also nonsingular.

Now let us consider the following matrix

$$\tilde{T}^T (sE - A - BK) \tilde{V} = \begin{bmatrix} T^T \\ T_\infty^T \end{bmatrix} (sE - A - BK) [V \ V_\infty].$$

Using (9.11) and (9.12) as well as relations (9.15) and (9.18), we have

$$\tilde{T}^T (sE - A - BK) \tilde{V}$$

$$= s \begin{bmatrix} T^T \\ T_\infty^T \end{bmatrix} E [V \ V_\infty] - \begin{bmatrix} T^T \\ T_\infty^T \end{bmatrix} (A + BK) [V \ V_\infty]$$

$$= s \begin{bmatrix} T^{\mathrm{T}} E V & 0 \\ 0 & 0 \end{bmatrix} - \begin{bmatrix} T^{\mathrm{T}}(AV + BW) & * \\ T_{\infty}^{\mathrm{T}}(AV + BW) & T_{\infty}^{\mathrm{T}}(A + BK)V_{\infty} \end{bmatrix}$$

$$= s \begin{bmatrix} T^{\mathrm{T}} E V & 0 \\ 0 & 0 \end{bmatrix} - \begin{bmatrix} T^{\mathrm{T}} E V J & * \\ T_{\infty}^{\mathrm{T}} E V J & T_{\infty}^{\mathrm{T}}(A + BK)V_{\infty} \end{bmatrix}$$

$$= s \begin{bmatrix} T^{\mathrm{T}} E V & 0 \\ 0 & 0 \end{bmatrix} - \begin{bmatrix} T^{\mathrm{T}} E V J & * \\ 0 & T_{\infty}^{\mathrm{T}}(A + BK)V_{\infty} \end{bmatrix}$$

$$= \begin{bmatrix} T^{\mathrm{T}} E V(sI - J) & * \\ 0 & -T_{\infty}^{\mathrm{T}}(A + BK)V_{\infty} \end{bmatrix}.$$

Therefore, we have

$$\det \tilde{T} \det(sE - A - BK) \det \tilde{V}$$
$$= \det(T^{\mathrm{T}} E V) \det(sI - J) \det(-T_{\infty}^{\mathrm{T}}(A + BK)V_{\infty}). \qquad (9.22)$$

Since $\det \tilde{T} \neq 0$, $\det \tilde{V} \neq 0$, and $\det(T^{\mathrm{T}} E V) \neq 0$ as shown above, $\det(sI - J)$ is not identically zero, it follows from the above (9.22) that $\det(sE - A - BK)$ is not identically zero if and only if condition (9.19) holds.   □

9.1.3 Problem Decomposition

In the above, we have given interpretations of the three requirements in Problem 9.1. Lemma 9.1 tells us that there exist a real matrix K and a group of vectors $v_{ij}^{k} \in \mathbb{C}^{n}$, $(k,\ j,\ i) \in \pi$, satisfying all the equations in (9.7) if and only if there exist a pair of matrices W and V and a matrix K satisfying the generalized Sylvester matrix equation (9.11), (9.12), and condition (9.13). Lemma 9.2 tells us that the second requirement holds if and only if the matrix EV has full-column rank. Lemma 9.3 states that the third requirement in Problem 9.1 is equivalent to condition (9.19). Based on these three lemmas, we can convert Problem 9.1 proposed in Sect. 9.1.1 equivalently into the following two subproblems.

Problem 9.2. [Solution of closed-loop eigenvectors] Given the descriptor linear system (9.1) satisfying Assumption 9.1, and the matrix J as described previously, find the general complete parametric expressions for the matrices V and W, which satisfy the generalized Sylvester matrix equation (9.11) and conditions (9.13) and (9.17).

Problem 9.3. [Solution of state feedback gain] Let the same conditions in Problem 9.2 hold, and matrices V_{∞} and T_{∞} be defined by (9.15) and (9.18), respectively. Furthermore, let V and W be the solutions to Problem 9.2. Find the general parametric form for the matrix K satisfying (9.12) and condition (9.19).

9.2 The Parametric Solution

It follows from the preceding section that the Problem 9.1 can be solved by solving the two separate Problems 9.2 and 9.3.

9.2.1 Solution of Closed-Loop Eigenvectors

Our objective in this subsection is to solve Problem 9.2. The development follows this natural and effective idea: first find the general complete parametric solutions V and W to the generalized Sylvester matrix equation (9.11), and then specify the degrees of freedom in the general solutions so as to let conditions (9.13) and (9.17) be satisfied.

General Solution to Equation (9.11)

By applying the direct, closed, complete parametric solution given in Theorem C.1 to the generalized Sylvester matrix equation (9.11) , under Assumption 9.1, the general complete parametric expressions for the closed-loop eigenvectors associated with the finite closed-loop eigenvalues, together with the corresponding vectors $\left\{w_{ij}^k\right\}$, are obtained as

$$\begin{bmatrix} v_{ij}^k \\ w_{ij}^k \end{bmatrix} = \begin{bmatrix} N(s_i) \\ D(s_i) \end{bmatrix} f_{ij}^k + \cdots + \frac{1}{(k-1)!} \frac{d^{k-1}}{ds^{k-1}} \begin{bmatrix} N(s_i) \\ D(s_i) \end{bmatrix} f_{ij}^1, \qquad (9.23)$$

$$(k,\ j,\ i) \in \pi,$$

where $N(s) \in \mathbb{R}^{n\times r}[s]$ and $D(s) \in \mathbb{R}^{r\times r}[s]$ are any pair of right coprime polynomial matrices satisfying

$$(A - sE)N(s) + BD(s) = 0, \qquad (9.24)$$

and $f_{ij}^k \in \mathbb{C}^r$, $(k,\ j,\ i) \in \pi$, is a group of arbitrarily chosen free parameter vectors, which represent the degree of freedom existing in the general solution to this equation.

Remark 9.1. In the special case that the Jordan matrix J is diagonal, we may, for simplicity, count the relative finite eigenvalues of the pair of matrices $(E,\ A_c)$ by multiplicities, in this case, the vectors f_{ij}^k, v_{ij}^k and w_{ij}^k can be denoted simply by f_i, v_i and w_i, respectively, and our solution (9.23) to the generalized Sylvester matrix equation in (9.11) becomes

$$\begin{bmatrix} v_i \\ w_i \end{bmatrix} = \begin{bmatrix} N(s_i) \\ D(s_i) \end{bmatrix} f_i, \quad i = 1, 2, \ldots, n_0. \qquad (9.25)$$

Remark 9.2. In contrast to Duan (1992b), the above general parametric expression (9.23) or (9.25) for the closed-loop eigenvectors associated with the finite closed-loop eigenvalues is in a direct closed explicit parametric form, and is thus simpler and more convenient to use. They can be immediately written out as soon as the pair of right coprime polynomial matrices $N(s)$ and $D(s)$ satisfying (9.24) are obtained. For solutions of such $N(s)$ and $D(s)$, please refer to Sect. C.3 in Appendix C (see also Duan 1992b, 1993b, 1996b, 1998a,b, 1999, 2000b, 2001, 2002; Duan and Patton 1998a ; Duan et al. 1999b).

Remark 9.3. Besides the above right coprime factorization method, another method for solving the generalized Sylvester matrix equation (9.11) using singular value decomposition, provided in Appendix C, can also be applied.

Conditions (9.13) and (9.17)

To obtain the solution to Problem 9.2, now let us further consider the conditions (9.13) and (9.17).

Condition (9.13), which ensures that the matrix K is real, can be equivalently turned into the following constraint on the group of parameter vectors $\{f_{ij}^k\}$ in view of the forms of our general solutions (9.23).

Constraint C1 $f_{ij}^k = \bar{f}_{lj}^k$ if and only if $s_i = \bar{s}_l$.

If we define the vectors

$$\hat{v}_{ij}^k = E' N(s_i) f_{ij}^k + \cdots + \frac{1}{(k-1)!} \frac{d^{k-1}}{ds^{k-1}} E' N(s_i) f_{ij}^1, \qquad (9.26)$$

$$(k, \ j, \ i) \in \pi,$$

with $E' \in \mathbb{R}^{n_0 \times n}$ defined by

$$TE = \begin{bmatrix} E' \\ 0 \end{bmatrix}, \quad \text{for some } T \in \mathbb{R}^{n \times n} \text{ and } \det T \neq 0, \qquad (9.27)$$

then condition (9.17) can be converted equivalently into the following constraint also on the group of parameter vectors $\{f_{ij}^k\}$ as follows.

Constraint C2 $\det \hat{V} \left(f_{ij}^k \right) \neq 0$.

Summing up the above two aspects gives the following conclusion concerning solution to Problem 9.2.

Theorem 9.1. *The solution to Problem 9.2, that is, the group of the closed-loop eigenvectors $\{v_{ij}^k\}$ associated with the closed-loop finite eigenvalues s_i, $i = 1, 2, \ldots, n_0'$, which ensures the existence of a real gain matrix K that gives closed-loop regularity, is given by (9.23) with the parameters $\{f_{ij}^k\}$ satisfying Constraints C1 and C2.*

In the special case that the Jordan matrix J is diagonal, in view of Remark 9.1, Constraint C2 can be simply expressed in the group of parameters f_i, $i = 1, 2, \ldots, n_0$, as follows.

Constraint C′2 $\det \left[E' N(s_1) f_1 \ E' N(s_2) f_2 \cdots E' N(s_{n_0}) f_{n_0} \right] \neq 0$.

9.2.2 Solution of the Gain Matrix K

Now suppose that the complete parametric forms of the matrices V and W are already obtained through formula (9.23) with the free parameters $\{f_{ij}^k\}$ chosen so that Constraints C1 and C2 are satisfied. In order to solve the matrix K satisfying (9.12) we introduce an auxiliary matrix equation in the following form

$$W_\infty = K V_\infty, \tag{9.28}$$

where W_∞ is a real matrix of dimension $r \times (n - n_0)$. Using this introduced auxiliary equation, we have

$$T_\infty^T (A + BK) V_\infty = T_\infty^T A V_\infty + T_\infty^T B W_\infty. \tag{9.29}$$

Therefore, condition (9.19), which ensures the regularity of the closed-loop system, can be turned into the following constraint on only the parameter matrix $W_\infty \in \mathbb{R}^{r \times (n - n_0)}$ as follows.

Constraint C3 $\det (T_\infty^T A V_\infty + T_\infty^T B W_\infty) \neq 0$.

For the general parametric solution of the feedback gain K, we have two expressions.

The First General Expression for K

By combining (9.12) with (9.28), we have

$$[W \ W_\infty] = K [V \ V_\infty]. \tag{9.30}$$

Due to Constraint C2 and Lemma 9.2, we know that the matrix $\tilde{V} = [V \ V_\infty]$ is nonsingular. Therefore, the general complete parametric solution to the gain matrix K can be given as follows

$$K = [W \ W_\infty][V \ V_\infty]^{-1}, \tag{9.31}$$

where $W_\infty \in \mathbb{R}^{r \times (n - n_0)}$ is a free parameter matrix which represents an extra degree of the design freedom existing in the solution of (9.12) other than the group of parameters $\{f_{ij}^k\}$.

Remark 9.4. Duan (1992b) and Fahmy and O'Reilly (1989) have both considered the solution of the gain matrix K directly from (9.12). Due to the nonsquareness of matrix V, their solutions appear to be much more complicated, and need prior specifications of some or all of the parameters contained in matrix V. In the above we have, by introducing the auxiliary equation (9.28), obtained an extended (9.30) with the composite matrix $[V \ V_\infty]$ being square and nonsingular, and thus derive the simple expression (9.31) for the feedback gain K. We call such a procedure "completing the square" (Duan 1998a, 1999).

The Second General Expression for K

Without loss of generality, let us further require that the columns of the matrix V_∞ are orthogonal to each other. Let $Q \in \mathbb{R}^{n \times n_0}$ be such a matrix that its columns form a set of normalized orthogonal base for $\mathscr{E}_r = \text{span}\{\text{rows of } E\}$, then from $EV_\infty = 0$, we know that

$$\tilde{Q} = [Q \ V_\infty] \tag{9.32}$$

forms an orthogonal matrix, and $Q^T V_\infty = 0$. Thus, we have

$$\tilde{Q}^T \tilde{V} = \begin{bmatrix} Q^T \\ V_\infty^T \end{bmatrix} [V \ V_\infty] = \begin{bmatrix} Q^T V & 0 \\ V_\infty^T V & I_{n_0} \end{bmatrix}.$$

Using the orthogonal property of the matrix $\tilde{Q}$, we have from the above relation

$$\tilde{V} = \tilde{Q} \begin{bmatrix} Q^T V & 0 \\ V_\infty^T V & I_{n_0} \end{bmatrix}. \tag{9.33}$$

Since $\tilde{V}$ is nonsingular, $\tilde{Q}$ is orthogonal, it follows from the above (9.33) that the matrix $Q^T V$ is nonsingular, and by applying the Matrix Inversion Lemma, Theorem A.2, we have

$$\tilde{V}^{-1} = \begin{bmatrix} (Q^T V)^{-1} & 0 \\ -V_\infty^T V(Q^T V)^{-1} & I_{n_0} \end{bmatrix} \begin{bmatrix} Q^T \\ V_\infty^T \end{bmatrix}, \tag{9.34}$$

which gives

$$\tilde{V}^{-1} = \begin{bmatrix} G \\ V_\infty^T (I - VG) \end{bmatrix}, \tag{9.35}$$

with G given by

$$G = (Q^T V)^{-1} Q^T. \tag{9.36}$$

Finally, using (9.31) and (9.35) we obtain the expression for the feedback gain K as follows:

$$K = WG + W_\infty V_\infty^T (I - VG), \qquad (9.37)$$

where the matrix G is given by (9.36), and the matrix Q is an arbitrary matrix whose columns form a set of normalized orthogonal bases for $\mathscr{E}_r$, which is the space spanned by the rows of the matrix E.

It is clear to see that the matrix Q in the second expression of the feedback gain K has nothing to do in nature with solution of K. It is introduced only for the computational purpose, that is, to realize solution of the inversion of the matrix $\tilde{V}$ by solving the inversion of some matrix with lower order, and to give some degrees of freedom to improve numerical stability. Due to this point, we call this matrix Q, or the whole matrix $\tilde{Q}$, the computational companion matrix associated with the solution process of K, and we call this method of solving inverses of matrices with high dimensions by introducing computational companions the "Method of Companion".

To sum up, we have the following conclusion for solution to Problem 9.3.

Theorem 9.2. *Regarding solution to Problem 9.3, the following statements hold:*

1. *The general parametric solution of the feedback gain K to Problem 9.3 is given by formula (9.31) with the parameter matrix $W_\infty \in \mathbb{R}^{r \times (n-n_0)}$ satisfying Constraint C3.*
2. *When the columns of the matrix V_∞ are chosen to be orthogonal to each other, the general parametric solution of the feedback gain K can be also given by (9.36)–(9.37) with the parameter matrix Q being an arbitrary matrix whose columns form a set of normalized orthogonal bases for $\mathscr{E}_r = \mathrm{span}\{\text{rows of } E\}$.*

Numerical Stability Considerations

From the numerical stability point of view, inverses of matrices, especially those of high dimensions, should be avoided in practical computations whenever possible. It is clear that both of our two expressions for the feedback gain matrix K involve inverses of matrices. In this subsection, we give some directions in computing numerically the gain matrix K.

There is one type of applications of eigenstructure assignment approaches, which impose special system design requirements on the structure of the right eigenvector matrix V and the corresponding matrix W. For this type of applications, the design parameters $\left\{ f_{ij}^k \right\}$ and W_∞ are sought by our constraints and the special system design requirements before the solution of K. Therefore, by the time to compute the gain matrix K, the matrices V and W are already known. In this case, the two expressions for the gain matrix K, namely, (9.31) and (9.36)–(9.37), may be only viewed as two different mathematical representations for the gain matrix K. Instead of using these two expressions directly, we may use the following two methods for solution of the gain matrix K, which avoid inverse of matrices and are more numerically stable.

Method 1 Instead of using the first expression (9.31), it is more preferable to solve the gain matrix K from the linear matrix equation presented by (9.30) using some numerically stable algorithms (for example, the QR algorithm) for linear matrix equations.

Method 2 Using (9.30) and (9.34), we have

$$K = \begin{bmatrix} W & W_\infty \end{bmatrix} \begin{bmatrix} (Q^T V)^{-1} & 0 \\ -V_\infty^T V (Q^T V)^{-1} & I_{n_0} \end{bmatrix} \begin{bmatrix} Q^T \\ V_\infty^T \end{bmatrix}.$$

Post-multiplying by $\tilde{Q}$ both sides of the above equation, and using the orthogonal property of the matrix $\tilde{Q}$, we obtain

$$K \begin{bmatrix} Q & V_\infty \end{bmatrix} = \begin{bmatrix} (W - W_\infty V_\infty^T V)(Q^T V)^{-1} & W_\infty \end{bmatrix}, \tag{9.38}$$

which gives

$$KQ = (W - W_\infty V_\infty^T V)(Q^T V)^{-1}. \tag{9.39}$$

Based on (9.38) and (9.39), we can give the following two steps for solution of the gain matrix K, which does not involve inverses of matrices.

Step 1 Solve the matrix W_0 from the following linear matrix equation

$$W_0(Q^T V) = W - W_\infty V_\infty^T V, \tag{9.40}$$

Using some numerically stable algorithm for solution of linear equations.
Step 2 Compute the matrix K by

$$K = W_0 Q^T + W_\infty V_\infty^T. \tag{9.41}$$

The above Method 2 is better than Method 1 because it involves a linear matrix equation of a lower order, and thus has less computational load and seemingly suffers less numerical problem.

Another type of applications of eigenstructure assignment approaches impose some special requirements directly on the gain matrix K (for example, the minimum gain problem). In such type of applications, it seems that the direct closed form for the gain matrix K, (9.31) or (9.36)–(9.37), has to be used, and the matrix inversions contained in these expressions generally cannot be avoided. In this case, the second expression (9.36)–(9.37) is also more preferable since it contains the same degree of design freedom and involves the inverse of a matrix with a lower order.

9.2.3 The Algorithm for Problem 9.1

Using Theorems 9.1 and 9.2, we have the following theorem for solution to Problem 9.1.

Theorem 9.3. *Regarding solution to Problem 9.1, the following statements hold.*

1. *Problem 9.1 has solutions if and only if there exist a group of parameter vectors $\{f_{ij}^k\}$ satisfying Constraints C1 and C2.*
2. *When the above condition is met, all the solutions to Problem 9.1 are given by (9.23) and (9.31) with Constraints C1, C2, and C3 satisfied.*
3. *When the columns of the matrix V_∞ are chosen orthogonal to each other, the general parametric solution of the feedback gain K can be also given by (9.36)–(9.37) with the parameter matrix Q being an arbitrary matrix whose columns form a set of normalized orthogonal bases for the space spanned by the rows of the matrix E.*

Based on the above, an algorithm for eigenstructure assignment in the S-controllable descriptor linear system (9.1) via the state feedback control (9.2) can be given as follows.

Algorithm 9.1. Eigenstructure assignment by state feedback.

Step 1 Solve the right coprime polynomial matrices $N(s)$ and $D(s)$ satisfying (9.24).

Step 2 Solve the infinite left and right eigenvector matrices T_∞ and V_∞ associated with the infinite closed-loop eigenvalue s_∞ defined by (9.15) and (9.18).

Step 3 Construct the matrices V and W by formula (9.23) and conventions Cv1–Cv3.

Step 4 Find a group of parameter vectors $\{f_{ij}^k\}$ satisfying Constraints C1 and C2. If such parameters do not exist, the specified eigenstructure is not assignable.

Step 5 Compute matrices V and W based on the parameters determined in Step 4.

Step 6 Choose a parameter matrix W_∞ satisfying Constraint C3.

Step 7 Compute the gain matrix K using (9.31) or (9.36)–(9.37), or the two numerical methods introduced in Sect. 9.2.2, based on the found matrices V_∞, W_∞, V, and W.

Remark 9.5. The above Theorem 9.3 and Algorithm 9.1 are, respectively, the generalizations of the Theorem 2 and Algorithm ESA for eigenstructure assignment in normal linear systems proposed in Duan (1993b).

Remark 9.6. When more requirements beyond the basic closed-loop eigenstructure are imposed on the closed-loop system, we can turn first these requirements into some additional constraints on the closed-loop eigenvalues or/and the parameters W_∞ and $\{f_{ij}^k\}$, and then solve from (9.23) and (9.31) (or (9.36)–(9.37)) the required solution to the problem by restricting parameters W_∞ and $\{f_{ij}^k\}$ to satisfy also the set of additional constraints. Corresponding changes in the above Algorithm 9.1 are only to bring the additional constraints related only to the group of parameter vectors $\{f_{ij}^k\}$ into consideration in Step 4, and those related to W_∞ or both W_∞ and $\{f_{ij}^k\}$ into consideration in Step 6. This is the basic idea lies in the applications of the approach (see, e.g., Duan and Patton 1999, 2001; Duan et al. 2000a, 2001a,b, 2002a,b).

9.3   The Left Eigenvector Matrix

Many applications of the eigenstructure assignment approaches, such as robust
pole assignment, disturbance decoupling, etc., require the solution of the closed-
loop left eigenvectors. This section will further give some ways to solve the
closed-loop left eigenvector matrix associated with the finite closed-loop relative
eigenvalues.

9.3.1   Preliminaries

For the closed-loop system (9.3), the left eigenvector matrix $T \in \mathbb{C}^{n \times n_0}$ associated
with the finite closed-loop eigenvalues is defined by

$$T^{\mathrm{T}}(A + BK) = JT^{\mathrm{T}}E, \quad \mathrm{rank} T = n_0. \tag{9.42}$$

Thus, the entire left eigenvector matrix for the closed-loop system (9.3) is given by

$$\tilde{T} = \begin{bmatrix} T & T_\infty \end{bmatrix},$$

with T_∞ being defined by (9.18). Since the closed-loop system is regular, it is easily
understandable that, like the case for right eigenvector matrix, the matrix $\tilde{T}$, as the
entire eigenvector matrix for the closed-loop system, should be nonsingular.

It is clear that the eigenvector matrix T, defined by (9.42), is not unique. Usually,
we are interested in normalized pairs of right and left eigenvector matrices defined
below.

Definition 9.1 (Stewart 1975). Let V and T be, respectively, the right and left
eigenvector matrices of the closed-loop system associated with the finite closed-
loop eigenvalues. They are said to be a normalized pair if they satisfy

$$T^{\mathrm{T}}EV = I. \tag{9.43}$$

Normalized pairs of right and left eigenvector matrices of the closed-loop system
associated with the finite closed-loop eigenvalues play an important role in closed-
loop eigenvalue sensitivity analysis (Stewart 1975; Kautsky et al. 1989; Syrmos
and Lewis 1992), and are essential in deriving the closed-loop canonical form and
the parametric expressions for the response of the closed-loop system (see the next
section). The following theorem gives a way to solve a normalized pair of right
and left eigenvector matrices based on a known pair of right and left eigenvector
matrices.

Theorem 9.4. *Let K and V be a pair of solutions to Problem 9.1, $T' \in \mathbb{C}^{n \times n_0}$ be
an arbitrary left eigenvector matrix of the closed-loop system associated with the
finite closed-loop eigenvalues. Then the following hold:*

1. The matrix $D = (T')^\mathrm{T} E V$ is nonsingular.
2. The matrix $T = T' D^{-1}$ and the matrix V form a normalized pair of eigenvector matrices for the closed-loop system associated with the finite closed-loop eigenvalues.

Proof. Since the matrix $T' \in \mathbb{C}^{n \times n_0}$ is a left eigenvector matrix of the matrix pencil $(E, A + BK)$ associated with the finite closed-loop eigenvalues, we have, for this T', the entire eigenvector matrix of the matrix pencil $(E, A + BK)$ as

$$\tilde{T}' = [T' \ T_\infty],$$

which is nonsingular. By the definitions of matrices V_∞ and T_∞, we can obtain

$$(\tilde{T}')^\mathrm{T} E \tilde{V} = \begin{bmatrix} (T')^\mathrm{T} E V & 0 \\ 0 & 0 \end{bmatrix}. \tag{9.44}$$

Since $\tilde{T}'$ and $\tilde{V}$, which are, respectively, the entire left and right eigenvector matrices of the closed-loop systems, are both nonsingular, further note $\mathrm{rank} E = n_0$, it follows from the above (9.44) that the matrix $D = (T')^\mathrm{T} E V$ is nonsingular.

Following from the definition of left eigenvector matrices and the given conditions, we have

$$(T')^\mathrm{T}(A + BK) = J(T')^\mathrm{T} E, \quad \mathrm{rank} T' = n_0.$$

With the above relations and the nonsingularity of matrix D, it is easy to verify that for the matrix $T = T' D^{-T}$, the matrix equation (9.42) holds. Therefore, this matrix T is also a left eigenvector matrix of the matrix pencil $(E, A + BK)$ associated with the finite closed-loop relative eigenvalues. Furthermore, by direct verification, it can be easily shown that this matrix T satisfies (9.43). □

9.3.2 The Parametric Expressions

The above Theorem 9.4 gives a way to turn a pair of right and left eigenvector matrices of the closed-loop system into a normalized pair. However, in many applications, we are interested in the complete parametric form of the left eigenvector matrix T, which forms a normalized pair with the general solution V to the eigenstructure assignment problem. The following theorem provides such a solution.

Theorem 9.5. *Let matrices K and V be the general solutions to Problem 9.1, and V_∞ and W_∞ be as defined previously. Then the following hold:*

1. *The compound matrix*

$$\Phi = [EV \ AV_\infty + BW_\infty] \tag{9.45}$$

is nonsingular.

2. *The left eigenvector matrix T of the closed-loop system associated with the finite closed-loop eigenvalues, which forms a normalized pair with the right eigenvector matrix V, is unique with respect to the parameters W_∞ and $\{f_{ij}^k\}$, and is given by*

$$T^T = [I \ 0][EV \ AV_\infty + BW_\infty]^{-1}. \tag{9.46}$$

Proof. Since T is a left eigenvector matrix of the closed-loop system associated with the finite closed-loop eigenvalues, (9.42) holds. Post-multiplying both sides of (9.42) by $\tilde{V}$, and using (9.15), we obtain

$$\begin{bmatrix} T^T \\ T_\infty^T \end{bmatrix} (A + BK)V_\infty = \begin{bmatrix} 0 \\ \Sigma_\infty \end{bmatrix}, \tag{9.47}$$

where

$$\Sigma_\infty = T_\infty^T (A + BK) V_\infty. \tag{9.48}$$

From (9.43) and the definition of T_∞, we also have

$$\begin{bmatrix} T^T \\ T_\infty^T \end{bmatrix} EV = \begin{bmatrix} I \\ 0 \end{bmatrix}. \tag{9.49}$$

Combining (9.47) with (9.49) yields

$$\begin{bmatrix} T^T \\ T_\infty^T \end{bmatrix} [EV \ (A + BK)V_\infty] = \begin{bmatrix} I & 0 \\ 0 & \Sigma_\infty \end{bmatrix}. \tag{9.50}$$

Since $[T \ T_\infty]$ forms an entire left eigenvector matrix and is nonsingular, and the right-hand side of the above equation is also nonsingular, in view of (9.29) and Constraint C3, the matrix $[EV \ (A + BK)V_\infty]$ must be nonsingular. Furthermore, noting

$$(A + BK)V_\infty = AV_\infty + BW_\infty$$

in view of (9.28), we have

$$\Phi = [EV \ AV_\infty + BW_\infty] = [EV \ (A + BK)V_\infty],$$

which is nonsingular. The first conclusion is proven.

Following from (9.50), we have

$$T^T [EV \ (A + BK)V_\infty] = [I \ 0]. \tag{9.51}$$

From this, we can obtain formula (9.46). The proof is then completed. □

The formula (9.46) for the left eigenvector matrix T involves the solution of the inverse of the matrix $\Phi = [EV \ AV_\infty + BW_\infty]$. In order to reduce the computational amount and also improve the conditioning to the solution of this inverse matrix, we here again apply the Method of Companion introduced in Sect. 9.2.2, and obtain the following result.

Theorem 9.6. *Let matrices K and V be the general solutions to Problem 9.1, and T_∞, V_∞, and W_∞ be as defined previously, but with the columns of the matrix T_∞ being orthogonal to each other. Then the following hold:*

1. *For an arbitrary matrix $P \in \mathbb{R}^{n \times n_0}$ with columns forming a set of normalized orthogonal base for $T_\infty^\perp$, the matrix $P^\mathrm{T} EV$ is nonsingular.*
2. *The left eigenvector matrix T of the closed-loop system, associated with the finite closed-loop eigenvalues, which forms a normalized pair with the right eigenvector matrix V, is unique with respect to the parameters W_∞ and $\left\{ f_{ij}^k \right\}$, and is given by*

$$T^\mathrm{T} = (P^\mathrm{T} EV)^{-1} P^\mathrm{T} \left(I - (AV_\infty + BW_\infty) \Sigma_\infty^{-1} T_\infty^\mathrm{T} \right), \qquad (9.52)$$

where Σ_∞ is defined as in (9.48).

Proof. Since the columns of P form a set of normalized orthogonal base for $T_\infty^\perp$, it is clear that $[P \ \ T_\infty]$ forms an orthogonal matrix. Choosing the computational companion matrix as $[P \ \ T_\infty]$, we have

$$\begin{bmatrix} P^\mathrm{T} \\ T_\infty^\mathrm{T} \end{bmatrix} \left[EV \ \ AV_\infty + BW_\infty \right] = \begin{bmatrix} P^\mathrm{T} EV & P^\mathrm{T}(AV_\infty + BW_\infty) \\ 0 & \Sigma_\infty \end{bmatrix}. \qquad (9.53)$$

Since $[P \ \ T_\infty]$ is orthogonal, $\left[EV \ \ AV_\infty + BW_\infty \right]$ is nonsingular in view of Theorem 9.5, and Σ_∞ is also nonsingular due to Constraint C3, it follows from the above (9.53) that the matrix $P^\mathrm{T} EV$ is nonsingular. The first conclusion is shown.

Applying the Matrix Inversion Lemma, Theorem A.2 in Appendix A, to the right-hand side of (9.53), and in view of the orthogonal property of the matrix $[P \ \ T_\infty]$, we obtain

$$\left[EV \ \ AV_\infty + BW_\infty \right]^{-1}$$
$$= \begin{bmatrix} (P^\mathrm{T} EV)^{-1} & -(P^\mathrm{T} EV)^{-1} P^\mathrm{T}(AV_\infty + BW_\infty) \Sigma_\infty^{-1} \\ 0 & \Sigma_\infty^{-1} \end{bmatrix} \begin{bmatrix} P^\mathrm{T} \\ T_\infty^\mathrm{T} \end{bmatrix}.$$

Therefore,

$$\left[EV \ \ AV_\infty + BW_\infty \right]^{-1}$$
$$= \begin{bmatrix} (P^\mathrm{T} EV)^{-1} P^\mathrm{T} \left(I - (AV_\infty + BW_\infty) \Sigma_\infty^{-1} T_\infty^\mathrm{T} \right) \\ \Sigma_\infty^{-1} T_\infty^\mathrm{T} \end{bmatrix}. \qquad (9.54)$$

With formula (9.54), the matrix T given by (9.46) can be easily shown to possess expression (9.52). $\qquad \square$

Remark 9.7. Since $\text{rank} E = n_0$, and $T_\infty^{\text{T}} E = 0$, it is easy to show that the matrix P may also be taken to be an arbitrary matrix with columns forming a set of normalized orthogonal base for the space spanned by the columns of the matrix E.

Remark 9.8. Similar to the treatment in Sect. 9.2.2, for numerical solution to the matrix T, we have two ways. One is to solve the matrix T^{T} from the linear matrix equation (9.51) using some numerically stable algorithm for solution of linear equations. The other is to solve, again using some numerically stable algorithm for linear equations, the matrix T^{T} from the following linear equation

$$(P^{\text{T}} E V) T^{\text{T}} = P^{\text{T}} \left(I - (A V_\infty + B W_\infty) \Sigma_\infty^{-1} T_\infty^{\text{T}} \right), \tag{9.55}$$

which is obtained from (9.52).

9.4 Response of the Closed-Loop System

In this section, we will explore, by establishing a canonical form for the closed-loop system, the response of the closed-loop system (9.3) with the gain matrix K given by our eigenstructure assignment approach proposed in Sect. 9.2. This problem can be precisely stated as follows.

Problem 9.4. [Closed-loop response analysis] Given system (9.1) satisfying Assumption 9.1, establish the analytical expression for the response of the closed-loop system (9.3), where the state feedback gain K is the general solution to Problem 9.3, that is, the matrix K is given by (9.31) with the parameter matrix W_∞ satisfying Constraint C3, and the matrices V and W are a pair of general solution to Problem 9.2.

9.4.1 The Canonical Form for the Closed-Loop System

In this subsection, we present a canonical form, in terms of the normalized pair of right and left closed-loop eigenvector matrices, for the closed-loop system (9.3), which is essential for deriving the general response formula for the closed-loop system (9.3).

Theorem 9.7. *Let the gain matrix K be given as in Problem 9.4, and T be a left eigenvector matrix of the closed-loop system associated with the finite closed-loop eigenvalues and satisfy (9.43). Then, under the relation*

$$x = \tilde{V} \tilde{x} \quad \text{or} \quad \tilde{x} = \tilde{V}^{-1} x, \tag{9.56}$$

the closed-loop system (9.3) can be equivalently transformed into the following canonical form

$$\begin{bmatrix} I & 0 \\ 0 & 0 \end{bmatrix} \dot{\tilde{x}} = \begin{bmatrix} J & 0 \\ 0 & T_\infty^T A V_\infty + T_\infty^T B W_\infty \end{bmatrix} \tilde{x} + \begin{bmatrix} T^T \\ T_\infty^T \end{bmatrix} B u_e, \qquad (9.57)$$

which can be divided, by letting

$$\tilde{x} = \begin{bmatrix} \tilde{x}_D \\ \tilde{x}_N \end{bmatrix}, \quad \tilde{x}_D \in \mathbb{R}^{n_0}, \ \tilde{x}_N \in \mathbb{R}^{n-n_0}, \qquad (9.58)$$

into the normal dynamical part

$$\dot{\tilde{x}}_D = J \tilde{x}_D + T^T B u_e \qquad (9.59)$$

and the nondynamical part

$$\tilde{x}_N = -(T_\infty^T A V_\infty + T_\infty^T B W_\infty)^{-1} T_\infty^T B u_e. \qquad (9.60)$$

Proof. Noting that $\tilde{T} = \begin{bmatrix} T & T_\infty \end{bmatrix}$ is nonsingular, pre-multiplying both sides of (9.3) by $\tilde{T}^T$, we have the following equivalent form of system (9.3):

$$\tilde{T}^T E \dot{x} = \tilde{T}^T A_c x + \tilde{T}^T B u_e.$$

Substituting (9.56) into the above system, we have

$$\tilde{T}^T E \tilde{V} \dot{\tilde{x}} = \tilde{T}^T A_c \tilde{V} \tilde{x} + \tilde{T}^T B u_e. \qquad (9.61)$$

In view of (9.15), (9.18), and (9.43), we have

$$\tilde{T}^T E \tilde{V} = \begin{bmatrix} T^T E V & 0 \\ 0 & 0 \end{bmatrix} = \begin{bmatrix} I & 0 \\ 0 & 0 \end{bmatrix}. \qquad (9.62)$$

In view of (9.10), (9.15), (9.18), (9.42), and (9.43) again, we have

$$\begin{aligned} \tilde{T}^T A_c \tilde{V} &= \begin{bmatrix} T^T(A+BK)V & T^T(A+BK)V_\infty \\ T_\infty^T(A+BK)V & T_\infty^T(A+BK)V_\infty \end{bmatrix} \\ &= \begin{bmatrix} J & 0 \\ 0 & T_\infty^T(A+BK)V_\infty \end{bmatrix}. \end{aligned} \qquad (9.63)$$

Substituting (9.62) and (9.63) into (9.61), and noticing (9.29) we obtain the canonical form (9.57), which can, by using Constraint C3, be clearly divided into dynamical and nondynamical parts as given by (9.59) and (9.60), respectively. □

Remark 9.9. The importance of the canonical form (9.57) for the closed-loop system lies in the following two aspects:

- It is composed clearly of a normal dynamical part of order n_0 and a nondynamical (algebraic) part of order $n - n_0$.

- The coefficient matrices of this canonical form are expressed by the closed-loop eigenvector matrices and the closed-loop Jordan matrix, which are both undetermined and may be chosen to achieve some desired properties of the system.

With the help of this canonical form, many problems for descriptor linear systems, for example, pole assignment, linear quadratic optimal control, etc., can be solved by converting them into the corresponding problems for normal linear systems.

9.4.2 The Closed-Loop Response

With the help of the canonical form for (9.3), introduced in Theorem 9.7, the response of the closed-loop system (9.3) can be immediately obtained as follows.

Theorem 9.8. *Let the conditions in Theorem 9.7 hold, then the response of the closed-loop system (9.3) is given by*

$$x(t) = V \tilde{x}_D - V_\infty \left(T_\infty^{\mathrm{T}} (A + BK) V_\infty \right)^{-1} T_\infty^{\mathrm{T}} B u_{\mathrm{e}}, \tag{9.64}$$

with

$$\tilde{x}_D(t) = \mathrm{e}^{Jt} T^{\mathrm{T}} E \, x(0) + \int_0^t \mathrm{e}^{J(t-\tau)} T^{\mathrm{T}} B u_{\mathrm{e}}(\tau) \mathrm{d}\tau. \tag{9.65}$$

Proof. Since
$$x = \begin{bmatrix} V & V_\infty \end{bmatrix} \tilde{x} = V \tilde{x}_D + V_\infty \tilde{x}_N, \tag{9.66}$$
$\tilde{x}_N$ can be easily obtained from (9.60) as

$$\tilde{x}_N = - \left(T_\infty^{\mathrm{T}} (A + BK) V_\infty \right)^{-1} T_\infty^{\mathrm{T}} B u_{\mathrm{e}}.$$

Substituting the above into (9.66) gives (9.64).

Since $\tilde{x}_D$ is governed by (9.59), in view of the well-known results for responses of continuous-time normal linear systems, it is clearly given by

$$\tilde{x}_D = \mathrm{e}^{Jt} \begin{bmatrix} I_{n_0} & 0 \end{bmatrix} \begin{bmatrix} V & V_\infty \end{bmatrix}^{-1} x(0) + \int_0^t \mathrm{e}^{J(t-\tau)} T^{\mathrm{T}} B u_{\mathrm{e}}(\tau) \mathrm{d}\tau. \tag{9.67}$$

Furthermore, using (9.43) and (9.15) we have

$$T^{\mathrm{T}} E \begin{bmatrix} V & V_\infty \end{bmatrix} = \begin{bmatrix} I_{n_0} & 0 \end{bmatrix}.$$

Thus, there holds
$$\begin{bmatrix} I_{n_0} & 0 \end{bmatrix} \begin{bmatrix} V & V_\infty \end{bmatrix}^{-1} = T^{\mathrm{T}} E.$$

Substituting the above relation into (9.67) gives (9.65).   □

The structure of the Jordan matrix J may be utilized in deriving $\tilde{x}_D$. In fact, in accordance with the structure of matrix J, let

$$\tilde{x}_D = \begin{bmatrix} x_d^1 \\ x_d^2 \\ \vdots \\ x_d^{n_0'} \end{bmatrix}, \quad x_d^i = \begin{bmatrix} x_d^{i1} \\ x_d^{i2} \\ \vdots \\ x_d^{iq_i} \end{bmatrix}, \quad x_d^{ij} \in \mathbb{R}^{p_{ij}}, \tag{9.68}$$

then the dynamical system (9.59) can be equivalently decomposed into the following group of dynamical systems with smaller dimensions:

$$\dot{x}_d^{ij} = J_{ij} x_d^{ij} + T_{ij}^{\mathrm{T}} B u_e, \tag{9.69}$$
$$j = 1, 2, \ldots, q_i, \quad i = 1, 2, \ldots, n_0'.$$

Thus, we have

$$x_d^{ij}(t) = \Phi_{ij}(t) x_d^{ij}(0) + \int_0^t \Phi_{ij}(t-\tau) T_{ij}^{\mathrm{T}} B u_e(\tau) d\tau, \tag{9.70}$$
$$j = 1, 2, \ldots, q_i, \quad i = 1, 2, \ldots, n_0',$$

with

$$\Phi_{ij}(t) = \begin{bmatrix} 1 & t & \frac{1}{2!}t^2 & \cdots & \frac{1}{(p_{ij}-1)!}t^{p_{ij}-1} \\ 0 & 1 & t & \cdots & \frac{1}{(p_{ij}-2)!}t^{p_{ij}-2} \\ \vdots & \vdots & \vdots & \ddots & \vdots \\ 0 & 0 & 0 & \cdots & 1 \end{bmatrix} e^{s_i t}. \tag{9.71}$$

The closed-loop response can also be stated in another usual way as in the following theorem.

Theorem 9.9. *Let x_{initial} and x_{external} be, respectively, the responses of the closed-loop system generated by the system initial value $x(0) = x_0$ and the external signal u_e, then under the given conditions of Theorem 9.7, the response of the closed-loop system (9.3) is given as*

$$x(t) = x_{\text{initial}}(t) + x_{\text{external}}(t) \tag{9.72}$$

with

$$x_{\text{initial}}(t) = V e^{Jt} T^{\mathrm{T}} E x_0, \tag{9.73}$$

and

$$x_{\text{external}}(t) = -V_\infty \left(T_\infty^{\mathrm{T}}(A+BK)V_\infty \right)^{-1} T_\infty^{\mathrm{T}} B u_e$$
$$+ \int_0^t V e^{J(t-\tau)} T^{\mathrm{T}} B u_e(\tau) d\tau. \tag{9.74}$$

Again, the structure of the Jordan matrix J can be utilized to express x_{initial} and x_{external}. In view of our conventions made in the beginning of Sect. 9.1, we have

$$x_{\text{initial}}(t) = \sum_{i,j} V_{ij}\, e^{J_{ij}t}\, T_{ij}^{\text{T}} E\, x_0, \tag{9.75}$$

and

$$x_{\text{external}}(t) = -V_\infty \left(T_\infty^{\text{T}}(A + BK)V_\infty\right)^{-1} T_\infty^{\text{T}} B u_{\text{e}}$$
$$+ \int_0^t \sum_{i,j} V_{ij} e^{J_{ij}(t-\tau)} T_{ij}^{\text{T}} B u_{\text{e}}(\tau)\mathrm{d}\tau. \tag{9.76}$$

For the special case of J being diagonal, we have

$$x_{\text{initial}}(t) = \sum_{i=1}^{n_0} v_i t_i^{\text{T}} e^{s_i t} E\, x_0, \tag{9.77}$$

$$x_{\text{external}}(t) = -V_\infty \left(T_\infty^{\text{T}}(A + BK)V_\infty\right)^{-1} T_\infty^{\text{T}} B u_{\text{e}}$$
$$+ \sum_{i=1}^{n_0} v_i t_i^{\text{T}} B \int_0^t e^{s_i(t-\tau)} u_{\text{e}}(\tau)\mathrm{d}\tau. \tag{9.78}$$

Remark 9.10. When complex eigenvalues are assigned to the closed-loop system, the Jordan form J and the associated eigenvector matrices T and V are all complex. Moreover, even the dynamical system (9.59) has complex coefficient matrices; however, the response given by our formula is still real. To avoid complex operations, a well-known technique can be applied, which converts, at the same time, all the three matrices T, V, and J into real ones. For simplicity, let us demonstrate this technique with the simple case of J being diagonal. Without loss of generality, let us assume that s_1 and $s_2 = \bar{s}_1 = \alpha + \beta j$ are a pair of conjugate eigenvalues, and all the other ones are real. In this case, we have $v_1 = \bar{v}_2$ and $w_1 = \bar{w}_2$, and the real substitutions of these three matrices can be taken as

$$J = \begin{bmatrix} \alpha & \beta & & & \\ -\beta & \alpha & & & \\ & & s_3 & & \\ & & & \ddots & \\ & & & & s_{n_0} \end{bmatrix},$$

$$V = \begin{bmatrix} \text{Re}(v_1) & \text{Im}(v_1) & v_3 & \cdots & v_{n_0} \end{bmatrix},$$
$$T = \begin{bmatrix} \text{Re}(t_1) & \text{Im}(t_1) & t_3 & \cdots & t_{n_0} \end{bmatrix}.$$

Remark 9.11. Our formulae for the closed-loop response are obtained based on the canonical form for the closed-loop system. As a consequence, these formulae are also expressed by the closed-loop eigenvector matrices and the closed-loop Jordan matrix, which are undetermined. By using these formulae, some of the important problems in the descriptor system context (for e.g., disturbance decoupling) can be easily solved.

9.5 An Example

To illustrate our approaches proposed in this chapter for eigenstructure assignment and response analysis, in this section we consider a system in the form of (9.1) with the following parameters (Chen and Chang 1993; Duan and Patton 1997)

$$E = \begin{bmatrix} 1 & 0 & 0 & 0 & 0 & 0 \\ 0 & 1 & 0 & 0 & 0 & 0 \\ 0 & 0 & 1 & 0 & 0 & 0 \\ 0 & 0 & 0 & 0 & 1 & 0 \\ 0 & 0 & 0 & 0 & 0 & 0 \\ 1 & 0 & 0 & 0 & 0 & 0 \end{bmatrix}, \quad A = \begin{bmatrix} 0 & 0 & 1 & 0 & 0 & 0 \\ 1 & 0 & 0 & 0 & 0 & 0 \\ 0 & 1 & 0 & 1 & 0 & 0 \\ 0 & 0 & 0 & 1 & 0 & 0 \\ 0 & 0 & 0 & 0 & 1 & 0 \\ 1 & 0 & 0 & 0 & 0 & 1 \end{bmatrix}, \quad B = \begin{bmatrix} 1 & 0 \\ 0 & 0 \\ 0 & 0 \\ 0 & 0 \\ 1 & 1 \\ 0 & 0 \end{bmatrix}.$$

For this system, we have $n = 6, n_0 = 4$ and $r = 2$. In the following, we consider the assignment of the following closed-loop eigenstructure

$$\Gamma = \{s_i, \ 0 \neq s_i \in \mathbb{R}, \ i = 1, 2, 3, 4\},$$

$$m_i = q_i = p_{ii} = 1, \ i = 1, 2, 3, 4,$$

which corresponds to

$$J = \text{diag}(s_1, \ s_2, \ s_3, \ s_4).$$

By a method given in Lemma C.3, we obtain

$$N(s) = \begin{bmatrix} s^2 & -s^2 \\ s & -s \\ 1 & 0 \\ 0 & s \\ 0 & 1 \\ s^2(s-1) & -s^2(s-1) \end{bmatrix}, \quad D(s) = \begin{bmatrix} -1+s^3 & -s^3 \\ 1-s^3 & -1+s^3 \end{bmatrix}.$$

The left and right eigenvector matrices associated with the infinite eigenvalue of the system are given by

$$V_\infty = \begin{bmatrix} 0 & 0 \\ 0 & 0 \\ 0 & 0 \\ 1 & 0 \\ 0 & 0 \\ 0 & 1 \end{bmatrix}, \quad T_\infty = \begin{bmatrix} 0 & -1 \\ 0 & 0 \\ 0 & 0 \\ 0 & 0 \\ 1 & 0 \\ 0 & 1 \end{bmatrix}.$$

9.5.1 The General Solutions

Let

$$f_i = \begin{bmatrix} c_i \\ d_i \end{bmatrix}, \quad c_i, \ d_i \in \mathbb{R}, \quad i = 1, 2, \ldots, 6,$$

then Constraint C1 holds automatically. Choose

$$E' = \begin{bmatrix} 1 & 0 & 0 & 0 & 0 & 0 \\ 0 & 1 & 0 & 0 & 0 & 0 \\ 0 & 0 & 1 & 0 & 0 & 0 \\ 0 & 0 & 0 & 0 & 1 & 0 \end{bmatrix},$$

then Constraint C2 can be expressed in the following form

$$\det \begin{bmatrix} (c_1 - d_1)s_1^2 & (c_2 - d_2)s_2^2 & (c_3 - d_3)s_3^2 & (c_4 - d_4)s_4^2 \\ (c_1 - d_1)s_1 & (c_2 - d_2)s_2 & (c_3 - d_3)s_3 & (c_4 - d_4)s_4 \\ c_1 & c_2 & c_3 & c_4 \\ d_1 & d_2 & d_3 & d_4 \end{bmatrix} \neq 0. \qquad (9.79)$$

Let

$$W_\infty = \begin{bmatrix} z_{11} & z_{12} \\ z_{21} & z_{22} \end{bmatrix},$$

then Constraint C3 is verified to be

$$\det \begin{bmatrix} z_{21} & z_{22} + 1 \\ z_{11} & z_{12} - 1 \end{bmatrix} \neq 0. \qquad (9.80)$$

Thus, the general parametric solutions of the closed-loop eigenvectors are given by

$$v_i = \begin{bmatrix} (c_i - d_i)s_i^2 \\ (c_i - d_i)s_i \\ c_i \\ s_i d_i \\ d_i \\ (c_i - d_i)(s_i - 1)s_i^2 \end{bmatrix}, \quad i = 1, 2, 3, 4, \qquad (9.81)$$

and the corresponding vectors w_i are given by

$$w_i = \begin{bmatrix} -c_i + (c_i - d_i)s_i^3 \\ (c_i - d_i)(1 - s_i^3) \end{bmatrix}, \quad i = 1, 2, 3, 4. \qquad (9.82)$$

The state feedback gain K to this eigenstructure assignment problem is given by

$$K = \tilde{W}\tilde{V}^{-1} = [\,W\quad W_\infty\,][\,V\quad V_\infty\,]^{-1}, \tag{9.83}$$

$$W = [\,w_1\quad w_2\quad w_3\quad w_4\,], \quad V = [\,v_1\ v_2\ v_3\ v_4\,], \tag{9.84}$$

where the parameters c_i, d_i, $i = 1, 2, 3, 4$, and z_{ij}, $i, j = 1, 2$, satisfy conditions (9.79) and (9.80).

According to Theorem 9.5, the left eigenvector matrix T is given by

$$T^{\mathrm{T}} = [\,I\quad 0\,][\,EV\quad AV_\infty + BW_\infty\,]^{-1}. \tag{9.85}$$

Corresponding to the above general solution, the closed-loop system can be transformed into the following dynamical part

$$\dot{\tilde{x}}_D = \mathrm{diag}(s_1,\ s_2,\ s_3,\ s_4)\tilde{x}_D + T^{\mathrm{T}}Bu_e, \tag{9.86}$$

and the following nondynamical part

$$\tilde{x}_N = -\begin{bmatrix} z_{11} + z_{21} & z_{12} + z_{22} \\ -z_{11} & -z_{12} + 1 \end{bmatrix}^{-1} \begin{bmatrix} 1 & 1 \\ -1 & 0 \end{bmatrix} u_e, \tag{9.87}$$

and the responses of the closed-loop system are given as in (9.77) and (9.78).

9.5.2 Special Solutions

Let us choose the finite closed-loop relative eigenvalues to be

$$s_i = -i, \quad i = 1, 2, 3, 4.$$

In addition, if we choose

$$c_1 = c_3 = c_4 = 1, \quad c_2 = 0, \quad d_1 = 2, \quad d_2 = d_4 = 1, \quad d_3 = 0,$$

and

$$z_{11} = z_{12} = z_{22} = 0, \quad z_{21} = -1,$$

then we have

$$K = \begin{bmatrix} -6 & -11 & -7 & 0 & 6 & 0 \\ 5 & 4 & -5 & -1 & 1 & 0 \end{bmatrix}, \quad W = \begin{bmatrix} 0 & 8 & -28 & -1 \\ -2 & -9 & 28 & 0 \end{bmatrix},$$

$$V = \begin{bmatrix} -1 & -4 & 9 & 0 \\ 1 & 2 & -3 & 0 \\ 1 & 0 & 1 & 1 \\ -2 & -2 & 0 & -4 \\ 2 & 1 & 0 & 1 \\ 2 & 12 & 0 & 0 \end{bmatrix}, \quad T = \begin{bmatrix} -0.5 & 1 & 0.5 & 0 \\ -2.5 & 4 & 1.5 & 10 \\ -3 & 3 & 1 & 30 \\ 3 & -3 & -1 & -20 \\ 0 & 0 & 0 & 1 \\ 0 & 0 & 0 & 0 \end{bmatrix}.$$

Corresponding to the above special solutions, the closed-loop system can be transformed into the following dynamical part

$$\dot{\tilde{x}}_D(t) = \begin{bmatrix} -1 \\ & -2 \\ & & -3 \\ & & & -4 \end{bmatrix} \tilde{x}_D(t) + \begin{bmatrix} -0.5 & 0 \\ 1 & 0 \\ 0.5 & 0 \\ 1 & 1 \end{bmatrix} u_e,$$

and the following nondynamical part

$$\tilde{x}_N(t) = \begin{bmatrix} 1 & 1 \\ -1 & 0 \end{bmatrix} u_e(t).$$

The responses of the closed-loop system are given as:

$x_{\text{initial}}(t)$

$$
= \left(\begin{bmatrix} 0.5 & 2.5 & 3 & 0 & -3 & 0 \\ -0.5 & -2.5 & -3 & 0 & 3 & 0 \\ -0.5 & -2.5 & -3 & 0 & 3 & 0 \\ 1 & 5 & 6 & 0 & -6 & 0 \\ -1 & -5 & -6 & 0 & 6 & 0 \\ -1 & -5 & -6 & 0 & 6 & 0 \end{bmatrix} e^{-t} + \begin{bmatrix} -4 & -16 & -12 & 0 & 12 & 0 \\ 2 & 8 & 6 & 0 & -6 & 0 \\ 0 & 0 & 0 & 0 & 0 & 0 \\ -2 & -8 & -6 & 0 & 6 & 0 \\ 1 & 4 & 3 & 0 & -3 & 0 \\ 12 & 48 & 36 & 0 & -36 & 0 \end{bmatrix} e^{-2t} \right.
$$

$$
\left. + \begin{bmatrix} 4.5 & 13.5 & 9 & 0 & -9 & 0 \\ -1.5 & -4.5 & -3 & 0 & 3 & 0 \\ 0.5 & 1.5 & 1 & 0 & -1 & 0 \\ 0 & 0 & 0 & 0 & 0 & 0 \\ 0 & 0 & 0 & 0 & 0 & 0 \\ -18 & -54 & -36 & 0 & 36 & 0 \end{bmatrix} e^{-3t} + \begin{bmatrix} 0 & 0 & 0 & 0 & 0 & 0 \\ 0 & 0 & 0 & 0 & 0 & 0 \\ 0 & 1 & 3 & 0 & -2 & 0 \\ 0 & -4 & -12 & 0 & 8 & 0 \\ 0 & 1 & 3 & 0 & -2 & 0 \\ 0 & 0 & 0 & 0 & 0 & 0 \end{bmatrix} e^{-4t} \right) x_0,
$$

and

$$x_{\text{external}}(t) = \begin{bmatrix} 0 & 0 \\ 0 & 0 \\ 0 & 0 \\ 1 & 1 \\ 0 & 0 \\ -1 & 0 \end{bmatrix} u_e + \begin{bmatrix} 0.5 & 0 \\ -0.5 & 0 \\ -0.5 & 0 \\ 1 & 0 \\ -1 & 0 \\ -1 & 0 \end{bmatrix} \int_0^t e^{-\tau} u_e(\tau) d\tau$$

$$+ \begin{bmatrix} -4 & 0 \\ 2 & 0 \\ 0 & 0 \\ -2 & 0 \\ 1 & 0 \\ 12 & 0 \end{bmatrix} \int_0^t e^{-2\tau} u_e(\tau) d\tau + \begin{bmatrix} 4.5 & 0 \\ -1.5 & 0 \\ 0.5 & 0 \\ 0 & 0 \\ 0 & 0 \\ -18 & 0 \end{bmatrix} \int_0^t e^{-3\tau} u_e(\tau) d\tau$$

$$+ \begin{bmatrix} 0 & 0 \\ 0 & 0 \\ 1 & 1 \\ -4 & -4 \\ 1 & 1 \\ 0 & 0 \end{bmatrix} \int_0^t e^{-4\tau} u_e(\tau) d\tau.$$

9.6 Notes and References

Eigenstructure assignment in descriptor linear systems is an important problem in the descriptor system context. In this chapter, the parametric approach in Duan (1998a) for the problem of eigenstructure assignment in descriptor linear systems via state feedback is presented.

Further Comments

The proposed approach presents simple general complete parametric expressions for both the closed-loop eigenvectors associated with the relative finite closed-loop eigenvalues and the state feedback gain matrix. It guarantees arbitrary assignment of rank E number finite closed-loop eigenvalues with arbitrary given algebraic and geometric multiplicities, realizes elimination of all possible initial time impulsive responses, and also removes most of the restrictions and assumptions of previous works, such as

- The distinctness of the finite closed-loop eigenvalues (for e.g., Ozcaldiran and Lewis 1987; Fahmy and O'Reilly 1989).

- The distinctness of the finite closed-loop eigenvalues from the open-loop ones (for e.g., Chen and Chang 1993; Fahmy and O'Reilly 1989).
- The distinctness of the closed-loop eigenvalues from zero (for e.g., Fahmy and O'Reilly 1989; Duan 1992b; Duan and Patton 1998a).
- The regularity assumption of the open-loop system (for e.g., Zagalak and Kucera 1995; Jing 1994; Chen and Chang 1993; Fahmy and O'Reilly 1989).

This approach also falls into the scope of the generalized Sylvester matrix equation approaches. However, different from Fletcher et al. (1986) and Fahmy and O'Reilly (1988), we use the complete parametric solution to the generalized Sylvester matrix equation rather than the generalized Sylvester matrix equation itself, hence complete parametric expressions for the closed-loop eigenvectors are obtained.

Different from state feedback control of normal linear systems, in a descriptor linear state feedback control system, the state feedback gain is not unique with respect to the closed-loop eigenvector matrix due to the rank deficiency of matrix E. This certainly gives more difficulty in deriving a general form for the feedback gain matrix K when dealing with eigenstructure assignment in descriptor linear systems. To obtain the general solution for the feedback gain matrix K, Fletcher (1988) and Duan (1995) used the inner inverses of matrices, Fahmy and O'Reilly (1989) and Duan (1992b) adopted elementary matrix operations. Both ways are complicated. To overcome this, in this approach the "completing the square" technique, introduced in Duan and Patton (1998a), is used. This technique introduces an auxiliary equation using the part of the eigenvector matrix V_∞ associated with the infinite closed-loop eigenvalues, and a parameter matrix W_∞ which represents an extra degree of freedom existing in solution of the gain matrix K other than the degree of freedom existing in the solution of the eigenvector matrix V. With this technique, two very simple, complete, and parametric solutions in direct closed forms for the feedback gain matrix K are easily obtained.

Closed-loop regularity is an essential requirement in the eigenstructure assignment problem for descriptor systems since it ensures the uniqueness of solutions of the systems. In Fahmy and O'Reilly (1989), Duan (1992b), and Duan and Patton (1998a), the closed-loop regularity problem is considered by using a condition, in terms of the determinant of the closed-loop system matrix $A_c = A + BK$, proposed by Fahmy and O'Reilly (1989). Since this condition, as Chen and Chang (1993) have pointed out, requires all the finite closed-loop relative eigenvalues to be distinct from zero, it excludes the important case of deadbeat control of discrete-time descriptor systems. To overcome this defect, in this chapter a new necessary and sufficient condition, in terms of the determinant of a matrix involving the parts of the left and the right eigenvector matrices associated with the infinite eigenvalue, is proposed. It not only allows assignment of zero closed-loop eigenvalues, but is also much simpler than the one which was used in Fahmy and O'Reilly (1989), Duan (1992b), and Duan and Patton (1998a), since it is the determinant of a matrix of much lower order.

Based on the general framework for eigenstructure assignment presented, several other related developments are also made in this chapter:

- First, solution of the closed-loop left eigenvector matrix T associated with the finite closed-loop eigenvalues is considered. A simple, general, complete parametric solution is presented for the closed-loop left eigenvector matrix T, which forms, together with the general right eigenvector matrix V associated with the finite closed-loop eigenvalues, a normalized pair. Based on this solution and with the help of the Method of Companion, another general solution containing the computational companion matrix is also obtained.
- Second, a canonical form for the closed-loop system is derived, which is expressed in terms of the closed-loop normalized pair of right and left eigenvector matrices, and clearly states that the closed-loop system, designed using the proposed eigenstructure assignment approach, can be equivalently transformed into the normal dynamical part and the nondynamical (algebraic) part.
- Third, with the help of this canonical form, general analytical fully parametric expression for the closed-loop system response is obtained in terms of the closed-loop eigenstructure, that is, the closed-loop eigenvector matrices and the closed-loop Jordan matrix.

Like the eigenstructure assignment result presented, both the canonical form and the response solution obtained for the closed-loop system may have some important applications in some descriptor systems design with specific properties.

A Brief Overview

Eigenstructure assignment in descriptor linear systems has attracted much attention in the last two decades. Main references in this aspect include Fletcher et al. (1986), Ozcaldiran and Lewis (1987), Fletcher (1988), Lewis and Ozcaldiran (1989), Fahmy and O'Reilly (1989), Sakr and Khalifa (1990), Georgiou and Krikelis 1992, Chen and Chang (1993), Jing (1994), Zagalak and Kucera (1995), Duan et al. (1991a, 2003), Duan and Patton (1997, 1998a), Duan and Wang (2003), and Duan (1992b, 1994, 1995, 1998a, 1999, 2003a,b, 2004d). As an introduction to this topic, this chapter only solves the problem of eigenstructure assignment in descriptor linear systems via state feedback. The content is mainly taken from Duan (1998a). Other work on eigenstructure assignment in descriptor linear systems via state feedback can be found in Fletcher et al. (1986), Fahmy and O'Reilly (1989), Georgiou and Krikelis (1992), Chen and Chang (1993), Zagalak and Kucera (1995), Duan and Patton (1998a), Duan and Wang (2003), and Duan (1992b, 1998a, 2004d). For eigenstructure assignment in descriptor linear systems via output feedback, one may refer to Fletcher (1988), Sakr and Khalifa (1990), and Duan (1999); for eigenstructure assignment in descriptor linear systems via state proportional-derivative feedback, one may refer to Jing (1994), Chen and Chang (1993), and Duan and Patton (1997).

By methodology, approaches to the problem of eigenstructure assignment via feedback presently proposed can mainly be classified into two categories, one is the geometric approaches (Ozcaldiran and Lewis 1987; Lewis and Ozcaldiran 1989), the other is the algebraic approaches (Fletcher et al. 1986; Fletcher 1988; Fahmy and O'Reilly 1989; Duan 1992b, 1995, 1998a, 1999; Duan and Patton 1998a). By the mathematical tools used, approaches can be also classified into the characteristic polynomial approaches (Fahmy and O'Reilly 1989; Chen and Chang 1993; Zagalak and Kucera 1995) and the generalized Sylvester matrix equation approaches (Fletcher et al. 1986; Fletcher 1988; Ozcaldiran and Lewis 1987; Lewis and Ozcaldiran 1989; Duan 1992b, 1995, 1998a, 1999; Duan and Patton 1998a).

Another type of approaches to eigenstructure assignment in descriptor linear systems are called parametric approaches since they parameterize all the solutions to the problem. A pioneering piece of work in this aspect was proposed by Fahmy and O'Reilly (1989). Later work can be found in Chen and Chang (1993), Duan (1992b, 1995, 1998a, 1999), and Duan and Patton (1998a). This type of approaches are important due to two reasons. On the one hand, as far as it is concerned with the general eigenstructure assignment problem, where only the basic eigenstructure requirements and the closed-loop regularity condition are imposed on the closed-loop system, the solution to the problem of eigenstructure assignment in a controllable descriptor linear system is generally not unique. Therefore, it is clearly better to give general parameter characterizations of all the closed-loop eigenvectors as well as the feedback gain than to establish existence conditions and provide a procedure with which only one single solution pair of the closed-loop eigenvector matrix and the feedback gain can be obtained. On the other hand, it has been well demonstrated, with both descriptor linear systems (see, e.g., Kautsky et al. 1989; Duan et al. 1999a, 2000b, 2001b, 2002a,b; Duan and Lam 2000), and normal linear systems (see Owens and O'Reilly 1987, 1989; Duan 1992a, 1993a; Duan and Patton 1998b; Duan and Liu 2002; Patton and Liu 1994; Duan et al. 2000a,d), that parametric approaches to eigenstructure assignment are very convenient for control systems design with specific design objectives and robustness requirements.

Recently, the author and his coauthors have generalized their previous results on eigenstructure assignment in descriptor linear systems to second-order and high-order systems. Specifically, Duan et al. (2005) considered eigenstructure assignment in a class of second-order descriptor linear systems, and presented a complete parametric approach, while Wang et al. (2006) solved the problem of partial eigenstructure assignment in second-order descriptor systems. Parametric approaches for eigenstructure assignment in high-order linear systems have been considered in Duan (2005).

Chapter 10
Optimal Control

This chapter studies optimal control of descriptor linear systems. After a brief introduction given in Sect. 10.1, the basic linear quadratic state regulation problem for descriptor linear systems is investigated in Sect. 10.2. Corresponding to the well-known bang–bang control for normal linear systems, the time-optimal control problem for descriptor linear systems is studied in Sect. 10.3. Some notes and references follow in Sect. 10.4.

10.1 Introduction

In Chaps. 5–9, we have examined several types of descriptor linear systems design problems. These design problems aim to achieve certain properties or structures of the closed-loop system, such as regularity, stability, normalizability, impulse elimination, dynamical order assignment, pole assignment, and eigenstructure assignment. In this chapter, we consider a different type of control problems—optimal control of descriptor linear systems. The purpose of optimal control is to optimize, via choosing a certain type of feedback controllers, some objective function associated with the considered system. Such type of control design problems often have clear practical implications.

Before presenting the concrete optimal control strategies, in this section we classify some of the basic elements involved in optimal control.

The Open-Loop System Regarding optimal control, the open-loop system to be controlled can be a general descriptor nonlinear system. However, in this chapter, we only consider the typical descriptor linear system:

$$E\dot{x}(t) = Ax(t) + Bu(t), \tag{10.1}$$

where $x(t) \in \mathbb{R}^n$ and $u(t) \in \mathbb{R}^r$ are the state and the control input, respectively; $E, A \in \mathbb{R}^{n \times n}$, and $B \in \mathbb{R}^{n \times r}$ are constant system coefficient matrices. It is assumed that the system (10.1) is regular and rank $E \le n$.

G.-R. Duan, *Analysis and Design of Descriptor Linear Systems*, Advances
in Mechanics and Mathematics 23, DOI 10.1007/978-1-4419-6397-0_10,
© Springer Science+Business Media, LLC 2010

Control Constraint and Admissible Control   In real applications, the control input $u(t)$ is often subjected to some constraint π. We use $\mathbb{U}$ to denote the collection set of $u(t)$ subject to constraint π, which is called the control constraint set, or simply, the constraint set. Any $u(t) \in \mathbb{U}$ is called an admissible control. For descriptor systems, admissible controls must be piecewise sufficiently and continuously differentiable functions since derivatives of control inputs are included in the state response of descriptor linear systems. The constraint π is often a vector function of state, control, and time t, that is,

$$\pi = \pi\left(x\left(t\right), u\left(t\right), t\right).$$

Initial and Terminal Times   Concerning control system design, an initial time has to be defined. The initial time can be an arbitrary time instant $t_0 \geq 0$. For simplicity, we often choose $t_0 = 0$. The terminal time is some time instant $t_1 > t_0$. It is the time point at which the control action is supposed to finish. Please note that the terminal time t_1 can also be infinity. The interval $[t_0 \quad t_1]$ gives the interested time period in the optimal control design.

Initial and Terminal States   Corresponding to the initial time $t_0 = 0$ and the terminal time t_1, the initial state of the system (10.1) is $x(0) = x_0$, and the terminal state is $x(t_1) = x_1$. For simplicity, in this chapter we often choose the terminal state to be zero, that is, $x(t_1) = 0$.

Control Target   For general optimal control problems, a target $\mathbb{S}$ is usually involved, which is usually a requirement on the state at the terminal time t_1, and thus is actually a constraint on the terminal time t_1 and/or the terminal state $x(t_1)$.

Cost Functional   For optimal control of system (10.1), an index, or a cost functional

$$J = J\left(x\left(t\right), u\left(t\right), t\right) \tag{10.2}$$

must be defined, which is a functional of the state and the control input of the system (10.1) as well as the time. The cost J is usually chosen with respect to the practical problem we are interested in. Different choices of J represent different meanings of the optimal control systems and thus classify the different types of optimal control problems, such as the problem of quadratic regulation, and the time-optimal control, the cheapest control (minimum energy consumption).

Optimal Controller   The optimal controller for system (10.1) with respect to the target $\mathbb{S}$, constraint π, initial time $t_0 = 0$, and initial state $x(0)$ is an admissible control $u^\star(t) \in \mathbb{U}$, with which the terminal state of the control system reaches the target $\mathbb{S}$ and the cost functional J is minimized, that is, $u^\star(t) \in \mathbb{U}$ satisfies

$$\min_{u \in \mathbb{U}} J\left(x\left(t\right), u\left(t\right), t\right) = J\left(x^\star\left(t\right), u^\star\left(t\right), t\right), \tag{10.3}$$

where $x^\star(t)$ represents the optimal trajectory which is the state response driven by optimal control $u^\star(t)$.

For normal systems, the optimal solution to a general optimal control problem may be obtained from the well-known maximum principle. However, for general descriptor systems there may not exist such a principle. Up to now, much effort has been made toward solving this problem (see, Lovass-Nagy et al. 1986b; Lewis 1985b; Bender and Laub 1987a; Dai 1988). Typically, problems concerning optimal control are often treated by first converting them into optimal problems for normal systems, and then solve them via techniques for optimal control in normal system theory.

10.2 Optimal Linear Quadratic State Regulation

It is well known in the normal linear systems theory that linear quadratic optimal regulation is the most important and basic part in the optimal control theory for normal linear systems. In this section, we introduce the problem of linear quadratic optimal regulation in descriptor linear systems. Due to length limit, we only concentrate on the infinite time state feedback problem. For a more comprehensive and relatively complete theory of linear quadratic control for descriptor linear systems, please refer to the celebrated book Mehrmann (1991).

10.2.1 Problem Formulation

Consider again the descriptor linear system

$$E\dot{x}(t) = Ax(t) + Bu(t), \quad x(0) = x_0, \tag{10.4}$$

where $x(t) \in \mathbb{R}^n$ and $u(t) \in \mathbb{R}^r$ are the state and the control input, respectively; $E, A \in \mathbb{R}^{n \times n}$, and $B \in \mathbb{R}^{n \times r}$ are constant matrices, (E, A) is regular. The cost functional takes the quadratic form:

$$J = \int_0^{\infty} \left[x^{\mathrm{T}}(t) Q x(t) + u^{\mathrm{T}}(t) R u(t) \right] dt, \tag{10.5}$$

where the matrices Q and R are symmetric positive definite and are called the weighting matrices. This is a problem with an infinite time interval and a quadratic cost functional and is thus called an infinite time state quadratic regulation problem. It is supposed that the admissible controls are piecewise sufficiently and continuously differentiable functions.

Due to its simple form, optimal control is often adopted in practice. Since imposing control means consuming energy, the optimal control with a quadratic cost functional generally require less control input effort due to the existence of the symmetric positive definite weighting matrix R in the cost functional.

Regarding the conditions for existence and uniqueness of solutions to the above optimal control problem, Cobb (1983a) has proven the following basic result.

Theorem 10.1 (Cobb 1983a). *For the regular descriptor linear system (10.4) with the quadratic cost functional (10.5), there exists an optimal control u^* (t) such that the cost functional (10.5) is minimized if and only if system (10.4) is stabilizable and I-controllable.*

Due to the above theorem, the following assumption is imposed in this section.

Assumption 10.1. System (10.4) is stabilizable and I-controllable.

Based on the above, the optimal control problem to be solved in this section can be described as follows.

Problem 10.1. [Optimal linear quadratic regulation] Given the regular descriptor linear system (10.4) satisfying Assumption 10.1, and the quadratic cost functional (10.5), find an optimal control u^* (t), $0 < t < \infty$, for the system (10.4) such that the cost functional (10.5) is minimized.

The basic idea underlying in our development for solving the above problem is to convert the problem into a linear optimal quadratic regulation problem for a normal linear system.

10.2.2 The Conversion

Under Assumption 10.1, it follows from Theorem 7.6 that there exists a matrix $K \in \mathbb{R}^{r \times n}$ satisfying

$$\deg \det (sE - (A + BK)) = \operatorname{rank} E. \tag{10.6}$$

Choose a preliminary feedback control

$$u = Kx + v, \tag{10.7}$$

where $v \in \mathbb{R}^r$ is a new input, then, when the controller (10.7) is applied to system (10.4), the closed-loop system is obtained as

$$E\dot{x} = (A + BK) x + Bv. \tag{10.8}$$

According to the selection of the matrix K, (10.6) holds. Thus, it follows from Theorem 7.1 that the system has a standard decomposition form with $N = 0$, that is, there exist a pair of nonsingular matrices Q_1 and P_1 such that

$$Q_1 E P_1 = \operatorname{diag}\left(I_{n_0}, 0\right), \quad Q_1 (A + BK) P_1 = \operatorname{diag}(A_1, I),$$

where $n_0 = \mathrm{rank}E$. By denoting

$$P_1^{-1}x = \begin{bmatrix} x_1 \\ x_2 \end{bmatrix}, \quad x_1 \in \mathbb{R}^{n_0}, \; x_2 \in \mathbb{R}^{n-n_0}, \tag{10.9}$$

system (10.8) is r.s.e. to

$$\begin{cases} \dot{x}_1 = A_1 x_1 + B_1 v, & x_1(0) = x_{10} \\ 0 = x_2 + B_2 v, & x_2(0) = x_{20} \end{cases}, \tag{10.10}$$

where x_{10} and x_{20} are defined by

$$\begin{bmatrix} x_{10} \\ x_{20} \end{bmatrix} = P_1^{-1}x_0.$$

Lemma 10.1. *Let the regular descriptor linear system (10.4) satisfy Assumption 10.1, K be a real matrix satisfying (10.6), and (10.5) be the quadratic cost functional. Furthermore, let (10.10) be a standard decomposition form of the system (10.8), and define*

$$\Phi = \begin{bmatrix} P_1 & 0 \\ KP_1 & I \end{bmatrix}^{\mathrm{T}} \begin{bmatrix} Q & 0 \\ 0 & R \end{bmatrix} \begin{bmatrix} P_1 & 0 \\ KP_1 & I \end{bmatrix}, \tag{10.11}$$

$$\begin{bmatrix} \hat{Q} & H \\ H^{\mathrm{T}} & \tilde{R} \end{bmatrix} = \begin{bmatrix} I & 0 \\ 0 & -B_2 \\ 0 & I \end{bmatrix}^{\mathrm{T}} \Phi \begin{bmatrix} I & 0 \\ 0 & -B_2 \\ 0 & I \end{bmatrix}, \tag{10.12}$$

$$\tilde{Q} = \hat{Q} - H\tilde{R}^{-1}H^{\mathrm{T}}, \tag{10.13}$$

$$\tilde{A}_1 = A_1 - B_1\tilde{R}^{-1}H^{\mathrm{T}}. \tag{10.14}$$

Then under the controller substitution

$$u(t) = Kx(t) - \tilde{R}^{-1}H^{\mathrm{T}}x_1(t) + w(t), \tag{10.15}$$

the optimal control for the descriptor system (10.4) with the quadratic functional (10.5) can be converted into the optimal control for the reduced-order normal linear system

$$\dot{x}_1 = \tilde{A}_1 x_1 + B_1 w, \tag{10.16}$$

with the cost functional

$$J = \int_0^\infty (x_1^{\mathrm{T}}\tilde{Q}x_1 + w^{\mathrm{T}}\tilde{R}w)\,dt. \tag{10.17}$$

Proof. From the second equation in (10.10), we have

$$x_2 = -B_2 v.$$

Combining this with (10.7) and (10.9) gives

$$
\begin{bmatrix} x \\ u \end{bmatrix} = \begin{bmatrix} I & 0 \\ K & I \end{bmatrix} \begin{bmatrix} x \\ v \end{bmatrix}
$$

$$
= \begin{bmatrix} P_1 & 0 \\ K P_1 & I \end{bmatrix} \begin{bmatrix} x_1 \\ x_2 \\ v \end{bmatrix}
$$

$$
= \begin{bmatrix} P_1 & 0 \\ K P_1 & I \end{bmatrix} \begin{bmatrix} I & 0 \\ 0 & -B_2 \\ 0 & I \end{bmatrix} \begin{bmatrix} x_1 \\ v \end{bmatrix}.
$$

Thus, with the matrices H, $\tilde{Q}$, $\tilde{R}$, and $\hat{Q}$ defined by (10.11), (10.12), and (10.13), the cost functional (10.5) can be expressed as

$$
\begin{aligned}
J &= \int_0^\infty \begin{bmatrix} x \\ u \end{bmatrix}^{\mathrm{T}} \begin{bmatrix} Q & 0 \\ 0 & R \end{bmatrix} \begin{bmatrix} x \\ u \end{bmatrix} \mathrm{d}t \\
&= \int_0^\infty \begin{bmatrix} x_1 \\ v \end{bmatrix}^{\mathrm{T}} \begin{bmatrix} \hat{Q} & H \\ H^{\mathrm{T}} & \tilde{R} \end{bmatrix} \begin{bmatrix} x_1 \\ v \end{bmatrix} \mathrm{d}t \\
&= \int_0^\infty \left(x_1^{\mathrm{T}} \hat{Q} x_1 + v^{\mathrm{T}} H^{\mathrm{T}} x_1 + x_1^{\mathrm{T}} H v + v^{\mathrm{T}} \tilde{R} v \right) \mathrm{d}t \\
&= \int_0^\infty \left(x_1^{\mathrm{T}} \hat{Q} x_1 - x_1^{\mathrm{T}} H \tilde{R}^{-1} H^{\mathrm{T}} x_1 + x_1^{\mathrm{T}} H \tilde{R}^{-1} H^{\mathrm{T}} x_1 \right) \mathrm{d}t \\
&\quad + \int_0^\infty \left(v^{\mathrm{T}} H^{\mathrm{T}} x_1 + x_1^{\mathrm{T}} H v + v^{\mathrm{T}} \tilde{R} v \right) \mathrm{d}t \\
&= \int_0^\infty \left(x_1^{\mathrm{T}} \hat{Q} x_1 + w^{\mathrm{T}} \tilde{R} w \right) \mathrm{d}t,
\end{aligned}
\tag{10.18}
$$

which is the same as (10.17), where

$$w = v + \tilde{R}^{-1} H^{\mathrm{T}} x_1. \tag{10.19}$$

From (10.19), we obtain

$$v = w - \tilde{R}^{-1} H^{\mathrm{T}} x_1.$$

Substituting this into (10.10) and (10.7) yields the system given by (10.14) and (10.16) and the control input relation (10.15).

To finish the proof, we finally show that the matrices $\tilde{Q}$ and $\tilde{R}$ are symmetric positive definite. Note that the matrix diag (Q, R) is symmetric positive definite and the matrix

$$\begin{bmatrix} P_1 & 0 \\ KP_1 & I \end{bmatrix}$$

is nonsingular. We know that the matrix Φ is also symmetric positive definite. Furthermore, noting that the matrix

$$\begin{bmatrix} I & 0 \\ 0 & -B_2 \\ 0 & I \end{bmatrix}$$

is of full-column rank, we know that the matrix given in (10.12) is symmetric positive definite, and so are the matrices $\hat{Q}$ and $\tilde{R}$. Paying attention to the relationship

$$\begin{bmatrix} \tilde{Q} & 0 \\ 0 & \tilde{R} \end{bmatrix} = \begin{bmatrix} I & -H\tilde{R}^{-1} \\ 0 & I \end{bmatrix} \begin{bmatrix} \hat{Q} & H \\ H^{\mathrm{T}} & \tilde{R} \end{bmatrix} \begin{bmatrix} I & -H\tilde{R}^{-1} \\ 0 & I \end{bmatrix}^{\mathrm{T}}, \tag{10.20}$$

and the nonsingularity of the matrix

$$\begin{bmatrix} I & -H\tilde{R}^{-1} \\ 0 & I \end{bmatrix},$$

we know that the matrix $\tilde{Q}$ is also symmetric positive definite. With this, the proof is completed.   □

It follows from the above lemma that the optimal control problem for descriptor system (10.4) is converted into the problem of finding the new control w for the normal linear system (10.16) such that the cost functional (10.17) is minimized.

10.2.3 The Optimal Regulator

To solve the optimal linear quadratic regulation problem for the converted normal linear system (10.16) using the result in normal system theory, we need to prove the following lemma.

Lemma 10.2. *The matrix pair* $(\tilde{A}_1, B_1)$ *is stabilizable.*

Proof. It follows from the stabilizability assumption of system (10.4) that the matrix pair (A_1, B_1) is also stabilizable. Furthermore, note that $(\tilde{A}_1, B_1)$ is the closed-loop system resulted in by applying a state feedback with gain matrix $-\tilde{R}^{-1}H^{\mathrm{T}}$ to the system (A_1, B_1). Recall the fact that state feedback does not change the stabilizability of the system, we know that $(\tilde{A}_1, B_1)$ is also stabilizable.   □

By applying the result for quadratic regulation of normal linear systems, the solution to the optimal control for the system (10.16) minimizing the cost functional (10.17) is given by

$$w^\star = -\tilde{R}^{-1} B_1^{\mathrm{T}} P x_1^\star, \tag{10.21}$$

where $x_1^\star$ is the optimal trajectory, the matrix P is the unique symmetric positive definite solution of the following Riccati matrix equation

$$P\tilde{A}_1 + \tilde{A}_1^{\mathrm{T}} P - PB_1 \tilde{R}^{-1} B_1^{\mathrm{T}} P + \tilde{Q} = 0, \tag{10.22}$$

and the minimum value of J is

$$\min J = x_1^{\mathrm{T}}(0) P x_1(0) = x_{10}^{\mathrm{T}} P x_{10}.$$

Furthermore, in view of (10.7) and (10.19), the control $u(t)$ for the original descriptor linear system is linked with the new control $w(t)$ by (10.15), thus the following result concerning solution to our Problem 10.1 holds.

Theorem 10.2. *For the regular descriptor linear system (10.4) which satisfies Assumption 10.1 and has the standard decomposition (10.10), the optimal control $u^\star(t)$ for the descriptor system (10.4), which minimizes the quadratic functional (10.5), is given by*

$$u^\star(t) = K^\star x^\star(t), \tag{10.23}$$

with

$$K^\star = K - \tilde{R}^{-1} \left(H^{\mathrm{T}} + B_1^{\mathrm{T}} P \right) \begin{bmatrix} I & 0 \end{bmatrix} P_1^{-1},$$

and the optimal trajectory $x^\star(t)$ is governed by

$$E\dot{x}^\star(t) = \left(A + BK^\star \right) x^\star(t), \tag{10.24}$$

which is stable and impulse-free. Furthermore, the minimum value of J is

$$\min J = x_0^{\mathrm{T}} \left(P_1^{\mathrm{T}} \right)^{-1} \begin{bmatrix} I \\ 0 \end{bmatrix} P \begin{bmatrix} I & 0 \end{bmatrix} P_1^{-1} x_0, \tag{10.25}$$

where the matrix P is the symmetric positive definite solution to the Riccati equation (10.22), while the matrices P_1, $\tilde{R}$ and $\tilde{Q}$ are as previously defined, and the matrix K is any matrix satisfying (10.6).

It should be noted that the optimal quadratic regulator (10.23) is in the form of state feedback.

By the above theorem, an algorithm for solving the optimal linear quadratic regulation problem for the descriptor linear system (10.4) with the quadratic cost functional (10.5) can be given as follows.

Algorithm 10.1. Optimal linear quadratic regulation by state feedback.

Step 1 Verify the stabilizability and I-controllability of system (10.4).

Step 2 Choose a matrix K such that $\deg \det (sE - (A + BK)) = \operatorname{rank} E$.

Step 3 Determine the matrices A_1, B_1, B_2 in the standard decomposition (10.10) as well as the transformation matrices Q_1 and P_1.

Step 4 Determine the matrices $\hat{Q}$, H, $\tilde{R}$, $\tilde{Q}$, and $\tilde{A}_1$ according to (10.12), (10.13), and (10.14).

Step 5 Find the unique solution P to the Riccati matrix equation (10.22).

Step 6 Finally, obtain the optimal control $u^*(t)$ and trajectory $x^*(t)$ using (10.23) and (10.24).

Remark 10.1. In the optimal regulation problem with quadratic cost functional for normal linear systems, to obtain the solution we need to solve a Riccati equation of order n, the same as the order of the system. However, as seen from (10.22), a Riccati equation of order $\operatorname{rank} E \leq n$ is needed for the descriptor system case. This produces much convenience when $\operatorname{rank} E$ is much smaller than n.

10.2.4 An Illustrative Example

Example 10.1. Consider again the system (4.80) in Example 4.7. Its state equation is

$$\begin{bmatrix} 1 & 0 & 0 \\ 0 & 1 & 0 \\ 0 & 0 & 0 \end{bmatrix} \dot{x} = \begin{bmatrix} 0 & 1 & 0 \\ 0 & 0 & 1 \\ 0 & 0 & -1 \end{bmatrix} x + \begin{bmatrix} 0 & 0 \\ 1 & 0 \\ 0 & 1 \end{bmatrix} u.$$

Our purpose is to design a state feedback controller to minimize the cost functional

$$\int_0^\infty \left(x^T x + u^T R u \right) dt.$$

In this problem, $\operatorname{rank} E = 2$, $n = 3$, $Q = I_3$, and

$$R = \begin{bmatrix} 2 & 0 \\ 0 & 1 \end{bmatrix}.$$

Furthermore, the system can be easily verified to be C-controllable, thus it is stabilizable and I-controllable. Therefore, the proposed Algorithm 10.1 can be applied to solve this linear quadratic regulation problem.

It can be easily checked that the matrix

$$K = \begin{bmatrix} -1 & -1 & -1 \\ 0 & 0 & -1 \end{bmatrix}$$

satisfies

$$\deg \det (sE - (A + BK)) = \operatorname{rank} E.$$

Direct verification shows that with the nonsingular matrices

$$Q_1 = \begin{bmatrix} 1 & 0 & 0 \\ 0 & 1 & 0 \\ 0 & 0 & 1 \end{bmatrix}, \ P_1 = \begin{bmatrix} 1 & 0 & 0 \\ 0 & 1 & 0 \\ 0 & 0 & -\frac{1}{2} \end{bmatrix},$$

we have

$$Q_1 E P_1 = \operatorname{diag} (I_2, 0), \ Q_1 (A + BK) P_1 = \operatorname{diag} (A_1, 1),$$

where

$$A_1 = \begin{bmatrix} 0 & 1 \\ -1 & -1 \end{bmatrix}, \ B_1 = \begin{bmatrix} 0 & 0 \\ 1 & 0 \end{bmatrix}, \ B_2 = \begin{bmatrix} 0 & 1 \end{bmatrix}.$$

For this system, solving (10.12)–(10.14), gives

$$\hat{Q} = \begin{bmatrix} 3 & 2 \\ 2 & 3 \end{bmatrix}, \ \tilde{R} = \begin{bmatrix} 2 & -1 \\ -1 & 1 \end{bmatrix}, \ \tilde{Q} = \begin{bmatrix} 1 & 0 \\ 0 & 1 \end{bmatrix},$$

$$\tilde{A}_1 = \begin{bmatrix} 0 & 1 \\ 0 & 0 \end{bmatrix}, \ H = \begin{bmatrix} -2 & 1 \\ -2 & 1 \end{bmatrix}.$$

Thus, the Riccati equation (10.22) is

$$P \begin{bmatrix} 0 & 1 \\ 0 & 0 \end{bmatrix} + \begin{bmatrix} 0 & 0 \\ 1 & 0 \end{bmatrix} P + \begin{bmatrix} 1 & 0 \\ 0 & 1 \end{bmatrix} - P \begin{bmatrix} 0 & 0 \\ 0 & 1 \end{bmatrix} P = 0. \qquad (10.26)$$

Let

$$P = \begin{bmatrix} p_1 & p_2 \\ p_2 & p_3 \end{bmatrix}.$$

By equating the corresponding elements on both sides of the matrix equation (10.26), we obtain the equation group

$$\begin{cases} 1 - p_2^2 = 0 \\ p_1 - p_2 p_3 = 0 \\ 2p_2 - p_3^2 + 1 = 0 \end{cases},$$

which has the unique solution

$$p_1 = \sqrt{3}, \ p_2 = 1, \ p_3 = \sqrt{3}.$$

Thus, the unique positive definite solution to the Riccati matrix equation (10.26) is

$$P = \begin{bmatrix} \sqrt{3} & 1 \\ 1 & \sqrt{3} \end{bmatrix}.$$

According to (10.23) and (10.24), the optimal control law for this problem is

$$u^\star(t) = \left[K - \tilde{R}^{-1} \left(H^T + B_1^T P \right) \left[I \ 0 \right] P_1^{-1} \right] x^\star(t)$$

$$= \begin{bmatrix} -1 & -\sqrt{3} & -1 \\ -1 & -\sqrt{3} & -1 \end{bmatrix} x^\star(t)$$

and the corresponding optimal trajectory is governed by

$$\begin{bmatrix} 1 & 0 & 0 \\ 0 & 1 & 0 \\ 0 & 0 & 0 \end{bmatrix} \dot{x}^\star(t) = \begin{bmatrix} 0 & 1 & 0 \\ -1 & -\sqrt{3} & 0 \\ -1 & -\sqrt{3} & -2 \end{bmatrix} x^\star(t).$$

It can be verified that this optimal closed-loop system has the finite pole set

$$\Gamma = \left\{ -2, -\frac{\sqrt{3}}{6} + \frac{1}{2}i, -\frac{\sqrt{3}}{6} - \frac{1}{2}i \right\},$$

hence is stable and impulse-free.

10.3 Time-Optimal Control

In this section, let us consider the time-optimal control of a descriptor linear system. This corresponds to the so-called bang–bang control for normal linear systems.

10.3.1 Problem Formulation

Consider again the system (10.4),

$$E\dot{x} = Ax + Bu, \quad x(0) = x_0, \tag{10.27}$$

where $x(t) \in \mathbb{R}^n$ and $u(t) \in \mathbb{R}^r$ are the state and the control input, respectively, $E, A \in \mathbb{R}^{n \times n}$, and $B \in \mathbb{R}^{n \times r}$ are constant matrices, (E, A) is regular, $x_0 \in \mathbb{R}^n$ is the system initial value.

The initial time is taken as $t_0 = 0$ and the cost functional takes the following special form

$$J = t,   (10.28)$$

which represents the time period taken to drive the state from one position to another. Let the terminal time be t_1, and the terminal state be $x_1 = x(t_1)$, then the problem is to find a control $u^*(t)$ for system (10.27) such that the state starting from $x_0 = x(0)$ is driven to $x_1 = x(t_1)$ within a minimum time period, i.e.,

$$t_1 = \min J.$$

Such a problem corresponds to the bang–bang control for normal linear systems.

According to the definition of controllability given in Chap. 4, if the control input is unconstrained, the state of a C-controllable system may be driven from an arbitrary initial state to any terminal state within an arbitrarily short period. In this case, the time-optimal problem does not make sense. Therefore, when the time-optimal problem is treated, the control is always constrained. In this section, we assume that the control

$$u = [u_1, u_2, \cdots, u_r]^{\mathrm{T}}   (10.29)$$

is bounded as follows:

$$|u_i| \leq \alpha_i, \quad i = 1, 2, \ldots, r,   (10.30)$$

and $u(t)$ is piecewise sufficiently and continuously differentiable. Furthermore, for the sake of simplicity, the terminal state is assumed to be zero, i.e., $x_1 = x(t_1) = 0$.

10.3.2 Time-Optimal Control of the Slow and Fast Subsystems

For any regular system (10.27), there exist two nonsingular matrices Q and P such that, under the transformation (P, Q), the system (10.27) is r.s.e. to the following standard decomposition form:

$$\begin{cases} \dot{x}_1 = A_1 x_1 + B_1 u \\ N \dot{x}_2 = x_2 + B_2 u \end{cases},   (10.31)$$

where $A_1 \in \mathbb{R}^{n_1 \times n_1}$, $N \in \mathbb{R}^{n_2 \times n_2}$, $n_1 + n_2 = n$, and N is nilpotent.

Now we examine the time-optimal control problems for these two subsystems in (10.31).

Time-Optimal Control of the Slow Subsystem

The slow subsystem

$$\dot{x}_1 = A_1 x_1 + B_1 u, \quad x_1(0) = x_{10}   (10.32)$$

is a normal linear system. From normal linear systems theory, we know that under constraint (10.30) the optimal control that drives any initial state of system (10.32) to a prescribed terminal state in the minimum time is the switching control or the so-called bang–bang control, which takes the extreme values of the constrained control elements.

Lemma 10.3. [Bang–bang control] *Consider the normal linear system (10.32), and let x_{10}, $x_{1f} \in \mathbb{R}^{n_1}$. Then the optimal control input which satisfies the constraints in (10.29)–(10.30) and drives within minimum time t_f the state of the system from $x_1(0) = x_{10}$ to $x_1(t_f) = x_{1f}$ is given by*

$$u_i^*(t, \lambda) = -\alpha_i \operatorname{sign}\left[\lambda^{\mathrm{T}} e^{A_1(t_f - t)} b_i\right], \quad i = 1, 2, \ldots, r, \tag{10.33}$$

where b_i is the i-th column of matrix B_1, and the optimal time t_f and the vector λ are determined by the following boundary conditions

$$x_{1f} = e^{A_1 t_f} x_{10} + \int_0^{t_f} e^{A_1(t_f - \tau)} B_1 u^*(\tau, \lambda) d\tau \tag{10.34}$$

and

$$\lambda^{\mathrm{T}}\left[A_1 x_f + B_1 u^*(t_f, \lambda)\right] = -1. \tag{10.35}$$

Time-Optimal Control of the Fast Subsystem

Consider the fast subsystem

$$N \dot{x}_2 = x_2 + B_2 u, \quad x_2(0) = x_{20}, \tag{10.36}$$

where the matrix N is with nilpotent index h.

For this system, the distributional solution is

$$x_2(t) = -\sum_{i=1}^{h-1} \delta^{(i-1)}(t) N^i x_{20} - \sum_{i=0}^{h-1} \sum_{k=0}^{i-1} N^i B_2 \delta^{(k)}(t) u^{(i-k-1)}(0)$$

$$- \sum_{i=0}^{h-1} N^i B_2 u^{(i)}(t), \tag{10.37}$$

the consistent initial value is

$$x_{20}(u) = -\sum_{i=0}^{h-1} N^i B_2 u^{(i)}(0), \tag{10.38}$$

and, with this initial value, the first two terms in (10.37) vanish. Thus,

$$x_2(t) = -\sum_{i=0}^{h-1} N^i B_2 u^{(i)}(t), \text{ when } x_{20}(0) = x_{20}(u),$$

and

$$\begin{cases} x_2(0) = \infty \\ x_2(t) = -\sum_{i=0}^{h-1} N^i B_2 u^{(i)}(t), \ t > 0 \end{cases}, \text{ when } x_{20}(0) \neq x_{20}(u).$$

Particularly, when $u(t) \equiv 0$, the consistent initial value turns to be $x_{20}(u) = 0$, in this case we have

$$x_2(t) \equiv 0, \text{ when } x_{20} = 0,$$

and

$$\begin{cases} x_2(0) = \infty \\ x_2(t) \equiv 0, \ t > 0 \end{cases}, \text{ when } x_{20} \neq 0.$$

The above fact shows that, with the control $u(t) \equiv 0$, the state can be driven to zero in any short period. However, the time period generally cannot be zero since impulse terms exist at the initial instant point $t_0 = 0$ when the system initial value is not taken zero. This shows that $u(t) \equiv 0$ is not an optimal solution to the time-optimal problem for a fast subsystem, but it can be arbitrarily close to an optimal one. This is because the corresponding minimum time, though not a fixed one, can be arbitrarily small.

10.3.3 The Solution

In view of the above discussion, we introduce the following definition.

Definition 10.1. Consider system (10.27) with the control constraints (10.30). Let $\varepsilon > 0$ be a prescribed scalar. If there exist an admissible control $u_\varepsilon^\star(t)$ and a time point t_1 such that the state $x(t)$ is driven to zero at time t_1, i.e., $x(t_1) = 0$, but there exists no admissible control such that the state $x(t)$ is driven to zero at $t_1 - \varepsilon$, then the control $u_\varepsilon^\star(t)$ is called an ε-suboptimal control and the time t_1 is called the ε-suboptimal time.

It is seen from the above definition that the ε-suboptimal control always exists for a fast subsystem. The ε-optimal control, although is not strict, is practically acceptable when ε is sufficiently small.

Based on the above results and definition, we have the following theorem about solution to the time-optimal problem for system (10.27).

Theorem 10.3. *Consider the regular descriptor linear system (10.27) with its slow subsystem (10.32). Let t_1 be the optimal time for the state of the slow subsystem*

(10.32) to reach zero, and $\bar{u}^\star(t)$ be the corresponding time-optimal control such that $x_1^\star(t_1) = 0$. Then for any $\varepsilon > 0$, the following control

$$u_\varepsilon^\star(t) = \begin{cases} \bar{u}^\star(t), & 0 \le t \le t_1, \\ 0, & t_1 < t \le t_1 + \varepsilon, \end{cases} \qquad (10.39)$$

is an ε-suboptimal control for system (10.27) such that the state of system (10.27) is driven to zero at time $t_1 + \varepsilon$, that is, $x(t_1 + \varepsilon) = 0$.

Proof. Let t_1 be the optimal time for the slow subsystem (10.32) and $\bar{u}^\star(t)$ be the corresponding time-optimal control, $0 \le t \le t_1$, such that $x_1^\star(t_1) = 0$. Recalling the state response of the slow subsystem, we have

$$x_1^\star(t_1) = e^{A_1 t_1} x_1(0) + \int_0^{t_1} e^{A_1(t_1-t)} B_1 \bar{u}^\star(t)\, dt = 0.$$

With the control defined in (10.39), we have

$$x_1^\star(t) = 0, \quad t \ge t_1.$$

Let (10.36) be a fast subsystem of (10.27), then there exists some $\varepsilon > 0$ such that

$$x_2^\star(t_1 + \varepsilon) = 0.$$

Combining the above two relations yields

$$x^\star(t_1 + \varepsilon) = 0.$$

Therefore, by definition $u^\star(t)$ is the ε-suboptimal control. $\qquad\qquad\square$

It clearly follows from the above theorem that the solvability of the time-optimal control for the descriptor system (10.27) is governed by that of its slow subsystem. Based on the above result and the proof of Theorem 10.3, the following algorithm can be given for solving the time-optimal control problem of the descriptor linear system (10.27).

Algorithm 10.2. Time-optimal control.

Step 1 Obtain the standard decomposition form (10.31) for system (10.27).

Step 2 Solve the time-optimal controller $\bar{u}^\star(t)$ for the slow subsystem according to Lemma 10.3.

Step 3 Solve the ε-suboptimal controller $u_\varepsilon^\star(t)$ for system (10.27) according to (10.39).

Example 10.2. Consider the following system

$$
\begin{bmatrix} 0 & 0 & 1 \\ 1 & 0 & 0 \\ 0 & 0 & 0 \end{bmatrix} \dot{x} = \begin{bmatrix} 0 & \frac{1}{2} & -\frac{1}{2} \\ -1 & 0 & 0 \\ 0 & 0 & \frac{1}{2} \end{bmatrix} x + \begin{bmatrix} -\frac{1}{2} \\ 1 \\ \frac{1}{2} \end{bmatrix} u, \qquad (10.40)
$$

with the initial state condition

$$
x(0) = \begin{bmatrix} -1 \\ 0 \\ 0 \end{bmatrix},
$$

and the control constraint

$$
|u| \le 2.
$$

A standard decomposition form for system (10.40) can be easily obtained as

$$
\begin{bmatrix} 1 & 0 & 0 \\ 0 & 0 & 1 \\ 0 & 0 & 0 \end{bmatrix} \begin{bmatrix} \dot{x}_1 \\ \dot{x}_2 \\ \dot{x}_3 \end{bmatrix} = \begin{bmatrix} -1 & 0 & 0 \\ 0 & 1 & 0 \\ 0 & 0 & 1 \end{bmatrix} \begin{bmatrix} x_1 \\ x_2 \\ x_3 \end{bmatrix} + \begin{bmatrix} 1 \\ 0 \\ 1 \end{bmatrix} u,
$$

for which the initial state condition is

$$
\begin{bmatrix} x_1(0) \\ x_2(0) \\ x_3(0) \end{bmatrix} = \begin{bmatrix} -1 \\ 0 \\ 0 \end{bmatrix}.
$$

Therefore, the slow subsystem is

$$
\dot{x}_1 = -x_1 + u, \quad x_1(0) = -1.
$$

Using bang–bang control theory for normal linear systems, the time optimal control for this subsystem can be obtained as

$$
\bar{u}^\star(t) = 2, \quad 0 \le t \le \ln\frac{3}{2},
$$

with the optimal terminal time

$$
t_1 = \ln\frac{3}{2}.
$$

Combining this with Theorem 10.3, we know that for any prescribed scalar $\varepsilon > 0$, the ε-suboptimal control is

$$
u_\varepsilon^\star(t) = \begin{cases} 2, & 0 \le t \le \ln\frac{3}{2} \\ 0, & \ln\frac{3}{2} < t \le \ln\frac{3}{2} + \varepsilon \end{cases},
$$

and under this control the state of the closed-loop system reaches zero at time $t_1 = \ln\frac{3}{2} + \varepsilon$.

10.4 Notes and References

In this chapter, optimal control of descriptor linear systems is considered, with emphases on the infinite time state quadratic regulation problem and the time-optimal control problem.

Further Comments

It is clearly seen that the basic idea underlying in the solutions to the infinite time state quadratic regulation problem and the time-optimal control problem is to first convert the problems into corresponding ones for normal linear systems and then solve the problems by adopting well-known techniques for normal linear systems. Again, standard decomposition has performed a fundamental role in realizing this idea. Here, we mention again that, for computation of a standard decomposition form of a given descriptor linear system, the Matlab Toolbox for linear descriptor systems introduced in Varga (2000) can be readily used, which was developed based on the basic techniques proposed in Misra et al. (1994).

Regarding the time-optimal control, we have seen from this chapter that, unlike the normal linear systems case, there generally does not exist an optimal solution in the descriptor linear systems case, but a suboptimal solution can drive the state of the system to the desired destination within a time period, which is arbitrarily close to the 'minimal one'. Regarding the optimal linear quadratic regulation problem, one should bear in mind that the condition for existence of a solution is that the system is stabilizable and I-controllable.

It is worth pointing out that deviations always exist in a real system model, thus causing the optimal control to lose its expected property. In some cases, a nonoptimal control law may have better practical effect than a theoretically optimal one due to these reasons. However, the optimal method provides an effective approach for control system design, which proves to be often reliable and useful in practice.

Finite-Time Quadratic Regulation

Following the same lines in Sect. 10.2, the finite-time state optimal quadratic regulation for the system (10.1) can be easily solved. The problem is to minimize the cost functional

$$ J = \frac{1}{2} x^{\mathrm{T}}(T) M x(T) + \frac{1}{2} \int_0^T \left[x^{\mathrm{T}}(t) Q x(t) + u^{\mathrm{T}}(t) R u(t) \right] \mathrm{d}t, $$

where $T > 0$ is a finite time, M is some real symmetric semipositive definite matrix. Similar to the treatment in Sect. 10.2, based on the standard decomposition the problem can be converted into a problem of finite-time quadratic regulation for a normal linear system. Details of the solution can be found in some published papers and books (e.g. Dai 1989b; Zhang and Yang 2003; Mehrmann 1991).

Due to length limit, the output quadratic regulation problem is not discussed, in which the cost functional takes the following form:

$$J = \frac{1}{2} y^{\mathrm{T}}(T) M y(T)$$
$$+ \frac{1}{2} \int_{t_0}^{T} \left[y^{\mathrm{T}}(t) Q y(t) + u^{\mathrm{T}}(t) R u(t) + u^{\mathrm{T}}(t) S^{\mathrm{T}} y(t) + y^{\mathrm{T}}(t) S u(t) \right] \mathrm{d}t,$$

where $S \in \mathbb{R}^{m \times r}$, $M = M^{\mathrm{T}} \in \mathbb{R}^{m \times m}$, $Q = Q^{\mathrm{T}} \in \mathbb{R}^{m \times m}$, $R = R^{\mathrm{T}} \in \mathbb{R}^{r \times r}$, and $y(t)$ is the system output given by the following output equation

$$y(t) = C x(t). \tag{10.41}$$

Mehrmann (1991) has given a thorough treatment on the above finite-time output regulation problem. The solution again adopts the standard decomposition of descriptor linear systems, and also depends on the concepts of strong stabilization and strong observation of the descriptor linear system (10.1) with the output equation (10.41).

A Brief Overview on Optimal Control

The materials in this chapter are closely related to Cheng et al. (1987), Campbell (1982) and Bender and Laub (1987b). Some of the other early relevant references are Bender and Laub (1987a), Cobb (1983a), Lewis (1985b), Lovass-Nagy et al. (1986b), Luenberger (1987), Kurina (1983), and Pandolfi (1981). More recently, there are also many reported results on the linear quadratic optimal control problem. The methods can be divided into, but not limited to, the following several categories.

Constant-Time Systems Regarding the optimal control of constant-time descriptor linear systems, there are numerous reported works. Besides the work of Cheng et al. (1987), Campbell (1982), and Bender and Laub (1987b), Mehrmann (1989) also analyzed the existence, uniqueness, and stability of solutions of singular, linear quadratic optimal control problems under different singularity assumptions and also showed that the continuous and discrete problems yield essentially the same results and can thus be studied analogously. Benner et al. (1999b) gave a numerical solution of linear quadratic optimal control problems for descriptor systems by extending the recently developed methods for Hamiltonian matrices to the general case of embedded pencils as they arise in descriptor systems. Muller (1999) considered linear quadratic optimal control of a type of constant-time linear mechanical descriptor systems. Zhu et al. (2002) studied a singular linear quadratic problem for nonregular descriptor systems. Kurina (2002) constructed optimal feedback control depending only on the system state for a control problem by the noncausal descriptor system using a nonsymmetric solution of the algebraic Riccati equation. Chen and Cheng (2004) dealt with singular linear quadratic suboptimal control problem with disturbance rejection for descriptor systems. Lu et al. (2004) also studied the singular linear quadratic problem for descriptor systems, and provided the optimal control such that the closed-loop system is regular, impulse-free, and stable. Stefanovski (2006) presented results concerning linear quadratic control, using the Silverman algorithm, of arbitrary descriptor systems and a numerical algorithm for achieving it.

Time-Varying Systems Several researchers have studied the linear quadratic optimal control problems for linear variable coefficient descriptor systems (Kunkel and Mehrmann 1997; Kurina 2003; Kurina and Marz 2004). Kurina (2003) obtained the controls in the feedback form and the optimal cost for the differential minimax linear quadratic control problems by descriptor systems with properly stated leading term in a Hilbert space while Kurina and Marz (2004) treated linear quadratic optimal control for time-varying descriptor systems in a Hilbert space setting.

Discrete-Time Systems Mehrmann (1988) introduced a new numerically stable, structure-preserving method for the discrete linear quadratic control problem with single input or single output while Kurina (1993) obtained the sufficient conditions of optimal control in the maximum principle form for linear discrete descriptor systems and considered the linear quadratic optimal control problem. Linear-quadratic control problems of discrete descriptor systems in a Hilbert space was obtained by Kurina (2001). Special solutions of the implicit operator Riccati equations were used, which act in all state space and, in general case, are not nonnegative definite. Recently, Kurina (2004) established the solvability of the linear quadratic discrete optimal control problem for descriptor systems with variable coefficients, also in Hilbert space.

H_∞/H_2 **Optimal Control** Besides the quadratic optimal control, H_∞/H_2 control has also attracted much attention. Ishihara and Terra (1999) examined the continuous-time H_2 optimal control for descriptor systems. Yang et al. (2005b) analyzed the H_2 suboptimal performance and studied the observer-based H_2 suboptimal control problem for linear continuous-time descriptor systems. Xing et al. (2005) investigated the H_∞ control problem based on a state observer for descriptor systems. Benner and his co-authors discussed the numerical solution of linear quadratic optimal control problems and H_∞ control problems (Benner et al. 1999a), and also proposed a robust numerical method for optimal H_∞ control (Benner et al. 2004). Optimal robust mixed H_2/H_∞ control for descriptor systems with time-varying uncertainties and distributed delays has been treated by Yue and Lam (2004), and the problem of a reliable observer-based controller with mixed H_2/H_∞ performance for descriptor systems using linear matrix inequality (LMI) approach has been addressed by Chen et al. (2005).

Others Besides the singular linear quadratic problem, some other problems related to optimal control have also been studied. These include sensitivity analysis for solutions of a parametric optimal control problem of a descriptor system (Kostyukova 2001), parametric optimal control for a descriptor system whose initial state depends on a parameter (Kostyukova 2003), optimal control for proper and nonproper descriptor dynamic systems (Muller 2003), guaranteed cost control of descriptor systems with time-delays (Fu and Duan 2006; Yue et al. 2005; Zhang and Yu 2006; Ren 2006a), nonlinear optimal control of descriptor systems (Glad and Sjoberg (2006)), and nonfragile control of nonlinear descriptor systems (Ren 2005; Ren 2006b).

To end this section, we finally mention that there is an important book addressing the quadratic control problem of descriptor linear systems (Mehrmann 1991). This book first gives a survey on the state of the art in the theory and the numerical solutions of general linear quadratic optimal control problems with differential algebraic equation constraints, and then provides new and extended theory and numerical methods for these problems. Quadratic optimal control problems occur in many practical applications, while this book not only presents a comprehensive theory and useful methods for quadratic optimal control of descriptor linear systems, but also helps increase the cooperation between pure mathematicians, numerical analysts, and practitioners.

Chapter 11
Observer Design

In this chapter, we consider the problem of observer design for descriptor linear systems, which is involved with reconstructing asymptotically the state vector or a linear combination of the state variables of a descriptor linear system.

After a general introduction about observers in Sect. 11.1, a type of full-order descriptor state observers studied in some detail in Sects. 11.2–11.4. Section 11.2 gives the condition for this type of observers and illustrates two types of design approaches for such observers. A particular design approach, which is called parametric eigenstructure assignment, is introduced in Sect. 11.3 for this type of full-order state observers. In Sect. 11.4, the design of such type of full-order observers with disturbance decoupling property is further investigated based on the parametric eigenstructure assignment approach proposed in Sect. 11.3. Section 11.5 introduces two types of normal reduced-order state observers, while Sect. 11.6 investigates the Luenberger-type observers for descriptor linear systems. A necessary and sufficient condition is first provided, and then a parametric design approach for such observers is proposed. Some notes and remarks follow in Sect. 11.7.

11.1 Introduction

We have seen from the last few chapters that state feedback control is very important in certain system designs. Under certain conditions, state feedback enables us to place the pole structure or the eigenstructure of a system such that the closed-loop system has some satisfactory properties. Many applications in deterministic systems design have shown that state feedback is very convenient and practical. However, state feedback is based on the assumption that the state is available, which is unfortunately not always true. In practice, the state is usually not directly available. What we can obtain is the control input $u(t)$ and the measured output $y(t)$ rather than the state itself. So the state feedback control usually cannot be realized directly. In general, there are two popular approaches to overcome this difficulty. The first one is to apply output feedback, that is, the measured output is used for feedback instead of the state. The second method first asymptotically estimates the state vector $x(t)$

G.-R. Duan, *Analysis and Design of Descriptor Linear Systems*, Advances
in Mechanics and Mathematics 23, DOI 10.1007/978-1-4419-6397-0_11,
© Springer Science+Business Media, LLC 2010

or the required $Kx(t)$ from the control input $u(t)$ and the measured output $y(t)$, and then use the estimate of $x(t)$ or $Kx(t)$ to construct the state feedback control.

Besides realizing state feedback control, observer design is also needed in some other applications. One of such applications is in active fault detection and isolation. There are many approaches for fault detection, among which the observer-based approaches have attracted much attention (see Duan et al. 2002a; Duan and Patton 2001, and the references therein). The basic idea behind the observer-based approaches is to construct a residual signal by using some types of observers. This residual is then examined for the likelihood of faults.

11.1.1 State Observers

Consider the descriptor linear system

$$\begin{cases} E\dot{x}(t) = Ax(t) + Bu(t) \\ y(t) = Cx(t) \end{cases}, \tag{11.1}$$

where $x(t) \in \mathbb{R}^n$, $u(t) \in \mathbb{R}^r$, and $y(t) \in \mathbb{R}^m$ are the state, the control input, and the measured output, respectively; E, A, B, and C are constant matrices of appropriate dimensions. It is assumed that system (11.1) is regular and $n_0 = \text{rank} E \leq n$.

In system (11.1), its initial state $x(0)$ is usually not known in advance, therefore it is difficult or impossible to reconstruct exactly the state vector $x(t)$. The so-called state observer is involved with reconstructing the state vector $x(t)$ of (11.1) asymptotically. The basic idea is to construct a system which takes the control input and the measured output of system (11.1) as inputs, and its output approaches the state of system (11.1) asymptotically for arbitrary given system initial values and control signal $u(t)$. Such a system is called a state observer for system (11.1). More strictly, we have the following definition for a state observer of a descriptor linear system.

Definition 11.1. Consider the regular system (11.1), and the constructed system

$$\begin{cases} \hat{E}\dot{z}(t) = Fz(t) + Su(t) + Ly(t) \\ w(t) = Mz(t) + Ny(t) + Hu(t) \end{cases}, \tag{11.2}$$

where $z(t) \in \mathbb{R}^p$, $w(t) \in \mathbb{R}^n$, $\hat{E}$, $F \in \mathbb{R}^{p \times p}$, and S, L, M, N, and H are constant matrices of appropriate dimensions. If

$$\lim_{t \to \infty} (w(t) - x(t)) = 0 \tag{11.3}$$

holds for arbitrarily initial conditions $z(0)$, $x(0)$, and arbitrarily given input signal $u(t)$, then system (11.2) is called a (linear) state observer for system (11.1). Particularly, if $\text{rank} \hat{E} < p$, (11.2) is called a descriptor state observer; otherwise, $\text{rank} \hat{E} = p$, then (11.2) is called a normal state observer. The integer p is called the order of the state observer. When $p = n$, (11.2) is called a full-order state observer, when $p < n$, (11.2) is called a reduced-order state observer. Furthermore, any pole of system (11.2) is called an observer pole, and $\sigma(\hat{E}, F)$ is called the pole set of observer (11.2).

In a normal-state observer, the matrix $\hat{E}$ is usually taken to be the identity matrix. In this case, a normal-state observer takes the following form:

$$\begin{cases} \dot{z}(t) = Fz(t) + Su(t) + Ly(t) \\ w(t) = Mz(t) + Ny(t) + Hu(t) \end{cases}. \tag{11.4}$$

In this chapter, only one type of full-order descriptor state observers, which takes the special form of (11.4), is considered (Sect. 11.2), and two types of reduced-order normal state observers of order $\text{rank}\,E$ and $n - m$, respectively, are studied (Sect. 11.5).

11.1.2 Function Kx Observers

As mentioned above, one of the main purposes of constructing an observer is to realize state feedback control. For this purpose, not all the state variables are required to be reconstructed, only a certain linear combination of the state variables would be sufficient for use. In this case, we may construct a system which produces directly this linear combination of the state variables. This gives rise to the so-called function Kx observer.

Definition 11.2. Consider the regular system (11.1), and let $K \in \mathbb{R}^{p \times n}$. If there exists a dynamic system in the form of (11.2) with $z \in \mathbb{R}^p$, $w \in \mathbb{R}^p$, and $\hat{E}$, F, S, M, L, N, and H being constant matrices of appropriate dimensions, such that

$$\lim_{t \to \infty} (Kx(t) - w(t)) = 0 \tag{11.5}$$

holds for arbitrarily given initial values $x(0)$, $z(0)$ and input signal $u(t)$, then the system (11.2) is called a function Kx observer for system (11.1). Specifically, when $\text{rank}\hat{E} < p$, system (11.2) is called a descriptor Kx observer, otherwise, $\text{rank}\hat{E} = p$, (11.2) is termed a normal Kx observer.

In a normal function observer, the matrix $\hat{E}$ is usually taken to be the identity matrix. Therefore, a normal function observer for system (11.1) also takes the form of (11.4), but with $w \in \mathbb{R}^p$ and satisfying (11.5) instead of (11.3).

Let $H = 0$, and $\hat{E} = I$, the function observer (11.2) reduces to the following form:

$$\begin{cases} \dot{z} = Fz + Su + Ly \\ w = Mz + Ny \end{cases}, \tag{11.6}$$

which is of the well-known Luenberger-type function observers. For simplicity, this chapter only considers the design of this Luenberger-type normal function observer (11.6). Conditions and design method are given in Sect. 11.6.

11.2 Descriptor State Observers

In this section, we are concerned with the design of a simple type of full-order descriptor state observers for system (11.1) in the following form:

$$
\begin{cases}
E\dot{z} = Az + Bu + L(y - Cz) \\
w = z
\end{cases}
, \tag{11.7}
$$

where $L \in \mathbb{R}^{n \times m}$ is a real design parameter matrix, which is called the observer gain matrix. The output w is expected to approach the state of system (11.1). This type of observers are often written in the following form:

$$
E\dot{\hat{x}} = A\hat{x} + Bu + L(y - C\hat{x}), \tag{11.8}
$$

where the state $\hat{x}$ is expected to reconstruct the state of system (11.1) asymptotically.

11.2.1 Existence Condition

The following theorem gives the condition for system (11.7) to be a simple type of full-order state observer for system (11.1).

Theorem 11.1. *The system (11.7), or (11.8) forms a full-order descriptor state observer for the regular system (11.1) if and only if the gain matrix L satisfies*

$$
\sigma(E, A - LC) \subset \mathbb{C}^-. \tag{11.9}
$$

Proof. We need only to show that the two combined systems (11.1) and (11.8) satisfy

$$
\lim_{t \to \infty} (\hat{x}(t) - x(t)) = 0, \ \forall x(0), \ \hat{x}(0) \text{ and } u(t) \tag{11.10}
$$

if and only if condition (11.9) is met.

Let

$$
e(t) = \hat{x}(t) - x(t)
$$

be the estimation error between the real and the estimated states. It is easy to verify that $e(t)$ satisfies

$$
E\dot{e}(t) = (A - LC)e(t), \tag{11.11}
$$

with the initial value

$$
e(0) = \hat{x}(0) - x(0).
$$

Therefore,

$$\lim_{t \to \infty} e(t) = \lim_{t \to \infty} (\hat{x}(t) - x(t)) = 0, \quad \forall x(0), \ \hat{x}(0) \ \text{and} \ u(t),$$

holds if and only if system (11.11) is stable, or equivalently, condition (11.9) holds. The proof is then completed. □

Regarding the state observer (11.8), apparently, $\hat{x}(t) = x(t)$ provided $\hat{x}(0) = x(0)$, which generally does not hold since $x(0)$ is unknown. However, since (11.10) holds, $\hat{x}(t)$ and $x(t)$ get very close to each other after sufficient long time.

In the state equation of the descriptor state observer (11.8), the first two terms are the same as the corresponding ones in system (11.1), and the third term $L(y(t) - C\hat{x}(t))$ acts as the error modification term, which modifies the estimation error brought by the difference between the initial conditions $\hat{x}(0)$ and $x(0)$, so that $\hat{x}(t)$ approaches $x(t)$ asymptotically.

The following theorem shows that detectability is a necessary and sufficient condition for system (11.1) to have a full-order descriptor state observer in the form of (11.8).

Theorem 11.2. *The regular system (11.1) has a full-order descriptor state observer in the form of (11.7) if and only if it is detectable.*

Proof. It immediately follows from Theorem 11.1 and the fact that a matrix L can be selected to satisfy condition (11.9) if and only if the system (11.1) is detectable. □

If $E = I_n$, Theorems 11.1 and 11.2 become the well-known results in normal linear system theory.

It is seen from Theorem 11.2 that detectability characterizes the ability to detect the state by the type of dynamic systems in the form of (11.7) or (11.8). Under the detectability assumption, the state $x(t)$ may be reconstructed asymptotically by an observer from the system input $u(t)$ and the system output $y(t)$.

Remark 11.1. The full-order state observer can be written in the following form

$$E\dot{\hat{x}} = (A - LC)\hat{x} + Bu + Ly.$$

Denote $K = -L^{\mathrm{T}}$, we have

$$A - LC = \left(A^{\mathrm{T}} + C^{\mathrm{T}}K\right)^{\mathrm{T}}.$$

Therefore, the problem of designing a full-order state observer in the form of (11.8) can be converted into a problem of designing a state feedback controller

$$v = Kx = -L^{\mathrm{T}}x$$

for the dual system

$$E^{\mathrm{T}} \dot{x} = A^{\mathrm{T}} x + C^{\mathrm{T}} v.$$

As a descriptor system, the descriptor state observer (11.8) may have impulse terms in its response. To guarantee that the descriptor state observer (11.7) is impulse-free, the matrix L must be so chosen that

$$\deg \det \left(sE - (A - LC) \right) = \operatorname{rank} E. \tag{11.12}$$

Recall results about impulse elimination by state feedback, and in view of the above Remark 11.1, we know immediately that there exists a matrix L such that (11.12) holds if and only if the matrix triple (E, A, C) is I-observable. Combining this fact with the above Theorem 11.2 immediately gives the following result.

Theorem 11.3. *The regular system (11.1) has a full-order descriptor state observer in the form of (11.8), which is impulse-free if and only if it is detectable and I-observable.*

Again, in view of the above Remark 11.1 and results for pole assignment, we see that the finite observer poles, that is, the finite eigenvalues of $(E, \ A - LC)$ can be arbitrarily assigned if and only if the matrix triple (E, A, C) is R-observable. Combining this fact with the above results, we obviously have the following result.

Theorem 11.4. *Consider the regular and detectable system (11.1), and let (11.8) be a full-order descriptor state observer for system (11.1). Then the following hold:*

1. *The set of finite poles of the observer system (11.8) can be replaced by an arbitrarily prescribed set if and only if the system (11.1) is R-observable.*
2. *The observer system (11.8) can possess an arbitrarily prescribed* $\operatorname{rank} E$ *finite poles if and only if (11.1) is R-observable and I-observable.*

11.2.2 Design Methods

The full-order descriptor state observer (11.8) has only one design parameter matrix L. Regarding the solution of this gain matrix L, there are basically two categories of approaches.

Indirect Approaches

Noticing that (11.9) is equivalent to

$$\sigma \left(E^{\mathrm{T}}, \ A^{\mathrm{T}} - C^{\mathrm{T}} L^{\mathrm{T}} \right) \subset \mathbb{C}^{-},$$

the problem can then be converted into solving a stabilizing state feedback controller

$$v = -L^{\mathrm{T}} x$$

for the dual system

$$E^{\mathrm{T}} \dot{x} = A^{\mathrm{T}} x + C^{\mathrm{T}} v.$$

Therefore, any state feedback stabilization technique can be readily applied. Moreover, it follows from the pole assignment result in Chap. 8 that when system (11.1) is R-observable, all the observer finite poles can be arbitrarily assigned. In this case, some pole assignment methods can also be readily applied.

Direct Approaches

It follows from Theorem 11.1 that system (11.7) forms a full-order descriptor observer for system (11.1) if and only if the gain matrix L satisfies (11.9). A direct approach directly seeks a matrix L satisfying condition (11.9).

The following example illustrates the basic idea of the direct approaches.

Example 11.1. Consider again the system (3.39) in Example 3.5. The state space representation is

$$\begin{cases} \begin{bmatrix} 1 & 0 & 0 & 0 \\ 0 & 1 & 0 & 0 \\ 0 & 0 & 1 & 0 \\ 0 & 0 & 0 & 0 \end{bmatrix} \dot{x} = \begin{bmatrix} 0 & 0 & 1 & 0 \\ 1 & 0 & 0 & 0 \\ 0 & 1 & 0 & 0 \\ 0 & 0 & 1 & 0 \end{bmatrix} x + \begin{bmatrix} 1 & 0 & 0 \\ 1 & -1 & 2 \\ 0 & 1 & 0 \\ 0 & 0 & 1 \end{bmatrix} u \\ y = \begin{bmatrix} 0 & 1 & 0 & 0 \\ 0 & 0 & 0 & 1 \end{bmatrix} x \end{cases} \tag{11.13}$$

It has been known in Example 4.5 that the system is C-observable. Thus, it is detectable. Denote

$$L^{\mathrm{T}} = \begin{bmatrix} l_{11} & l_{21} & l_{31} & l_{41} \\ l_{12} & l_{22} & l_{32} & l_{42} \end{bmatrix},$$

we have

$$\det(sE - (A - LC))$$
$$= l_{42} s^3 + (l_{32} - l_{22} l_{41} + l_{21} l_{42} - 1) s^2$$
$$+ (l_{21} l_{32} + l_{11} l_{42} + l_{22} - l_{22} l_{31} - l_{21} - l_{12} l_{41}) s$$
$$+ (l_{12} + l_{31} l_{42} - l_{31} l_{12} + l_{41} - l_{41} l_{32} - l_{11} + l_{11} l_{32} - l_{42}) \tag{11.14}$$

Setting

$$l_{11} = 1, \ l_{12} = 2, \ l_{21} = 3, \ l_{22} = 1, \tag{11.15}$$

$$l_{31} = 0, \ l_{32} = 2, \ l_{41} = 1, \ l_{42} = 1, \qquad\qquad (11.16)$$

(11.14) becomes

$$\det{(sE - (A - LC))} = (s + 1)^3.$$

Thus,

$$\sigma\,(E, A - LC) = \{-1, -1, -1\} \subset \mathbb{C}^-.$$

According to Theorem 11.1, we can give the full-order descriptor state observer for system (11.13) as follows:

$$\begin{cases} E\dot{z} = (A - LC)\,z + Bu + Ly \\ w = z \end{cases},$$

where l_{ij}, $i = 1, 2, 3, 4$, $j = 1, 2$, are given by (11.15) and (11.16).

11.3   Eigenstructure Assignment Design

Let us again consider the descriptor linear system

$$\begin{cases} E\dot{x} = Ax\,(t) + Bu\,(t) \\ y\,(t) = Cx\,(t) \end{cases}, \qquad\qquad (11.17)$$

where $x\,(t) \in \mathbb{R}^n$, $u\,(t) \in \mathbb{R}^r$, and $y\,(t) \in \mathbb{R}^m$ are the state, the control input, and the measured output, respectively; E, A, B, and C are constant matrices of appropriate dimensions. It is assumed that system (11.17) is regular and $n_0 = \mathrm{rank}E \leq n$. For this system, we consider the design of a type of full-order descriptor state observers in the form of

$$E\dot{\hat{x}} = A\hat{x} + Bu - L\,(y - C\hat{x}). \qquad\qquad (11.18)$$

It follows from the pole assignment result in Chap. 8 that when system (11.17) is R-observable and I-observable, all the observer poles can be arbitrarily assigned and the observer system (11.18) can also be made impulse-free. In this section, we propose a parametric eigenstructure assignment approach for designing the full-order descriptor observer (11.18) under the following assumptions.

Assumption 11.1. $\mathrm{rank}C = m$.

Assumption 11.2. System (11.17) is regular and S-observable.

The full-order descriptor state observer (11.18) for system (11.17) can also be written in the form of

$$E\dot{\hat{x}} = A_o\hat{x} + Bu - Ly, \qquad\qquad (11.19)$$

with

$$A_o = A + LC. \tag{11.20}$$

Suppose that the observer system (11.19) has distinct finite eigenvalues, or poles, denoted by s_i, $i = 1, 2, \ldots, n_0$. Furthermore, denote the left eigenvector of the above system, or the matrix pair (E, A_o), associated with the eigenvalue s_i by t_i, then by definition we have

$$t_i^T(A + LC) = s_i t_i^T E, \quad i = 1, 2, \ldots, n_0. \tag{11.21}$$

Therefore, the eigenstructure assignment problem for the observer system (11.19) can be stated as follows.

Problem 11.1. Given the descriptor linear system (11.17) satisfying Assumptions 11.1 and 11.2, and a group of self-conjugate distinct complex numbers s_i, $i = 1, 2, \ldots, n_0$, on the left half complex plane, determine a real matrix L and a group of vectors t_i, $i = 1, 2, \ldots, n_0$, such that the following requirements are met:

1. All the equations in (11.21) hold.
2. The group of eigenvectors t_i, $i = 1, 2, \ldots, n_0$, are linearly independent.
3. The matrix pair (E, A_o) is regular.

11.3.1 Eigenstructure Assignment Result

Taking the transpose on both sides of (11.21) gives

$$(A^T + C^T L^T)t_i = s_i E^T t_i, \quad i = 1, 2, \ldots, n_0. \tag{11.22}$$

Furthermore, since the regularity of the matrix pair (E, A_o) is equivalent to that of (E^T, A_o^T), it is easy to see that the above Problem 11.1 is simply a special case of the right eigenstructure assignment problem for the dual system

$$E^T \dot{x} = A^T x + C^T v \tag{11.23}$$

via the state feedback control

$$v = Kx, \quad K = L^T. \tag{11.24}$$

Therefore, the solution to Problem 11.1 can be readily obtained by using some right eigenstructure assignment result via state feedback control.

Under Assumptions 11.1 and 11.2, there exist a pair of left coprime polynomial matrices $H(s) \in \mathbb{R}^{m \times n} [s]$ and $L(s) \in \mathbb{R}^{m \times m} [s]$ satisfying

$$H(s)(A - sE) + L(s)C = 0. \tag{11.25}$$

Furthermore, define matrices $T_\infty \in \mathbb{R}^{n \times (n - n_0)}$ and $V_\infty \in \mathbb{R}^{n \times (n - n_0)}$ by

$$T_\infty^T E = 0, \quad \mathrm{rank} T_\infty = n - n_0, \tag{11.26}$$

and

$$E V_\infty = 0, \quad \mathrm{rank} V_\infty = n - n_0. \tag{11.27}$$

Then, by using the eigenstructure assignment result in Sect. 9.2, the following theorem for solution to Problem 11.1 can be derived.

Theorem 11.5. *Consider the descriptor linear system (11.17) satisfying Assumptions 11.1 and 11.2.*

1. *Problem 11.1 has a solution if and only if there exist a group of parameter vectors* $g_i \in \mathbb{R}^m$, $i = 1, 2, \ldots, n_0$, *satisfying the following constraints:*

Constraint C1 $g_i = \bar{g}_l$ *if* $s_i = \bar{s}_l$.

Constraint C2 $\mathrm{rank} \begin{bmatrix} g_1^T H(s_1) E \\ \vdots \\ g_{n_0}^T H(s_{n_0}) E \end{bmatrix} = n_0.$

2. *Let*

$$\begin{cases} T = [\, t_1 \; t_2 \; \cdots \; t_{n_0} \,] \\ t_i^T = g_i^T H(s_i), \quad i = 1, 2, \ldots, n_0 \end{cases}, \tag{11.28}$$

and

$$\begin{cases} Z = [\, z_1 \; z_2 \; \cdots \; z_{n_0} \,] \\ z_i^T = g_i^T L(s_i), \quad i = 1, 2, \ldots, n_0 \end{cases}, \tag{11.29}$$

with g_i, $i = 1, 2, \ldots, n_0$, *satisfying Constraints C1 and C2. Then there holds*

$$\det [\, T \; T_\infty \,] \neq 0, \tag{11.30}$$

and all the solutions to Problem 11.1 are given by (11.28) and

$$L = [\, T \; T_\infty \,]^{-T} [\, Z \; Z_\infty \,]^T, \tag{11.31}$$

with Z *given by (11.29) and* $Z_\infty \in \mathbb{R}^{m \times (n - n_0)}$ *being a parameter matrix satisfying*

Constraint C3 $\det \left(T_\infty^T A V_\infty + Z_\infty^T C V_\infty \right) \neq 0.$

The above result in fact gives a general parametric approach for the design of the full-order descriptor state observers in the form of (11.19). This result can also be easily generalized, using the general state feedback eigenstructure assignment results in Sect. 9.2, into the case that the observer eigenvalues have arbitrary algebraic and geometric multiplicities.

Remark 11.2. The pair of left coprime polynomial matrices $H(s)$ and $L(s)$ satisfying (11.25) performs an important role in the parameterization of the observer gain and the observer eigenvectors. Since (11.25) is equivalent to

$$(A^{\mathrm{T}} - sE^{\mathrm{T}})H^{\mathrm{T}}(s) + C^{\mathrm{T}}L^{\mathrm{T}}(s) = 0,$$

the pair of left coprime polynomial matrices $H(s)$ and $L(s)$ can be obtained using any method mentioned in Chap. 9 for solving a pair of right coprime polynomial matrices $N(s)$ and $D(s)$ satisfying

$$(A - sE)N(s) + BD(s) = 0.$$

Remark 11.3. It follows from the above Theorem 11.5 that the key step in solving the observer eigenstructure assignment problem is to find the parameter vectors g_i, $i = 1, 2, \ldots, n_0$, and Z_∞ satisfying Constraints C1, C2, and C3. According to pole assignment theory for descriptor linear systems, such parameters always exist when Assumptions 11.1 and 11.2 are valid and also when the observer eigenvalues are restricted, as required in the problem statement, to be distinct.

Remark 11.4. When there are complex eigenvalues assigned to the observer system (11.19), the matrices T and Z are both complex. However, the gain matrix given by (11.31) is still real. If one wishes to avoid complex computation, a well-known technique can be applied, which converts, at the same time, both the two matrices T and Z into real ones. This technique can be demonstrated as follows. Without loss of generality, let us assume that $s_2 = \bar{s}_1 = \alpha + \beta j$ are a pair of conjugate eigenvalues, and all the other ones are real. In this case, if we choose $g_2 = \bar{g}_1 = \xi + \eta j$, with ξ and η real, as required by Constraint C1, we have $t_1 = \bar{t}_2$ and $z_1 = \bar{z}_2$, and the real substitutions of these matrices can be taken as

$$T = \begin{bmatrix} \mathrm{Re}(t_1) & \mathrm{Im}(t_1) & t_3 & \cdots & t_p \end{bmatrix}$$

and

$$Z = \begin{bmatrix} \mathrm{Re}(z_1) & \mathrm{Im}(z_1) & z_3 & \cdots & z_p \end{bmatrix}.$$

11.3.2 The Algorithm and Example

One of the main advantages of the proposed parametric eigenstructure assignment approach for the type of full-order descriptor state observers in the form of (11.19) is that it provides all the design degrees of freedom represented by the design parameters g_i, $i = 1, 2, \ldots, n_0$, and Z_∞. Constraints C1, C2, and C3 are the basic ones for eigenstructure assignment and observer regularity. Once there are additional requirements on the observer system, we can link these requirements with the design parameters, and convert them into certain extra constraints on the design parameters. This aspect is well demonstrated in the next section with disturbance decoupling.

Based on the above discussion and Theorem 11.5, an algorithm for solving the observer eigenstructure assignment problem can be given as follows.

Algorithm 11.1. Observer design based on eigenstructure assignment.

Step 1 Solve the matrices T_∞ and V_∞, defined by (11.26) and (11.27), respectively.
Step 2 Solve a pair of left coprime polynomial matrices $H(s)$ and $L(s)$ satisfying (11.25) (refer to Remark 11.2).
Step 3 Specify a group of distinct finite observer eigenvalues s_i, $i = 1, 2, \ldots, n_0$.
Step 4 Express Constraints C1–C3, as well as any additional constraints (if exist), into forms represented by the entries of the parameter vectors g_i, $i = 1, 2, \ldots, n_0$, and the parameter matrix Z_∞.
Step 5 Find a set of parameters g_i, $i = 1, 2, \ldots, n_0$, and Z_∞ satisfying Constraints C1–C3, as well as any additional system design constraints. If such parameters do not exist, this approach is invalid, or the problem does not have a solution.
Step 6 Compute the observer gain L according to (11.28), (11.29), and (11.31) based on the parameters g_i, $i = 1, 2, \ldots, n_0$, and Z_∞ obtained in Step 5.

Remark 11.5. As is well known, the eigenvector of a matrix pair associated with a certain eigenvalue is not unique. Such a property allows us to simplify the solution to Step 6 in the above Algorithm 11.1 by setting one nonzero element in each g_i to 1. This treatment does not reduce the degrees of design freedom but can usually simplify the solution procedure.

Example 11.2. Consider a system in the form of (11.17) with the following coefficient matrices:

$$E = \begin{bmatrix} 1 & 0 & 0 & 0 \\ 0 & 1 & 0 & 0 \\ 0 & 0 & 1 & 0 \\ 0 & 0 & 0 & 0 \end{bmatrix}, \quad A = \begin{bmatrix} 0 & 0 & 1 & 0 \\ 1 & 0 & 0 & 0 \\ 0 & 1 & 0 & 1 \\ 0 & 0 & 1 & 0 \end{bmatrix},$$

$$B = \begin{bmatrix} 1 & 0 & 0 \\ 1 & -1 & 2 \\ 0 & 1 & 0 \\ 0 & 0 & 1 \end{bmatrix}, \quad C = \begin{bmatrix} 0 & 1 & 0 & 0 \\ 0 & 0 & 0 & 1 \end{bmatrix},$$

where the matrices A, B, and C are taken from the example in Fletcher (1988) and Duan (1995, 1999). It is easy to verify that with this example system Assumptions 11.1 and 11.2 hold. In the following, we solve the observer design problem for this example system following Algorithm 11.1.

Step 1 The left and right infinite eigenvector matrices T_∞ and V_∞ , defined by (11.26) and (11.27), respectively, are easily obtained as

$$V_\infty = T_\infty = [0 \ \ 0 \ \ 0 \ \ 1]^\mathsf{T}.$$

Step 2 By the method given in Lemma C.3, the pair of right coprime polynomial matrices $H(s)$ and $L(s)$ can be obtained as

$$H(s) = \begin{bmatrix} 1 & s & 0 & -1 \\ 0 & 0 & 1 & s \end{bmatrix}, \quad L(s) = \begin{bmatrix} s^2 & 0 \\ -1 & -1 \end{bmatrix}.$$

Step 3 To present a more general result, the closed-loop eigenvalues s_i, $i = 1, 2, 3$, are not specified, but are only restricted to be real, negative, and distinct from each other.

Step 4 Denote

$$g_i = \begin{bmatrix} \beta_{i1} \\ \beta_{i2} \end{bmatrix}, \quad i = 1, 2, 3,$$

with β_{ij}, $i = 1, 2, 3$; $j = 1, 2$, being a group of real scalars. Then Constraint C1 holds automatically, and Constraint C2 can be easily verified to be

$$
\begin{aligned}
\Delta_1 = \ & (s_2 - s_1)\beta_{11}\beta_{21}\beta_{32} + (s_3 - s_2)\beta_{21}\beta_{12}\beta_{31} \\
& + (s_1 - s_3)\beta_{11}\beta_{31}\beta_{22} \neq 0.
\end{aligned}
\tag{11.32}
$$

Furthermore, let

$$Z_\infty = z_1^\infty = \begin{bmatrix} \eta_1 \\ \eta_2 \end{bmatrix},$$

then, Constraint C3 can be easily obtained as $\eta_2 \neq 0$.

Step 5 It follows from the above that the design parameters are composed of the following three sets:

1. The observer poles s_i, $i = 1, 2, 3$, which are real and distinct.
2. The set of real scalars β_{ij}, $i = 1, 2, 3$, $j = 1, 2$, which satisfy (11.32).
3. The set of real scalars η_i, $i = 1, 2$, with $\eta_2 \neq 0$.

Step 6 It can be obtained according to (11.28) and (11.29) that

$$
T = \begin{bmatrix}
\beta_{11} & \beta_{21} & \beta_{31} \\
s_1\beta_{11} & s_2\beta_{21} & s_3\beta_{31} \\
\beta_{12} & \beta_{22} & \beta_{32} \\
s_1\beta_{12} - \beta_{11} & s_2\beta_{22} - \beta_{21} & s_3\beta_{32} - \beta_{31}
\end{bmatrix}
\tag{11.33}
$$

and

$$Z = \begin{bmatrix} s_1^2 \beta_{11} - \beta_{12} & s_2^2 \beta_{21} - \beta_{22} & s_3^2 \beta_{31} - \beta_{32} \\ -\beta_{12} & -\beta_{22} & -\beta_{32} \end{bmatrix}. \qquad (11.34)$$

The observer gain matrix L is then given by (11.31).

The degrees of freedom provided by this design are represented by s_j, β_{ij}, and η_i, $i = 1, 2, 3$; $j = 1, 2$. Specially choosing

$$s_1 = -1, s_2 = -2, s_3 = -3,$$

and

$$\beta_{11} = \beta_{21} = \beta_{32} = 1, \ \beta_{12} = \beta_{22} = \beta_{31} = 0,$$

we have

$$T = \begin{bmatrix} 1 & 1 & 0 \\ -1 & -2 & 0 \\ 0 & 0 & 1 \\ -1 & -1 & -3 \end{bmatrix}, \ Z = \begin{bmatrix} 1 & 4 & -1 \\ 0 & 0 & -1 \end{bmatrix}.$$

If we further choose $\eta_1 = 0$, $\eta_2 = 1$, then we obtain

$$L = \begin{bmatrix} -2 & 1 \\ -3 & 0 \\ -1 & 2 \\ 0 & 1 \end{bmatrix}.$$

It is clearly seen from this example that the proposed eigenstructure assignment approach is simple and effective, and can provide solutions with certain degrees of freedom, which can be further utilized. In the next section, this example will be examined again. The degrees of freedom will be utilized to achieve disturbance decoupling.

11.4   Observer Design with Disturbance Decoupling

A great advantage of the parametric design approach proposed in the preceding section for the full-order descriptor state observer (11.19) is that additional design requirements can be easily handled. The basic idea is to link any additional design requirement with the design parameters s_i, g_i, $i = 1, 2, \ldots, n_0$, and Z_∞, and convert them into certain constraints on these parameters. To illustrate this point, in this section, let us consider the design of state observers in the form of (11.19)–(11.20) with disturbance decoupling property.

11.4.1 Problem Formulation

The problem can be stated as follows.

Problem 11.2. Given the descriptor linear system

$$\begin{cases} E\dot{x} = Ax + Bu + Dd \\ y = Cx \end{cases}, \tag{11.35}$$

where $d \in \mathbb{R}^q$ is an unknown disturbance, $D \in \mathbb{R}^{n \times q}$ is a known matrix, all the other variables are as stated previously, find a state observer in the form of (11.19)–(11.20) satisfying the following requirements:

1. It is regular and has $n_0 = \text{rank}\, E$ distinct observer poles.
2. The weighted estimation error

$$r = W\,(y - C\hat{x}) \tag{11.36}$$

is free from the effect of the disturbance $d\,(t)$, where $W \in \mathbb{R}^{t \times m}$ is a weighting matrix.

It is obvious that the three requirements in Problem 11.1 imply the first requirement in Problem 11.2. Therefore, the main task in solving Problem 11.2 is to convert its second requirement into constraints on the design parameters s_i, g_i, $i = 1, 2, \ldots, n_0$, where the matrices and Z_∞.

It follows from (11.19) and (11.36) that the effect of the disturbance $d(t)$ to the error signal $r(t)$ is characterized, in the frequency domain, by

$$R(s) = WC(sE - A_o)^{-1}Dd(t).$$

Therefore, to meet the second requirement in Problem 11.2, it is necessary and sufficient to achieve

$$WC(sE - A_o)^{-1}D = 0, \quad \forall s \in \mathbb{C}. \tag{11.37}$$

11.4.2 Preliminaries

To derive a parametric condition for (11.37), the concept of normalized left and right eigenvectors needs to be introduced.

Let t_i and v_i be the left and right eigenvectors of the matrix pair (E, A_o) associated with the eigenvalue s_i, respectively. Then, by definition, they are determined by

$$t_i^T A_o = s_i t_i^T E, \qquad A_o v_i = s_i E v_i. \tag{11.38}$$

Define the left and right eigenvector matrices for the matrix pair (E, A_o) as

$$T = [t_1 \ t_2 \ \cdots \ t_{n_0}],$$ (11.39)

and

$$V = [v_1 \ v_2 \ \cdots \ v_{n_0}],$$ (11.40)

respectively, then they clearly satisfy the following in view of the two equations in (11.38):

$$A_o V = EV\Lambda, \qquad T^{\mathrm{T}} A_o = \Lambda T^{\mathrm{T}} E,$$ (11.41)

where

$$\Lambda = \mathrm{diag}\,(s_1, s_2, \ldots, s_{n_0}).$$ (11.42)

The groups of left eigenvectors t_i, $i = 1, 2, \ldots, n_0$, and right eigenvectors v_i, $i = 1, 2, \ldots, n_0$, are said to be a normalized pair of groups if they further satisfy

$$T^{\mathrm{T}} EV = I.$$ (11.43)

Theorem 11.6. *Let the matrix pair (E, A_o) be regular and have distinct finite eigenvalues s_i, $i = 1, 2, \ldots, n_0$; T_∞ and V_∞ be as defined in (11.26) and (11.27); t_i and v_i, $i = 1, 2, \ldots, n_0$, be a normalized pair of groups of the left and right eigenvectors of the matrix pair (E, A_o). Then the following hold:*

1. *The matrix $\Sigma_\infty = T_\infty^{\mathrm{T}} A_o V_\infty$ is nonsingular.*
2. *Condition (11.37) holds if and only if*

$$R_\infty = WCV_\infty (T_\infty^{\mathrm{T}} A_o V_\infty)^{-1} T_\infty^{\mathrm{T}} D = 0,$$ (11.44)

and

$$R_i = WCv_i t_i^{\mathrm{T}} D = 0, \quad i = 1, 2, \ldots, n_0.$$ (11.45)

Proof. Since the matrix pair (E, A_o) is regular, it follows from the Lemma 9.3 that the first conclusion holds. Thus, in the following, it suffices only to prove the second one.

Put

$$\tilde{T} = [\,T \quad T_\infty\,], \quad \tilde{V} = [\,V \quad V_\infty\,].$$ (11.46)

Then using (11.26), (11.27), (11.41), and (11.43) gives

$$\tilde{T}^{\mathrm{T}} E \tilde{V} = \begin{bmatrix} I & 0 \\ 0 & 0 \end{bmatrix}, \quad \tilde{T}^{\mathrm{T}} A_o \tilde{V} = \begin{bmatrix} \Lambda & 0 \\ 0 & T_\infty^{\mathrm{T}} A_o V_\infty \end{bmatrix}.$$ (11.47)

Combining the above two relations in (11.47) produces

$$\tilde{T}^{\mathrm{T}} (sE - A_o) \tilde{V} = \begin{bmatrix} sI - \Lambda & 0 \\ 0 & -T_\infty^{\mathrm{T}} A_o V_\infty \end{bmatrix}.$$ (11.48)

In view of the first conclusion of the theorem, we know from (11.48) that both matrices $\tilde{T}$ and $\tilde{V}$ are nonsingular. Therefore, it follows from (11.48) and the non-singularity of matrices $T_\infty^T A_o V_\infty$, $\tilde{T}$ and $\tilde{V}$ that

$$
\begin{aligned}
WC(sE - A_o)^{-1}D &= WC\tilde{V}[\tilde{T}^T(sE - A_o)\tilde{V}]^{-1}\tilde{T}^T D \\
&= WC\tilde{V}\begin{bmatrix} sI - \Lambda & 0 \\ 0 & -T_\infty^T A_o V_\infty \end{bmatrix}^{-1}\tilde{T}^T D \\
&= WC\tilde{V}\begin{bmatrix} (sI - \Lambda)^{-1} & 0 \\ 0 & -(T_\infty^T A_o V_\infty)^{-1} \end{bmatrix}\tilde{T}^T D \\
&= WC(V(sI - \Lambda)^{-1}T^T - V_\infty(T_\infty^T A_o V_\infty)^{-1}T_\infty^T)D \\
&= \sum_{i=1}^{n_0} \frac{WCv_i t_i^T D}{s - s_i} - WCV_\infty(T_\infty^T A_o V_\infty)^{-1}T_\infty^T D \\
&= \sum_{i=1}^{n_0} \frac{R_i}{s - s_i} - R_\infty.
\end{aligned}
$$

This immediately gives the second conclusion of the theorem in view of the arbitrariness of the variable s and the distinctness of the finite closed-loop eigenvalues s_i, $i = 1, 2, \ldots, n_0$. □

The above theorem immediately gives the following corollaries.

Corollary 11.1. *Let the matrix pair (E, A_o) be regular and have distinct finite eigenvalues s_i, $i = 1, 2, \ldots, n_0$; T_∞ and V_∞ be as defined in (11.26) and (11.27); and t_i, $i = 1, 2, \ldots, n_0$, be a group of the left eigenvectors of the matrix pair (E, A_o). Then condition (11.37) holds if (11.44) is valid and*

$$
R_i' = t_i^T D = 0, \quad i = 1, 2, \ldots, n_0. \tag{11.49}
$$

Corollary 11.2. *Let the matrix pair (E, A_o) be regular and have distinct finite eigenvalues s_i, $i = 1, 2, \ldots, n_0$; T_∞ and V_∞ be as defined in (11.26) and (11.27); and t_i, $i = 1, 2, \ldots, n_0$, and v_i, $i = 1, 2, \ldots, n_0$, be a normalized pair of groups of the left and right eigenvectors of the matrix pair (E, A_o). Then*

$$
WC(sE - A_o)^{-1}D \neq 0 \tag{11.50}
$$

for some $s \in \mathbb{C}$ if and only if one of the matrices R_∞ and R_i, $i = 1, 2, \ldots, n_0$, defined by (11.44)–(11.45) are different from zero.

Remark 11.6. In the case of $n - n_0 = 1$, the matrix $\Sigma_\infty = T_\infty^T A_o V_\infty$ becomes a scalar and is nonzero under the conditions of Theorem 11.6. In this case, it is clearly seen that condition (11.44) is equivalent to the following

$$
R_\infty' = WCV_\infty T_\infty^T D = 0. \tag{11.51}
$$

11.4.3　Constraints for Disturbance Decoupling

Theorem 11.6 can be readily applied to treat condition (11.37). However, using Theorem 11.6, the parametric representation of the right observer eigenvectors v_i, $i = 1, 2, \ldots, n$, has to be given based on the left eigenvector matrix T and the required observer structure. This is technically feasible (see Sect. 9.3), but requires a more complicated procedure. For simplicity, the sufficient condition given in Corollary 11.1 is used here instead of the necessary and sufficient condition given in Theorem 11.6. It follows from Corollary 11.1 that condition (11.37) holds if

$$t_i^T D = 0, \quad i = 1, 2, \ldots, n_0, \tag{11.52}$$

and

$$WCV_\infty (T_\infty^T A_o V_\infty)^{-1} T_\infty^T D = 0. \tag{11.53}$$

By using (11.28), condition (11.52) can be converted into the following constraint on the parameters g_i, $i = 1, 2, \ldots, n_0$.

Constraint C4 $g_i^T H(s_i) D = 0$, $\quad i = 1, 2, \ldots, n_0$.

It can be obtained from (11.31) that $Z_\infty^T = T_\infty^T L$. Utilizing this relation gives

$$\begin{aligned} T_\infty^T A_o V_\infty &= T_\infty^T (A + LC) V_\infty \\ &= T_\infty^T A V_\infty + Z_\infty^T C V_\infty, \end{aligned} \tag{11.54}$$

where Z_∞ is a parameter matrix satisfying Constraint C3. By using (11.54), condition (11.53) can be converted into the following constraint on the parameter matrices Z_∞ and W:

Constraint C5 $WCV_\infty (T_\infty^T A V_\infty + Z_\infty^T C V_\infty)^{-1} T_\infty^T D = 0$.

In view of Remark 11.6, in the case of $n - n_0 = 1$, the matrix

$$\Sigma_\infty = T_\infty^T A_o V_\infty = T_\infty^T A V_\infty + Z_\infty^T C V_\infty$$

becomes a nonzero scalar under the conditions of Theorem 11.6. In this case, the above Constraint C5 is equivalent to the following:

Constraint C'5 $WCV_\infty T_\infty^T D = 0$.

It follows from the above reasoning that the following theorem for the solution to Problem 11.2 holds.

Theorem 11.7. *Let Assumptions 11.1 and 11.2 hold. Then, Problem 11.2 has a solution if there exists a group of parameter vectors g_i, $i = 1, 2, \ldots, n_0$, and two parameter matrices Z_∞ and W satisfying Constraints C1–C5. In this case, the observer gain is given by (11.31), and the weighting matrix is given by the parameter matrix W.*

11.4.4 The Example

Regarding solution to Problem 11.2, we do not need to propose a new algorithm. The one proposed in the preceding Sect. 11.3 can be readily applied. The only change that needs to be made is that the disturbance decoupling Constraints C4 and C5 need to be treated besides the basic Constraints C1–C3 in the Steps 4 and 5 in Algorithm 11.1. In the following, let us demonstrate this idea with the example system considered in Sect. 11.3.

Example 11.3. Consider again the example system treated in Sect. 11.3. The coefficient matrices are as follows:

$$
E = \begin{bmatrix} 1 & 0 & 0 & 0 \\ 0 & 1 & 0 & 0 \\ 0 & 0 & 1 & 0 \\ 0 & 0 & 0 & 0 \end{bmatrix}, \quad
A = \begin{bmatrix} 0 & 0 & 1 & 0 \\ 1 & 0 & 0 & 0 \\ 0 & 1 & 0 & 1 \\ 0 & 0 & 1 & 0 \end{bmatrix},
$$

$$
B = \begin{bmatrix} 1 & 0 & 0 \\ 1 & -1 & 2 \\ 0 & 1 & 0 \\ 0 & 0 & 1 \end{bmatrix}, \quad
C = \begin{bmatrix} 0 & 1 & 0 & 0 \\ 0 & 0 & 0 & 1 \end{bmatrix}, \quad
D = \begin{bmatrix} 0 \\ 0 \\ 0 \\ 1 \end{bmatrix}.
$$

In the following, we solve the problem of observer design with disturbance decoupling for this example system following Algorithm 11.1. The results for the first three steps are exactly the same as those in Example 11.2, and hence are omitted.

Step 4 Denote

$$
g_i = \begin{bmatrix} \beta_{i1} \\ \beta_{i2} \end{bmatrix}, \quad i = 1, 2, 3,
$$

with β_{ij}, $i = 1, 2$; $j = 1, 2, 3$, being a group of real scalars. Then Constraint C1 holds automatically, Constraint C2 can be easily verified to be

$$
\Delta_1 = (s_2 - s_1)\beta_{11}\beta_{21}\beta_{32} + (s_3 - s_2)\beta_{21}\beta_{12}\beta_{31}
$$
$$
+ (s_1 - s_3)\beta_{11}\beta_{31}\beta_{22} \neq 0, \tag{11.55}
$$

and Constraint C4 can be obtained as

$$
\beta_{11} = s_1\beta_{12}, \quad \beta_{21} = s_2\beta_{22}, \quad \beta_{31} = s_3\beta_{32}. \tag{11.56}
$$

Furthermore, let

$$
Z_\infty = z_1^\infty = \begin{bmatrix} \eta_1 \\ \eta_2 \end{bmatrix},
$$

Constraint C3 can be easily obtained as $\eta_2 \neq 0$. Finally, it is easy to check that Constraint C5 is $w_2 = 0$, where $W = \begin{bmatrix} w_1 & w_2 \end{bmatrix}$.

Step 5 Combining (11.55) and (11.56) gives

$$\beta_{12} \neq 0, \ \beta_{22} \neq 0 \text{ and } \beta_{32} \neq 0.$$

Therefore, in view of Remark 11.5, we can set $\beta_{12} = \beta_{22} = \beta_{32} = 1$, and hence obtain from (11.56),

$$\beta_{11} = s_1, \ \beta_{22} = s_2 \text{ and } \beta_{32} = s_3.$$

Moreover, we have η_1 arbitrary, $\eta_2 \neq 0$ and $w_2 = 0$ as reasoned above, and $w_1 \neq 0$ since W has to be different from the zero matrix.

Step 6 It can be obtained according to (11.28) and (11.29) that

$$
T = \begin{bmatrix}
\beta_{11} & \beta_{21} & \beta_{31} \\
s_1\beta_{11} & s_2\beta_{21} & s_3\beta_{31} \\
\beta_{12} & \beta_{22} & \beta_{32} \\
s_1\beta_{12} - \beta_{11} & s_2\beta_{22} - \beta_{21} & s_3\beta_{32} - \beta_{31}
\end{bmatrix}
$$

$$
= \begin{bmatrix}
s_1 & s_2 & s_3 \\
s_1^2 & s_2^2 & s_3^2 \\
1 & 1 & 1 \\
0 & 0 & 0
\end{bmatrix}, \tag{11.57}
$$

and

$$
Z = \begin{bmatrix}
s_1^2\beta_{11} - \beta_{12} & s_2^2\beta_{21} - \beta_{22} & s_3^2\beta_{31} - \beta_{32} \\
-\beta_{12} & -\beta_{22} & -\beta_{32}
\end{bmatrix}
$$

$$
= \begin{bmatrix}
s_1^3 - 1 & s_2^3 - 1 & s_3^3 - 1 \\
-1 & -1 & -1
\end{bmatrix}. \tag{11.58}
$$

Thus, it follows from (11.31) that the observer gain matrix L is given by

$$
L = \frac{1}{\Delta} \begin{bmatrix}
s_1^3(s_2^2 - s_3^2) + s_2^3(s_3^2 - s_1^2) + s_3^3(s_1^2 - s_2^2) & 0 \\
s_1^3(s_3 - s_2) + s_2^3(s_1 - s_3) + s_3^3(s - s_1) & 0 \\
(s_1 s_2 s_3 - 1)\Delta & -\Delta \\
\eta_1 & \eta_2
\end{bmatrix}, \tag{11.59}
$$

with $\eta_2 \neq 0$ and

$$\Delta = s_1^2(s_3 - s_2) + s_2^2(s_1 - s_3) + s_3^2(s_2 - s_1). \tag{11.60}$$

In practical applications, the observer poles s_i, $i = 1, 2, 3$, can be properly chosen to meet some performance requirements. If we simply choose $s_1 = -1$, $s_2 = -2$, $s_3 = -3$, the following observer gain is obtained:

$$L = \frac{1}{2} \begin{bmatrix} 22 & 0 \\ 12 & 0 \\ 14 & 2 \\ \eta_1 & \eta_2 \end{bmatrix}, \quad \eta_2 \neq 0.$$

It is clearly seen from the above that the proposed approach is simple and effective, and can provide solutions with certain degrees of freedom, which can be still further utilized.

11.5 Normal Reduced-Order State Observers

In the preceding two sections, we studied the design of a type of full-order descriptor state observers for descriptor linear systems. However, as pointed out before, input derivatives are usually involved in the state response of descriptor systems, that is, in addition to $u(t)$ and $y(t)$, their derivatives may also be involved in the state response of descriptor state observers. Since these derivative terms are usually not available, descriptor state observers are usually difficult to be realized. Furthermore, the strength of a high-frequency noise is often amplified by its derivative. Thus, a descriptor state observer may be sensitive to high-frequency noises. Due to these facts, normal-state observers are much more preferable practically.

In this section, we consider the design of normal reduced-order observers. First, let us introduce a simple type of normal observers with a reduced-order of rank$E$.

11.5.1 Normal Rank E-Order State Observers

Regarding the existence of reduced-order normal-state observers for system (11.17), we have the following result.

Theorem 11.8. *Let the regular system (11.17) be both detectable and I-observable. Then it has a normal-state observer of order* rank$E$ *in the following form:*

$$\begin{cases} \dot{z} = Fz + Su + Ly \\ w = Mz + Ny + Hu \end{cases}. \tag{11.61}$$

Proof. Under the assumption of detectability and I-observability, we know that there exists a matrix $\bar{L} \in \mathbb{R}^{n \times m}$ such that

$$\deg \det \left(sE - (A - \bar{L}C) \right) = \operatorname{rank} E, \tag{11.62}$$

and

$$\sigma(E, A - \bar{L}C) \subset \mathbb{C}^-. \tag{11.63}$$

Thus, by Theorem 11.1, the system

$$\begin{cases} E\dot{\bar{z}} = (A - \bar{L}C)\bar{z} + Bu + \bar{L}y \\ w = \bar{z} \end{cases} \tag{11.64}$$

forms a full-order state observer for system (11.17). Therefore,

$$\lim_{t \to \infty} (w(t) - x(t)) = 0, \ \forall \bar{z}(0), x(0). \tag{11.65}$$

Furthermore, (11.62) implies that there exist two nonsingular matrices Q and P such that

$$QEP = \operatorname{diag}\left(I_{n_0}, 0\right), \ Q(A - \bar{L}C)P = \operatorname{diag}\left(\bar{A}_1, I\right), \tag{11.66}$$

where $n_0 = \operatorname{rank} E$ and $\bar{A}_1 \in \mathbb{R}^{n_0 \times n_0}$. Moreover, in view of (11.63),

$$\sigma\left(\bar{A}_1\right) = \sigma\left(E, A - \bar{L}C\right) \subset \mathbb{C}^-.$$

Denote

$$\begin{cases} QB = \begin{bmatrix} B_1 \\ B_2 \end{bmatrix}, & B_1 \in \mathbb{R}^{n_0 \times r}, \ B_2 \in \mathbb{R}^{(n-n_0) \times r} \\ Q\bar{L} = \begin{bmatrix} L_1 \\ L_2 \end{bmatrix}, & L_1 \in \mathbb{R}^{n_0 \times m}, \ L_2 \in \mathbb{R}^{(n-n_0) \times m} \end{cases}, \tag{11.67}$$

and

$$P^{-1}\bar{z} = \begin{bmatrix} z_1 \\ z_2 \end{bmatrix}, \ z_1 \in \mathbb{R}^{n_0}, \ z_2 \in \mathbb{R}^{n-n_0}.$$

Then the state observer (11.64) is r.s.e. to

$$\begin{cases} \dot{z}_1 = \bar{A}_1 z_1 + B_1 u + L_1 y \\ 0 = z_2 + B_2 u + L_2 y \\ w = P \begin{bmatrix} z_1 \\ z_2 \end{bmatrix} \end{cases}. \tag{11.68}$$

Solving z_2 from the second equation in (11.68), and substituting it into the third one gives

$$
\begin{cases}
\dot{z} = \bar{A}_1 z + B_1 u + L_1 y \\
w = P \begin{bmatrix} I \\ 0 \end{bmatrix} z - P \begin{bmatrix} 0 \\ I \end{bmatrix} B_2 u - P \begin{bmatrix} 0 \\ I \end{bmatrix} L_2 y
\end{cases}
\tag{11.69}
$$

where $z = z_1 \in \mathbb{R}^{n_0}$. System (11.69) is in the form of (11.61), which is still a state observer of system (11.17) since (11.65) holds. □

It follows from the above proof that the following algorithm can be given for designing a normal rank$E$-order state observer for a detectable and I-observable descriptor linear system.

Algorithm 11.2. Design of normal rank$E$-order state observers.

Step 1 Find the matrix $\bar{L}$ satisfying (11.62) and (11.63).
Step 2 Find the nonsingular matrices P and Q, as well as the matrix $\bar{A}_1$ satisfying (11.66).
Step 3 Solve for the matrices B_i and L_i, $i = 1, 2$, according to (11.67).
Step 4 Form the observer according to (11.69).

Example 11.4. Consider the following system

$$
\begin{cases}
\begin{bmatrix} 0 & 1 & 0 \\ 0 & 0 & 1 \\ 0 & 0 & 0 \end{bmatrix} \dot{x} = \begin{bmatrix} 1 & 0 & 0 \\ 0 & 1 & 0 \\ 0 & 0 & 1 \end{bmatrix} x + \begin{bmatrix} 0 & 0 \\ 1 & 0 \\ 0 & 1 \end{bmatrix} u \\
y = \begin{bmatrix} 1 & 0 & 0 \end{bmatrix} x
\end{cases}
\tag{11.70}
$$

It is easy to verify that the system is detectable and I-observable. In fact, choosing the matrix

$$
\bar{L} = \begin{bmatrix} 2 & 1 & 1 \end{bmatrix}^{\mathrm{T}},
$$

we have

$$
\deg \det \left(sE - (A - \bar{L}C) \right) = \mathrm{rank}E = 2,
$$

and

$$
\sigma(E, A - \bar{L}C) = \left\{ -\frac{1}{2} + \frac{\sqrt{3}i}{2}, -\frac{1}{2} - \frac{\sqrt{3}i}{2} \right\} \subset \mathbb{C}^-.
$$

Taking the following transformation matrices

$$
Q = \begin{bmatrix} 1 & 0 & -1 \\ 0 & 1 & -1 \\ 0 & 0 & -1 \end{bmatrix}, \quad P = \begin{bmatrix} 0 & 1 & 1 \\ 1 & 0 & 0 \\ 0 & 1 & 0 \end{bmatrix},
$$

we can show that (11.66) holds with

$$
\bar{A}_1 = \begin{bmatrix} 0 & -1 \\ 1 & -1 \end{bmatrix}, \ B_1 = \begin{bmatrix} 0 & -1 \\ 1 & -1 \end{bmatrix}, \ B_2 = \begin{bmatrix} 0 & -1 \end{bmatrix}
$$

and

$$
L_1 = \begin{bmatrix} 1 \\ 0 \end{bmatrix}, \ L_2 = -1.
$$

Therefore, according to the method given in Theorem 11.8, system (11.70) has the following normal rank$E$-order state observer:

$$
\begin{cases} \dot{z} = \begin{bmatrix} 0 & -1 \\ 1 & -1 \end{bmatrix} z + \begin{bmatrix} 0 & -1 \\ 1 & -1 \end{bmatrix} u + \begin{bmatrix} 1 \\ 0 \end{bmatrix} y \\[4mm] w = \begin{bmatrix} 0 & 1 \\ 1 & 0 \\ 0 & 1 \end{bmatrix} z + \begin{bmatrix} 1 \\ 0 \\ 0 \end{bmatrix} y + \begin{bmatrix} 0 & 1 \\ 0 & 0 \\ 0 & 0 \end{bmatrix} u \end{cases}.
\tag{11.71}
$$

11.5.2 Normal-State Observers of Order $n - m$

The reduced-order state observer designed in this subsection is a straight-forward generalization of the reduced-order observer for normal linear systems.

Theorem 11.9. *Assume that the regular system (11.17) is detectable, dual normalizable, and* rank$C = m$. *Then it has a reduced-order normal observer of order $n-m$ of the following form:*

$$
\begin{cases} \dot{z} = Fz + Su + Ly \\ w = Mz + Ny \end{cases}.
\tag{11.72}
$$

Proof. Since rank$C = m$, without loss of generality we can assume

$$
C = [C_1 \ C_2], \ C_1 \in \mathbb{R}^{m \times m}, \ \text{rank} C_1 = m.
\tag{11.73}
$$

Let

$$
P = \begin{bmatrix} C_1^{-1} & -C_1^{-1}C_2 \\ 0 & I_{n-m} \end{bmatrix},
\tag{11.74}
$$

then P is nonsingular and $CP = [I_m \ 0]$.

Denote

$$
EP = \begin{bmatrix} \tilde{E}_{11} & \tilde{E}_{12} \\ \tilde{E}_{21} & \tilde{E}_{22} \end{bmatrix}, \ \tilde{E}_{11} \in \mathbb{R}^{m \times m},
\tag{11.75}
$$

then it follows from the dual normalizability assumption for system (11.17) that

$$n = \text{rank} \begin{bmatrix} E \\ C \end{bmatrix} = \text{rank} \begin{bmatrix} EP \\ CP \end{bmatrix} = \text{rank} \begin{bmatrix} \tilde{E}_{11} & \tilde{E}_{12} \\ \tilde{E}_{21} & \tilde{E}_{22} \\ I_m & 0 \end{bmatrix},$$

which implies

$$\text{rank} \begin{bmatrix} \tilde{E}_{12} \\ \tilde{E}_{22} \end{bmatrix} = n - m.$$

Thus, there exists matrix $\begin{bmatrix} Q_{11} \\ Q_{21} \end{bmatrix} \in \mathbb{R}^{n \times m}$ such that

$$\text{rank} \begin{bmatrix} Q_{11} & \tilde{E}_{12} \\ Q_{21} & \tilde{E}_{22} \end{bmatrix} = n. \tag{11.76}$$

Let

$$Q = \begin{bmatrix} Q_{11} & \tilde{E}_{12} \\ Q_{21} & \tilde{E}_{22} \end{bmatrix}^{-1}, \tag{11.77}$$

then Q is nonsingular and

$$Q \begin{bmatrix} \tilde{E}_{12} \\ \tilde{E}_{22} \end{bmatrix} = \begin{bmatrix} 0 \\ I_{n-m} \end{bmatrix}.$$

For the nonsingular matrices Q and P defined above, we have

$$QEP = Q \begin{bmatrix} \tilde{E}_{11} & \tilde{E}_{12} \\ \tilde{E}_{21} & \tilde{E}_{22} \end{bmatrix} = \begin{bmatrix} E_{11} & 0 \\ E_{21} & I_{n-m} \end{bmatrix}, \quad CP = [I_m \ 0],$$

where

$$Q \begin{bmatrix} \tilde{E}_{11} \\ \tilde{E}_{21} \end{bmatrix} = \begin{bmatrix} E_{11} \\ E_{21} \end{bmatrix}. \tag{11.78}$$

Denote

$$QAP = \begin{bmatrix} A_{11} & A_{12} \\ A_{21} & A_{22} \end{bmatrix}, \quad QB = \begin{bmatrix} B_1 \\ B_2 \end{bmatrix}. \tag{11.79}$$

From the above discussion, it is clearly seen that under the coordinate transformation

$$x = P \begin{bmatrix} x_1 \\ x_2 \end{bmatrix},$$

system (11.17) is r.s.e. to

$$\begin{cases} E_{11}\dot{x}_1 = A_{11}x_1 + A_{12}x_2 + B_1u \\ E_{21}\dot{x}_1 + \dot{x}_2 = A_{21}x_1 + A_{22}x_2 + B_2u, \\ y = x_1 \end{cases} \tag{11.80}$$

where $x_1 \in \mathbb{R}^m$, $x_2 \in \mathbb{R}^{n-m}$.

From (11.80), we see that the substate x_1 may be obtained directly from the measured output $y(t)$, and the substate x_2 satisfies the following dynamic equation:

$$\begin{cases} \dot{x}_2 = A_{22}x_2 + A_{21}y - E_{21}\dot{y} + B_2u \\ \tilde{y} = E_{11}\dot{y} - A_{11}y - B_1u = A_{12}x_2 \end{cases} \tag{11.81}$$

Moreover, since the descriptor linear system (11.17) is assumed to be detectable, we have

$$\text{rank}\begin{bmatrix} sE - A \\ C \end{bmatrix}$$

$$= \text{rank}\begin{bmatrix} sQEP - QAP \\ CP \end{bmatrix}$$

$$= \text{rank}\begin{bmatrix} sE_{11} - A_{11} & -A_{12} \\ sE_{21} - A_{21} & sI - A_{22} \\ I_m & 0 \end{bmatrix}$$

$$= n, \ \forall s \in \bar{\mathbb{C}}^+, s \text{ finite.}$$

Therefore,

$$\text{rank}\begin{bmatrix} sI - A_{22} \\ A_{12} \end{bmatrix} = n - m, \ \forall s \in \bar{\mathbb{C}}^+, s \text{ finite,}$$

showing that (A_{22}, A_{12}) is detectable. Thus, there exists a matrix $L_2 \in \mathbb{R}^{(n-m)\times m}$ such that $\sigma(A_{22} - L_2A_{12}) \subset \mathbb{C}^-$. For this matrix L_2, from Theorem 11.1 (the special case of $E = I_n$) we know that the descriptor linear system (11.81) has the following state observer:

$$\dot{\hat{x}}_2 = (A_{22} - L_2A_{12})\hat{x}_2 + A_{21}y - E_{21}\dot{y} + B_2u + L_2\tilde{y}, \tag{11.82}$$

which guarantees

$$\lim_{t\to\infty}(\hat{x}_2(t) - x_2(t)) = 0, \ \forall x_2(0), \ \hat{x}_2(0).$$

Let

$$z = \hat{x}_2 + (E_{21} - L_2E_{11})y,$$

then it follows from (11.82) that z satisfies

$$\begin{cases} \dot{z} = (A_{22} - L_2 A_{12}) z + (B_2 - L_2 B_1) u \\ \quad + [A_{21} - L_2 A_{11} - (A_{22} - L_2 A_{12}) (E_{21} - L_2 E_{11})] y. \\ \hat{x}_2 = z - (E_{21} - L_2 E_{11}) y \end{cases} \tag{11.83}$$

Since $x_1 = y$ and $\hat{x}_2$ is an asymptotic estimation of the state x_2 for system (11.81), a state observer in the form of (11.72) of order $n - m$ for system (11.81) can be obtained, with $z \in \mathbb{R}^{n-m}$, $w \in \mathbb{R}^n$, and

$$\begin{cases} F = A_{22} - L_2 A_{12} \\ S = B_2 - L_2 B_1 \\ L = A_{21} - L_2 A_{11} - (A_{22} - L_2 A_{12}) (E_{21} - L_2 E_{11}) \\ M = P \begin{bmatrix} 0 \\ I_{n-m} \end{bmatrix} \\ N = P \begin{bmatrix} I_m \\ -E_{21} + L_2 E_{11} \end{bmatrix} \end{cases} \tag{11.84}$$

The proof is then done.   □

The state observer (11.72) is a normal one and its output is a linear combination of z and y. Such observers are easier to realize physically than descriptor ones.

Based on the above proof, an algorithm for designing a normal-state observer of order $n - m$ can be given as follows. Please note that in this algorithm condition (11.73) is assumed to be true. This does not reduce the generality since otherwise some state transformation can be applied to the system such that this condition holds for the transformed system. We can then apply this algorithm to the transformed system.

Algorithm 11.3. *Design of normal-state observers of order $n - m$.*

Step 1 Compute the nonsingular matrix P according to (11.74).
Step 2 Find the matrices $\tilde{E}_{12}$ and $\tilde{E}_{22}$ satisfying (11.75).
Step 3 Find the matrices Q_{11} and Q_{21} satisfying (11.76), and compute the nonsingular matrix Q according to (11.77).
Step 4 Find the matrices E_{11}, E_{21}, A_{ij}, and B_i, $i, j = 1, 2$, according to (11.78) and (11.79).
Step 5 Find a matrix L_2 such that $\sigma (A_{22} - L_2 A_{12}) \subset \mathbb{C}^-$.
Step 6 Compute the coefficient matrices of the observer (11.72) according to (11.84).

Example 11.5. Consider a descriptor system in the form of (11.17) with the following coefficient matrices

$$E = \begin{bmatrix} 0 & 1 & 0 \\ 0 & 0 & 1 \\ 0 & 0 & 0 \end{bmatrix}, \quad A = \begin{bmatrix} 1 & 0 & 0 \\ 0 & 1 & 0 \\ 0 & 0 & 1 \end{bmatrix},$$

$$B = \begin{bmatrix} 0 & 0 \\ 1 & 0 \\ 0 & 1 \end{bmatrix}, \ C = \begin{bmatrix} 1 & 0 & 0 \end{bmatrix}.$$

In this system, $n = 3$, $m = 1$, $\text{rank} C = 1 = m$, and this system can be easily checked to be observable. Thus, it has an observer of order 2 in the form of (11.72). Following the first three steps in Algorithm 11.3, we obtain the transformation matrices

$$P = \begin{bmatrix} 1 & 0 & 0 \\ 0 & 1 & 0 \\ 0 & 0 & 1 \end{bmatrix}, \ Q = \begin{bmatrix} 0 & 0 & 1 \\ 1 & 0 & 0 \\ 0 & 1 & 0 \end{bmatrix}.$$

Furthermore, performing the Step 4 in Algorithm 11.3 gives

$$E_{11} = 0, \ E_{21} = \begin{bmatrix} 0 \\ 0 \end{bmatrix},$$

$$A_{11} = 0, \ A_{12} = \begin{bmatrix} 0 & 1 \end{bmatrix}, \ A_{21} = \begin{bmatrix} 1 \\ 0 \end{bmatrix}, \ A_{22} = \begin{bmatrix} 0 & 0 \\ 1 & 0 \end{bmatrix},$$

and

$$B_1 = \begin{bmatrix} 0 & 1 \end{bmatrix}, \ B_2 = \begin{bmatrix} 0 & 0 \\ 1 & 0 \end{bmatrix}.$$

Since $(A_{22} \ A_{12})$ is observable, in the Step 5 of Algorithm 11.3 we find the matrix $L_2 = \begin{bmatrix} 1 & 1 \end{bmatrix}^T$, which satisfies

$$\sigma (A_{22} - L_2 A_{12}) = \left\{ -\frac{1}{2} + \frac{\sqrt{3}}{2}i, -\frac{1}{2} - \frac{\sqrt{3}}{2}i \right\} \subset \mathbb{C}^-.$$

Following the Step 6 of Algorithm 11.3, we can finally construct the observer as follows:

$$\begin{cases} \dot{z} = \begin{bmatrix} 0 & -1 \\ 1 & -1 \end{bmatrix} z + \begin{bmatrix} 1 \\ 0 \end{bmatrix} y + \begin{bmatrix} 0 & -1 \\ 1 & -1 \end{bmatrix} u \\ w = \begin{bmatrix} 1 \\ 0 \\ 0 \end{bmatrix} y + \begin{bmatrix} 0 & 0 \\ 1 & 0 \\ 0 & 1 \end{bmatrix} z \end{cases} \tag{11.85}$$

It can be obtained that the state response $x(t)$ of this example system is

$$x(t) = \begin{bmatrix} 0 & 0 \\ -1 & 0 \\ 0 & -1 \end{bmatrix} u(t) + \begin{bmatrix} -1 & 0 \\ 0 & -1 \\ 0 & 0 \end{bmatrix} \dot{u}(t) + \begin{bmatrix} 0 & -1 \\ 0 & 0 \\ 0 & 0 \end{bmatrix} \ddot{u}(t),$$

which includes input derivatives of orders up to 2. However, it is implied from (11.85) that the state $x(t)$ may be asymptotically approximated by a value w, which is the output of a normal system. This interesting phenomenon reveals a common fact in normal-state observers for descriptor systems. A variable which apparently includes the input derivatives may be sometimes asymptotically tracked by the output of a system that does not involve apparently the input derivatives. The core point for this fact lies in that input derivatives are included invisibly in the measured output $y(t)$.

11.6 Normal Function Kx Observers

In the preceding three sections, we studied the design of state observers for the descriptor system

$$\begin{cases} E\dot{x} = Ax + Bu \\ y = Cx \end{cases}, \tag{11.86}$$

where $x(t) \in \mathbb{R}^n$, $u(t) \in \mathbb{R}^r$, and $y(t) \in \mathbb{R}^m$ are the state, the control input, and the measured output, respectively; E, A, B, and C are constant matrices of appropriate dimensions. In this section, we turn to study the normal function Kx observers for system (11.86). For simplicity, we only concentrate on the following Luenberger-type normal function observers:

$$\begin{cases} \dot{z} = Fz + Su + Ly \\ w = Mz + Ny \end{cases}, \tag{11.87}$$

where $z \in \mathbb{R}^p$ is the observer state vector and $w \in \mathbb{R}^q$ is the observer output vector. The design purpose is to seek the matrix parameters F, S, L, M, and N such that

$$\lim_{t \to \infty} (Kx(t) - w(t)) = 0 \tag{11.88}$$

holds for some given matrix $K \in \mathbb{R}^{q \times n}$ and arbitrarily given initial values $x(0)$, $z(0)$, and control input $u(t)$.

The main task of this section is to provide the conditions for normal function Kx observers for descriptor linear systems, and to give a parametric design approach to this Luenberger-type of function Kx observers.

11.6.1 Conditions for Normal Function Kx Observers

Let $T \in \mathbb{R}^{p \times n}$, and define

$$\varepsilon = z - TEx, \quad e = w - Kx, \tag{11.89}$$

then it follows from (11.86) and (11.87) that

$$\begin{cases} \dot{\varepsilon} = F\varepsilon + (LC - TA + FTE)x + (S - TB)u \\ e = M\varepsilon + (MTE + NC - K)x \end{cases}. \qquad (11.90)$$

When F is stable, and

$$S = TB, \qquad (11.91)$$

$$TA - FTE = LC, \qquad (11.92)$$

$$K = MTE + NC, \qquad (11.93)$$

system (11.90) becomes

$$\begin{cases} \dot{\varepsilon} = F\varepsilon \\ e = M\varepsilon \end{cases}.$$

Therefore, for arbitrary admissible initial values $x(0)$, $z(0)$ and control input $u(t)$ there holds

$$\lim_{t \to \infty} e(t) = 0. \qquad (11.94)$$

This states that (11.87) is a function Kx Luenberger-type observer for system (11.86).

Conversely, it can also be shown that when (11.94) holds for arbitrary admissible initial values $x(0)$, $z(0)$ and control input $u(t)$, the matrix F must be stable and the (11.91)–(11.93) hold (see Dai 1989b, and also Lin and Wang 1997; Kawaji and Kim 1995).

The above discussion clearly gives the following theorem.

Theorem 11.10. *Assume that the regular system (11.86) is R-observable and system (11.87) is observable. Then (11.87) is a normal Kx observer for system (11.86) if and only if $\sigma(F) \subset \mathbb{C}^-$, and there exists a matrix $T \in \mathbb{R}^{p \times n}$ satisfying (11.91)–(11.93).*

Obviously, the above function Kx observer (11.87) reduces to a normal-state observer when the given matrix K is an identity one. Based on this observation and the above theorem, we can prove the following result on the minimal order of normal-state observers for system (11.86).

Corollary 11.3. *Let the regular system (11.86) be R-observable and $\mathrm{rank}C = m$. If the observable system (11.87) is a normal-state observer such that*

$$\lim_{t \to \infty} (x(t) - w(t)) = 0, \ \forall z(0), \ x(0), \ \text{and} \ u(t),$$

then the minimal order of observer (11.87) is $p = n - m$.

Proof. Assume that (11.87) is a normal-state observer for system (11.86). Then $K = I_n$ in (11.88). By Theorem 11.10, there exists a matrix $T \in \mathbb{R}^{p \times n}$ such that

$$I_n = MTE + NC = \begin{bmatrix} M & N \end{bmatrix} \begin{bmatrix} TE \\ C \end{bmatrix},$$

indicating

$$\text{rank}\begin{bmatrix} TE \\ C \end{bmatrix} = n.$$

Hence,

$$\text{rank}\,(TE) \geq n - m.$$

Therefore, $p \geq n - m$ by the fact of $T \in \mathbb{R}^{p \times n}$. Furthermore, recall Theorem 11.9 that a state observer of order $n - m$ exists for system (11.86), the conclusion thus holds. □

Remark 11.7. It is shown in the preceding section that a normal-state observer of order $n_0 = \text{rank}E$ exists for system (11.86) under certain conditions. Yet with many practical systems, there may hold $n_0 < n - m$. Please note that this fact does not contradict with the above result because the normal-state observer of order n_0 considered in the preceding section is in the form of (11.61), which does not fall into the scope of the observers described by (11.87).

11.6.2 Parametric Design for Normal Function Observers

In the above, we have presented conditions for the Luenberger-type normal function Kx observers for descriptor linear systems. In this subsection, we propose, based on the conditions given in Theorem 11.10, a parametric approach to the design of the Luenberger-type function Kx observers in the form of (11.87). Relevant work can be found in Duan et al. (1991b, 1993).

Parametric Expressions for Observer Parameters

To derive the general parametric form of the Luenberger-type function Kx observers for system (11.86), we introduce the following assumptions on system (11.86).

Assumption 11.3. $\text{rank}B = r$, $\text{rank}C = m$.

Assumption 11.4. The matrix triple (E, A, C) is R-observable.

Parametric Expression for the Matrix F The only requirement on the coefficient matrix F of the Luenberger function Kx observer (11.87) is that it has eigenvalues with negative real parts. For convenience and simplicity, let us restrict the matrix F to be nondefective, it thus has a diagonal Jordan matrix. Therefore, a general form for the matrix F can be given, based on the Jordan form decomposition theory, as

$$F = Q\Lambda Q^{-1}, \quad \Lambda = \text{diag}(s_1, s_2, \ldots, s_p), \tag{11.95}$$

where, s_i, $i = 1, 2, \ldots, p$, are clearly the eigenvalues of the matrix F. They are self-conjugate and satisfy the following constraint.

Constraint C1 $\mathrm{Re}\,(s_i) < 0$, $i = 1, 2, \ldots, p$.
The matrix

$$Q = \begin{bmatrix} q_1 & q_2 & \cdots & q_p \end{bmatrix} \in \mathbb{C}^{p \times p}$$

is obviously the right eigenvector matrix of the matrix F, which satisfies

Constraint C2 $\det Q \neq 0$ and $q_i = \bar{q}_l$ if $s_i = \bar{s}_l$.
In the case of $p = 1$, this matrix Q can be chosen to be 1 without loss of generality.

Parametric Expressions for the Matrices T and L   According to Theorem 11.10, the matrices T and L are determined by the matrix equation

$$TA - FTE = LC. \tag{11.96}$$

Equation (11.96) is a generalized Sylvester matrix equation. It is in the dual form of the one considered in Appendix C (see also Duan 1992b, 1993b, 1996b). By applying the complete parametric solution proposed in Theorem C.1 for generalized Sylvester matrix equations, the parametric expressions for the matrices T and L satisfying (11.96) can be readily obtained as

$$T = QV, \quad L = -QW, \tag{11.97}$$

with

$$V = \begin{bmatrix} v_1 & v_2 & \cdots & v_p \end{bmatrix}^{\mathrm{T}}, \quad v_i = N(s_i)f_i, \tag{11.98}$$

$$W = \begin{bmatrix} w_1 & w_2 & \cdots & w_p \end{bmatrix}^{\mathrm{T}}, \quad w_i = D(s_i)f_i, \tag{11.99}$$

where $N(s)$ and $D(s)$ are a pair of right coprime polynomial matrices satisfying the following right coprime factorization

$$\left(sE^{\mathrm{T}} - A^{\mathrm{T}}\right)^{-1} C^{\mathrm{T}} = N\,(s)\,D^{-1}\,(s), \tag{11.100}$$

and $f_i \in \mathbb{C}^m$, $i = 1, 2, \ldots, p$, are a group of design parameters satisfying the following constraint:

Constraint C3 $f_i = \bar{f}_l$ if $s_i = \bar{s}_l$.

Remark 11.8. The right coprime factorization (11.100) performs an important role in the parameterization of the matrices T and L. For solution of the right coprime factorization (11.100), refer to Remark 11.2.

Remark 11.9. When there are complex eigenvalues chosen for the matrix F, the matrices T and L are complex. This can be prevented by a well-known technique, which can convert, at the same time, all the three matrices Λ, T, and L into real ones (see Remark 11.4).

Parametric Expressions for the Matrices N and M   According to Theorem 11.10, the matrices N and M are determined by the matrix equation

$$K = MTE + NC, \tag{11.101}$$

which can be equivalently written as

$$K = \begin{bmatrix} M & N \end{bmatrix} \begin{bmatrix} TE \\ C \end{bmatrix}. \tag{11.102}$$

It follows from (11.102) that there exist a pair of matrices M and N satisfying (11.101) if and only if the following constraint is met.

Constraint C4: $\text{rank} \begin{bmatrix} TE \\ C \end{bmatrix} = \text{rank} \begin{bmatrix} TE \\ C \\ K \end{bmatrix}$.

Under the above constraint, the parametric solutions to the observer gain matrices M and N can be presented.

Lemma 11.1. *Let Assumptions 11.3 and 11.4 be met, the matrix T be given by (11.97)–(11.98). Then there exist, when Constrain C4 is met, a matrix $K_0 \in \mathbb{R}^{q \times r^*}$ and a nonsingular matrix P satisfying*

$$P \begin{bmatrix} TE \\ C \end{bmatrix} Q_1 = \begin{bmatrix} T_0 & 0 \\ 0 & 0 \end{bmatrix}, K Q_1 = \begin{bmatrix} K_0 & 0 \end{bmatrix}, \tag{11.103}$$

where $T_0 \in \mathbb{R}^{r^* \times r^*}$ is nonsingular, r^* is the common rank in Constraint C4. In this case, the matrices N and M are given by

$$\begin{bmatrix} M & N \end{bmatrix} = \begin{bmatrix} K_0 T_0^{-1} & N' \end{bmatrix} P, \tag{11.104}$$

with $N' \in \mathbb{R}^{q \times (m+p-r^*)}$ being an arbitrary real parameter matrix.

Proof. Under the condition of Constraint C4, there exist nonsingular matrices P and Q_1 satisfying (11.103). Post-multiplying both sides of (11.102) by Q_1, and using (11.103) gives

$$\begin{bmatrix} K_0 & 0 \end{bmatrix} = \begin{bmatrix} M & N \end{bmatrix} P^{-1} \begin{bmatrix} T_0 & 0 \\ 0 & 0 \end{bmatrix},$$

which is equivalent to

$$K_0 = \begin{bmatrix} M & N \end{bmatrix} P^{-1} \begin{bmatrix} T_0 \\ 0 \end{bmatrix}. \tag{11.105}$$

Let

$$[M \ N] \, P^{-1} = [M' \ N'], \ N' \in \mathbb{R}^{q \times (m+p-r^*)}, \qquad (11.106)$$

then (11.105) becomes

$$K_0 = M'T_0. \qquad (11.107)$$

Combining the above relation with (11.106) gives the parametric expression (11.104) for the matrices M and N. □

By summarizing the above, the following theorem about the parametric solution of the Luenberger function Kx observers in the form of (11.87) can be given as follows:

Theorem 11.11. *Let Assumptions 11.3 and 11.4 be met.*

1. *A Luenberger-type observer in the form of (11.87), with the matrix $F \in \mathbb{R}^{p \times p}$ nondefective, exists for system (11.86) if there exist parameters s_i, q_i, and f_i, $i = 1, 2, \ldots, p$, satisfying Constraints C1–C4.*
2. *When the above condition is met, all the Luenberger observers in the form of (11.87) for system (11.86), with the matrix F nondefective, are parameterized by (11.95)–(11.99) and (11.103)–(11.104) with the parameters s_i, q_i, and f_i, $i = 1, 2, \ldots, p$, satisfying Constraints C1-C4.*

Based on the above theorem, we have the following algorithm for the design of the Luenberger-type of normal observers in the form of (11.87) for system (11.86).

Algorithm 11.4. Luenberger observer design.

Step 1 Solve a pair of right coprime polynomial matrices $N(s)$ and $D(s)$ satisfying the right coprime factorization (11.100).
Step 2 Specify the observer order p, and obtain the specific forms for Constraints C1–C4.
Step 3 Find a set of parameters s_i, f_i, and q_i, $i = 1, 2, \ldots, p$, satisfying Constraints C1–C4.
Step 4 Compute the coefficient matrices F, T, L, N, and M according to (11.95), (11.97)–(11.99) and (11.103)–(11.104).

Remark 11.10. In the case that the parameters s_i, q_i, f_i, $i = 1, 2, \ldots, p$, and N' satisfying Constraints C1–C4 do not exist, the order of the observer may be increased to provide sufficient design freedom for existence of a solution.

Remark 11.11. An important advantage of the above approach is that it provides all the design freedom. These degrees of freedom can be further utilized to achieve additional system specifications.

An Illustrative Example

Consider a linear system in the form of (11.86) with the following parameters

$$E = \begin{bmatrix} 1 & 0 & 0 \\ 0 & 0 & 1 \\ 0 & 0 & 0 \end{bmatrix}, \quad A = \begin{bmatrix} -5 & 0 & 0 \\ 0 & 1 & 0 \\ 0 & 0 & 1 \end{bmatrix},$$

$$B = \begin{bmatrix} 0 \\ 1 \\ 0 \end{bmatrix}, \quad C = \begin{bmatrix} 1 & 0 & 0 \\ 0 & 1 & 0 \end{bmatrix}.$$

We will design a Luenberger observer which tracks asymptotically the function Kx, with

$$K = \begin{bmatrix} 1 & 1 & 1 \end{bmatrix},$$

which can stabilize the given system.

In the following, we will obtain the observer gain matrices using Algorithm 11.4.

Step 1 By using a method given by Duan (1996b), the solution to the right coprime factorization (11.100) can be obtained as follows:

$$N(s) = \begin{bmatrix} 0 & 1 \\ 1 & 0 \\ s & 0 \end{bmatrix}, \quad D(s) = \begin{bmatrix} 0 & 5+s \\ -1 & 0 \end{bmatrix}. \tag{11.108}$$

Step 2 For this example, a first-order observer in the form of (11.87) is considered. In this case, the matrix Q can be chosen to be 1 without loss of generality. In order that Constraint C1 is met, the observer pole s is restricted to be a negative real scalar, then the matrix $F = s$. Choosing

$$f = \begin{bmatrix} \alpha_1 \\ \alpha_2 \end{bmatrix} \tag{11.109}$$

then Constraint C3 holds automatically, and the matrices T and L are given, in view of (11.97)–(11.99), as

$$T = \begin{bmatrix} \alpha_2 & \alpha_1 & s\alpha_1 \end{bmatrix}, \quad L = -\begin{bmatrix} (5+s)\alpha_2 & -\alpha_1 \end{bmatrix}.$$

With this matrix T, it is obvious that Constraint C4 is $\alpha_1 \neq 0$ and $r^* = 3$.

Step 3 Under the condition $\alpha_1 \neq 0$, the matrices P and Q_1 can both be chosen to be identity matrices when the matrices T_0 and K_0 are expressed as follows:

$$T_0 = \begin{bmatrix} \alpha_2 & \alpha_1 & s\alpha_1 \\ 1 & 0 & 0 \\ 0 & 1 & 0 \end{bmatrix}, \quad K_0 = \begin{bmatrix} 1 & 1 & 1 \end{bmatrix}.$$

Step 4 Summing up the above, the design freedom involved in this design is composed of the closed-loop eigenvalue s and parameters α_1 and α_2.

Step 5 The matrices M and N can be obtained as

$$\begin{bmatrix} M & N \end{bmatrix} = \begin{bmatrix} 1 & 1 & 1 \end{bmatrix} \begin{bmatrix} \alpha_2 & \alpha_1 & s\alpha_1 \\ 1 & 0 & 0 \\ 0 & 1 & 0 \end{bmatrix}^{-1}$$

$$= \begin{bmatrix} \dfrac{1}{s\alpha_1} & -\dfrac{\alpha_2}{s\alpha_1} + 1 & -\dfrac{1}{s} + 1 \end{bmatrix}.$$

This gives

$$M = \frac{1}{s\alpha_1}, \quad N = \begin{bmatrix} -\dfrac{\alpha_2}{s\alpha_1} + 1 & -\dfrac{1}{s} + 1 \end{bmatrix}.$$

Specially choosing $s = -1$, $\alpha_1 = \alpha_2 = 1$ gives the following specific solution:

$$F = -1, \ T = \begin{bmatrix} 1 & 1 & -1 \end{bmatrix}, \ L = \begin{bmatrix} -4 & 1 \end{bmatrix},$$
$$M = -1, \ N = \begin{bmatrix} 2 & 2 \end{bmatrix}.$$

11.7 Notes and References

In this chapter, we studied the problem of observer design for descriptor linear systems, and considered designs of several types of observers.

It is worth pointing out that the rank E order reduced-order state observer studied in Sect. 11.5 can be realized by the Luenberger observer introduced in Sect. 11.6 by properly selecting the coefficient matrices. Readers are encouraged to establish such a relation.

For design of the full-order descriptor state observer and the Luenberger-type function observer, we have introduced a parametric method, which utilizes the parametric solution to a type of generalized Sylvester matrix equation proposed by Duan (1992b, 1993b, 1996b, 2004c).

This chapter is closely related to the concepts of observability and detectability, which are the dual concept of controllability and stability. Furthermore, it can be also easily seen that some of the problems in observer design are also the dual ones in controller design.

The observer problem for descriptor linear systems has been considered by quite a few researchers. Early works on state observers for continuous-time descriptor systems can be found in Wang and Dai (1987a), El-Tohami et al. (1983), Shafai and Carroll (1987), Saidahmed (1985),and Wang (1984). More recent results can be found in Hou and Muller (1999), Minamide et al. (1997), Wang and Zou (2001), Sun (2006), and Yang et al. (2006). In El-Tohami et al. (1983), a method based on pseudoinverse was used to study this problem. In Hou and Muller (1999), the problem of observer design is studied by introducing two detectability notions for the most general class of linear descriptor systems. While Minamide et al. (1997) gave a parametrization of all linear function observers for descriptor systems and converted the problem of finding all function observers to that of solving a proper stable linear matrix equation, and Wang and Zou (2001) presented a method for the design of Luenberger observers for descriptor systems, and revealed that the input-generatable impulsive modes are very crucial to the existence of normal Luenberger observers of the systems. In Yang et al. (2006), the H_2 observer design for descriptor systems is examined, and a parameterization of all H_2 observer is proposed. In addition, Wu and his coauthor have considered the design of generalized PI/PD observers for descriptor linear systems (Wu and Duan 2006a,b, 2007a), and Gao and Ho (2004) also dealt with proportional multiple-integral observer design, but for descriptor systems with measurement output disturbances. Different from most other researchers, Sun (2006) treated the problem of inputs decoupled observer design in finite time for nonregular descriptor systems with unknown inputs.

Besides the above, many researchers have paid attention to the designs of many different types of observers for more complicated systems. Zhang et al. (1997) studied state observers of interconnected descriptor systems, Biehn and his/her coauthors examined the designs of observers for linear time-varying descriptor systems (Biehn et al. 1998, 2001), Fu and Duan (2005) treated robust guaranteed cost observer design for uncertain descriptor time-delay systems with Markovian jumping parameters. For nonlinear systems, Aslund and Frisk (2006) addressed the problem of observer design for nonlinear models represented by differential-algebraic equations, while Lu et al. (2004) and Lu and Ho (2006a) considered observer design for a class of continuous-time Lipschitz descriptor systems.

This is the last chapter in this book. By now we have introduced quite a few important concepts related to descriptor linear systems. In comparison with the concepts in the normal linear system theory, these concepts are more complicated. To give the reader an overall picture of these concepts, a fairly complete summarization of their inter-relations has been provided in Fig. 11.1.

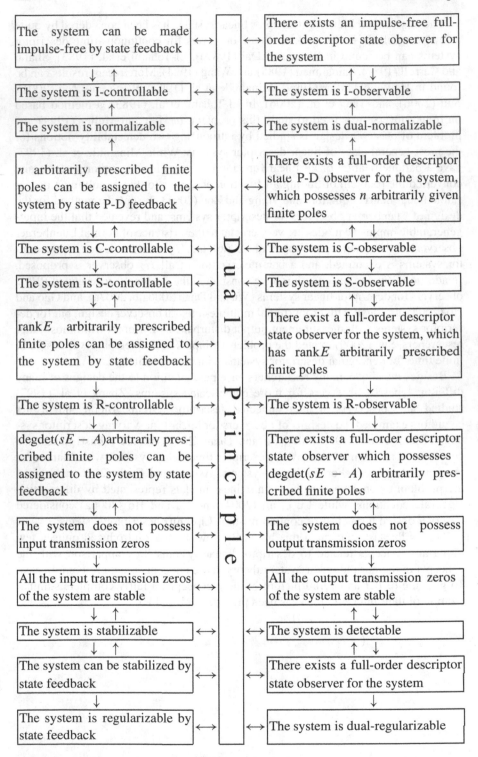

Fig. 11.1. Relations governed by dual principle

Part III
Appendices

Appendix A
Some Mathematical Results

This appendix presents some mathematical results which are repeatedly used in the book.

A.1 Delta Function $\delta(\cdot)$

Definition A.1. A function $\varphi(x)$ defined on $(-\infty, +\infty)$ is called a test function, if it is infinite times differentiable and has zero value outside a bounded closed set. Furthermore, the closure of the set $\{x \mid \varphi(x) \neq 0\}$, which is then a bounded closed set, is called the support of the test function $\varphi(x)$.

In the following, we use $\mathbb{D}$ to represent the set of all test functions, which is a linear space obviously.

Definition A.2. Let $\{\varphi_i(x)\}_{i=1}^{\infty}$ be a sequence in $\mathbb{D}$, and $\varphi(x) \in \mathbb{D}$. If

1. There exists a bounded closed set such that all the supports of $\varphi_i(x), i=1, 2, \ldots,$ are subsets of this set.
2. For any $k = 0, 1, 2, \ldots,$ the sequence $\{\varphi_i^{(k)}(x)\}_{i=1}^{\infty}$ converges to $\varphi^{(k)}(x)$ uniformly.

Then we say that the sequence of test functions $\{\varphi_i(x)\}_{i=1}^{\infty}$ converges to $\varphi(x)$ in $\mathbb{D}$.

Definition A.3. A linear and continuous (relative to the convergence defined above) functional on $\mathbb{D}$ is called a generalized function, or distribution.

Let δ be a functional on $\mathbb{D}$ defined as follows:

$$\int_{-\infty}^{+\infty} \delta(x) f(x) \mathrm{d}x = f(0), \ \forall f \in \mathbb{D}.$$

Obviously, δ is a linear and continuous functional on $\mathbb{D}$. It thus is a generalized function, which is often called delta (or Dirac) function.

G.-R. Duan, *Analysis and Design of Descriptor Linear Systems*, Advances
in Mechanics and Mathematics 23, DOI 10.1007/978-1-4419-6397-0_12,
© Springer Science+Business Media, LLC 2010

If $\mathscr{I}$ is an interval on $(-\infty, +\infty)$ and $f(x)$ is defined only on $\mathscr{I}$, then we can extend $f(\cdot)$ naturally by

$$\widetilde{f}(x) = \begin{cases} f(x), & \text{if } x \in \mathscr{I} \\ 0, & \text{otherwise} \end{cases},$$

and this makes the following convenience

$$\int_{\mathscr{I}} \delta(x) f(x) \mathrm{d}x = \int_{-\infty}^{+\infty} \delta(x) \widetilde{f}(x) \mathrm{d}x$$

$$= \widetilde{f}(0)$$

$$= \begin{cases} f(0), & \text{if } 0 \in \mathscr{I} \\ 0, & \text{otherwise} \end{cases}. \tag{A.1}$$

Now we can define $\delta^{(i)}(\cdot)$, $i = 1, 2, \ldots$, recursively as follows:

$$\int_{\mathscr{I}} \delta^{(i)}(x) f(x) \mathrm{d}x = -\int_{\mathscr{I}} \delta^{(i-1)}(x) f'(x) \mathrm{d}x,$$

and further have

$$\int_{\mathscr{I}} \delta^{(i)}(x) f(x) \mathrm{d}x = \begin{cases} (-1)^i f^{(i)}(0), & \text{if } 0 \in \mathscr{I} \\ 0, & \text{otherwise} \end{cases}. \tag{A.2}$$

Lemma A.1. *Let $f(x)$ be a usual continuous function, and k be a positive integer. If for an arbitrary function $u(x)$, which is k times continuously differentiable and satisfies $u^i(0) = 0$, $i = 0, 1, \ldots, k$, the following condition*

$$\int_{-\infty}^{+\infty} f(x) u(x) \mathrm{d}x = 0$$

always holds, then $f(x)$ is identically zero.

Proof. Otherwise, there exists $x_0 \neq 0$ such that $f(x_0) \neq 0$. Without loss of generality, we suppose $f(x_0) > 0$. Since $f(x)$ is continuous at x_0, we can take $\delta > 0$ such that

$$0 \notin \{x \mid |x - x_0| \leq \delta\},$$

and

$$f(x) > \frac{1}{2} f(x_0), \text{ if } |x - x_0| \leq \delta.$$

Let

$$u(x) = \begin{cases} \exp\left(\frac{\delta^2}{(x-x_0)^2 - \delta^2}\right), & \text{if } |x - x_0| < \delta \\ 0, & \text{if } |x - x_0| \geq \delta \end{cases},$$

which is k times continuously differentiable and satisfies $u^i(0) = 0, i = 0, 1, \ldots, k$, (in fact, u is a test function, and the support of u is $\{x \mid |x - x_0| \leq \delta\}$, which then does not contain 0). Then we have

$$\int_{-\infty}^{+\infty} f(x)u(x)dx = \int_{x_0-\delta}^{x_0+\delta} f(x)u(x)dx$$

$$\geq \frac{1}{2}f(x_0)\int_{x_0-\delta}^{x_0+\delta} \exp\left(\frac{\delta^2}{(x-x_0)^2 - \delta^2}\right)dx$$

$$> 0.$$

This contradicts the assumption.   □

Theorem A.1 (Linear Independence Theorem). *Let $f(x)$ be a usual continuous function, which is not identically zero, and k be a positive integer. Then the functions $f(x), \delta(x), \delta^{(1)}(x), \ldots, \delta^{(k)}(x)$ are linearly independent.*

Proof. Suppose that

$$cf(x) + c_0\delta(x) + c_1\delta^{(1)}(x) + \cdots + c_k\delta^{(k)}(x) = 0 \tag{A.3}$$

holds for constants $c, c_0, c_1, \ldots, c_k$, we need to prove all these constants are zero.

Take an arbitrary function $u(x)$, which is k times continuously differentiable and satisfies $u^{(i)}(0) = 0, i = 0, 1, \ldots, k$. Multiplying both sides of (A.3) by this function, and then taking integration, gives

$$\int_{-\infty}^{+\infty} cf(x)u(x)dx + \sum_{i=0}^{k} c_i \int_{-\infty}^{+\infty} \delta^{(i)}(x)u(x)dx = 0.$$

Since

$$\int_{-\infty}^{+\infty} \delta^{(i)}(x)u(x)dx = (-1)^i u^{(i)}(0)$$

$$= 0, \quad i = 0, 1, 2, \ldots, k,$$

we get

$$\int_{-\infty}^{+\infty} cf(x)u(x)dx = 0.$$

Then it follows from Lemma A.1 that $cf(x)$ is identically zero and we have further $c = 0$ by the assumption.

Now we take an arbitrary function $u(x)$, which is k times continuously differentiable and satisfies $u(0) = 1$ but $u^{(i)}(0) = 0, \quad i = 1, \ldots, k$. Multiplying both sides of (A.3) by $u(x)$ and taking integration give

$$c_0 \int_{-\infty}^{+\infty} \delta(x)u(x)dx + \sum_{i=1}^{k} c_i \int_{-\infty}^{+\infty} \delta^{(i)}(x)u(x)dx = 0.$$

Since

$$\int_{-\infty}^{+\infty} \delta(x)u(x)\mathrm{d}x = u(0)$$
$$= 1,$$

and

$$\int_{-\infty}^{+\infty} \delta^{(i)}(x)u(x)\mathrm{d}x = (-1)^i u^{(i)}(0)$$
$$= 0, \quad i = 1, 2, \cdots, k,$$

we get

$$c_0 = 0.$$

It is easily seen that by taking different test functions $u(x)$ we can get successively $c_i = 0, i = 1, 2, \ldots, k$. This completes the proof. $\qquad\square$

Corollary A.1. *Let $g(x)$ be a (generalized) function which has the following decomposition*

$$g(x) = f(x) + c_0\delta(x) + c_1\delta^{(1)}(x) + \cdots + c_k\delta^{(k)}(x),$$

where $f(x)$ is a usual continuous function. Then the decomposition is unique, that is, we can determine the data $\{f(x), c_0, c_1, \ldots, c_k\}$ uniquely from $g(x)$.

A.2 Laplace Transform

Laplace transform is a very basic mathematical tool and is used in a few sections of the book.

Definition A.4. *Let $f(t) \in \mathbb{R}^n$ be a vector function defined on the interval $[0, +\infty)$, and $|\int_0^\infty e^{-st} f(t)\mathrm{d}t| < \infty$, $\forall s \in \mathscr{C}$. The Laplace transform of the function $f(t)$, denoted by $\mathscr{L}[f]$, is a vector function of the complex variable s defined as*

$$F(s) = \mathscr{L}[f(t)] = \int_0^\infty e^{-st} f(t)\mathrm{d}t. \tag{A.4}$$

The following proposition gives two important properties of Laplace transform.

Proposition A.1. *Let $f(t), f_1(t), f_2(t) \in \mathbb{R}^n$ be three arbitrary functions defined on the interval $[0, +\infty)$, α, β be any two constants. Then*

$$\mathscr{L}[\alpha f_1(t) + \beta f_2(t)] = \alpha\mathscr{L}[f_1(t)] + \beta\mathscr{L}[f_2(t)], \tag{A.5}$$

and

$$\mathscr{L}[f^{(k)}(t)] = s^k \mathscr{L}[f(t)] - \sum_{i=0}^{k-1} s^i f^{(k-1-i)}(0), \quad k = 1, 2, \ldots. \tag{A.6}$$

Proof. Since integration admits a linear property, (A.5) is obvious. Formula (A.6) can be seen from the following deduction

$$\mathcal{L}[f^{(k)}(t)] = \int_0^\infty e^{-st} f^{(k)}(t)dt$$

$$= e^{-st} f^{(k-1)}(t)|_0^\infty - \int_0^\infty (-s)e^{-st} f^{(k-1)}(t)dt$$

$$= s\mathcal{L}[f^{(k-1)}(t)] - f^{(k-1)}(0)$$

$$= s\left(s\mathcal{L}[f^{(k-2)}(t)] - f^{(k-2)}(0)\right) - f^{(k-1)}(0)$$

$$= s^2 \mathcal{L}[f^{(k-2)}(t)] - \sum_{i=0}^{1} s^i f^{(k-1-i)}(0)$$

$$\vdots$$

$$= s^k \mathcal{L}[f(t)] - \sum_{i=0}^{k-1} s^i f^{(k-1-i)}(0).$$

The proof is then completed. □

Taking $k = 1, 2$ in (A.6) gives

$$\mathcal{L}[f'(t)] = s\mathcal{L}[f(t)] - f(0)$$

and

$$\mathcal{L}[f''(t)] = s^2 \mathcal{L}[f(t)] - sf(0) - f'(0).$$

Using the above definition and properties, we can obtain the Laplace transform of a given function.

Proposition A.2. *The Laplace transforms of* $\delta^{(i)}(t)$, $i = 0, 1, 2, \ldots$, *are given by*

$$\mathcal{L}[\delta^{(i)}](s) = s^i, \ i = 0, 1, 2, \ldots,$$

where $\delta^{(0)}(t)$ *represents* $\delta(t)$ *as usual.*

Proof. By Definitions A.4 and A.2, we have

$$\mathcal{L}[\delta^{(i)}](s) = \int_{[0,+\infty)} \delta^{(i)}(t)e^{-st}dt$$

$$= (-1)^i (-s)^i e^{-s \cdot 0}$$

$$= s^i.$$

This completes the proof. □

Definition A.5. Let $F(s) \in \mathbb{C}^n$ be a complex vector function which is analytical on $\{s | \text{Re}(s) \geq \sigma\}$ for some $\sigma \in \mathbb{R}$. The inverse Laplace transform of the function $F(s)$, denoted by $\mathcal{L}^{-1}[F(s)]$, is a vector function of time t defined as

$$f(t) = \mathcal{L}^{-1}[F(s)] = \frac{1}{2\pi i} \int_{\sigma-j\infty}^{\sigma+j\infty} F(s)e^{st} ds. \tag{A.7}$$

Note that under very mild conditions,

$$
\begin{aligned}
\mathcal{L}^{-1}[\mathcal{L}[f(t)]] &= \mathcal{L}^{-1}\left(\int_0^\infty e^{-st} f(t) dt \right) \\
&= \frac{1}{2\pi i} \int_{\sigma-j\infty}^{\sigma+j\infty} \left(\int_0^\infty e^{-s\tau} f(\tau) d\tau \right) e^{st} ds \\
&= \frac{1}{2\pi i} \int_{\sigma-j\infty}^{\sigma+j\infty} \int_0^\infty e^{-s(\tau-t)} f(\tau) d\tau ds \\
&= \frac{1}{2\pi i} \int_0^\infty f(\tau) \int_{\sigma-j\infty}^{\sigma+j\infty} e^{-s(\tau-t)} ds d\tau \\
&= \int_0^\infty f(\tau) \delta(\tau - t) d\tau \\
&= f(t).
\end{aligned}
$$

So, generally the inverse Laplace transform of $F(s)$ is $f(t)$ if the Laplace transform of $f(t)$ is $F(s)$, and vice versa. The Laplace transforms of some basic functions are given below:

$$\mathcal{L}[\delta(t-a)] = e^{-as},$$

$$\mathcal{L}\left[\frac{t^n}{n!}\right] = \frac{1}{s^{n+1}}, \quad \mathcal{L}[e^{-nt}] = \frac{1}{s+n},$$

$$\mathcal{L}[\sin kt] = \frac{k}{s^2+k^2}, \quad \mathcal{L}[\cos kt] = \frac{s}{s^2+k^2}.$$

A.3   Determinants and Inverses of Block Matrices

This section provides the formulae for the determinants and inverses of block matrices. They are often used in this book.

Theorem A.2. *Let*

$$A = \begin{bmatrix} A_{11} & A_{12} \\ A_{21} & A_{22} \end{bmatrix} \in \mathbb{C}^{n\times n}$$

with $A_{ij} \in \mathbb{C}^{n_i \times n_j}, i, j = 1, 2$, and $n_1 + n_2 = n$.

1. *Suppose A_{11} is nonsingular, and define*

$$\Phi = A_{22} - A_{21}A_{11}^{-1}A_{12}. \tag{A.8}$$

Then

$$\det A = \det A_{11} \det \Phi = \det A_{11} \det(A_{22} - A_{21}A_{11}^{-1}A_{12}), \tag{A.9}$$

and hence the matrix A is nonsingular if and only if Φ is nonsingular, and in this case,

$$\begin{bmatrix} A_{11} & A_{12} \\ A_{21} & A_{22} \end{bmatrix}^{-1} = \begin{bmatrix} A_{11}^{-1} + A_{11}^{-1}A_{12}\Phi^{-1}A_{21}A_{11}^{-1} & -A_{11}^{-1}A_{12}\Phi^{-1} \\ -\Phi^{-1}A_{21}A_{11}^{-1} & \Phi^{-1} \end{bmatrix}. \tag{A.10}$$

2. *Suppose A_{22} is nonsingular, and define*

$$\Psi = A_{11} - A_{12}A_{22}^{-1}A_{21}. \tag{A.11}$$

Then

$$\det A = \det A_{22} \det \Psi = \det A_{22} \det(A_{11} - A_{12}A_{22}^{-1}A_{21}), \tag{A.12}$$

and hence the matrix A is nonsingular if and only if Ψ is nonsingular, and in this case,

$$\begin{bmatrix} A_{11} & A_{12} \\ A_{21} & A_{22} \end{bmatrix}^{-1} = \begin{bmatrix} \Psi^{-1} & -\Psi^{-1}A_{12}A_{22}^{-1} \\ -A_{22}^{-1}A_{21}\Psi^{-1} & A_{22}^{-1} + A_{22}^{-1}A_{21}\Psi^{-1}A_{12}A_{22}^{-1} \end{bmatrix}. \tag{A.13}$$

Proof. Proof of conclusion 1. When A_{11} is nonsingular, it can be easily checked that the matrix A has the following decomposition:

$$\begin{bmatrix} A_{11} & A_{12} \\ A_{21} & A_{22} \end{bmatrix} = \begin{bmatrix} I & 0 \\ A_{21}A_{11}^{-1} & I \end{bmatrix} \begin{bmatrix} A_{11} & 0 \\ 0 & \Phi \end{bmatrix} \begin{bmatrix} I & A_{11}^{-1}A_{12} \\ 0 & I \end{bmatrix}. \tag{A.14}$$

Taking determinants over both sides, and using

$$\det \begin{bmatrix} I & 0 \\ A_{21}A_{11}^{-1} & I \end{bmatrix} = \det \begin{bmatrix} I & A_{11}^{-1}A_{12} \\ 0 & I \end{bmatrix} = 1,$$

yields (A.9).

It clearly follows from (A.9) that the matrix A is nonsingular if and only if Φ is nonsingular. In the case, following (A.14) we have

$$\begin{bmatrix} A_{11} & A_{12} \\ A_{21} & A_{22} \end{bmatrix}^{-1} = \begin{bmatrix} I & A_{11}^{-1}A_{12} \\ 0 & I \end{bmatrix}^{-1} \begin{bmatrix} A_{11}^{-1} & 0 \\ 0 & \Phi^{-1} \end{bmatrix} \begin{bmatrix} I & 0 \\ A_{21}A_{11}^{-1} & I \end{bmatrix}^{-1}$$

$$= \begin{bmatrix} I & -A_{11}^{-1}A_{12} \\ 0 & I \end{bmatrix} \begin{bmatrix} A_{11}^{-1} & 0 \\ 0 & \Phi^{-1} \end{bmatrix} \begin{bmatrix} I & 0 \\ -A_{21}A_{11}^{-1} & I \end{bmatrix}$$

$$= \begin{bmatrix} A_{11}^{-1} + A_{11}^{-1}A_{12}\Phi^{-1}A_{21}A_{11}^{-1} & -A_{11}^{-1}A_{12}\Phi^{-1} \\ -\Phi^{-1}A_{21}A_{11}^{-1} & \Phi^{-1} \end{bmatrix},$$

that is, (A.10) holds.

Proof of conclusion 2. When A_{22} is nonsingular, the matrix A can be decomposed as

$$\begin{bmatrix} A_{11} & A_{12} \\ A_{21} & A_{22} \end{bmatrix} = \begin{bmatrix} I & A_{12}A_{22}^{-1} \\ 0 & I \end{bmatrix} \begin{bmatrix} \Psi & 0 \\ 0 & A_{22} \end{bmatrix} \begin{bmatrix} I & 0 \\ A_{22}^{-1}A_{21} & I \end{bmatrix}.$$

Therefore, the proof can be carried out similarly.   □

In the case of $A_{12} = 0$ or $A_{21} = 0$, we immediately have the following corollary of above theorem.

Corollary A.2. *Let* $A_{ij} \in \mathbb{C}^{n_i \times n_j}$, $i, j = 1, 2$. *Then*

$$\begin{bmatrix} A_{11} & 0 \\ A_{21} & A_{22} \end{bmatrix}^{-1} = \begin{bmatrix} A_{11}^{-1} & 0 \\ -A_{22}^{-1}A_{21}A_{11}^{-1} & A_{22}^{-1} \end{bmatrix},$$

$$\begin{bmatrix} A_{11} & A_{12} \\ 0 & A_{22} \end{bmatrix}^{-1} = \begin{bmatrix} A_{11}^{-1} & -A_{11}^{-1}A_{12}A_{22}^{-1} \\ 0 & A_{22}^{-1} \end{bmatrix}.$$

Comparing the $(1,1)$ block of A^{-1} given in (A.10) and (A.13) produces the following corollary.

Corollary A.3. *Let* $A_{ij} \in \mathbb{C}^{n_i \times n_j}$, $i, j = 1, 2$, A_{11} *and* A_{22} *are both nonsingular. Then*

$$(A_{11} - A_{12}A_{22}^{-1}A_{21})^{-1} = A_{11}^{-1} + A_{11}^{-1}A_{12}(A_{22} - A_{21}A_{11}^{-1}A_{12})^{-1}A_{21}A_{11}^{-1}.$$

By applying Theorem A.2 to a matrix of the following form

$$A = \begin{bmatrix} I_m & X \\ -Y & I_n \end{bmatrix},$$

we immediately have the following corollary.

Corollary A.4. *Let* $X \in \mathbb{C}^{m \times n}$ *and* $Y \in \mathbb{C}^{n \times m}$. *Then*

$$\det \begin{bmatrix} I_m & X \\ -Y & I_n \end{bmatrix} = \det(I_n + YX) = \det(I_m + XY), \qquad (A.15)$$

and when $\Phi = I_n + YX$ is nonsingular,

$$\begin{bmatrix} I_m & X \\ -Y & I_n \end{bmatrix}^{-1} = \begin{bmatrix} I_m - X\Phi^{-1}Y & -X\Phi^{-1} \\ \Phi^{-1}Y & \Phi^{-1} \end{bmatrix}, \tag{A.16}$$

and when $\Psi = I_m + XY$ is nonsingular,

$$\begin{bmatrix} I_m & X \\ -Y & I_n \end{bmatrix}^{-1} = \begin{bmatrix} \Psi^{-1} & -\Psi^{-1}X \\ Y\Psi^{-1} & I_n - Y\Psi^{-1}X \end{bmatrix}. \tag{A.17}$$

Specially choosing in the above Corollary A.4 $n = 1$, $X = x \in \mathbb{C}^m$, and $Y = y^* \in \mathbb{C}^{1 \times m}$, we have

$$\Phi = 1 + y^*x \text{ and } \Psi = I_m + xy^*,$$

and (A.15), (A.16), and (A.17) reduce to

$$\det\begin{bmatrix} I_m & x \\ -y^* & 1 \end{bmatrix} = \det(I_m + xy^*) = 1 + y^*x,$$

$$\begin{bmatrix} I_m & x \\ -y^* & 1 \end{bmatrix}^{-1} = \begin{bmatrix} I_m - (1 + y^*x)^{-1}xy^* & -(1 + y^*x)^{-1}x \\ (1 + y^*x)^{-1}y^* & (1 + y^*x)^{-1} \end{bmatrix},$$

and

$$\begin{bmatrix} I_m & x \\ -y^* & 1 \end{bmatrix}^{-1} = \begin{bmatrix} (I_m + xy^*)^{-1} & -(I_m + xy^*)^{-1}x \\ y^*(I_m + xy^*)^{-1} & I_n - y^*(I_m + xy^*)^{-1}x \end{bmatrix}.$$

A.4 Nilpotent Matrices

Nilpotent matrices and their properties perform a very basic function in this book. For readers' convenience, here we present the definition and some basic properties, which are used in the book.

Perhaps, the nilpotent matrix we know best is the following one:

$$A = \begin{bmatrix} 0 & 1 & & \\ & \ddots & \ddots & \\ & & 0 & 1 \\ & & & 0 \end{bmatrix} \in \mathbb{R}^{n \times n}. \tag{A.18}$$

The definition for a general nilpotent matrix is as follows.

Definition A.6. A square matrix $N \in \mathbb{R}^{n \times n}$ is called nilpotent if there exists a positive integer k such that $N^k = 0$. The smallest k satisfying $N^k = 0$ is called the nilpotent index of N.

According to the above definition, one class of nilpotent matrices are the strictly lower or upper triangular matrices. Particularly, the matrix A given in (A.18) is clearly a nilpotent matrix with index n. For this nilpotent matrix, it is obvious that

$$\det(sN - I) = (-1)^n, \quad \forall s \in \mathscr{C}. \tag{A.19}$$

The following theorem gives a very basic necessary and sufficient condition for nilpotent matrices.

Theorem A.3. *A matrix $N \in \mathbb{R}^{n \times n}$ is nilpotent if and only if all its eigenvalues are zero.*

Proof. *Necessity*: Assume that N is nilpotent and has index k. Then, $N^k = 0$, for some positive integer k. Let λ be an eigenvalue of N. Then $Nx = \lambda x$ for a nonzero vector x. By induction, we have

$$0 = N^k x = \lambda^k x.$$

Since $x \neq 0$, we obtain that $\lambda = 0$. Therefore, all the eigenvalues of a nilpotent matrix are zero.

Sufficiency: Suppose all the eigenvalues $\lambda_1, \lambda_2, \ldots, \lambda_n$ of the matrix N are zero. Then the characteristic polynomial

$$\varphi(\lambda) = \det(\lambda I - N) = (\lambda - \lambda_1)(\lambda - \lambda_2) \cdots (\lambda - \lambda_n) = \lambda^n.$$

According to the Cayley–Hamilton theorem, we have

$$\varphi(N) = N^n = 0,$$

that is, N is nilpotent.   □

The above theorem has the following corollary.

Corollary A.5. *Let $N \in \mathbb{R}^{n \times n}$ be a nilpotent matrix. Then the following hold:*

1. $\det N = 0$.
2. $\operatorname{tr}(N) = 0$.
3. $\det(sI - N) = s^n$.
4. $\det(sN - I) = (-1)^n, \quad \forall s \in \mathbb{C}$.

Proof. Since the determinant of a matrix is the product of its eigenvalues, it follows from Theorem A.3 that the first conclusion holds. Similarly, since the trace of a square matrix is the sum of its eigenvalues, it follows from Theorem A.3 that the second conclusion holds. The third conclusion is shown in the sufficiency part of the above Theorem A.3.

The fourth condition is trivial when $s = 0$. When $s \neq 0$, denoting $\lambda = \frac{1}{s}$, and using the third conclusion we have

$$\det(sN - I) = \det\left[s\left(N - \frac{1}{s}I\right)\right]$$
$$= (-1)^n s^n \det(\lambda I - N)$$
$$= (-1)^n s^n \lambda^n$$
$$= (-1)^n.$$

Therefore, the fourth conclusion holds. $\square$

The following theorem gives some further properties of nilpotent matrices.

Theorem A.4. *Nilpotent matrices admit the following properties.*

1. *Any matrix that is similar to a nilpotent matrix is also nilpotent. Similar nilpotent matrices have the same nilpotent index.*
2. *A nonzero nilpotent matrix cannot be symmetric.*
3. *The product of a nilpotent matrix N with a square matrix commutative to N is also nilpotent.*

Proof. Proof of conclusion 1. Assume the matrix A is similar to a nilpotent matrix N, that is, there exists a nonsingular matrix P such that

$$A = PNP^{-1}.$$

Since N is nilpotent, there exists a positive integer k such that $N^k = 0$. It follows that

$$A^k = PN^k P^{-1} = 0.$$

Thus, A is also nilpotent.

Proof of conclusion 2. Suppose that N is a symmetric nonzero nilpotent matrix. Then there exists an orthogonal matrix P such that

$$N = P^{-1} DP$$

for some diagonal matrix D. Because N is nilpotent, there exists a positive integer k such that

$$0 = N^k = P^{-1} D^k P.$$

It follows that $D = 0$. So

$$N = P^{-1} DP = 0.$$

We obtain a contradiction. Therefore, N cannot be symmetric.

Proof of conclusion 3. Assume that the nilpotent matrix N has index k, and C is a square matrix, which is commutative with N, then

$$CN = NC.$$

Therefore,

$$(CN)^k = C^k N^k = 0,$$
$$(NC)^k = N^k C^k = 0.$$

Therefore, the conclusion holds.   □

The following theorem gives some decompositions related to nilpotent matrices.

Theorem A.5. *Let* $A \in \mathbb{R}^{n \times n}$ *be an arbitrarily given matrix. Then the following hold:*

1. *There exists a diagonalizable matrix* $D \in \mathbb{R}^{n \times n}$ *and a nilpotent matrix* $N \in \mathbb{R}^{n \times n}$ *such that*

$$A = D + N.$$

2. *There exist a nilpotent matrix* N_1, *and two nonsingular matrices* A_1 *and* T, *of appropriate dimensions, such that*

$$T^{-1} A T = \begin{bmatrix} A_1 & 0 \\ 0 & N_1 \end{bmatrix}. \tag{A.20}$$

Proof. Let the Jordan matrix J of the given matrix A be in the following form

$$J = \mathrm{diag}(J_1(\lambda_1), J_2(\lambda_2), \ldots, J_r(\lambda_r)), \tag{A.21}$$

with

$$J_i = \begin{bmatrix} \lambda_i & 1 & & \\ & \ddots & \ddots & \\ & & \lambda_i & 1 \\ & & & \lambda_i \end{bmatrix}. \tag{A.22}$$

Then it follows from the well-known Jordan decomposition that there exists a nonsingular matrix P such that

$$P^{-1} A P = J. \tag{A.23}$$

Proof of conclusion 1. Denote

$$\begin{cases} D' = \mathrm{diag}(D_1, \ldots, D_r) \\ D_i = \mathrm{diag}(\lambda_i, \ldots, \lambda_i) \end{cases},$$

and

$$\begin{cases} N' = \mathrm{diag}(N_1, \ldots, N_r) \\ N_i = \begin{bmatrix} 0 & 1 & & \\ & \ddots & \ddots & \\ & & 0 & 1 \\ & & & 0 \end{bmatrix} \end{cases}$$

Then the matrix N' is nilpotent, and it is clear to see that

$$J_i = D_i + N_i,$$

and hence following from the above relation and (A.21)–(A.22),

$$J = D' + N'.$$

Combining the above relation with (A.23) gives

$$A = PD'P^{-1} + PN'P^{-1}.$$

Finally, choose $D = PD'P^{-1}$ and $N = PN'P^{-1}$. Because N is still nilpotent in view of Theorem A.4, the conclusion is proven.

Proof of conclusion 2. By applying elementary transformations to the Jordan matrix J given in (A.21)–(A.22), we can find a nonsingular matrix Q satisfying

$$Q^{-1}JQ = \begin{bmatrix} A_1 & 0 \\ 0 & N_1 \end{bmatrix}, \qquad (A.24)$$

with

$$A_1 = \operatorname{diag}(J'_1, J'_2, \ldots, J'_p),$$

and

$$N_1 = \operatorname{diag}(J'_{p+1}, J'_{p+2}, \ldots, J'_r),$$

where J'_i, $i = 1, 2, \ldots, p$, are those Jordan blocks in J, which are associated with nonzero eigenvalues of A, while $J'_j j = p + 1, p + 2, \ldots, r$, are those Jordan blocks in J, which are associated with zero eigenvalues of A. As a consequence, A_1 is nonsingular, and N_1 is nilpotent. Finally, letting $T = PQ$, and combining (A.24) with (A.23) gives (A.20). □

A.5 Some Operations of Linear Subspaces

In this and the following sections, we use $\mathbb{F}$ to represent $\mathbb{R}$ or $\mathbb{C}$, the set of real or complex numbers, respectively.

Definition A.7. Let V, V_1, and V_2 be subspaces of $\mathbb{F}^n$, and $x \in \mathbb{F}^n$.

1. The orthogonal complement of V, denoted by $V^\perp$, is defined as

$$V^\perp = \{x | x \in \mathbb{F}^n, x^T v = 0 \text{ for all } v \in V\}.$$

2. The sum of the vector x and the subspace, denoted by $x + V$, is

$$x + V = \{x + v | v \in V\}.$$

3. The sum of the two subspaces V_1 and V_2, denoted by $V_1 + V_2$, is

$$V_1 + V_2 = \{v_1 + v_2 | v_1 \in V_1, v_2 \in V_2\}.$$

4. The intersection of the two subspaces V_1 and V_2, denoted by $V_1 \cap V_2$, is

$$V_1 \cap V_2 = \{v | v \in V_1 \text{ and } v \in V_2\}.$$

The following proposition gives some properties of the operations introduced in the above definition.

Proposition A.3. *Let V, V_1, and V_2 be subspaces of $\mathbb{F}^n$. Then the following hold:*

1. *$V^\perp, V_1 + V_2$, and $V_1 \cap V_2$ are all subspaces of $\mathbb{F}^n$.*
2. *$(V^\perp)^\perp = V$.*
3. *$V_1 = V_2 \iff (V_1)^\perp = (V_2)^\perp$.*

Proof. The first conclusion can be directly verified from the definition of subspaces.

For the second one, let $\{v_1, v_2, \ldots, v_{n_1}\}$ and $\{u_1, u_2, \ldots, u_{n_2}\}$ be bases of V and $V^\perp$, respectively. Because $V^\perp$ can be seen as the solution space of the following linear equations group

$$\begin{bmatrix} v_1 & v_2 & \cdots & v_{n_1} \end{bmatrix}^\mathrm{T} x = 0,$$

we have $n_2 = n - n_1$ from the theory of linear algebra. Therefore, as the solution space of the following linear equations group

$$\begin{bmatrix} u_1 & u_2 & \cdots & u_{n_2} \end{bmatrix}^\mathrm{T} x = 0,$$

$(V^\perp)^\perp$ has dimension $n - n_2 = n_1$, that is

$$\dim(V^\perp)^\perp = \dim(V).$$

On the other hand, it is obvious that $V \subset (V^\perp)^\perp$, so the conclusion follows.

The third conclusion follows immediately from the second one.   □

The following definition introduces another operation, which is used in Chap. 4 of the book.

Definition A.8. Let V_1 be a subspace of $\mathbb{F}^{n_1}$, and V_2 be a subspace of $\mathbb{F}^{n_2}$. Then the external direct sum of V_1 and V_2, denoted by $V_1 \oplus V_2$, is defined as

$$V_1 \oplus V_2 = \left\{ \begin{bmatrix} v_1 \\ v_2 \end{bmatrix} \middle| v_1 \in V_1, v_2 \in V_2 \right\}.$$

The following proposition reveals two important and basic properties of the external direct sum of subspaces.

Proposition A.4. *Let V_1 and V_2 be two subspaces of $\mathbb{F}^{n_1}$ and $\mathbb{F}^{n_2}$, respectively. Then the following hold:*

1. *$V_1 \oplus V_2$ is a subspace of $\mathbb{F}^{n_1+n_2}$.*
2. *$\dim(V_1 \oplus V_2) = \dim(V_1) + \dim(V_2)$.*

Proof. The first conclusion can be checked by using the definition of subspace.

For the second one, suppose that $\{v_1, v_2, \ldots, v_k\}$ and $\{u_1, u_2, \ldots, u_l\}$ are bases of V_1 and V_2, respectively. Then it is easy to see that the following set

$$\left\{ \begin{bmatrix} v_1 \\ 0 \end{bmatrix}, \begin{bmatrix} v_2 \\ 0 \end{bmatrix}, \ldots, \begin{bmatrix} v_k \\ 0 \end{bmatrix}, \begin{bmatrix} 0 \\ u_1 \end{bmatrix}, \begin{bmatrix} 0 \\ u_2 \end{bmatrix}, \ldots, \begin{bmatrix} 0 \\ u_l \end{bmatrix} \right\}$$

is a basis of $V_1 \oplus V_2$. Thus, the conclusion follows. $\qquad \square$

A.6 Kernels and Images of Matrices

Definition A.9. Let $A \in \mathbb{F}^{m \times n}$.

1. The set $\{x \mid x \in \mathbb{F}^n, Ax = 0\}$ is a subspace of $\mathbb{F}^n$, which is called the kernel of A and denoted by $\ker A$, that is

$$\ker A = \{x \mid x \in \mathbb{F}^n, Ax = 0\}.$$

2. The set $\{Ax \mid x \in \mathbb{F}^n\}$ is a subspace of $\mathbb{F}^m$, which is called the image of A and denoted by Image A. That is,

$$\text{Image } A = \{Ax \mid x \in \mathbb{F}^n\}.$$

The following proposition summarizes some basic properties of the kernel and the image of a given matrix.

Proposition A.5. *Let $A \in \mathbb{F}^{m \times n}$, $B \in \mathbb{F}^{n \times k}$, and $C \in \mathbb{F}^{m \times l}$.*

1. *$\dim(\text{Image } A) = \text{rank} A$.*
2. *$\text{rank} A + \dim(\ker A) = n$.*
3. *Image $(AB) = A$Image $B = \{Ax \mid x \in \text{Image } B\}$.*
4. *When B is nonsingular,*

$$\ker(AB) = B^{-1} \ker A = \{z \mid z \in \mathbb{C}^k, Bz \in \ker A\}.$$

5. *Image A + Image C = Image $[A\ C]$.*

Proof. Let $\{v_1, v_2, \ldots, v_p\}$ be a maximal linear independent subset of the set of all column vectors of A. Then $\text{rank} A = p$,

$$\text{Image } A = \text{span}\{v_1, v_2, \ldots, v_p\},$$

and the first conclusion follows. The second conclusion is a well-known result in linear algebra, while the third and the fourth conclusions can be directly verified by using the above Definition A.9.

Note

$$\text{Image } A + \text{Image } C = \{Ax + Cy | x \in \mathbb{F}^n, y \in \mathbb{F}^l\}$$

$$= \left\{ [A\ C] \begin{bmatrix} x \\ y \end{bmatrix} \middle| \begin{bmatrix} x \\ y \end{bmatrix} \in \mathbb{F}^{n+l} \right\}$$

$$= \text{Image } [A\ C].$$

The last conclusion also holds true.   □

The following proposition further gives some equivalent relations related to kernels and images of matrices.

Proposition A.6. *Let $A \in \mathbb{F}^{m \times n}$, and $B \in \mathbb{F}^{n \times k}$.*

1. Image $(AB) = $ Image $A \iff$ Image $B + \ker A = \mathbb{F}^n$.
2. $\ker(AB) = \ker B \iff$ Image $B \cap \ker A = \emptyset$, *where $\emptyset$ is the null subspace of* $\mathbb{F}^n$.
3. *Let $x \in \mathbb{F}^m$. Then $x \in$ (Image $A)^\perp \iff x^T A = 0$.*

Proof. Proof of conclusion 1. Necessity: Since Image $(AB) = $ Image A, then for any $x \in \mathbb{F}^n$, there exists $y \in \mathbb{F}^k$ such that

$$Ax = ABy, \tag{A.25}$$

which is equivalent to

$$A(x - By) = 0.$$

Therefore,

$$x - By \in \ker A. \tag{A.26}$$

Further noting that $By \in$ Image B, we have

$$x = (x - By) + By \in \text{Image } B + \ker A.$$

Sufficiency: Suppose Image $B + \ker A = \mathbb{F}^n$. Then for $\forall x \in \mathbb{F}^n$, there exists $y \in \mathbb{F}^k$, such that (A.26) holds. This implies that for any $x \in \mathbb{F}^n$, there exists $y \in \mathbb{F}^k$, such that (A.25) holds, and this means Image $(AB) = $ Image A holds.

Proof of conclusion 2. Necessity: Otherwise, there exists $y \in \mathbb{F}^n$ such that $y \neq 0$ and $y \in$ Image $B \cap \ker A$. Therefore, $Ay = 0$ and $y = Bz$ for some $z \in \mathbb{F}^k$. Thus, we have $z \in \ker(AB)$, but $z \notin \ker B$, and these relations contradict to $\ker(AB) = \ker B$.

Sufficiency: It is obvious that $\ker B \subset \ker(AB)$. Suppose that there exists $z \in \mathbb{F}^k$ such that $z \in \ker(AB)$, but $z \notin \ker B$. Then we have $(AB)z = A(Bz) = 0$ but

$Bz \neq 0$. Put $y = Bz$, which then satisfies $y \neq 0, y \in$ Image B, and $y \in \ker A$. These relations contradict to Image $B \cap \ker A = 0$.

Proof of conclusion 3. This conclusion follows immediately from the Definitions A.7 and A.9. □

A.7 Singular Value Decomposition

Singular value decomposition (SVD) is a very useful tool in matrix analysis. Since it involves only unitary or orthogonal matrices, its solution is regarded to be numerically very simple and reliable. Because of this nice feature, singular value decomposition has been repeatedly used in this book.

In this section, the results are valid in either $\mathbb{R}^{n \times m}$ or $\mathbb{C}^{n \times m}$. For convenience, we denote by $\mathbb{F}^{n \times m}$ the matrix space, which may be either $\mathbb{R}^{n \times m}$ or $\mathbb{C}^{n \times m}$.

To introduce SVD, let us first give a very important basic fact about unitary matrices.

Lemma A.2. *Let $A \in \mathbb{F}^{n \times m}$. Then*

$$\|UAV\|_2 = \|A\|_2 \quad and \quad \|UAV\|_F = \|A\|_F$$

hold for any appropriately dimensioned unitary matrices U and V.

Proofs of this lemma can be found in certain algebra textbooks.

Based on the above lemma, we can now prove the following singular value decomposition.

Theorem A.6. *For an arbitrarily given matrix $A \in \mathbb{F}_p^{m \times n}$, there always exist unitary matrices $U \in \mathbb{F}^{m \times m}$ and $V \in \mathbb{F}^{n \times n}$ such that*

$$A = U \begin{bmatrix} \Sigma & 0 \\ 0 & 0 \end{bmatrix} V^*, \tag{A.27}$$

where

$$\Sigma = \text{diag}(\sigma_1, \sigma_2, \ldots, \sigma_p) \tag{A.28}$$

and

$$\sigma_1 \geq \sigma_2 \geq \cdots \geq \sigma_p \geq 0. \tag{A.29}$$

Proof. Let $\sigma_1 = \|A\|$ and assume $m \geq n$ (otherwise, we consider A^* instead). Then, from the definition of $\|A\|$, there exists a nonzero $z \in \mathbb{F}^n$ such that

$$\|Az\|_2 = \sigma_1 \|z\|_2. \tag{A.30}$$

Define

$$x = \frac{z}{\|z\|_2} \in \mathbb{F}^n, \quad y = \frac{1}{\sigma_1} Ax \in \mathbb{F}^m. \tag{A.31}$$

Then, both x and y are unitary vectors. Based on these two unitary vectors, we can construct the following unitary matrices

$$V_1 = \begin{bmatrix} x & V_0 \end{bmatrix} \in \mathbb{F}^{n \times n}, \quad U_1 = \begin{bmatrix} y & U_0 \end{bmatrix} \in \mathbb{F}^{m \times m}, \tag{A.32}$$

and consequently,

$$y^* A x = \frac{1}{\sigma_1} (Ax)^* Ax = \frac{1}{\sigma_1} \|Ax\|_2^2 = \sigma_1, \tag{A.33}$$

$$U_0^* A x = \sigma_1 U_0^* \left(\frac{1}{\sigma_1} Ax \right) = \sigma_1 U_0^* y = 0, \tag{A.34}$$

and

$$U_0^* y = 0. \tag{A.35}$$

Using (A.31)–(A.35), we have

$$
\begin{aligned}
B_1 &= U_1^* A V_1 \\
&= \begin{bmatrix} y^* A x & y^* A V_0 \\ U_0^* A x & U_0^* A V_0 \end{bmatrix} \\
&= \begin{bmatrix} \sigma_1 y^* y & y^* A V_0 \\ \sigma_1 U_0^* y & U_0^* A V_0 \end{bmatrix} \\
&= \begin{bmatrix} \sigma_1 & w^* \\ 0 & A_1 \end{bmatrix},
\end{aligned}
\tag{A.36}
$$

where

$$w = V_0^* A^* y \in \mathbb{F}^{n-1} \quad \text{and} \quad A_1 = U_0^* A V_0 \in \mathbb{F}^{(m-1) \times (n-1)}.$$

Let $e_1 = \begin{bmatrix} 1 & 0 & \cdots & 0 \end{bmatrix}^{\mathrm{T}}$. Since

$$\|B_1^* e_1\|_2^2 = \left\| \begin{bmatrix} \sigma_1 \\ w \end{bmatrix} \right\|_2^2 = (\sigma_1^2 + w^* w),$$

it follows from the above equation that

$$\|B_1\|_2^2 \geq \sigma_1^2 + w^* w.$$

On the other side, we have from Lemma A.2 and (A.36) that

$$\|A\|_2 = \|B_1\|_2 = \sigma_1.$$

Comparing the above two relations gives $w = 0$. Therefore, (A.36) finally turns into

$$B_1 = U_1^* A V_1 = \begin{bmatrix} \sigma_1 & 0 \\ 0 & A_1 \end{bmatrix}. \tag{A.37}$$

Applying the above process to the matrix A_1, we can obtain two unitary matrices $V_2 \in \mathbb{F}^{(n-1)\times(n-1)}$ and $U_2 \in \mathbb{F}^{(m-1)\times(m-1)}$ such that

$$B_2 = U_2^* A_1 V_2 = \begin{bmatrix} \sigma_2 & 0 \\ 0 & A_2 \end{bmatrix},$$

with

$$\begin{aligned}
\sigma_2 &= \|A_1\|_2 \\
&\leq \left\| \begin{bmatrix} \sigma_1 & 0 \\ 0 & A_1 \end{bmatrix} \right\|_2 \\
&= \|U_1^* A V_1\|_2 \\
&= \|A\|_2 \\
&= \sigma_1.
\end{aligned}$$

Proceeding with this process, we obtain the unitary matrices $V_i \in \mathbb{F}^{(n-i+1)\times(n-i+1)}$ and $U_i \in \mathbb{F}^{(m-i+1)\times(m-i+1)}$, $i = 1, 2, \ldots, p$, such that

$$B_i = U_i^* A_{i-1} V_i = \begin{bmatrix} \sigma_i & 0 \\ 0 & A_i \end{bmatrix}, \ i = 1, 2, \ldots, p,$$

with

$$\sigma_i = \|A_{i-1}\|_2 \leq \sigma_{i-1}, \ \text{ and } \ A_p = 0.$$

Put

$$\hat{V}_i = \begin{bmatrix} I_{i-1} & 0 \\ 0 & V_i \end{bmatrix}, \ \hat{U}_i = \begin{bmatrix} I_{i-1} & 0 \\ 0 & U_i \end{bmatrix}, \ i = 1, 2, \ldots, p,$$

and

$$V = \prod_{i=1}^{p} \hat{V}_i, \ \ U = \prod_{i=1}^{p} \hat{U}_i.$$

Then it is easy to know that both matrices V and U are unitary and satisfy

$$U^* A V = \mathrm{diag}(\sigma_1, \sigma_2, \ldots, \sigma_p, 0, \ldots, 0).$$

This completes the proof.   □

Denote the two unitary matrices in Theorem A.6 as

$$U = \begin{bmatrix} u_1 & u_2 & \cdots & u_m \end{bmatrix} \in \mathbb{F}^{m\times m}, \tag{A.38}$$

$$V = \begin{bmatrix} v_1 & v_2 & \cdots & v_n \end{bmatrix} \in \mathbb{F}^{n\times n}, \tag{A.39}$$

then it follows from (A.27) to (A.29) that

$$Av_i = \sigma_i u_i, \quad A^* u_i = \sigma_i v_i, \quad i = 1, 2, \ldots, p. \tag{A.40}$$

Due to the above fact, let us introduce the following definition.

Definition A.10. Let matrix $A \in \mathbb{F}_p^{m \times n}$. Then a group of positive scalars σ_i, $i = 1, 2, \ldots, p$, are called the singular values of the matrix A, denoted by $\sigma_i = \sigma_i(A)$, $i = 1, 2, \ldots, p$, if there exist two unitary matrices $U \in \mathbb{F}^{m \times m}$ and $V \in \mathbb{F}^{n \times n}$ satisfying (A.27)–(A.29). Furthermore, if the matrices U and V are partitioned as in (A.38)–(A.39), then the vectors u_i and v_j are, respectively, called the left singular vector and the right singular vector associated with the singular value σ_i.

Some useful properties of SVD are collected in the following theorem.

Theorem A.7. Let $A \in \mathbb{F}_p^{m \times n}$ and σ_i, $i = 1, 2, \ldots, p$, be the singular values of the matrix A, the vectors u_i and v_j be, respectively, the left and the right singular vectors associated with the singular value σ_i. Then the following hold:

1. $\|A\|_2 = \sigma_1 = \max\{\sigma_i, i = 1, 2, \ldots, p\}$.
2. $\|A\|_F^2 = \sigma_1^2 + \sigma_2^2 + \cdots + \sigma_p^2$.
3. $\ker A = \mathrm{span}\{v_{p+1}, \ldots, v_n\}$ and $(\ker A)^\perp = \mathrm{span}\{v_1, \ldots, v_p\}$.
4. $\mathrm{Image}\ A = \mathrm{span}\{u_1, \ldots, u_p\}$ and $(\mathrm{Image}\ A)^\perp = \mathrm{span}\{u_{p+1}, \ldots, u_m\}$.
5. σ_i^2 is an eigenvalue of AA^* or A^*A, u_i is an associated eigenvector of AA^*, and v_i is an associated eigenvector of A^*A.
6. For any appropriately dimensioned unitary matrices U_0 and V_0, there hold

$$\sigma_i(U_0 A V_0) = \sigma_i(A), \quad i = 1, \ldots, p.$$

7. $A \in \mathbb{F}^{m \times n}$ has a dyadic expansion:

$$A = \sum_{i=1}^{p} \sigma_i u_i v_i^* = U_p \Sigma_p V_p^*,$$

where
$$U_p = \begin{bmatrix} u_1 & u_2 & \cdots & u_p \end{bmatrix}, V_r = \begin{bmatrix} v_1 & v_2 & \cdots & v_p \end{bmatrix},$$

and
$$\Sigma_r = \mathrm{diag}(\sigma_1, \sigma_2, \ldots, \sigma_p).$$

Since all the seven conclusions can be easily shown, the proof of the theorem is omitted.

To end this section, let us finally make a remark about the numerical property of singular value decomposition. This has two aspects. First, singular value decomposition involves only two unitary or orthogonal matrices. This makes it a very favorable tool in many applications because the inverses of unitary or orthogonal matrices are simply their transposes and can be obtained accurately. Second, the process of obtaining a singular value decomposition of a given matrix is known to be numerically very reliable, and hence the obtained singular value decomposition of a given matrix is generally of a very high precision. Computationally, a singular value decomposition of a given matrix can be easily obtained by the Matlab command svd.

Appendix B
Rank-Constrained Matrix Matching and Least Square Problems

The rank-constrained least square problem seeks the matrix X with an assigned rank p such that the Frobenius norm of $AX - B$ is minimized for some given matrices A and B. This problem reduces to the so-called rank-constrained matrix matching problem when the matrix A is an identity one. It is first shown that the solution to the rank-constrained least square problem can be converted to the solution to the rank-constrained matrix matching problem, and then a complete parametric solution to the rank-constrained matrix matching problem is presented. The proposed approach is convenient to use and possesses good numerical reliability since it mainly involves a singular value decomposition. To the best knowledge of the authors, such results are novel in the literature.

The results about the rank-constrained matrix matching problem is used in Chap. 6 with the solution of all the derivative feedback controllers which have minimum Frobenius norm and assigns the desired dynamical order to the closed loop system.

B.1 The Problems

In this appendix, two types of minimum norm problems are investigated: the rank-constrained matrix matching problem and the rank-constrained least square problem. Recalling the notation $\mathbb{R}_r^{m \times n} = \{M \mid M \in \mathbb{R}^{m \times n}, \text{rank} M = r\}$, we can state these two problems as follows.

Problem B.1. [The rank-constrained matrix matching problem] Let $M \in \mathbb{R}_r^{m \times n}$ be a given matrix, and p be a positive integer satisfying $p \leq \min\{m, n\}$.

1. Determine a necessary and sufficient condition for the existence of the following number

$$\gamma(p, M) = \min_{X \in \mathbb{R}_p^{m \times n}} \|X - M\|_{\text{F}}, \tag{B.1}$$

where $\|\cdot\|_{\text{F}}$ represents the Frobenius norm.

G.-R. Duan, *Analysis and Design of Descriptor Linear Systems*, Advances
in Mechanics and Mathematics 23, DOI 10.1007/978-1-4419-6397-0_13,
© Springer Science+Business Media, LLC 2010

2. When $\gamma_{(p,M)}$ defined in (B.1) exists, determine the following set

$$\mathscr{S}_{(p,M)} = \left\{ X \mid X \in \mathbb{R}_p^{m \times n} \text{ and } \|X - M\|_{\mathrm{F}} = \gamma_{(p,M)} \right\}. \tag{B.2}$$

Problem B.2. [The rank-constrained least square problem] Let $A \in \mathbb{R}^{m \times n}$ and $B \in \mathbb{R}^{m \times t}$ be two given matrices, and p be a positive integer satisfying $p \leq \min\{n, t\}$.

1. Determine a necessary and sufficient condition for the existence of the following number

$$\eta_p = \min_{X \in \mathbb{R}_p^{n \times t}} \|AX - B\|_{\mathrm{F}}. \tag{B.3}$$

2. When η_p defined in (B.3) exists, determine the following set

$$\Omega_p = \left\{ X \mid X \in \mathbb{R}_p^{n \times t} \text{ and } \|AX - B\|_{\mathrm{F}} = \eta_p \right\}. \tag{B.4}$$

It should be pointed out that the case of $p = 0$ is not included in the above rank-constrained matrix matching and least square problems because in this case both problems obviously have a unique zero solution.

Obviously, the rank-constrained matrix matching problem is a special case of the rank-constrained least square problem. It has a direct application in dynamical order assignment in descriptor linear systems via state derivative feedback (see Duan and Zhang 2002b). The main result of the rank-constrained matrix matching problem is used in Chap. 6 to solve dynamical order assignment via derivative feedback controllers with minimum Frobenius norm.

The rank-constrained least square problem can be completely solved based on the solutions to the rank-constrained matrix matching problem. Instead of a single solution, all the solutions to the Problem B.1 are characterized.

B.2 Preliminaries

Let matrices A and B and integer p be given as in Problem B.2. Applying singular value decomposition to the matrix A gives

$$A = U \begin{bmatrix} \Sigma & 0 \\ 0 & 0 \end{bmatrix} V^{\mathrm{T}}, \tag{B.5}$$

where $U \in \mathbb{R}^{m \times m}$ and $V \in \mathbb{R}^{n \times n}$ are orthogonal, and $\Sigma \in \mathbb{R}^{\alpha \times \alpha}$ is diagonal positive definite. Further, let

$$U^{\mathrm{T}} B = \begin{bmatrix} B_1 \\ B_2 \end{bmatrix}, \quad B_1 \in \mathbb{R}^{\alpha \times t}, \tag{B.6}$$

and define

$$\begin{cases} \Pi = \{p + \alpha - n, \ p + \alpha - n + 1, \cdots, \min\{\alpha, p\}\} \\ \Pi_0 = \{k \in \Pi \ |\gamma_{(k,B_1)} \text{ exists}\} \end{cases} \tag{B.7}$$

and

$$\Lambda = \{k \ |k \in \Pi_0, \ \gamma_{(k,B_1)} = \mu_0\}, \ \mu_0 = \min_{k \in \Pi_0} \gamma_{(k,B_1)}. \tag{B.8}$$

Then the following two lemmas, which will be used in the next section, can be easily shown.

Lemma B.1. *Let $A \in \mathbb{R}_\alpha^{m \times n}$ and $B \in \mathbb{R}^{m \times t}$, and matrices U, V, Σ, B_1, and B_2 satisfy (B.5) and (B.6). Then the following hold:*

1. For any $X \in \mathbb{R}^{n \times t}$, there exist $Y \in \mathbb{R}^{\alpha \times t}$ and $Z \in \mathbb{R}^{(n-\alpha) \times t}$ such that

$$X = V \begin{bmatrix} \Sigma^{-1} Y \\ Z \end{bmatrix}. \tag{B.9}$$

2. For an arbitrary matrix $X \in \mathbb{R}^{n \times t}$ possessing the form of (B.9), there holds

$$\|AX - B\|_F^2 = \|Y - B_1\|_F^2 + \|B_2\|_F^2.$$

3. Equation $\|AX - B\|_F = \|B_2\|_F$ holds if and only if the matrix X possesses the form

$$X = V \begin{bmatrix} \Sigma^{-1} B_1 \\ Z \end{bmatrix}, \quad Z \in \mathbb{R}^{(n-\alpha) \times t}. \tag{B.10}$$

Lemma B.2. *Let $Y \in \mathbb{R}_k^{\alpha \times t}$, Π be defined as in (B.7), and integer p be given as in Problem B.2. Then*

$$\begin{bmatrix} Y \\ Z \end{bmatrix} \in \mathbb{R}_p^{n \times t} \tag{B.11}$$

for some $Z \in \mathbb{R}^{(n-\alpha) \times t}$ if and only if $k \in \Pi$.

The following two lemmas are also to be used in Sect. B.3.

Lemma B.3. *Define the diagonal positive definite matrix*

$$D = \text{diag}(d_1, d_2, \dots, d_r), \ 0 < d_1 \leq \cdots \leq d_r. \tag{B.12}$$

Then the following hold:

1. There exist an integer s and two series of scalars λ_i and m_i, $i = 1, 2, \dots, s$, such that

$$0 < \lambda_1 < \lambda_2 < \cdots < \lambda_s, \tag{B.13}$$

$$m_1 + m_2 + \cdots + m_s = r, \tag{B.14}$$

and

$$D = \mathrm{diag}(d_1, \ldots, d_r) = \begin{bmatrix} \lambda_1 I_{m_1} & & \\ & \ddots & \\ & & \lambda_s I_{m_s} \end{bmatrix}. \tag{B.15}$$

2. Further define

$$l_k = m_1 + m_2 + \cdots + m_k, \quad k = 1, 2, \ldots, s, \tag{B.16}$$

and

$$l_0 = 0, \tag{B.17}$$

then

$$0 = l_0 < l_1 < l_2 < \cdots < l_s = r, \tag{B.18}$$

and for any integer t such that $0 < t \le r$ there exists a positive integer k_t satisfying

$$l_{k_t - 1} < t \le l_{k_t}. \tag{B.19}$$

Proof. The first conclusion is obvious. We here only prove the second one.

Noting that the relation (B.18) clearly follows from (B.16) and (B.14), we thus suffice only to show (B.19).

If $0 < t \le m_1$, we obviously have

$$l_0 < t \le l_1. \tag{B.20}$$

If, on the other hand, $m_1 < t \le r$, then there clearly exists a positive integer k_t ($k_t \ge 2$) such that

$$m_1 + m_2 + \cdots + m_{k_t - 1} < t \le m_1 + m_2 + \cdots + m_{k_t},$$

or equivalently,

$$l_{k_t - 1} < t \le l_{k_t} \text{ for some } k_t \ge 2. \tag{B.21}$$

Combining (B.20) and (B.21) produces (B.19). $\qquad\square$

Now we give another lemma whose proof is provided in Sect. B.6.

Lemma B.4. *Suppose the matrix D is defined as in (B.12) and t is an integer satisfying $0 < t < r$. Then for any matrix $Z \in \mathbb{R}^{r \times r}_{r-t}$ there holds*

$$\|Z - D\|_{\mathrm{F}}^2 \ge d_1^2 + d_2^2 + \cdots + d_t^2. \tag{B.22}$$

Furthermore, the equality in (B.22) holds if and only if

$$Z = D + \Theta \begin{bmatrix} 0 & 0 \\ \Gamma & 0 \end{bmatrix} D, \tag{B.23}$$

with

$$\Gamma = \mathrm{diag}\left(Y_1, Y_2, \ldots, Y_{k_t}\right), \tag{B.24}$$

where $\Theta \in \mathbb{R}^{r \times r}$ *is an arbitrary orthogonal matrix,* $Y_i \in \mathbb{R}^{m_i \times m_i}$, $i = 1, 2, \ldots,$ $k_t - 1$, *and* $Y_{k_t} \in \mathbb{R}^{(t-l_{k_t}-1) \times m_{k_t}}$ *are full-row rank upper triangular matrices satisfying*

$$Y_i Y_i^T = I, \quad i = 1, 2, \ldots, k_t, \tag{B.25}$$

and k_t *and* l_{k_t} *are integers defined as in Lemma B.3. In particular, if the matrix D satisfies*

$$0 < d_1 < d_2 < \cdots < d_r, \tag{B.26}$$

then the matrix Γ *in (B.24) becomes a diagonal involutory matrix (i.e., with all the diagonal elements being 1 or* -1) *in* $\mathbb{R}^{t \times t}$.

Remark B.1. The term of full-row rank upper triangular matrix appeared in the above lemma refers to any matrix $A \in \mathbb{R}^{m \times n}$ possessing the form $A = \begin{bmatrix} A_1 & A_2 \end{bmatrix}$, where $A_1 \in \mathbb{R}^{m \times m}$ is nonsingular upper triangular.

B.3   Solution to Problem B.1

Now let us consider solution to the rank-constrained matrix matching problem (Problem B.1). For convenience, the notations $\gamma_{(p,M)}$ and $\mathscr{S}_{(p,M)}$ in Problem B.1 are simply denoted by γ_p and $\mathscr{S}_p$ from now on.

Recall that $\mathrm{rank} M = r$, by conducting the singular value decomposition of the matrix M, we can obtain two orthogonal matrices $U_M \in \mathbb{R}^{m \times m}$ and $V_M \in \mathbb{R}^{n \times n}$ and a diagonal positive definite matrix $D_M \in \mathbb{R}^{r \times r}$ satisfying

$$M = U_M \begin{bmatrix} D_M & 0 \\ 0 & 0 \end{bmatrix} V_M^T, \tag{B.27}$$

where the matrix D_M possesses the form of D represented by (B.12).

Existence Condition for γ_p

The following theorem provides a necessary and sufficient condition for the existence of the number γ_p defined in (B.1).

Theorem B.1. *Let the conditions in Problem B.1 hold.*

1. When $p > r$, *for an arbitrary* ε *there exists* $X_\varepsilon \in \mathbb{R}_p^{m \times n}$ *such that*

$$\|X_\varepsilon - M\|_F = \varepsilon.$$

2. The γ_p *defined in (B.1) exists if and only if* $p \leq r$.

Proof. Proof of conclusion 1. Note that the singular value decomposition (B.27) holds. For any given matrix $X \in \mathbb{R}_p^{m \times n}$, let

$$U_M^{\mathrm{T}} X V_M = \begin{bmatrix} X_1 & X_2 \\ X_3 & X_4 \end{bmatrix}, \quad X_1 \in \mathbb{R}^{r \times r}. \tag{B.28}$$

Then, by specially choosing in (B.28)

$$X_1 = D_M, \quad X_2 = 0_{r \times (n-r)}, \quad X_3 = 0_{(m-r) \times r},$$

and

$$X_4 = \begin{bmatrix} \frac{\varepsilon}{\sqrt{p-r}} I_{p-r} & 0 \\ 0 & 0 \end{bmatrix}, \quad \varepsilon > 0,$$

we obtain the following special matrix in $\mathbb{R}_p^{m \times n}$:

$$X_\varepsilon = U_M \begin{bmatrix} D_M & 0 \\ 0 & \begin{bmatrix} \frac{\varepsilon}{\sqrt{p-r}} I_{p-r} & 0 \\ 0 & 0 \end{bmatrix} \end{bmatrix} V_M^{\mathrm{T}}. \tag{B.29}$$

Further, in view of the fact that for any orthogonal matrices Y and Z, there holds

$$\|XZ\|_{\mathrm{F}} = \|YX\|_{\mathrm{F}} = \|X\|_{\mathrm{F}},$$

we can derive from (B.27) and (B.29) that

$$\|X_\varepsilon - M\|_{\mathrm{F}} = \left\| \frac{\varepsilon}{\sqrt{p-r}} I_{p-r} \right\|_{\mathrm{F}} = \varepsilon.$$

Proof of conclusion 2. The necessity follows immediately from the first conclusion. The sufficiency lies in Theorems B.2 and B.3 given in the following.   □

The Complete Solution

In this subsection, we present, under the condition of $p \le r$, a characterization of the set $\mathscr{S}_p$ defined in (B.2), and thus provide a complete solution to the rank-constrained matrix matching problem. First, let us consider the special case of $p = r$.

Theorem B.2. *Let the conditions in Problem B.1 hold. If $p = r$, then $\gamma_p = 0$ and $\mathscr{S}_p = \{M\}$.*

Proof. Since $p = r$, we can choose $X = M$. This results in $\gamma_p = 0$. Therefore, $M \in \mathscr{S}_p$. To further show $\mathscr{S}_p = \{M\}$, let us take an arbitrary element $X \in \mathscr{S}_p$, and show that $X = M$.

It follows from (B.27) and (B.3) that

$$\|X - M\|_F^2 = \left\| \begin{bmatrix} X_1 - D_M & X_2 \\ X_3 & X_4 \end{bmatrix} \right\|_F^2$$

$$= \|X_1 - D_M\|_F^2 + \|X_2\|_F^2 + \|X_3\|_F^2 + \|X_4\|_F^2. \tag{B.30}$$

This implies that $X \in \mathscr{S}_p$ if and only if

$$X_1 = D_M, \quad X_2 = 0, \quad X_3 = 0, \quad X_4 = 0. \tag{B.31}$$

It can then be easily verified that the combination of (B.27), (B.28), and (B.31) gives $X = M$. □

The following theorem deals with the case of $p < r$.

Theorem B.3. *Let the conditions in Problem B.1 hold. Further, assume that (B.27) holds, and $p < r$. Then*

$$\gamma_p = (d_1^2 + d_2^2 + \cdots + d_{r-p}^2)^{1/2}, \tag{B.32}$$

and $X \in \mathscr{S}_p$ if and only if

$$X = U_M \begin{bmatrix} D_M + \Theta \begin{bmatrix} 0 & 0 \\ \widetilde{\Gamma} & 0 \end{bmatrix} D_M & 0 \\ 0_{(m-r) \times r} & 0 \end{bmatrix} V_M^{\mathrm{T}}, \tag{B.33}$$

with

$$\widetilde{\Gamma} = \mathrm{diag} \left(Y_1, Y_2, \ldots, Y_{k_{r-p}} \right), \tag{B.34}$$

where $\Theta \in \mathbb{R}^{r \times r}$ is an arbitrary orthogonal matrix, $Y_{k_{r-p}} \in \mathbb{R}^{(r-p-l_{k_{r-p}-1}) \times m_{k_{r-p}}}$ and $Y_i \in \mathbb{R}^{m_i \times m_i}$, $i = 1, 2, \ldots, k_{r-p} - 1$, are full-row rank upper triangular matrices satisfying

$$Y_i Y_i^{\mathrm{T}} = I, \quad i = 1, 2, \ldots, k_{r-p}, \tag{B.35}$$

and k_{r-p} and $l_{k_{r-p}}$ are defined as in Lemma B.3. In particular, if (B.26) is satisfied, then the matrix $\widetilde{\Gamma}$ in (B.34) becomes a diagonal involutory matrix in $\mathbb{R}^{(r-p) \times (r-p)}$.

Proof. It follows from (B.30) that

$$\|X - M\|_F^2 \geq \|X_1 - D_M\|_F^2.$$

This, together with Lemma B.4, implies

$$\|X - M\|_F^2 \geq d_1^2 + d_2^2 + \cdots + d_{r-\mathrm{rank}X_1}^2 \geq d_1^2 + d_2^2 + \cdots + d_{r-p}^2, \tag{B.36}$$

and that the inequalities in (B.36) turn simultaneously into equalities if and only if

$$X_2 = 0, \quad X_3 = 0, \quad X_4 = 0, \quad X_1 = D_M + \Theta \begin{bmatrix} 0 & 0_{p \times (r - l_{kr} - p)} \\ \widetilde{\Gamma} & 0 \end{bmatrix} D_M, \quad (B.37)$$

where the matrix $\widetilde{\Gamma}$ is defined as in (B.34) and $\Theta \in \mathbb{R}^{r \times r}$ is an arbitrary orthogonal matrix. Combining (B.28) and (B.37) completes the proof.   □

B.4 Solution to Problem B.2

Instead of giving a direct solution to Problem B.2, in this section we present a relation between Problems B.1 and B.2, which allows us to solve Problem B.2 via the solution to Problem B.1.

Theorem B.4. *Let* $A \in \mathbb{R}_\alpha^{m \times n}$ *and* $B \in \mathbb{R}^{m \times t}$ *be two given matrices, and* $p \leq \min\{n, t\}$ *be a positive integer. Further, let* V, B_1, B_2, Π, μ_0, *and* Λ *be defined by* (B.5)–(B.8). *Then the following hold:*

1. *The number* η_p *defined in* (B.3) *exists if and only if* $\operatorname{rank} B_1 \in \Pi$ *or* $\gamma_{(k, B_1)}$ *exists for all* $k \in \Pi$, *and in this case there holds*

$$\eta_p = \left(\mu_0^2 + \| B_2 \|_F^2 \right)^{\frac{1}{2}}. \quad (B.38)$$

2. *When* η_p *exists, the solution set* Ω_p *to the least square problem is given by*

$$\Omega_p = \bigcup_{k \in \Lambda} \left\{ V \begin{bmatrix} \Sigma^{-1} Y \\ Z \end{bmatrix} \middle| Y \in \mathscr{S}_{(k, B_1)}, \ \begin{bmatrix} Y \\ Z \end{bmatrix} \in \mathbb{R}_p^{n \times t} \right\}, \quad (B.39)$$

which reduces, when $\operatorname{rank} B_1 \in \Pi$ *and* $\gamma_{(k, B_1)}$ *exists for all* $k \in \Pi$, *to*

$$\Omega_p = \left\{ V \begin{bmatrix} \Sigma^{-1} B_1 \\ Z \end{bmatrix} \middle| \begin{bmatrix} B_1 \\ Z \end{bmatrix} \in \mathbb{R}_p^{n \times t} \right\}. \quad (B.40)$$

Proof. The proof is divided into the following three steps.

Step 1. Showing the necessity of the first conclusion, that is, $\operatorname{rank} B_1 \in \Pi$ or $\gamma_{(k, B_1)}$ exists for all $k \in \Pi$ if the number η_p defined in (B.3) exists.

Suppose the number η_p defined in (B.3) exists and $\gamma_{(k_0, B_1)}$ does not exist for some $k_0 \in \Pi$. Then it suffices only to show $\operatorname{rank} B_1 \in \Pi$.

It follows from the Theorem B.1 in Sect. B.3 and the fact that $\gamma_{(k_0, B_1)}$ does not exist for some $k_0 \in \Pi$ that for any $\varepsilon > 0$ there exists $Y_\varepsilon \in \mathbb{R}_{k_0}^{\alpha \times t}$ such that

$$\| Y_\varepsilon - B_1 \|_F = \varepsilon. \quad (B.41)$$

It further follows from $k_0 \in \Pi$, $Y_\varepsilon \in \mathbb{R}_{k_0}^{\alpha \times t}$ and Lemma B.2 that there exists $Z_\varepsilon \in \mathbb{R}^{(n-\alpha) \times t}$ such that

$$X_\varepsilon = V \begin{bmatrix} \Sigma^{-1} Y_\varepsilon \\ Z_\varepsilon \end{bmatrix} \in \mathbb{R}_p^{n \times t}. \tag{B.42}$$

Combining (B.42) and the second conclusion in Lemma B.1 gives

$$\|A X_\varepsilon - B\|_F^2 = \|Y_\varepsilon - B_1\|_F^2 + \|B_2\|_F^2. \tag{B.43}$$

Substituting (B.41) into (B.43) yields

$$\|A X_\varepsilon - B\|_F^2 = \varepsilon^2 + \|B_2\|_F^2. \tag{B.44}$$

On the other hand, we have from the first two conclusions in Lemma B.1 that

$$\|A X - B\|_F \geq \|B_2\|_F, \quad \forall X \in \mathbb{R}_p^{n \times t}. \tag{B.45}$$

Noting that the number η_p defined in (B.3) exists, it thus follows from (B.44) and (B.45) that there exists $X_0 \in \mathbb{R}_p^{n \times t}$ such that

$$\|A X_0 - B\|_F = \|B_2\|_F.$$

This, together with the third conclusion of Lemma B.1, implies that X_0 possesses the form (B.10). Further, using Lemma B.2, we have $\mathrm{rank}(\Sigma^{-1} B_1) \in \Pi$, or equivalently, $\mathrm{rank}\, B_1 \in \Pi$. The proof is then done.

Step 2. Proving, when $\gamma_{(k, B_1)}$ exists for all $k \in \Pi$, that the number η_p defined in (B.3) exists and is given by (B.38), and in this case the solution set Ω_p is given by (B.39). Recalling the definitions of Π and Π_0, we clearly have $\Pi = \Pi_0$ when $\gamma_{(k, B_1)}$ exists for all $k \in \Pi$. Let

$$\Delta = \left\{ Y \,\middle|\, \exists\, Z \in \mathbb{R}^{(n-\alpha) \times t} \text{ s.t. } \begin{bmatrix} Y \\ Z \end{bmatrix} \in \mathbb{R}_p^{n \times t} \right\}.$$

Then, it follows from Lemma B.2 that $Y \in \Delta$ if and only if $Y \in \mathbb{R}_k^{\alpha \times t}$ for some $k \in \Pi$. This implies

$$\Delta = \bigcup_{k \in \Pi} \left\{ Y \,\middle|\, Y \in \mathbb{R}_k^{\alpha \times t} \right\}. \tag{B.46}$$

Since $\gamma_{(k, B_1)}$ exists for all $k \in \Pi$, we further have

$$\min_{Y \in \Delta} \|Y - B_1\|_F = \min_{k \in \Pi} \min_{Y \in \mathbb{R}_k^{\alpha \times t}} \|Y - B_1\|_F = \min_{k \in \Pi} \gamma_{(k, B_1)}. \tag{B.47}$$

Combining (B.47), (B.8), and the relation $\Pi = \Pi_0$ gives

$$\min_{Y \in \Delta} \|Y - B_1\|_F = \mu_0.$$

This, together with the definition of Δ and the first two conclusions in Lemma B.1, implies that the number η_p defined in (B.3) exists and is given by (B.38). Combining (B.38) and (B.4) gives

$$\Omega_p = \left\{ X \, \middle| \, X \in \mathbb{R}_p^{n \times t}, \; \|AX - B\|_F = \left(\mu_0^2 + \|B_2\|_F^2 \right)^{\frac{1}{2}} \right\}.$$

Using the first two conclusions in Lemma B.1 again gives

$$\Omega_p = \left\{ V \begin{bmatrix} \Sigma^{-1} Y \\ Z \end{bmatrix} \, \middle| \, \|Y - B_1\|_F = \mu_0, \; \begin{bmatrix} Y \\ Z \end{bmatrix} \in \mathbb{R}_p^{n \times t} \right\}, \qquad \text{(B.48)}$$

which can be written, with the help of Lemma B.2, as

$$\Omega_p = \bigcup_{k \in \Pi} \left\{ V \begin{bmatrix} \Sigma^{-1} Y \\ Z \end{bmatrix} \, \middle| \, \|Y - B_1\|_F = \mu_0, \; Y \in \mathbb{R}_k^{\alpha \times t}, \; \begin{bmatrix} Y \\ Z \end{bmatrix} \in \mathbb{R}_p^{n \times t} \right\}.$$

Further, in view of $\Pi = \Pi_0$ and (B.8), we have

$$\Omega_p = \bigcup_{k \in \Lambda} \left\{ V \begin{bmatrix} \Sigma^{-1} Y \\ Z \end{bmatrix} \, \middle| \, \|Y - B_1\|_F = \gamma_{(k,B_1)}, \; Y \in \mathbb{R}_k^{\alpha \times t}, \; \begin{bmatrix} Y \\ Z \end{bmatrix} \in \mathbb{R}_p^{n \times t} \right\}.$$

This, together with (B.2), implies the relation (B.39). Hence, the proof is completed.

Step 3. Proving, when $\operatorname{rank} B_1 \in \Pi$, that the number η_p defined in (B.3) exists and is given by (B.38), and in this case the solution set Ω_p is given by (B.40).

First, it follows from (B.1) that $\gamma_{(\operatorname{rank} B_1, B_1)} = 0$, and hence $\operatorname{rank} B_1 \in \Pi_0$ in view of $\operatorname{rank} B_1 \in \Pi$. Thus, it follows from $\gamma_{(\operatorname{rank} B_1, B_1)} = 0$ and $\operatorname{rank} B_1 \in \Pi_0$ as well as the second equation of (B.8) that $\mu_0 = 0$.

Second, it follows from $\operatorname{rank} B_1 \in \Pi$, Lemma B.2 and the definition of the set Δ in (B.46) that $B_1 \in \Delta$, and hence

$$\min_{Y \in \Delta} \|Y - B_1\|_F = 0.$$

This, together with the first two conclusions in Lemma B.1, gives $\eta_p = \|B_2\|_F$.

Combining the above two aspects yields the relation (B.38). It again follows from $\eta_p = \|B_2\|_F$ and (B.4) that

$$\Omega_p = \left\{ X \in \mathbb{R}_p^{n \times t} \, \middle| \, \|AX - B\|_F = \|B_2\|_F \right\}.$$

This, together with the third conclusion of Lemma B.1, obviously produces (B.40). With the above three steps, the whole proof of the theorem is completed.   □

Remark B.2. If the number η_p defined in (B.3) does not exist, then, it follows from the first conclusion of Theorem B.4 that $\operatorname{rank} B_1 \notin \Pi$ and $\gamma_{(k_0, B_1)}$ does not exist for some $k_0 \in \Pi$. On the one hand, it is seen from Step 1 in the proof of Theorem B.4

that the relation (B.44) holds when $\gamma_{(k_0, B_1)}$ does not exist for $k_0 \in \Pi$. On the other hand, it follows from $\text{rank} B_1 \notin \Pi$ and Lemma B.2 that

$$X = V \begin{bmatrix} \Sigma^{-1} B_1 \\ Z \end{bmatrix} \notin \mathbb{R}_p^{n \times t}, \quad \forall Z \in \mathbb{R}^{(n-\alpha) \times t}.$$

This, together with the second and the third conclusions in Lemma B.1, *clearly gives*

$$\|AX - B\|_F > \|B_2\|_F, \quad \forall X \in \mathbb{R}_p^{n \times t}. \tag{B.49}$$

It thus follows from (B.44) and (B.49) that when *the number* η_p *defined in (B.3) does not exist*, $\|AX - B\|_F$ can arbitrarily approach but can never reach $\|B_2\|_F$.

It is clearly seen from the above theorem that Problem B.2 is completely solved once Problem B.1 is completely solved.

B.5 An Illustrative Example

Let

$$M = \begin{bmatrix} 2 & 1 & 1 & -2 \\ -1 & -1 & 0 & 2 \\ 2 & 3 & -1 & -6 \end{bmatrix},$$

then, $m = 3$, $n = 4$, and $r = 2$. By applying Theorems B.1 and B.2 in Sect. B.3, we can conclude that the associated rank-constrained matrix matching problem has no solution when $p = 3$ and has a unique solution $X = M$ when $p = 2$.

To cope with the case of $p = 1$, we obtain, using Matlab, the matrices U_M, V_M, and D_M in (B.27) as follows:

$$U_M = \begin{bmatrix} -0.9093 & -0.3430 & 0.2357 \\ 0.1283 & 0.3076 & 0.9428 \\ 0.3959 & -0.8875 & 0.2357 \end{bmatrix}, D_M = \begin{bmatrix} 1.7749 & 0 \\ 0 & 7.9278 \end{bmatrix},$$

$$V_M = \begin{bmatrix} -0.6508 & -0.3492 & 0.5383 & 0.4059 \\ 0.0846 & -0.4179 & -0.6461 & 0.6330 \\ -0.7353 & 0.0687 & -0.5383 & -0.4059 \\ -0.1692 & 0.8359 & -0.0539 & 0.5194 \end{bmatrix}.$$

Thus, by applying Theorem B.3 in Sect. B.3, we have $\gamma_1 = 1.7749$, and the general element in set $\mathscr{S}_1$ as follows:

$$X = U_M \begin{bmatrix} \begin{bmatrix} 1 + p_1 & 0 \\ p_2 & 1 \end{bmatrix} D_M & 0_{2 \times 2} \\ 0_{1 \times 2} & 0_{1 \times 2} \end{bmatrix} V_M^T, \tag{B.50}$$

where p_1, $p_2 \in \mathbb{R}$ are any parameters satisfying $p_1^2 + p_2^2 = 1$.

B.6 Proof of Lemma B.4

To prove Lemma B.4, the following lemma is needed.

Lemma B.5. *Let $Q \in \mathbb{R}^{t \times r}$ be of full-row rank. Then there exists an orthogonal matrix T such that*

$$TQ = \begin{bmatrix} Q_1 \\ \vdots \\ Q_t \end{bmatrix} \tag{B.51}$$

with

$$Q_i = \begin{bmatrix} 0 \cdots 0 \ q_{ic_i} \cdots q_{ir} \end{bmatrix} \in \mathbb{R}^{1 \times r}, \tag{B.52}$$

where

$$q_{ic_i} > 0, 1 \leq i \leq t \quad \text{and} \quad 1 \leq c_1 < \cdots < c_t \leq r. \tag{B.53}$$

Denote the singular value decomposition of ZD^{-1} by

$$ZD^{-1} = U \begin{bmatrix} \tilde{\Sigma} & 0 \\ 0 & 0 \end{bmatrix} V^{\mathrm{T}}, \tag{B.54}$$

where $U \in \mathbb{R}^{r \times r}$ and $V \in \mathbb{R}^{r \times r}$ are two orthogonal matrices, and $\tilde{\Sigma} \in \mathbb{R}^{(r-t) \times (r-t)}$ is a positive definite diagonal matrix in view of $Z \in \mathbb{R}_{r-t}^{r \times r}$. Then

$$
\begin{aligned}
Z - D &= U \begin{bmatrix} \tilde{\Sigma} & 0 \\ 0 & 0 \end{bmatrix} V^{\mathrm{T}} D - D \\
&= U \left(\begin{bmatrix} [\ \tilde{\Sigma} \ 0\ ] V^{\mathrm{T}} \\ 0 \end{bmatrix} - U^{\mathrm{T}} \right) D.
\end{aligned} \tag{B.55}
$$

Let

$$U = \begin{bmatrix} P \\ Q \end{bmatrix}^{\mathrm{T}}, \quad P \in \mathbb{R}^{(r-t) \times r}. \tag{B.56}$$

Then

$$Z - D = U \begin{bmatrix} [\ \tilde{\Sigma} \ 0\ ] V^{\mathrm{T}} - P \\ -Q \end{bmatrix} D, \tag{B.57}$$

and hence

$$\|Z - D\|_{\mathrm{F}}^2 = \|[\ \tilde{\Sigma} \ 0\ ] V^{\mathrm{T}} D - PD\|_{\mathrm{F}}^2 + \|QD\|_{\mathrm{F}}^2 \geq \|QD\|_{\mathrm{F}}^2. \tag{B.58}$$

and the inequality in (B.58) turns into an equality if and only if

$$[\ \tilde{\Sigma} \ \ 0\ ]V^{\mathrm{T}} = P. \tag{B.59}$$

It follows from (B.56) that $Q\,Q^{\mathrm{T}} = I_t$ and $\mathrm{rank}Q = t$, hence we can assume, in view of Lemma B.5, that (B.52) holds for some orthogonal matrix T. This implies

$$\|QD\|_{\mathrm{F}}^2 = \|TQD\|_{\mathrm{F}}^2 = \|Q_1D\|_{\mathrm{F}}^2 + \cdots + \|Q_tD\|_{\mathrm{F}}^2. \tag{B.60}$$

Again using $Q\,Q^{\mathrm{T}} = I_t$ and the orthogonality of T, we can derive that

$$(TQ)(TQ)^{\mathrm{T}} = I_t. \tag{B.61}$$

In view of (B.52), the above equation can be equivalently written as

$$Q_iQ_i^{\mathrm{T}} = 1, \quad i = 1, 2, \ldots, t,$$

which is equivalent to, further in view of (B.52),

$$\sum_{j=c_i}^{r} q_{ij}^2 = 1, \quad i = 1, 2, \ldots, t.$$

This, together with (B.12), implies that

$$\begin{aligned}
\|Q_iD\|_{\mathrm{F}}^2 &= q_{ic_i}^2 d_{c_i}^2 + \cdots + q_{ir}^2 d_r^2 \\
&\geq (q_{ic_i}^2 + \cdots + q_{ir}^2)d_{c_i}^2 \\
&= d_{c_i}^2 \\
&\geq d_i^2, \quad i = 1, 2, \ldots, t. \tag{B.62}
\end{aligned}$$

With (B.60) and (B.62), the relation

$$\|QD\|_{\mathrm{F}}^2 \geq d_1^2 + d_2^2 + \cdots + d_t^2 \tag{B.63}$$

is easily obtained. The inequalities in (B.62) turn into equalities if and only if

$$d_{c_i}^2 = d_i^2, \quad i = 1, 2, \ldots, t, \tag{B.64}$$

and

$$q_{ij_i} = 0 \quad \text{whenever} \quad d_{j_i} > d_i, \quad i = 1, 2, \ldots, t. \tag{B.65}$$

It follows from (B.64) that (B.52) becomes

$$TQ = [\ G \ \ H\ ], \tag{B.66}$$

where $G \in \mathbb{R}^{t \times t}$ is nonsingular upper triangular and $H \in \mathbb{R}^{t \times (r-t)}$. This, together with Lemma B.5 and (B.65), gives

$$G = \begin{bmatrix} Y_1 & 0 & 0 & \cdots \\ 0 & \ddots & \ddots & \ddots \\ \vdots & \ddots & Y_{k_t-1} & 0 \\ 0 & \cdots & 0 & G_1 \end{bmatrix}, \quad H = \begin{bmatrix} 0 & 0 \\ Y & 0 \end{bmatrix}, \tag{B.67}$$

where $Y_i \in \mathbb{R}^{m_i \times m_i}$, $i = 1, 2, \ldots, k_t - 1$, and $G_1 \in \mathbb{R}^{(t-l_{k_t}-1) \times (t-l_{k_t}-1)}$ are nonsingular upper triangular, and $Y \in \mathbb{R}^{(t-l_{k_t}-1) \times (l_{k_t}-t)}$. Substituting (B.67) into (B.66) yields

$$TQ = \begin{bmatrix} \Gamma & 0_{t \times (r-l_{k_t})} \end{bmatrix}, \tag{B.68}$$

where Γ is defined as in (B.24). Combining (B.61), (B.68), and (B.67) gives (B.25).

Now (B.22) follows from (B.58) and (B.63). Further, combining (B.57), (B.59), and (B.68), we see that the equality holds in (B.22) if and only if

$$Z - D = U \begin{bmatrix} 0 \\ -Q \end{bmatrix} D = U \begin{bmatrix} 0 \\ -T^T \begin{bmatrix} \Gamma & 0 \end{bmatrix} \end{bmatrix} D. \tag{B.69}$$

By putting

$$\Theta = U \begin{bmatrix} I & 0 \\ 0 & -T^T \end{bmatrix},$$

then (B.23) can easily be obtained from (B.69).

In particular, if D satisfies (B.26), then (B.65) becomes $q_{ij} = 0$, $\forall i = 1, 2, \ldots, t$, $j > i$. In this case, the matrix Γ in (B.68), and (B.24) as well, obviously become diagonal involutory matrices in $\mathbb{R}^{t \times t}$. The proof is then completed. □

Appendix C
Generalized Sylvester Matrix Equations

This appendix presents the general complete parametric solution to a type of generalized Sylvester matrix equations. The results are essential for solving the problems of eigenstructure assignment (Chapter 9) and parametric observer design (Chap. 11).

C.1   Introduction

Consider the generalized Sylvester matrix equation

$$AV + BW = EVJ, \tag{C.1}$$

where $A, E \in \mathbb{R}^{n \times n}, B \in \mathbb{R}^{n \times r}, J \in \mathbb{C}^{p \times p}$ $(p \leq \operatorname{rank}E \leq n)$ are known, and $V \in \mathbb{C}^{n \times p}$ and $W \in \mathbb{C}^{r \times p}$ are matrices to be determined. As it is shown in Tsui (1987), the special case of this equation

$$AV + BW = VJ \tag{C.2}$$

is closely related to the problem of eigenvalue assignment, eigenstructure assignment, and observer design in normal linear systems (Tsui 1985, 1987; Tsui 1988a; Tsui 1988b; Duan 1992b; Duan et al. 1999a), while the more generalized Sylvester matrix equation (C.1), with the matrix E usually singular, is fundamental in descriptor linear system theory and has found applications in problems of observer design (Verhaegen and Van Dooren 1986; Dai 1989b) and eigenstructure assignment (Fletcher et al. 1986; Duan 1992b, 1995, 1998a, 1999; Duan and Patton 1998a) in linear descriptor systems.

In solving (C.1) and (C.2), finding the complete parametric solutions, that is, parametric solutions consisting of the maximum number of free parameters, is of extreme importance, since many problems such as robustness in control system design require full use of the design freedom (see Tsui 1988a,b; Duan 1992a, 1993a; Duan and Patton 1999; Duan et al. 2000d, 2001b; Duan and Lam 2000; Duan and Ma 1995).

G.-R. Duan, *Analysis and Design of Descriptor Linear Systems*, Advances in Mechanics and Mathematics 23, DOI 10.1007/978-1-4419-6397-0_14, © Springer Science+Business Media, LLC 2010

For (C.2), several explicit parametric solutions have been proposed for matrix J in the companion form (Luenberger 1967) and Jordan form (Tsui 1987; Duan 1993b). While for the general equation (C.1), solutions have been considered by several researchers (Verhaegen and Van Dooren 1986; Ozcaldiran and Lewis 1987; Duan 1992b, 1996b, 2004c), but the approaches are not completely parametric except the ones given in Duan (1992b, 1996b, 2004c). In fact, Verhaegen and Van Dooren (1986) and Lewis and Ozcaldiran (1989) have studied a different problem of finding a special triangular matrix V for some W and a matrix J possessing desired eigenvalues. Duan (1992b) presented a complete parametric solution to this equation with no presence of the R-controllability condition ($\mathrm{rank}\begin{bmatrix} sE - A & B \end{bmatrix} = n$ for all $s \in \mathbb{C}$) for matrix J in arbitrary Jordan form. This solution has the weakness that it is not in direct, explicit form but in recursive form, while a direct, explicit solution usually provides much convenience in some system design problem (see Duan 1992b, 1993a; Duan et al. 2000d, 2001b; Duan and Lam 2000; Duan and Ma 1995). In this appendix, a general, complete parametric solution in a direct, explicit, clear, and neat form of (C.1) proposed in Duan (1996b) is introduced, where the matrix J is in Jordan form. Our recent advances on these equations can be found in Duan (2004b), Duan and Zhou (2006), Zhou and Duan (2006, 2007), and Wu and Duan (2007c,d).

C.2 Preliminaries

Equation (C.1) is directly concerned with eigenstructure assignment and observer design for descriptor linear systems. When it is applied to eigenstructure assignment in descriptor systems, the matrix J in the equation is directly required to be in Jordan form (Fletcher et al. 1986; Duan 1992b). In applications of observer design for descriptor systems, the matrix J can be arbitrarily chosen as long as it is stable (Verhaegen and Van Dooren 1986). This equals arbitrarily choosing the Jordan form and a corresponding eigenvector matrix, and with a transformation the equation can be changed into one with a given matrix J in Jordan form. For (C.1) with matrix J not in Jordan form, we suggest converting it into the case with matrix J in Jordan form with the help of some similarity transformations whenever a numerical stable solution can be obtained (Golub and Wilkinson 1976; Kagstrom and Ruhe 1980). Such a conversion greatly simplifies the problem in the sense that the converted matrix equation may be immediately decomposed into a number of matrix equations of the same form with reduced dimensions as shown in the following proposition.

Proposition C.1. *Let* $J = \mathrm{diag}\,(J_1, J_2, \ldots, J_m)$. *Then (C.1) can be equivalently decomposed into the following m matrix equations:*

$$AV_i + BW_i = EV_iJ_i, \quad i = 1, 2, \ldots, m, \tag{C.3}$$

with V_i and W_i, $i = 1, 2, \ldots, m$, of proper dimensions defined by

$$V = \begin{bmatrix} V_1 & V_2 & \cdots & V_m \end{bmatrix},$$
$$W = \begin{bmatrix} W_1 & W_2 & \cdots & W_m \end{bmatrix}.$$

Therefore, without loss of generality, we may require the matrix J in (C.1) to be a single upper triangle Jordan block of order p with eigenvalue σ. That is, the following assumption is in requirement.

Assumption C.1. $J = \sigma I_p + N$, where N is some nilpotent matrix with index $p-1$.

In accordance with this assumption, we have

$$J = \begin{bmatrix} \sigma & 1 & & \\ & \sigma & \ddots & \\ & & \ddots & 1 \\ & & & \sigma \end{bmatrix}.$$

Corresponding to the structure of this Jordan block J, we make the following convention throughout this section.

Convention Cv4 The matrices V and W in (C.1) are in the following forms:

$$V = \begin{bmatrix} v_1 & v_2 & \cdots & v_p \end{bmatrix},$$
$$W = \begin{bmatrix} w_1 & w_2 & \cdots & w_p \end{bmatrix},$$

where v_i and w_i are vectors of dimension n.

With Convention Cv4, the following fact can be immediately obtained.

Lemma C.1. *Equation (C.1) is equivalent to the following group of vector equations:*

$$\begin{bmatrix} A - \sigma E & B \end{bmatrix} \begin{bmatrix} v_k \\ w_k \end{bmatrix} = E v_{k-1}, \ v_0 = 0, \ k = 1, 2, \ldots, p, \qquad (C.4)$$

which can be further equivalently arranged into the following matrix form

$$\Phi x = 0 \qquad (C.5)$$

with

$$\Phi = \begin{bmatrix} [A - \sigma E \ \ B] & & & \\ [-E \ \ 0] & [A - \sigma E \ \ B] & & \\ & \ddots & \ddots & \\ & & [-E \ \ 0] & [A - \sigma E \ \ B] \end{bmatrix} \qquad (C.6)$$

and

$$x = \begin{bmatrix} v_1^T & w_1^T & v_2^T & w_2^T & \cdots & v_p^T & w_p^T \end{bmatrix}^T. \tag{C.7}$$

It follows from the above lemma that the problem of seeking the matrix pair (V, W) satisfying (C.1) is equivalent to the problem of finding the vector chains v_k and w_k, $k = 1, 2, \ldots, p$, satisfying (C.4) or (C.5).

Besides Assumption C.1, we also require the following assumption.

Assumption C.2. rank $\begin{bmatrix} A - sE & B \end{bmatrix} = n$ for all $s \in \mathbb{C}$.

Assumption C.2 is equivalent to the R-controllability condition of the linear descriptor system $E\dot{x} = Ax + Bu$. It is well known that subject to Assumption C.2, there exist unimodular matrices $P(s)$ and $Q(s)$ satisfying

$$P(s)\begin{bmatrix} A - sE & B \end{bmatrix} Q(s) = \begin{bmatrix} 0 & I \end{bmatrix}. \tag{C.8}$$

Duan (1992b) has given a parametric complete solution to the matrix equation (C.1) in an iteration form. Under Assumption C.2, this solution appears as follows.

Lemma C.2. *Let Assumptions C.1 and C.2 be valid, then all the solutions to the matrix equation (C.1) are given by*

$$\begin{bmatrix} v_k \\ w_k \end{bmatrix} = Q(\sigma) \begin{bmatrix} f_k \\ P(\sigma) Ev_{k-1} \end{bmatrix},$$

where $f_k \in \mathbb{C}^r$, $k = 1, 2, \ldots, p$, are any group of parameter vectors in $\mathbb{C}^r$, and $P(s)$ and $Q(s)$ are two unimodular matrices satisfying (C.8).

C.3 Solution Based on Right Coprime Factorization

In this section, we consider the solution to (C.1) based on right coprime factorization. Please note that this solution allows the eigenvalue σ to be unknown and set undetermined.

The General Solution

Our main result in this subsection is as follows.

Theorem C.1. *Subject to Assumptions C.1 and C.2, there always exist solutions to the matrix equation (C.1), and all of the solutions are given by*

$$\begin{bmatrix} v_k \\ w_k \end{bmatrix} = \begin{bmatrix} N(\sigma) \\ D(\sigma) \end{bmatrix} f_k + \cdots + \frac{1}{(k-1)!} \frac{d^{k-1}}{ds^{k-1}} \begin{bmatrix} N(\sigma) \\ D(\sigma) \end{bmatrix} f_1, \tag{C.9}$$

$$k = 1, 2, \ldots, p,$$

where $f_k \in \mathbb{C}^r, k = 1, 2, \ldots, p$, *are arbitrarily chosen free parameter vectors, and* $N(s) \in \mathbb{R}^{n \times r}, D(s) \in \mathbb{R}^{r \times r}$ *are any pair of right coprime matrix polynomials satisfying*

$$(A - sE) N(s) + BD(s) = 0. \tag{C.10}$$

Proof. First, we show that the vector chains $v_k, k = 1, 2, \ldots, p$, given by Theorem C.1, are solutions of (C.1).

Taking the differential of order l in both sides of (C.10), we obtain

$$(A - sE) \frac{d^l}{ds^l} N(s) + B \frac{d^l}{ds^l} D(s) = lE \frac{d^{l-1}}{ds^{l-1}} N(s), \tag{C.11}$$

$$l = 0, 1, 2, \ldots, k - 1.$$

Substituting s with σ in (C.11) and post-multiplying by vector $\frac{1}{l!} f_{k-l}$ on both sides of (C.11) gives

$$(A - \sigma E) \frac{1}{l!} \frac{d^l}{ds^l} N(\sigma) f_{k-l} + B \frac{1}{l!} \frac{d^l}{ds^l} D(\sigma) f_{k-l}$$

$$= E \frac{1}{(l-1)!} \frac{d^{l-1}}{ds^{l-1}} N(\sigma) f_{k-l}, \tag{C.12}$$

$$l = 0, 1, 2, \ldots, k - 1, \quad k = 1, 2, \ldots, p.$$

Summing up the equations in (C.12) for $l = 0, 1, 2, \ldots, k - 1$, and using (C.9), we can obtain

$$(A - \sigma E) v_k + B w_k = E v_{k-1}, \quad v_0 = 0, \quad k = 1, 2, \ldots, p, \tag{C.13}$$

which is equivalent to (C.4). Thus, we have proven that the vectors given by Theorem C.1 are solutions to (C.1) in view of Lemma C.1.

Second, we prove the completeness of the parametric solution given by (C.9) and (C.10). Let $\mathscr{S}$ be the set of vectors $\begin{bmatrix} v_k \\ w_k \end{bmatrix}$, $k = 1, 2, \ldots, p$, given by Theorem C.1, and $\mathscr{K}$ be the kernel of the linear map

$$\begin{bmatrix} V \\ W \end{bmatrix} \Rightarrow AV - EVF + BW.$$

Then we have $\mathscr{S} \subset \mathscr{K}$, and hence $\dim(\mathscr{K}) \geq \dim(\mathscr{S})$, following the sufficiency part of this proof. To prove the completeness of our solution (C.9), we need only to show

$$\dim(\mathscr{S}) = \dim(\mathscr{K}) = p \times r.$$

Let

$$\xi = \begin{bmatrix} v_p & w_p & \cdots & v_1 & w_1 \end{bmatrix}^{\mathrm{T}},$$

$$f = \begin{bmatrix} f_p & f_{p-1} & \cdots & f_1 \end{bmatrix}^{\mathrm{T}},$$

$$M = \begin{bmatrix} N(\sigma) & N'(\sigma) & \cdots & \frac{1}{(p-1)!} \frac{d^{p-1}}{ds^{p-1}} N(\sigma) \\ D(\sigma) & D'(\sigma) & \cdots & \frac{1}{(p-1)!} \frac{d^{p-1}}{ds^{p-1}} D(\sigma) \\ & N(\sigma) & \cdots & \frac{1}{(p-2)!} \frac{d^{p-2}}{ds^{p-2}} N(\sigma) \\ & D(\sigma) & \cdots & \frac{1}{(p-2)!} \frac{d^{p-2}}{ds^{p-2}} D(\sigma) \\ & & \ddots & \vdots \\ & & & N(\sigma) \\ & & & D(\sigma) \end{bmatrix},$$

then the equations in (C.9) can be rewritten in the compact matrix form

$$\xi = Mf.$$

It follows from the right coprime property of $N(s)$ and $D(s)$ that the above matrix M has rank $p \times r$. Thus, we have $\dim(\mathcal{K}) \geq \dim(\mathcal{S}) = p \times r$. On the other hand, it can be easily deduced from Lemma C.2 that the degree of freedom in the complete general solutions v_k and w_k to (C.1), with Assumptions C.1 and C.2 satisfied, is at most $p \times r$. Therefore, we have $\dim(\mathcal{K}) \leq p \times r$. With these two aspects, we arrive at the conclusion. □

Theorem C.1 is a general result. It may have important applications in analysis and design of descriptor linear systems. Solution (C.9) naturally reduces to the one obtained in Duan (1993b) when $E = I$. In the special case of $p = 1$, it becomes

$$\begin{bmatrix} v \\ w \end{bmatrix} = \begin{bmatrix} N(\sigma) \\ D(\sigma) \end{bmatrix} f.$$

Solution of $N(s)$ and $D(s)$

Solution (C.9) can be obtained as soon as the pair of right coprime polynomial matrices $N(s)$ and $D(s)$ are derived. Here, we present several ways for solving the pair of right-coprime polynomial matrices $N(s)$ and $D(s)$ satisfying (C.10).

Let $\mathrm{adj}(sE - A)$ be the adjoint matrix of $(sE - A)$. Then it is obvious that, under the regularity of (E, A), a special pair of polynomial matrices $N(s)$ and $D(s)$ satisfying (C.10) are

$$N(s) = \mathrm{adj}(sE - A)B, \quad D(s) = \det(sE - A)I_r. \tag{C.14}$$

Moreover, this pair of $N(s)$ and $D(s)$ are easily proven to be right coprime in view of Assumption C.2. It can be further reasoned that in the special case of $\det(\sigma E - A) \neq 0$, the above pairs of $N(s)$ and $D(s)$ in (C.14) may be replaced by

$$N(s) = (sE - A)^{-1}B, \quad D(s) = I_r. \tag{C.15}$$

Besides the above two solutions, the following lemma provides another constructive method for solution of the pair of right coprime polynomial matrices $N(s)$ and $D(s)$ required in Theorem C.1.

Lemma C.3. *Let Assumption C.2 hold and $Q(s)$ be any unimodular polynomial matrix satisfying (C.8). Then the pair of polynomial matrices $N(s)$ and $D(s)$ satisfying (C.10) are given by follows:*

$$\begin{bmatrix} N(s) \\ D(s) \end{bmatrix} = \alpha Q(s) \begin{bmatrix} I_r \\ 0_{n \times r} \end{bmatrix}, \tag{C.16}$$

where α is an arbitrary nonzero scalar.

Proof. Partition the unimodular matrix $Q(s)$ as

$$Q(s) = \begin{bmatrix} Q_{11}(s) & Q_{12}(s) \\ Q_{21}(s) & Q_{22}(s) \end{bmatrix} \tag{C.17}$$

with

$$Q_{11}(s) \in \mathbb{R}^{n \times r}[s], \quad Q_{21}(s) \in \mathbb{R}^{r \times r}[s]. \tag{C.18}$$

Then (C.16) is equivalent to

$$\begin{cases} N(s) = \alpha Q_{11}(s) \\ D(s) = \alpha Q_{21}(s) \end{cases}. \tag{C.19}$$

Since $Q(s)$ is an unimodular matrix, there holds

$$\operatorname{rank} \begin{bmatrix} Q_{11}(s) \\ Q_{21}(s) \end{bmatrix} = r, \quad \forall s \in \mathbb{C}. \tag{C.20}$$

This indicates that $Q_{11}(s)$ and $Q_{21}(s)$ are right coprime, hence so are the $N(s)$ and $D(s)$ given in (C.19). Further, it follows from (C.8), (C.17), and (C.19) that

$$(sE - A)N(s) = BD(s). $$

This is equivalent to (C.10). □

Remark C.1. Lemma C.3 has converted the solution of the pair of right coprime polynomial matrices $N(s)$ and $D(s)$ into a problem of solving the unimodular polynomial matrix $Q(s)$ satisfying the Smith form reduction (C.8). The Smith form reduction (C.8) can be easily realized by hand using some simple elementary matrix transformations for relatively lower order cases. For higher order cases, the Maple function smith can be readily used, and the Matlab toolbox, DeLiSBasics, developed by the author using singular value decompositions and the Matlab symbolic computation toolbox can also be applied. Regarding solution of the pair of right coprime polynomial matrices $N(s)$ and $D(s)$, several ways have been given in Duan (1996b) under the controllability condition of the open-loop system (9.1), and computational methods can also be found in Beelen and Veltkamp (1987), Bongers and Heuberger (1990), Van Dooren (1990), Duan (2000a), and Duan et al. (2002c).

Remark C.2. Besides simplicity and neatness, the solution (C.9) has another advantage, that is, it allows the eigenvalue σ to be unknown. This offers a great advantage in certain control applications where the closed-loop eigenvalues can be set undetermined and used as a part of design parameters.

C.4 Solution Based on SVD

In the case that the eigenvalue σ of the Jordan block J is known a priori, Assumption C.2 can be reduced to the following:

Assumption C.3. $\mathrm{rank}\begin{bmatrix} A - \sigma E & B \end{bmatrix} = n$.

Under this assumption, the extended matrix Φ defined in (C.6) is clearly seen to be of full-row rank. Applying singular value decomposition to the matrix Φ gives

$$P\Phi U = \begin{bmatrix} \Sigma & 0 \end{bmatrix}, \tag{C.21}$$

where P and U are two unitary matrices of appropriate dimensions, and Σ is a diagonal matrix with positive diagonal elements since Φ is of full-row rank.

Based on Lemma C.1 and the above singular value decomposition (C.21), the following result regarding the solution to the generalized Sylvester matrix equation (C.1) can be easily obtained (proof omitted).

Theorem C.2. *Suppose Assumptions C.1 and C.3 hold, and let U be an orthogonal matrix satisfying the singular value decomposition (C.21), with Φ being defined as in (C.6). Then the general solution to the generalized Sylvester matrix equation (C.1) is given by*

$$x = U \begin{bmatrix} 0 \\ f \end{bmatrix}, \tag{C.22}$$

where x is defined by (C.7), and $f \in \mathbb{R}^{rp}$ is an arbitrary parameter vector.

Remark C.3. In the special case of $p = 1$, we have $J = \sigma$ (a scalar) and the matrices V and W are all vectors of n dimension. This special case is important because it is commonly encountered in applications of pole assignment in linear systems design. In this case, the singular value decomposition (C.21) reduces to

$$P \begin{bmatrix} A - \sigma E & B \end{bmatrix} U = \begin{bmatrix} \Sigma & 0 \end{bmatrix}. \tag{C.23}$$

Remark C.4. Unlike the solution (C.9), the solution (C.22) given in the above Theorem C.2 requires the eigenvalue σ to be known a priori, but it adopts singular value decomposition and is thus both numerically simple and reliable.

C.5 An Example

Consider an equation in the form of (C.1) with the following parameters (refer to Duan 1992b)

$$A = \begin{bmatrix} 0 & 1 & 0 \\ 0 & 0 & 1 \\ 0 & 0 & -1 \end{bmatrix}, \quad E = \begin{bmatrix} 1 & 0 & 0 \\ 0 & 1 & 0 \\ 0 & 0 & 0 \end{bmatrix}, \tag{C.24}$$

$$B = \begin{bmatrix} 0 & 0 \\ 1 & 0 \\ 0 & 1 \end{bmatrix}, \quad F = \begin{bmatrix} \lambda & 0 & 0 \\ 0 & \sigma & 1 \\ 0 & 0 & \sigma \end{bmatrix}, \quad \lambda, \sigma \in \mathbb{C}. \tag{C.25}$$

It follows from Proposition C.1 that this equation can be decomposed into the following two equations:

$$AV_1 + BW_1 = EVJ_1, \tag{C.26}$$

$$AV_2 + BW_2 = EVJ_2, \tag{C.27}$$

with

$$J_1 = \lambda, \quad J_2 = \begin{bmatrix} \sigma & 1 \\ 0 & \sigma \end{bmatrix},$$

and matrices A, B, E as in (C.24)–(C.25).

It is easily verified that Assumption C.2 holds. By applying matrix elementary transformations to matrix $\begin{bmatrix} A - sE & B \end{bmatrix}$, we obtain $P(s) = I_3$ and

$$Q(s) = \begin{bmatrix} 1 & 0 & 0 & 0 & 0 \\ s & 0 & 1 & 0 & 0 \\ 0 & 1 & 0 & 0 & 0 \\ s^2 & -1 & s & 1 & 0 \\ 0 & 1 & 0 & 0 & 1 \end{bmatrix}.$$

Therefore, using Lemma C.3 we obtain

$$N(s) = \begin{bmatrix} 1 & 0 \\ s & 0 \\ 0 & 1 \end{bmatrix}, \quad D(s) = \begin{bmatrix} s^2 & -1 \\ 0 & 1 \end{bmatrix}.$$

Now applying our Theorem C.1 to (C.26), we have

$$\begin{bmatrix} v_1 \\ w_1 \end{bmatrix} = \begin{bmatrix} N(\lambda) \\ D(\lambda) \end{bmatrix} f.$$

Denoting

$$f = \begin{bmatrix} x_1 & x_2 \end{bmatrix}^{\mathrm{T}} \in \mathbb{C}^2,$$

then the general solution of (C.26) is obtained as

$$V_1 = [v_1] = \begin{bmatrix} x_1 \\ \lambda x_1 \\ x_2 \end{bmatrix}, \quad W_1 = [w_1] = \begin{bmatrix} \lambda^2 x_1 - x_2 \\ x_2 \end{bmatrix}. \tag{C.28}$$

Again applying Theorem C.1 to (C.27), we have

$$\begin{bmatrix} v_2 \\ w_2 \end{bmatrix} = \begin{bmatrix} N(\sigma) \\ D(\sigma) \end{bmatrix} f_1,$$

$$\begin{bmatrix} v_3 \\ w_3 \end{bmatrix} = \begin{bmatrix} N(\sigma) \\ D(\sigma) \end{bmatrix} f_2 + \frac{\mathrm{d}}{\mathrm{d}s} \begin{bmatrix} N(\sigma) \\ D(\sigma) \end{bmatrix} f_1.$$

Letting

$$f_i = \begin{bmatrix} x_{1i} & x_{2i} \end{bmatrix}^{\mathrm{T}} \in \mathbb{C}^2, \quad i = 1, 2,$$

then the general solution of (C.27) is obtained as

$$V_2 = \begin{bmatrix} x_{11} & x_{12} \\ \sigma x_{11} & \sigma x_{12} + x_{11} \\ x_{21} & x_{22} \end{bmatrix} \tag{C.29}$$

and

$$W_2 = \begin{bmatrix} \sigma^2 x_{11} - x_{21} & \sigma^2 x_{12} + 2\sigma x_{11} - x_{12} \\ x_{12} & x_{22} \end{bmatrix}. \tag{C.30}$$

Finally, it follows from Proposition C.1, (C.28)–(C.30), that the general solution of the equation with parameters in (C.25)–(C.26) is given by

$$V = \begin{bmatrix} x_1 & x_{11} & x_{12} \\ \lambda x_{11} & \sigma x_{11} & \sigma x_{12} + x_{11} \\ x_{12} & x_{21} & x_{22} \end{bmatrix}$$

and

$$W = \begin{bmatrix} \lambda^2 x_1 - x_2 & \sigma^2 x_{11} - x_{12} & \sigma^2 x_{12} + 2\sigma x_{11} - x_{22} \\ x_2 & x_{21} & x_{22} \end{bmatrix}$$

with x_i, x_{jk}, i, j, $k = 1, 2$, being arbitrary complex scalars.

References

Al-Nasr, N., Lovass-Nagy, V., O'Connor, D., and Powers, A. L. (1983a). Output function control in general state space systems containing the first derivative of the input vector. *Int. J. Syst. Sci.*, 14(9):1029–1042.

Al-Nasr, N., Lovass-Nagy, V., and Rabson, G. (1983b). General eigenvalue placement in linear control systems by output feedback. *Int. J. Syst. Sci.*, 14(5):519–528.

Aoki, T., Hosoe, S., and Hayakawa, S. (1983). Structural controllability for linear systems in descriptor form. *Trans. Soc. Instrum. Contr. Eng. (Japan)*, 19(8):628–635.

Armentano, V. A. (1984). Eigenvalue placement for generalized linear systems. *Syst. Contr. Lett.*, 4(4):199–202.

Aslund, J. and Frisk, E. (2006). An observer for non-linear differential-algebraic systems. *Automatica*, 42(6):959–965.

Bajic, V. B. (1986). Partial exponential stability of semi-state systems. *Int. J. Contr.*, 44(3):1383–1394.

Bavafa-Toosi, Y., Ohmori, H., and Labibi, B. (2006). Note on finite eigenvalues of regular descriptor systems. *IEE Proc. Contr. Theor. Appl.*, 153(4):502–503.

Beelen, T. and Van Dooren, P. (1988). An improved algorithm for the computation of Kronecker's canonical form of a singular pencil. *Linear Algebra Appl.*, 105:9–65.

Beelen, T., Van Dooren, P., and Verhaegen, M. (1986). A class of fast staircase algorithms for generalized state space systems. In *Proceedings of 1986 American Control Conference*, pages 425–426, Seattle, WA.

Beelen, T. G. J. and Veltkamp, G. W. (1987). Numerical computation of a coprime factorization of a transfer-function matrix. *Syst. Contr. Lett.*, 9(4):281–288.

Bender, D. J. (1985). *Descriptor systems and geometric control theory*. PhD thesis, Dept. of Elect. Comp. Eng., Univ. of California, Santa Barbara, Calif., Sept.

Bender, D. J. and Laub, A. J. (1985). Controllability and observability at infinite of multivariable linear second-order models. *IEEE Trans. Automat. Contr.*, 30(12):1234–1237.

Bender, D. J. and Laub, A. J. (1987a). The linear-quadratic optimal regulation for descriptor systems: Discrete-time case. *Automatica*, 23(1):71–86.

Bender, D. J. and Laub, A. J. (1987b). The linear-quadratic optimal regulator for descriptor systems. *IEEE Trans. Automat. Contr.*, 32(8):627–688.

Benner, P., Byers, R., Mehrmann, V., and Xu, H. G. (1999a). Numerical methods for linear quadratic and H_∞ control problems. *Dyn. Syst. Contr. Coding Comput. Vis. Prog. Syst. Contr. Theor.*, 25:203–222.

Benner, P., Byers, R., Mehrmann, V., and Xu, H. G. (1999b). Numerical solution of linear-quadratic control problems for descriptor systems. In *Proceedings of the IEEE International Symposium on Computer-Aided Control System Design*, pages 46–51.

Benner, P., Byers, R., Mehrmann, V., and Xu, H. G. (2004). A robust numerical method for optimal H_∞ control. In *Proceedings of the 43rd Conference on Decision and Control*, pages 424–425.

Benner, P. and Sokolov, V. I. (2006). Partial realization of descriptor systems. *Syst. Contr. Lett.*, 55(11):929–938.

Biehn, N., Campbell, S. L., Delebecque, F., and Nikoukhah, R. (1998). Observer design for linear time varying descriptor systems: Numerical algorithms. In *Proceedings of the IEEE Conference on Decision and Control*, volume 4, pages 3801–3806, Tampa, USA.

Biehn, N., Campbell, S. L., Nikoukhah, R., and Delebecque, F. (2001). Numerically constructible observers for linear time-varying descriptor systems. *Automatica*, 37(3):445–452.

Blackman, P. F. (1977). *Introduction to state-variable analysis*. The Macmillan, London.

Bongers, P. M. M. and Heuberger, P. S. C. (1990). Discrete normalized coprime factorization. *Lect. Notes Contr. Inform. Sci.*, 144:307–313.

Boukas, E. K. (2007). Singular linear systems with delay: H_∞ stabilization. *Optim. Contr. Appl. Meth.*, 28(4):259–274.

Brogan, W. L. (1991). *Modern control theory, 3rd edition*. Prentice-Hall, Englewood Cliffs, N. J.

Bunse-Gerstner, A., Byers, R., Mehrmann, V., and Nichols, N. K. (1999). Feedback design for regularizing descriptor systems. *Linear Algebra Appl.*, 299(1–3):119–151.

Bunse-Gerstner, A., Mehrmann, V., and Nichols, N. K. (1992). Regularization of descriptor systems by derivative and proportional state feedback. *SIAM J. Matrix Anal. Appl.*, 13(1):46–67.

Bunse-Gerstner, A., Mehrmann, V., and Nichols, N. K. (1994). Regularization of descriptor systems by output feedback. *IEEE Trans. Automat. Contr.*, 39(8):1742–1748.

Byers, R., Kunkel, P., and Mehrmann, V. (1997). Regularization of linear descriptor systems with variable coefficients. *SIAM J. Contr. Optim.*, 35(1):117–133.

Byers, R. and Mehrmann, V. (1997). Descriptor systems without controllability at infinity. *SIAM J. Contr. Optim.*, 35(2):462–479.

Campbell, S. L. (1980). *Singular systems of differential equations*. Pitman Publishing Co., London

Campbell, S. L. (1982). *Singular systems of differential equations II*. Pitman, New York.

Campbell, S. L. and Rose, N. J. (1982). A second-order singular linear system arising in electric power systems analysis. *Int. J. Syst. Sci.*, 13(1):101–108.

Cao, Y. Y. and Lin, Z. L. (2004). A descriptor system approach to robust stability analysis and controller synthesis. *IEEE Trans. Automat. Contr.*, 49(11):2081–2084.

Chaabane, M., Bachelier, O., and Souissi, M. (2006). Stability and stabilization of continuous descriptor systems:an lmi approach. In *Proceedings of the Mathematical Problem in Engineering, volume 2006*, (Article ID: 39367).

Chen, C. T. (1970). *Introduction to linear systems theory*. Holt Rinhart and Winston, New York.

Chen, H. C. and Chang, F. R. (1993). Chained eigenstructure assignment for constant-ratio proportional and derivative (CRPD) control law in controllable singular systems. *Syst. Contr. Lett.*, 21(5):405–411.

Chen, L. and Cheng, Z. L. (2004). Singular LQ suboptimal control problem with disturbance rejection for descriptor systems. In *Proceedings of the American Control Conference*, volume 5, pages 4595–4600, Boston, USA.

Chen, N., Gui, W. H., and Zhai, G. S. (2006). Design of robust decentralized H_∞ control for interconnected descriptor systems with norm-bounded parametric uncertainties. In *Proceedings of the IEEE International Symposium on Intelligent Control*, pages 537–542.

Chen, Y. and Duan, G. R. (2006). Conditions for c-controllability and c-observability of rectangular descriptor systems. In *Proceedings of the 6th World Congress on Intelligent Control and Automation*, pages 628–630.

Chen, Y. P., Zhang, Q. L., and Zhai, D. (2005). Reliable controller design with mixed H_2/H_∞ performance for descriptor systems. *Acta Automat. Sinica*, 31(2):262–266.

Cheng, Z., Hong, H., and Zhang, J. (1987). The optimal state regulation of generalized dynamical systems with quadratic cost functional. In *Proceedings of the Triennial World Congress of the International Federation of Automatic Control*, 9:127–131.

Chou, J. H., Chen, S. H., and Fung, R. F. (2001). Sufficient conditions for the controllability of linear descriptor systems with both time-varying structured and unstructured parameter uncertainties. *IMA J. Math. Contr. Inform.*, 18(4):469–477.

Chou, J. H., Chen, S. H., and Zhang, Q. L. (2006). Robust controllability for linear uncertain descriptor systems. *Linear Algebra Appl.*, 414(2–3):632–651.

Christodoulou, M. A. and Paraskevopoulos, P. N. (1984). Formula for feedback gains in eigenvalue control of singular systems. *Electron Lett. (GB)*, 20(1):18–19.

Christodoulou, M. A. and Paraskevopoulos, P. N. (1985). Solvability, controllability and observability of singular systems. *J. Opt. Theor. Appl.*, 45(1):53–72.

Chu, D. L. and Cai, D. Y. (2000). Regularization of singular systems by output feedback. *J. Comput. Math.*, 18(1):43–60.

Chu, D. L., Chan, H. C., and Ho, D. W. C. (1997). A general framework for state feedback pole assignment of singular systems. *Int. J. Contr.*, 67(2):135–152.

Chu, D. L., Chan, H. C., and Ho, D. W. C. (1998). Regularization of singular systems by derivative and proportional output feedback. *SIAM J. Matrix Anal. Appl.*, 19(1):21–38.

Chu, D. L. and Golub, G. H. (2006). On a generalized eigenvalue problem for nonsquare pencils. *SIAM J. Matrix Anal. Appl.*, 28(3):770–787.

Chu, D. L. and Ho, D. W. C. (1999). Necessary and sufficient conditions for the output feedback regularization of descriptor systems. *IEEE Trans. Automat. Contr.*, 44(2):405–412.

Chu, D. L., Mehrmann, V., and Nichols, N. K. (1999). Minimum norm regularization of descriptor systems by output feedback. *Linear Algebra Appl.*, 296:39–77.

Cobb, J. D. (1981). Feedback and pole placement in descriptor variable systems. *Int. J. Contr.*, 33(6):1135–1146.

Cobb, J. D. (1983a). Descriptor variable systems and optimal state regulation. *IEEE Trans. Automat. Contr.*, 28(5):601–611.

Cobb, J. D. (1983b). A further interpretation of inconsistent initial conditions in descriptor variable systems. *IEEE Trans. Automat. Contr.*, 28(9):920–922.

Cobb, J. D. (1984). Controllability, observability and duality in singular systems. *IEEE Trans. Automat. Contr.*, 29(12):1076–1082.

Craig, J. J. (1986). *Introduction to robotics: mechanices and control*. Addison-Wesley Publishing Company, Reading, Mass.

Cullen, D. J. (1984). *The equivalence of linear systems*. PhD thesis, Australian Natl. Univ., Sydney, Australia.

Cullen, D. J. (1986a). An algebraic condition for controllability at infinite. *Syst. Contr. Lett.*, 6(5):321–324.

Cullen, D. J. (1986b). State-space realizations at infinity. *Int. J. Contr.*, 43(4):1075–1088.

Dai, L. (1987). System equivalence, controllability and observability in singular systems. *J. Grad. School. USTC. Acad. Sinica*, (1):42–50.

Dai, L. (1988). An H_∞ method for decentralized stabilization in singular systems. In *Proceedings of the IEEE International Conference on System, Man, and Cybernetics*, pages 722–725, Aug. 8–13, Shenyang, China.

Dai, L. (1989a). Impulsive modes and causality in singular systems. *Int. J. Contr.*, 50(4): 1267–1281.

Dai, L. (1989b). *Singular control systems. Lecture Notes in Control and Information Sciences*. Springer, Berlin.

Debeljkovic, D. L., Koruga, D., Milinkovic, S. A., Jovanovic, M. B., and Jacic, L. A. (1998). Finite time stability of linear discrete descriptor systems. In *Proceedings of the Mediterranean Electrotechnical Conference – MELECON*, volume 1, pages 504–508, Tel-Aviv, Israel.

Duan, G. R. (1992a). Simple algorithm for robust eigenvalue assignment in linear output feedback. *IEE Proc. Contr. Theor. Appl.*, 139(5):465–469.

Duan, G. R. (1992b). Solution to matrix equation $AV + BW = EVF$ and eigenstructure assignment for descriptor systems. *Automatica*, 28(3):639–643.

Duan, G. R. (1993a). Robust eigenstructure assignment via dynamical compensators. *Automatica*, 29(2):469–474.

Duan, G. R. (1993b). Solutions to matrix equation $AV + BW = VF$ and their application to eigenstructure assignment in linear systems. *IEEE Trans. Automat. Contr.*, 38(2):276–280.

Duan, G. R. (1994). Eigenstructure assignment by decentralized output feedback–a complete parametric approach. *IEEE Trans. Automat. Contr.*, 39(5):1009–1014.

Duan, G. R. (1995). Parametric eigenstructure assignment in descriptor linear systems via output feedback. *IEE Proc. Contr. Theor. Appl.*, 142(6):611–616.

Duan, G. R. (1996a). *Linear systems theory, 1st edition (in Chinese)*. Harbin Institute of Technology Press, Harbin, China.

Duan, G. R. (1996b). On the solution to Sylvester matrix equation $AV + BW = EVF$. *IEEE Trans. Automat. Contr.*, 41(4):612–614.

Duan, G. R. (1998a). Eigenstructure assignment and response analysis in descriptor linear systems with state feedback control. *Int. J. Contr.*, 69(5):663–694.

Duan, G. R. (1998b). Right coprime factorization for single input systems using Hessenberg forms. In *Proceedings of the 6th IEEE Mediterranean Conference on Control and Systems*, pages 573–577, June 9–11, Italy.

Duan, G. R. (1999). Eigenstructure assignment in descriptor linear systems via output feedback: A new complete parametric approach. *Int. J. Contr.*, 72(4):345–364.

Duan, G. R. (2000a). Right coprime factorization using system upper Hessenberg forms – the multi-input system case. In *Proceedings of the 39th IEEE Conference on Decision and Control*, pages 1960–1965, Sydney, Australia.

Duan, G. R. (2000b). A simple iterative solution to right coprime factorisations of single input linear systems. In *Proceedings of the International Conference on Control, (Control'2000)*, Cambridge, UK.

Duan, G. R. (2001). Right coprime factorization using system upper hessenberg forms – the multi-input system case. *IEE Proc. Contr. Theor. Appl.*, 148(6):433–441.

Duan, G. R. (2002). Right coprime factorizations for multi-input descriptor linear systems–a simple numerically stable solution. *Asian J. Contr.*, 4(2):146–158.

Duan, G. R. (2003a). Parametric eigenstructure assignment via output feedback based on singular value decompositions. *IEE Proc. Contr. Theor. Appl.*, 150(1):93–100.

Duan, G. R. (2003b). Two parametric approaches for eigenstructure assignment in second-order linear systems. *J. Contr. Theor. Appl.*, 1(1):59–64.

Duan, G. R. (2004a). *Linear systems theory, 2nd edition (in Chinese)*. Harbin Institute of Technology Press, Harbin, China.

Duan, G. R. (2004b). A note on combined generalized Sylvester matrix equations. *J. Contr. Theor. Appl.*, 2(4):397–400.

Duan, G. R. (2004c). On solution to the matrix equation $AV + BW = EVJ + R$. *Appl. Math. Lett.*, 17:1197–1202.

Duan, G. R. (2004d). Parametric eigenstructure assignment in second-order descriptor linear systems. *IEEE Trans. Automat. Contr.*, 49(10):1789–1794.

Duan, G. R. (2005). Parametric approaches for eigenstructure assignment in high-order linear systems. *Int. J. Contr. Autom. Syst.*, 3(2):419–429.

Duan, G. R., Howe, D., and Patton, R. J. (1999a). Robust fault detection in descriptor linear systems via generalized unknown input observers. In *Proceedings of the IFAC World Congress'99*, volume P, pages 43–48, July 5–8, Beijing, China.

Duan, G. R., Howe, D., and Patton, R. J. (2002a). Robust fault detection in descriptor linear systems via generalized unknown-input observers. *Int. J. Syst. Sci.*, 33(5):369–377.

Duan, G. R., Irwin, G. W., and Liu, G. P. (2000a). Disturbance attenuation in linear systems via dynamical compensators–a parametric eigenstructure assignment approach. *IEE Proc. Contr. Theor. Appl.*, 146(2):129–136.

Duan, G. R. and Lam, J. (2000). Robust eigenvalue assignment in descriptor systems via output feedback. In *the 8th IFAC Symposium on 'Computer Aided Control System Design'*, Salford, UK.

Duan, G. R. and Lam, J. (2004). Robust eigenvalue assignment in descriptor systems via output feedback. *Asian J. Contr.*, 6(1):145–154.

Duan, G. R., Li, J. H., and Zhou, L. S. (1993). Design of robust Luenberger observers. *Chin. J. Autom. (English edition)*, 5(1):27–32.

Duan, G. R. and Liu, G. P. (2002). Complete parametric approach for eigenstructure assignment in a class of second-order linear systems. *Automatica*, 38(4):725–729.

Duan, G. R., Liu, G. P., and Thompso, S. (2000b). Disturbance decoupling in descriptor systems via output feedback – a parametric eigenstructure assignment approach. In *Proceedings of the 39th IEEE Conference on Decision and Control*, pages 3660–3665, Sydney, Australia.

Duan, G. R., Liu, G. P., and Thompson, S. (2000c). Eigenstructure assignment design for proportional-integral observers–the discrete-time case. In *Proceedings of the 8th IFAC Symposium on 'Computer Aided Control System Design'*, Salford, UK.

Duan, G. R., Liu, G. P., and Thompson, S. (2001a). Eigenstructure assignment design for proportional-integral observers – the continuous-time case. *IEE Proc. Contr. Theor. Appl.*, 148(3):263–267.

Duan, G. R., Liu, G. P., and Thompson, S. (2003). Eigenstructure assignment design for proportional-integral observers – the discrete-time case. *Int. J. Syst. Sci.*, 5(34):357–363.

Duan, G. R., Liu, W. Q., and Liu, G. P. (2000d). Robust model reference control for multivariable linear systems: A parametric approach. In *Proceedings of the IFAC Symposium on Robust control (Rocond 2000)*, pages 83–88, June 21–23, Prague, Czech Republic.

Duan, G. R. and Ma, K. M. (1995). Robust Luenberger function observers for linear systems. In *Pre-prints of IFAC Youth Automatic Conference*, pages 382–387, Aug. 22–25, Beijing, China.

Duan, G. R., Nichols, N. K., and Liu, G. P. (2001b). Robust pole assignment in descriptor linear systems via state feedback. In *Proceedings of the European Control Conference*, pages 2386–2391.

Duan, G. R., Nickols, N. K., and Liu, G. P. (2002b). Robust pole assignment in descriptor linear systems via state feedback. *Eur. J. Contr.*, 8(2):136–149.

Duan, G. R. and Patton, R. J. (1997). Eigenstructure assignment in descriptor systems via proportional plus derivative state feedback. *Int. J. Contr.*, 68(5):1147–1162.

Duan, G. R. and Patton, R. J. (1998a). Eigenstructure assignment in descriptor systems via state feedback–a new complete parametric approach. *Int. J. Syst. Sci.*, 29(2):167–178.

Duan, G. R. and Patton, R. J. (1998b). Robust fault detection in linear systems using Luenberger observers. In *Proceedings of International Conference on CONTROL'98*, pages 1468–1473, Sep. 1–4, Swansea, UK.

Duan, G. R. and Patton, R. J. (1999). Robust pole assignment in descriptor linear systems via proportional plus partial derivative state feedback. *Int. J. Contr.*, 72(13):1193–1203.

Duan, G. R. and Patton, R. J. (2001). Robust fault detection in linear systems using Luenberger-type unknown-input observers – a parametric approach. *Int. J. Syst. Sci.*, 32(4):533–540.

Duan, G. R., Thompson, S., and Liu, G. P. (1999b). On solution to the matrix equation $AV + EVJ = BW + G$. In *Proceedings of the 38th IEEE Conference on Decision and Control*, pages 2742–2743, Phoenix, Arizona, USA.

Duan, G. R. and Wang, G. S. (2003). Partial eigenstructure assignment for descriptor linear systems: a complete parametric approach. In *Proceedings of the IEEE 2003 Conference on Decision and Control*, volume 4, pages 3402–3407, Dec. 9–12, Hawaii, USA.

Duan, G. R., Wang, G. S., and Choi, J. W. (2002c). Eigenstructure assignment in a class of second-order linear systems: A complete parametric approach. In *Proceedings of the 8th Annual Chinese Automation and Computer Society Conference*, pages 89–96, Manchester, UK.

Duan, G. R., Wang, G. S., and Liu, G. P. (2005). Eigenstructure assignment in a class of second-order descriptor linear systems: a complete parametric approach. *Int. J. Autom. Comput.*, 2(1):1–5.

Duan, G. R. and Wu, A. G. (2005a). Design of observers with disturbance decoupling in descriptor systems (in Chinese). *Contr. Theor. Appl.*, 22(1):123–126.

Duan, G. R. and Wu, A. G. (2005b). I-controllablizability of descriptor linear systems. *Dyn. Contin. Discrete Impuls. Syst. Series A Math. Anal.*, 3:1197–1204.

Duan, G. R. and Wu, A. G. (2005c). Impulse elimination via P-D feedback in descriptor linear systems. *Dyn. Contin. Discrete Impuls. Syst. Series A Math. Anal.*, 3:714–721.

Duan, G. R. and Wu, A. G. (2005d). Impulse elimination via state feedback in descriptor linear systems. *Dyn. Contin. Discrete Impuls. Syst. Series A Math. Anal.*, 3:722–729.

Duan, G. R., Wu, G. Y., and Chen, S. J. (1992). An algorithm for robust control system design. In *Proceedings of IECON'92*, pages 1153–1156, Nov. 9–13, San Diego, California, USA.

Duan, G. R., Wu, G. Y., and Huang, W. H. (1991a). Eigenstructure assignment for time-varying systems. *Sci. China (English edition)*, 34(2):246–256.

Duan, G. R. and Wu, Y. L. (2005e). Robust pole assignment in matrix descriptor second-order linear systems. *Trans. Inst. Meas. Contr.*, 27(4):279–295.

Duan, G. R. and Zhang, X. (2002a). Dynamical order assignment in linear descriptor systems via partial state derivative feedback. In *Proceedings of the Annual Conference of the UK-China Automation Society*, Beijing, China.

Duan, G. R. and Zhang, X. (2002b). Dynamical order assignment in linear descriptor systems via state derivative feedback. In *Proceedings of the 41th IEEE Conference on Decision and Control*, volume 4, pages 4533–4538, Las Vegas, USA.

Duan, G. R. and Zhang, X. (2002c). Regularization of linear descriptor systems via output plus partial state derivative feedback. In *Proceedings of the 5th Asia-Pacific Conference on Control and Measurement*, pages 105–110, July 8–12, Yunnan, China.

Duan, G. R. and Zhang, X. (2003). Regularization of linear descriptor systems via output plus partial state derivative feedback. *Asian J. Contr.*, 5(3):334–340.

Duan, G. R. and Zhou, B. (2006). Solutions to the second-order Sylvester matrix equation. *IEEE Trans. Autom. Contr.*, 51(5):805–809.

Duan, G. R., Zhou, L. S., and Xu, Y. M. (1991b). A parametric approach for observer- based control system design. In *Proceedings of Asia-Pacific Conference on Measurement and Control*, pages 295–300, Aug. 21–23, Guangzhou, China.

El-Tohami, M., Lovass-Nagy, V., and Mukundan, R. (1983). On the design of observers for generalized state space systems using singular value decomposition. *Int. J. Contr.*, 38(3):673–685.

Fahmy, M. M. and O'Reilly, J. (1988). Parametric eigenstructure assignment by output-feedback control – the case of multiple-eigenvalues. *Int. J. Contr.*, 48(4):1519–1535.

Fahmy, M. M. and O'Reilly, J. M. (1989). Parametric eigenstructure assignment for continuous-time descriptor systems. *Int. J. Contr.*, 49(1):129–143.

Fahmy, M. M. and Tantawy, H. S. (1990). Dynamical order assignment for linear descriptor systems. *Int. J. Contr.*, 52:175–190.

Fairman, F. W. (1998). *Linear control theory: The state space approach*. Wiley, Chichester.

Feng, J. E., Zhang, W. H., Cheng, Z. L., and Cui, P. (2005). H_∞ output feedback control for descriptor systems with delayed-state. In *Proceedings of the American Control Conference*, 7:4892–4896.

Fletcher, L. R. (1986). Regularizability of descriptor systems. *Int. J. Syst. Sci.*, 47(6):843–847.

Fletcher, L. R. (1988). Eigenstructure assignment by output feedback in descriptor systems. *IEE Proc. Contr. Theor. Appl.*, 135:302–308.

Fletcher, L. R., Kautsky, J., and Nichols, N. K. (1986). Eigenstructure assignment in descriptor systems. *IEEE Trans. Automat. Contr.*, 31(12):1138–1141.

Fridman, E. M. (2001). A Lyapunov-based approach to stability of descriptor systems with delay. In *Proceedings of the IEEE Conference on Decision and Control*, volume 3, pages 2850–2855, Orlando, Florida, USA.

Fu, Y. M. and Duan, G. R. (2005). Robust guaranteed cost observer design for uncertain descriptor time-delay systems with markovian jumping parameters. *Acta Automat. Sinica*, 31(3):479–483.

Fu, Y. M. and Duan, G. R. (2006). Robust guaranteed cost control for descriptor systems with markov jumping parameters and states delays. *ANZIAM J.*, (47):569–580.

Fuhrmann, P. A. (1977). On strict system equivalence and similarity. *Int. J. Contr.*, 25(1):5–10.

Furuta, K., Sano, S., and Atherton, D. (1988). *State variable methods in automatic control*. Wiley, Chichester.

Gantmacher, F. R. (1974). *The theory of matrices*. Chelsea, New York.

Gao, Z. and Ding, S. X. (2007a). Fault estimation and fault-tolerant control for descriptor systems via proportional, multiple-integral and derivative observer design. *IET Contr. Theor. Appl.*, 1(5):1208–1218.

Gao, Z. and Ding, S. X. (2007b). Sensor fault reconstruction and sensor compensation for a class of nonlinear state-space systems via a descriptor system approach. *IET Contr. Theor. Appl.*, 1(3):578–585.

Gao, Z. and Ho, D. W. C. (2004). Proportional multiple-integral observer design for descriptor systems with measurement output disturbances. *IEE Proc. Contr. Theor. Appl.*, 151(3):279–288.

Gao, Z. W. and Ding, S. X. (2007c). Actuator fault robust estimation and fault-tolerant control for a class of nonlinear descriptor systems. *Automatica*, 43(5):912–920.

Geerts, T. (1993). Solvability conditions, consistency, and weak consistency for linear differential-algebraic equations and time-invariant singular systems: the general case. *Linear Algebra Appl.*, 181:111–130.

Georgiou, C. and Krikelis, N. J. (1992). Eigenstructure assignment for descriptor systems via state variable feedback. *Int. J. Syst. Sci.*, 23(1):99–108.

Glad, T. and Sjoberg, J. (2006). Hamilton-jacobi equations for nonlinear descriptor systems. *Am. Contr. Conf.*, 1–12:1027–1031.

Golub, G. H. and Wilkinson, J. H. (1976). Ill-conditioned eigensystems and the computation of the Jordan canonical form. *SIAM Rev.*, 18(4):578–619.

Gontian, Y. and Tarn, T. J. (1982). Strong controllability and strong observability of generalized dynamic systems. In *Proceedings of the Allerton Conference on Communication, Control and Computing*, pages 834–842, Monticello, USA.

Haggman, B. C. and Bryant, P. R. (1984). Solution of singular constrained differential equations: A generalization of circuits containing capacitor – only loops and inductor – only cutsets. *IEEE Trans. Circuits Syst.*, 31(12):1015–1028.

Hayton, G. E., Fretwell, P., and Pugh, A. C. (1986). Fundamental equivalence of generalized state space systems. *IEEE Trans. Automat. Contr.*, 31(5):431–439.

Hiller, M. (1994). What technical developments are on the horizon? In *Proc. 11th ISARC'94*, Brighton England.

Ho, D. W. C., Shi, X. Y., and Wang, Z. D. (2005). Filtering for a class of stochastic descriptor systems. *Dyn. Contin. Discrete Impul. Syst. Series B Appl. Algorithm*, (2):848–853.

Hou, M. (1995). *Descriptor systems: observers and fault diagnosis*. VDI Verlag, Wuppertal.

Hou, M. (2004). Controllability and elimination of impulsive modes in descriptor systems. *IEEE Trans. Automat. Contr.*, 49(10):1723–1727.

Hou, M. and Muller, P. C. (1999). Causal observability of descriptor systems. *IEEE Trans. Automat. Contr.*, 44(1):158–163.

Huang, X. and Huang, B. (2005). Fixed-order controller design for linear time-invariant descriptor systems: A bmi approach. *Int. J. Syst. Sci.*, 36(1):13–18.

Ishihara, J. Y. and Terra, M. H. (1999). Continuous-time H_2 optimal control problem for descriptor systems. In *Proceedings of the American Control Conference*, volume 4, pages 4098–4099, San Diego, California, USA.

Ishihara, J. Y. and Terra, M. H. (2001). Generalized Lyapunov theorems for rectangular descriptor systems. In *Proceedings of the IEEE Conference on Decision and Control*, volume 3, pages 2858–2859, Orlando, Florida, USA.

Jing, H. Y. (1994). Eigenstructure assignment by proportional-derivative state- feedback in singular systems. *Syst. Contr. Lett.*, 22(1):47–52.

Kagstrom, B. and Ruhe, A. (1980). An algorithm for numerical computation of the Jordan normal form of a complex matrix. *ACM Trans. Math. Software*, 6(3):398–419.

Kailath, T. (1980). *Linear systems*. Prentic-Hall, Englewood Cliffs, N. J.

Kalyuthnaya, T. S. (1978). Controllability of general differential systems. *Diff. Eq.*, 14:314–320.

Kautsky, J. and Nichols, N. K. (1986). Algorithms for robust pole assignment in singular systems. In *Proceedings of American Control Confrence*, volume 1, pages 433–436, Seattle, USA.

Kautsky, J. and Nichols, N. K. (1990). Robust pole assignment in systems subject to structured perturbations. *Syst. Contr. Lett.*, 15(5):373–380.

Kautsky, J., Nichols, N. K., and Chu, E. K. W. (1989). Robust pole assignment in singular control-systems. *Linear Algebra Appl.*, 121:9–37.

Kautsky, J., Nichols, N. K., and Van Dooren, P. (1985). Robust pole assignment in linear state feedback. *Int. J. Contr.*, 41(5):1129–1155.

Kawaji, S. and Kim, H. S. (1995). Full order observer for linear descriptor systems with unknown-inputs. In *Proceedings of the 34th Conference on Decision and Control*, pages 2366–2368, Dec., New Orleans, LA.

Kiruthi, G., Yazdani, H., and Newcomb, R. W. (1980). A hysteresis circuit seen through semi-state equations. In *Proceedings of the Midwest Symposium on Circuit Theory*, Toledo, Ohio.

Kostyukova, O. I. (2001). Sensitivity analysis for solutions of a parametric optimal control problem of a descriptor system. *Int. J. Comput. Syst. Sci.*, 41(1):45–56.

Kostyukova, O. I. (2003). Investigation of properties of solutions of a parametric optimal control problem in a singular point. *Int. J. Comput. Syst. Sci.*, 42(3):379–390.

Kunkel, P. and Mehrmann, V. (1997). The linear quadratic optimal control problem for linear descriptor systems with variable coefficients. *Math. Contr. Signals Syst.*, 10(3):247–264.

Kunkel, P. and Mehrmann, V. (2006). *Differential-algebraic equations. Analysis and numerical solution*. EMS Publishing House, Zurich.

Kunkel, P., Mehrmann, V., and Rath, W. (2001). Analysis and numerical solution of control problems in descriptor form. *Math. Contr. Signals Syst.*, 14:29–61.

Kurina, G. A. (1983). Asymptotic solution of one class of singularly perturbed optimal control problems. *Appl. Math. Mech.*, 47(3):309–315.

Kurina, G. A. (1993). On the sufficient conditions of optimal control for discrete descriptive systems. *Avtomatika i Telemekhanika*, (8):52–55.

Kurina, G. A. (2001). Linear-quadratic optimal control problems for discrete descriptor systems. *Syst. Struct. Contr.*, 1:261–266.

Kurina, G. A. (2002). Optimal feedback control proportional to the systems state can be found for non-causal descriptor systems (a remark on a paper by p. c. muller). *Int. J. Appl. Math. Comput. Sci*, 12(4):591–593.

Kurina, G. A. (2003). Minimax linear-quadratic control problems for descriptor systems with properly stated leading term. In *Proceedings of 11th Mediterranean Conference on Control and Automation*.

Kurina, G. A. (2004). Linear-quadratic discrete optimal control problems for descriptor systems in Hilbert space. *J. Dyn. Contr. Syst.*, 10(3):365–375.

Kurina, G. A. and Marz, R. (2004). On linear-quadratic optimal control problems for time-varying descriptor systems. *SIAM J. Contr. Optim.*, 42(6):2062–2077.

Lam, J. and Tam, H. K. (2000). Robust output feedback pole assignment using genetic algorithm. *Proc. Inst. Mech. Eng. I J. Syst. Contr. Eng.*, 214(15):327–334.

Lam, J., Tso, H. K., and Tsing, N. K. (1997). Robust deadbeat regulation. *Int. J. Contr.*, 67(4):587–602.

Lewis, F. L. (1985a). Fundamental, reachability and observability matrices for discrete descriptor systems. *IEEE Trans. Automat. Contr.*, 30(5):502–505.

Lewis, F. L. (1985b). Preliminary notes on optimal control for sigular systems. In *Proceedings of the IEEE Conference on Decision and Control*, pages 266–272, Fort Lauderdale, Fla.

Lewis, F. L. (1986). A survey of linear singular systems. *Circuits Syst. Signal Process*, 5(1):3–36.

Lewis, F. L. and Ozcaldiran, K. (1984). Reachability and controllability for descriptor systems. In *Proceedings of the 27th Midwest Symposium on Circuits and Systems*, volume 2, pages 690–695, Morgantown, W.V.

Lewis, F. L. and Ozcaldiran, K. (1985). On the eigenstructure assignment of singular systems. In *Proceedings of the IEEE Conference on Decision and Control*, pages 179–182, Fort Lauderdale, Fla.

Lewis, F. L. and Ozcaldiran, K. (1989). Geometric structure and feedback in singular systems. *IEEE Trans. Automat. Contr.*, 34(4):450–455.

Liang, B. and Duan, G. R. (2004). Robust H_∞ fault-tolerant control for uncertain descriptor systems by dynamical compensators. *J. Contr. Theor. Appl.*, 2(3):228–292.

Lin, C., Lam, J., Wang, J., and Yang, G. H. (2001). Analysis on robust stability for interval descriptor systems. *Syst. Contr. Lett.*, 42(4):267–278.

Lin, C., Wang, J. L., and Soh, C. B. (1997). Robust stability of linear interval descriptor systems. In *Proceedings of the IEEE Conference on Decision and Control*, volume 4, pages 3822–3823, San Diego, USA.

Lin, S. F. and Wang, A. P. (1997). Unknown input observers for singular systems designed by eigenstructure assignment. *J. Franklin Inst.*, 340:43–61.

Liu, G. P. and Patton, R. J. (1996). On eigenvectors and generalized eigenvectors in multi-variable descriptor control systems. In *Proceedings of the Institution of Mechanical Engineers Part I-Jounal of System and Control Engineering*, volume 210, pages 183–188.

Liu, G. P. and Patton, R. J. (1998). Robust eigenstructure assignment for descriptor systems. *Int. J. Syst. Sci.*, 29(1):75–84.

Liu, H. L. and Duan, G. R. (2006). Robust state-feedback impulse elimination design of uncertain time-varying descriptor linear systems. *Dyn. Contin. Discrete Impul. Syst. Series A Math. Anal.*, (13):691–698.

Liu, Y. Q. and Wen, C. X. (1997). *Flexible structure control of descriptor linear systems (in Chinese)*. Press of the South China University of Technology, Guangzhou, China.

Lohmann, B. and Labibi, B. (2000). Decentralized stabilization using descriptor systems. In *Proceedings of the IEEE International Conference on Systems, Man and Cybernetics*, volume 3, pages 2252–2256, Nashville, USA.

Lovass-Nagy, V., Powers, D. L., and Schilling, R. J. (1994). On the regularizing descriptor systems by output feedback. *IEEE Trans. Automat. Contr.*, 39(7):1507–1509.

Lovass-Nagy, V., Powers, D. L., and Schilling, R. J. (1996). On regularizing descriptor systems by output feedback. *IEEE Trans. Automat. Contr.*, 41(11):1689–1690.

Lovass-Nagy, V., Powers, D. L., and Yan, H. C. (1986a). On controlling generalized stale-space (descriptor) systems. *Int. J. Contr.*, 43(4):1271–1282.

Lovass-Nagy, V., Schilling, R., and Yan, H. C. (1986b). A note on optimal control of generalized state space (descriptor) systems. *Int. J. Contr.*, 44(3):613–624.

Lu, G. P. and Ho, D. W. C. (2006a). Full-order and reduced-order observers for lipschitz descriptor systems: the unified lmi approach. *IEEE Trans. Circuits Syst. II Exp. Briefs*, 53(7):563–567.

Lu, G. P. and Ho, D. W. C. (2006b). Solution existence and stabilization for bilinear descriptor systems with time-delay. In *Proceedings of the 9th International Conference on Control, Automation, Robotics and Vision*, (1–5):996–1000.

Lu, G. P., Ho, D. W. C., and Zheng, Y. F. (2004). Observers for a class of descriptor systems with Lipschitz constraint. In *Proceedings of the American Control Conference*, pages 3474–3479, Boston, USA.

Luenberger, D. G. (1967). Canonical forms for linear multivariable systems. *IEEE Trans. Automat. Contr.*, 12(3):290–293.

Luenberger, D. G. (1977). Dynamic equation in descriptor form. *IEEE Trans. Automat. Contr.*, 22(3):312–321.

Luenberger, D. G. (1987). Dynamic equilibria for linear systems and quadratic costs. *Automatica*, 23(1):117–122.

Martens, J. P., De, M. G., and Vanwormhoudt, M. C. (1984). CADLEM: A simulation program for nonlinear electronic networks. *Int. J. Electron.*, 57(1):97–110.

Mehrmann, V. (1988). A symplectic orthogonal method for single input or single output discrete time linear quadratic control problems. *SIAM J. Matrix Anal. Appl.*, 9:221–248.

Mehrmann, V. (1989). Existence, uniqueness and stability of solutions to singular linear quadratic optimal control problems. *Linear Algebra Appl.*, 121:291–331.

Mehrmann, V. (1991). *The autonomous linear quadratic control problem: theory and numerical solution. No. 163 Lecture Notes in Control and Information Sciences*. Springer, Berlin.

Minamide, N., Fujisaki, Y., and Shimizu, A. (1997). Parametrization of all observers for descriptor systems. *Int. J. Contr.*, 66(5):767–777.

Misra, P., Van Dooren, P., and Varga, A. (1994). Computation of structural invariants of generalized state-space systems. *Automatica*, 30(12):1921–1936.

Moor, B. D. and Golub, G. H. (1991). The restricted singular value decomposition: Properties and applications. *SIAM J. Matrix Anal. Appl.*, 12:401–425.

Mukundan, R. and Dayawansa, W. (1983). Feedback control of singular systems-proportional and derivative feedback of the state. *Int. J. Syst. Sci.*, 14(6):615–632.

Muller, P. C. (1999). Linear-quadratic optimal control of descriptor systems. *J. Braz. Soc. Mech. Sci.*, 21(3):423–432.

Muller, P. C. (2003). Optimal control of proper and nonproper descriptor systems. *Arch. Appl. Mech.*, 72(11–12):875–884.

Muller, P. C. (2006). Modified Lyapunov equations for LTI descriptor systems. *J. Braz. Soc. Mech. Sci. Eng.*, 28(4):448–452.

Murota, K. (1983). Structural controllability of a system with some fixed coefficients. *Trans. Soc. Instrum. Contr. Eng. (Japan)*, 19(9):683–690.

Newcomb, R. (1981). The semi-state description of nonlinear time variable circuits. *IEEE Trans. Circuit Syst.*, 28(1):62–71.

Newcomb, R. W. (1982). Semi-state design theory: binary and swept hysteresis. *Circuits. Syst. Signal Process.*, (1):203.

Olsson, K. H. A. and Ruhe, A. (2006). A rational krylov for eigenvalue computation and model order reduction. *Bit Numer. Math.*, (46):S99–S111 Suppl.

Owens, J. and Askarpour, S. (2001). Integrated approach to dynamical order assignment in linear descriptor systems. *IEE Proc. Contr. Theor. Appl.*, 148(4):329–332.

Owens, T. J. and O'Reilly, J. (1987). Parametric state feedback control with response insensitivity. *Int. J. Contr.*, 45(3):791–809.

Owens, T. J. and O'Reilly, J. (1989). Parametric state feedback control for arbitrary eigenvalue assignment with minimum sensitivity. *IEE Proc. D Contr. Theor. Appl.*, 136(6):307–313.

Ozcaldiran, K. and Lewis, F. (1987). A geometric approach to eigenstructure assignment for singular systems. *IEEE Trans. Automat. Contr.*, 32(7):629–631.

Ozcaldiran, K. and Lewis, F. L. (1990). On the regularizability of singular systems. *IEEE Trans. Automat. Contr.*, 35(10):1156–1160.

Pandolfi, L. (1980). Controllability and stabilizability for linear systems of algebraic and differential equations. *J. Opt. Theor. Appl.*, 30(4):601–620.

Pandolfi, L. (1981). On the regulator problem for linear degenerate control systems. *J. Opt. Theor. Appl.*, 33(2):241–252.

Patton, R. J. and Liu, G. P. (1994). Robust-control design via eigenstructure assignment, genetic algorithms and gradient-based optimization. *IEE Proc. Contr. Theor. Appl.*, 141(3):202–208.

Petzold, L. R. (1982). Differential / algebraic equations are not ODEs. *SIAM J. Sci. Stat. Comp.*, 3(2):367–384.

Piao, F. X., Zhang, Q. L., and Ma, X. Z. (2006). Robust H_∞ control for uncertain descriptor systems with state and control delay. *J. Syst. Eng. Electron.*, 17(3):571–575.

Potter, J. M., Anderson, B. D. O., and Morse, A. S. (1979). Single-channel control of a two-channel system. *IEEE Trans. Automat. Contr.*, 24(3):491–492.

Pugh, A. C. and Shelton, A. K. (1978). On a new definition of strict system equivalence. *Int. J. Contr.*, 27(5):657–672.

Ren, J. (2006a). Guaranteed cost control for nonlinear descriptor system with time delays via t-s fuzzy model. *Dyn. Contin. Discrete Impul. Syst. Series B Appl. Algorithm*, (13E):406–411, Part 1 Suppl.

Ren, J. (2006b). Non-fragile robust H_∞ fuzzy controller design for a class of nonlinear descriptor systems with time-varying delays in states. In *Proceedings of the 3rd International Conference on Fuzzy Systems and Systems and Knowledge Discovery*, pages 119–128.

Ren, J. S. (2005). Non-fragile lq fuzzy control for a class of nonlinear descriptor system with time delays. In *Proceedings of 2005 International Conference on Machine Learning and Cybernetics*, 1–9:797–802.

Saidahmed, M. T. F. (1985). Generalized reduced order observer for singular systems. In *Proc. 28th Midwest Symp. on Circuits and Systems*, pages 581–585, Louisville, Ky.

Sakr, A. F. and Khalifa, I. (1990). Eigenstructure assignment for descriptor systems by output-feedback compensation. *Syst. Contr. Lett.*, 14(2):139–144.

Schmidt, T. (1994). *Parametrschaetzung bei mehrkoerpersystemen mit zwangsbedingungen.* VDI-Verlag, Dusseldorf.

Schraft, R. D., Degenhart, E., and Hagele, M. (1993). Service robots: the appropriate level of automation and the role of users, operators in the task execution. In *Proceedings of Symposium on System, Man and Cybernetics*, volume 4, pages 164–169, Le Touuet, France.

Schraft, R. D. and Wanner, M. C. (1993). The aircraft cleaning robot skywash. *Industrial Robot*, (20):21–24.

Shafai, B. and Carroll, R. L. (1987). Design of a minimal-order observer for singular systems. *Int. J. Contr.*, 45(3):1075–1082.

Shaked, U. and Fridman, E. (2002). A descriptor system approach to H_∞ control of linear time-delay systems. *IEEE Trans. Automat. Contr.*, 47(2):253–270.

Singh, S. P. and Liu, R. W. (1973). Existence of state equation representation of linear large-scale dynamic systems. *IEEE Trans. Circuit Theor.*, 20(5):239–246.

Smith, R. J. (1966). *Circuits, devices, and system.* Wiley, New York.

Stefanovski, J. (2006). Lq control of descriptor systems by cancelling structure at infinity. *Int. J. Contr.*, 79(3):224–238.

Stengel, D. N., Larson, R. E., Luenberger, D. G., and Cline, T. B. (1979). A descriptor variable approach to modeling and optimization of large-scale systems. In *Proc. Eng.*, volume 7, Switzland.

Stewart, G. W. (1975). Gerschgorin theory for the generalized eigenvalue problem $Ax = \lambda Bx$. *Math. Comput.*, 29(130):600–606.

Stykel, T. (2002a). *Analysis and numerical solution of generalized Lyapunov equations.* PhD thesis, TU Berlin.

Stykel, T. (2002b). On criteria for asymptotic stability of differential-algebraic equations. *ZAMM Z. Angew. Math.*, 82:147–158.

Stykel, T. (2002c). Stability and inertia theorems for generalized Lyapunov equations. *Linear Algebra Appl.*, 355:297–314.

Stykel, T. (2006a). Balanced truncation model reduction for semidiscretized stokes equation. *Linear Algebra Appl.*, 415(2–3):262–289.

Stykel, T. (2006b). On some norms for descriptor systems. *IEEE Trans. Automat. Contr.*, 51(5):842–847.

Su, X. M., Shi, H. Y., Lv, M. Z., and Zhang, Q. L. (2004). On impulsive controllability of linear periodic descriptor systems. In *Proceedings of the 5th World Congress on Intelligent Control and Automation*, volume 1, pages 1–5, Hangzhou, China.

Su, X. M., Zhang, Q. L., and Jin, J. Q. (2002). Stability analysis of interval descriptor systems: A matrix inequalities approach. In *Proceedings of the World Congress on Intelligent Control and Automation (WCICA)*, volume 2, pages 1007–1011, Shanghai, China.

Sun, J. G. (1987). On numerical-methods for robust pole assignment in control- system design .2. *J. Comput. Math.*, 5(4):352–363.

Sun, L. Y. (2006). Observer design for nonregular descriptor systems with unknown inputs in finite time. In *Proceedings of the 6th World Congress on Intelligent Control and Automation*, volume 1, pages 591–595.

Syrmos, V. L. and Lewis, F. L. (1992). Robust eigenvalue assignment for generalized systems. *Automatica*, 28(6):1223–1228.

Takaba, K., Morihira, N., and Katayama, T. (1995). A generalized lyapunov theorem for descriptor system. *Syst. Contr. Lett.*, 24(1):49–51.

Tanaka, K., Ohtake, H., and Wang, H. O. (2007). A descriptor system approach to fuzzy control system design via fuzzy Lyapunov functions. *IEEE Trans. Fuzzy Syst.*, 15(3):333–341.

Tsui, C. C. (1985). A new algorithm for the design of multifunctional observers. *IEEE Trans. Automat. Contr.*, 30(1):89–93.

Tsui, C. C. (1987). A complete analytical solution to the equation $TA - FT = LC$ and its applications. *IEEE Trans. Automat. Contr.*, 32(8):742–744.

Tsui, C. C. (1988a). New approach to robust observer design. *Int. J. Contr.*, 47(3):745–751.

Tsui, C. C. (1988b). On robust observer compensator design. *Automatica*, 24(5):687–692.

488   References

Vafiadis, D. and Karcanias, N. (1997). Canonical forms for descriptor systems under restricted system equivalence. *Automatica*, 33(5):955–958.

Van Dooren, P. (1981). The generalized eigenstructure problem in linear system theory. *IEEE Trans. Automat. Contr.*, 26(1):111–129.

Van Dooren, P. (1990). Rational and polynomial matrix factorizations via recursive pole zero cancelation. *Linear Algebra Appl.*, (137):663–697.

Varga, A. (2000). A descriptor systems toolbox for MATLAB. In *Proceedings of the IEEE Symposium on System, Man and Cybernetics*, Anchorage, Alaska, USA.

Verghese, G. C. (1981). Further notes on singular descriptions. In *Proceedings of the 1981 JAC Conference*, volume 1, Article ID: TA-4, Charlottesville, USA.

Verghese, G. C., Levy, B. C., and Kailath, T. (1981). A generalized state space for singular systems. *IEEE Trans. Automat. Contr.*, 26(4):811–831.

Verhaegen, M. and Van Dooren, P. (1986). A reduced observer for descriptor systems. *Syst. Contr. Lett.*, 8(11):29–38.

Wang, C. and Dai, L. (1986). Singular dynamic systems – a survey. *Contr. Theor. Appl.*, (1):2–12.

Wang, C. and Dai, L. (1987a). State observation decoupling disturbance for singular systems. *Contr. Theor. Appl.*, (3):23–30.

Wang, C. and Dai, L. (1987b). State observer structure in singular systems. *Acta. Math. Appl. Sinica*, 10(1):121–124.

Wang, D. H. and Soh, C. B. (1999). On regularizing singular systems by decentralized output feedback. *IEEE Trans. Automat. Contr.*, 44(1):148–152.

Wang, E. and Wang, C. (1986). Optimal recurrent filtering for linear discrete stochastic singular systems (I). In *Preprints Ann. Natl. Conf. on Control Theory and Its Appl.*, volume 1, pages 118–121, Heilongjiang, China.

Wang, G. S., Lv, Q., Liang, B., and Duan, G. R. (2006). Partial eigenstructure assignment in second-order descriptor systems. *Dyn. Contin. Discrete Impuls. Syst. Ser. A Math. Anal.*, (13):1022–1029 Part 2 Suppl.

Wang, J., Zhang, Q. L., Liu, W. Q., and Yuan, Z. H. (2004). Infinite-eigenvalue assignment for descriptor systems via state variable feedback (in Chinese). In *Proceedings of the 5th World Congress on Intelligent Control and Automation*, volume 2, pages 1022–1026, Hangzhou, China.

Wang, W. and Zou, Y. (2001). Analysis of impulsive modes and luenberger observers for descriptor systems. *Syst. Contr. Lett.*, 44(5):347–353.

Wang, Y. (1984). *Disturbance resistance and output regulation for linear singular systems*. PhD thesis, Inst. of Systems Science, Academia Sinica, June, Beijing, China.

Wang, Z. D. (2000). Parametric circular eigenvalue assignment for descriptor systems via state feedback. *IMA J. Math. Contr. Inform.*, 17(1):57–66.

Wanner, M. C., Baumeister, K., Kohler, G. W., and Walze, H. (1986). Hochflexible handhabungssysteme. *Ergebnisse Einer Einsatzfalluntersuchung Robotersysteme*, 2(4):217–224.

Wanner, M. C. and Kong, R. (1990). *Roboter auberhalb der fertigungstechnik, in industrieroboter: handbuch fur industrie und wissenschaft*. H.-J. Warnecke and R. D. Schraft (Eds), Springer, Berlin.

Wilkinson, J. H. (1965). *The algebraic eigenvalue problem*. Oxford University Press, Oxford.

Wu, A. G. and Duan, G. R. (2006a). Design of generalized PI observers for continuous-time descriptor linear systems. *IEEE Trans. Circuits Syst. I Regul. Pap.*, 53(12):2828–2837.

Wu, A. G. and Duan, G. R. (2006b). Design of PI observers for continuous-time descriptor linear systems. *IEEE Trans. Syst. Man Cybern. B Cybern.*, 36(6):1423–1431.

Wu, A. G. and Duan, G. R. (2007a). Design of PD observers in descriptor linear systems. *Int. J. Contr.*, 5(1):93–98.

Wu, A. G. and Duan, G. R. (2007b). Impulsive-mode controllablisability in descriptor linear systems. *IET Contr. Theor. Appl.*, 1(3):558–563.

Wu, A. G. and Duan, G. R. (2007c). Kronecker maps and Sylvester-polynomial matrix equations. *IEEE Trans. Automa. Contr.*, 52(5):905–910.

Wu, A. G. and Duan, G. R. (2007d). Solution to the generalised Sylvester matrix equation $AV + BW = EVF$. *IET Contr. Theor. Appl.*, 1(1):402–408.

Xie, G. M. and Wang, L. (2002). Controllability of linear descriptor systems with multiple time-delays in control. In *Proceedings of the IEEE International Conference on Systems, Man and Cybernetics*, volume 2, pages 671–676, Yasmine Hammamet, Tunisia.

Xie, G. M. and Wang, L. (2003). Controllability of linear descriptor systems. *IEEE Trans. Circuit Syst. I Fundam. Theor. Appl.*, 50(3):455–460.

Xing, W., Zhang, Q., and Liu, W. (2005). H_∞ control based on state observer for descriptor systems. In *Volumes of the Optimization and Control with Applications*, pages 481–493.

Yamada, T. and Luenberger, D. G. (1985a). Algorithm to verify generic causality and controllability of descriptor systems. *IEEE Trans. Automat. Contr.*, 30(9):874–880.

Yamada, T. and Luenberger, D. G. (1985b). Generic controllability theorems for descriptor systems. *IEEE Trans. Automat. Contr.*, 30(2):144–152.

Yan, Z. B. (2007). Geometric analysis of impulse controllability for descriptor system. *Syst. Contr. Lett.*, 56(1):1–6.

Yan, Z. B. and Duan, G. R. (2005). Time-domain solution to descriptor variable systems. *IEEE Trans. Automat. Contr.*, 50(11):1796–1799.

Yan, Z. B. and Duan, G. R. (2006). Impulse controllability and impulse observability in descriptor system. *Dyn. Contin. Discrete Impul. Syst. Ser. A Math. Anal.*, 13:617–624.

Yang, C., Zhang, Q., and Zhou, L. (2007a). Generalised absolute stability analysis and synthesis for lure-type descriptor systems. *IET Contr. Theor. Appl.*, 1(3):617–623.

Yang, C. Y., Zhang, Q. L., and Lin, Y. P. (2007b). Positive realness and absolute stability problem of descriptor systems. *IEEE Trans. Circuits Syst. I Regul. Pap.*, 54(5):1142–1149.

Yang, C. Y., Zhang, Q. L., and Zhou, L. N. (2005a). Practical stability of descriptor systems. *Dyn. Contin. Discrete Impul. Syst. Ser. B Appl. Algorithm*, (12):44–57 Suppl.

Yang, C. Y., Zhang, Q. L., and Zhou, L. N. (2006). Practical stability of descriptor systems with time delays in terms of two measurements. *J. Franklin Inst. Eng. Appl. Math.*, 343(6):635–646.

Yang, D. M., Zhang, Q. L., and Sha, C. M. (2005b). Observer-based H_2 suboptimal control for descriptor systems. *Dyn. Contin. Discrete Impul. Syst. Ser. B Appl. Algorithm*, (12):186–196 Suppl.

Yeu, T. K., Kim, H. S., and Kawaji, S. (2005). Fault detection, isolation and reconstruction for descriptor systems. *Asian J. Contr.*, 7(4):356–367.

Yip, E. L. and Sincovec, R. F. (1981). Solvability, controllability and observability of continuous descriptor systems. *IEEE Trans. Automat. Contr.*, 26(3):702–706.

Yue, D. and Lam, J. (2004). Suboptimal robust mixed H_2/H_∞ controller design for uncertain descriptor systems with distributed delays. *Comput. Math. Appl.*, 47(6–7):1041–1055.

Yue, D., Lam, J., and Ho, D. W. C. (2005). Delay-dependent robust exponential stability of uncertain descriptor systems with time-varying delays. *Dyn. Contin. Discrete Impul. Syst. Ser. B Appl. Algorithm*, 12(1):129–149.

Zagalak, P. and Kucera, V. (1995). Eigenstructure assignment in linear descriptor systems. *IEEE Trans. Automat. Contr.*, 40(1):144–147.

Zhai, D., Zhang, Q. L., Zhang, G. S., Su, X. M., and Liu, N. Y. (2002). Robust stability analysis for descriptor systems. In *Proceedings of the World Congress on Intelligent Control and Automation (WCICA)*, volume 2, pages 931–935, Shanghai, China.

Zhai, G. (2005). Decentralized H_∞ controller design for descriptor systems. *Contr. Intell. Syst.*, 33(3):158–165.

Zhang, D. and Yu, L. (2006). Delay-dependent robust guaranteed cost control for generalized uncertain neutral systems. *Dyn. Contin. Discrete Impul. Syst. Ser. A Math. Anal.*, (13):974–982 Part 2 Suppl.

Zhang, Q. L. (1997). *Decentralized and robust control of large-scale descriptor systems (in Chinese)*. The North-west University of Technology Press, Xian, China.

Zhang, Q. L. and Yang, D. M. (2003). *Analysis and synthesis of uncertain descriptor linear systems (in Chinese)*. The Northeastern University Press, Shenyang, China.

Zhang, Q. L., Zheng, L. H., and Fan, Z. P. (1997). State observers of interconnected descriptor systems. *Adv. Model. Anal. C*, 49(1):41–48.

Zhou, B. and Duan, G. R. (2006). An explicit solution to right coprime factorization with application in eigenstructure assignment. *J. Contr. Theor. Appl.*, 4(2):147–154.

Zhou, B. and Duan, G. R. (2007). Solutions to generalized Sylvester matrix equation by schur decomposition. *Int. J. Syst. Sci.*, 38(5):369–375.

Zhou, Z., Shaymann, M. A., and Tarn., T. J. (1987). Singular systems: a new approach in the time domain. *IEEE Trans. Automat. Contr.*, 32(1):42–50.

Zhu, J. D., Ma, S. P., and Cheng, Z. L. (2002). Singular LQ problem for nonregular descriptor systems. *IEEE Trans. Automat. Contr.*, 47(7):1128–1133.

Zhu, J. D. and Tian, Y. P. (2004). A canonical form for non-square linear descriptor systems (in Chinese). In *Proceedings of the 5th World Congress on Intelligent Control and Automation (WCICA)*, volume 2, pages 975–979, Hangzhou, China.

Index